Special Volume:
Foundations of Computational Mathematics
Guest Editor: F. Cucker

Handbook of Numerical Analysis

General Editor:

P.G. Ciarlet

Laboratoire Jacques-Louis Lions
Université Pierre et Marie Curie
4 Place Jussieu
75005 PARIS, France

and

Department of Mathematics
City University of Hong Kong
Tat Chee Avenue
KOWLOON, Hong Kong

ELSEVIER

Amsterdam • Boston • London • New York • Oxford • Paris
San Diego • San Francisco • Singapore • Sydney • Tokyo

Volume XI

Special Volume: Foundations of Computational Mathematics

Guest Editor:

F. Cucker

Department of Mathematics
City University of Hong Kong
KOWLOON, Hong Kong

2003
ELSEVIER
Amsterdam • Boston • London • New York • Oxford • Paris
San Diego • San Francisco • Singapore • Sydney • Tokyo

ELSEVIER SCIENCE B.V.
Sara Burgerhartstraat 25
P.O. Box 211, 1000 AE Amsterdam, The Netherlands

First edition 2003

Library of Congress Cataloging in Publication Data
A catalog record from Library of Congress has been applied for.

British Library Cataloguing in Publication Data
Foundations of computational mathematics: special volume. –
(Handbook of numerical analysis; 11)
1. Numerical analysis 2. Mathematics – Data processing
I. Cucker, F.
519.4

ISBN: 0-444-51247-0
ISSN (Series): 1570-8659

⊗ The paper used in this publication meets the requirements of ANSI/NISO Z39.48-1992 (Permanence of Paper).

Printed in the United Kingdom

General Preface

In the early eighties, when Jacques-Louis Lions and I considered the idea of a *Handbook of Numerical Analysis*, we carefully laid out specific objectives, outlined in the following excerpts from the "General Preface" which has appeared at the beginning of each of the volumes published so far:

> During the past decades, giant needs for ever more sophisticated mathematical models and increasingly complex and extensive computer simulations have arisen. In this fashion, two indissociable activities, *mathematical modeling* and *computer simulation*, have gained a major status in all aspects of science, technology and industry.
>
> In order that these two sciences be established on the safest possible grounds, mathematical rigor is indispensable. For this reason, two companion sciences, *Numerical Analysis* and *Scientific Software*, have emerged as essential steps for validating the mathematical models and the computer simulations that are based on them.
>
> *Numerical Analysis* is here understood as the part of *Mathematics* that describes and analyzes all the numerical schemes that are used on computers; its objective consists in obtaining a clear, precise, and faithful, representation of all the "information" contained in a mathematical model; as such, it is the natural extension of more classical tools, such as analytic solutions, special transforms, functional analysis, as well as stability and asymptotic analysis.
>
> The various volumes comprising the *Handbook of Numerical Analysis* will thoroughly cover all the major aspects of Numerical Analysis, by presenting accessible and in-depth surveys, which include the most recent trends.
>
> More precisely, the Handbook will cover the *basic methods of Numerical Analysis*, gathered under the following general headings:
>
> - Solution of Equations in $\mathbb{R}^n$,
> - Finite Difference Methods,
> - Finite Element Methods,
> - Techniques of Scientific Computing.

It will also cover the *numerical solution of actual problems of contemporary interest in Applied Mathematics*, gathered under the following general headings:

- Numerical Methods for Fluids,
- Numerical Methods for Solids.

In retrospect, it can be safely asserted that Volumes I to IX, which were edited by both of us, fulfilled most of these objectives, thanks to the eminence of the authors and the quality of their contributions.

After Jacques-Louis Lions' tragic loss in 2001, it became clear that Volume IX would be the last one of the type published so far, i.e., edited by both of us and devoted to some of the general headings defined above. It was then decided, in consultation with the publisher, that each future volume will instead be devoted to a single "*specific application*" and called for this reason a "*Special Volume*". "*Specific applications*" will include Mathematical Finance, Meteorology, Celestial Mechanics, Computational Chemistry, Living Systems, Electromagnetism, Computational Mathematics etc. It is worth noting that the inclusion of such "specific applications" in the *Handbook of Numerical Analysis* was part of our initial project.

To ensure the continuity of this enterprise, I will continue to act as Editor of each Special Volume, whose conception will be jointly coordinated and supervised by a Guest Editor.

P.G. CIARLET
July 2002

Contents of Volume XI

Contents of the Handbook

VOLUME XI

SPECIAL VOLUME: FOUNDATIONS OF COMPUTATIONAL MATHEMATICS

Special Volume:
Foundations of Computational Mathematics

On the Foundations of Computational Mathematics

B.J.C. Baxter

School of Economics, Mathematics and Statistics, Birkbeck College, University of London, Malet Street, London WC1E 7HX, UK

A. Iserles

Department of Applied Mathematics and Theoretical Physics, University of Cambridge, Silver Street, Cambridge CB3 9EW, UK

"... there is no permanent place in the world for ugly mathematics."
G.H. HARDY [1992]

1. On the place of numerical analysis in the mathematical universe

A long time ago, when younger and rasher mathematicians, we both momentarily harboured the ambition that one day, older and wiser, we might write a multivolume treatise titled "On the Mathematical Foundations of Numerical Analysis". And then it dawned that such a creation already exists: it is called 'a mathematics library'. Indeed, it is almost impossible to identify a mathematical theme, no matter how 'pure', that has never influenced numerical reasoning: analysis of all kinds, of course, but also algebra, graph theory, number theory, probability theory, differential geometry, combinatorial topology, category theory ... The mainspring of numerical analysis is, indeed, the entire aquifer of mathematics, pure and applied, and this is an enduring attraction of the discipline. Good luck to those content to spend their entire mathematical existence toiling in a narrow specialism; our choice is an activity that borrows eclectically, indeed with abandon, from all across the mathematical landscape.

Foundations of Computational Mathematics
Special Volume (F. Cucker, Guest Editor) of
HANDBOOK OF NUMERICAL ANALYSIS, VOL. XI
P.G. Ciarlet (Editor)

This is emphatically *not* the popular mathematical perception of numerical analysis. The conventional view is captured in a memorable footnote of HODGES [1983], bravely defining what numerical analysis was *circa* 1945 (from the standpoint of a pure mathematician) and then adding "As a branch of mathematics, however, numerical analysis probably ranked the lowest, even below the theory of statistics, in terms of what most university mathematicians found interesting". (To which many contemporary numerical analysts will add "so, not much has changed".) Numerical analysis as a subject has had a bad name and it makes sense to tackle this perception head on, before any further elaboration of the discipline, its future and its interaction with other areas of scholarly endeavour.

Numerical analysis lies at the meeting point of pure mathematics, computer sciences and application areas. It often attracts some degree of hostility from all three. Yet it is important to note that the complaints of pure mathematicians, computer scientists and the broad constituency of scientists and engineers are very different and often contradictory.

To pure mathematicians, numerical analysis fails the Hardy test: it is (supposedly) ugly. The beauty of mathematics, described so lovingly by G.H. HARDY [1992], is in the quest for pattern and rigour: precise definition and well-formulated theorems with perfect proofs. Falling back on numerical calculation is, almost axiomatically, the moment mathematics fails: the end of rigour and precision, the abandonment of traditional theorem-and-proof, merely tedious discretization and number crunching. This is nonsense. A mathematical problem does not cease to be a *mathematical* problem once we resort to discretization: in principle, we replace a differential equation with a difference equation. The mathematical problem is, if anything, more demanding and intricate. Numerical formulation allows a computer – a mindless contraption – to reproduce strings of numbers and figures that might (or might not) bear some resemblance to the true solution of the underlying problem. The work of a theoretical numerical analyst is precisely to verify the provenance and precision of the above numbers and figures, using exactly the same rules of mathematical engagement as, say, a topologist or an algebraic geometer.

The opposition of computer scientists and of the applications' community is different in kind. Computer scientists are historically wedded to the discrete and display disregard, and even disdain, for computations involving real numbers or floating point arithmetic (SMALE [1990]); it is particularly illuminating that C, the *lingua franca* of programming languages, was not originally provided with complex numbers. Applied mathematicians, scientists and engineers dislike numerical analysis because they view numerical computation as a necessary evil, distracting them from their true focus. For example, suppose that you are an applied mathematician. Your aim is to further the understanding of a natural or technological phenomenon. You have obtained experimental data, formulated a mathematical model (almost certainly in terms of partial differential equations), analysed it with perturbation techniques, ever dreading the moment of that awful trigraph 'DNS': direct numerical simulation. For, sooner or later, you will be forced to spend weeks of your (or your graduate students') time programming, debugging, and fine-tuning algorithms, rather than absorbing yourself in your genuine area of interest.

The focus of this essay is on the interaction of numerical with pure mathematics. Firstly, we will dispense briefly with the other protagonists, not because they are less important or less interesting but because they are a matter for another discourse. As far as computer scientists are concerned, their genuine 'interface of interaction' with numerical thinking is in two distinct areas: complexity theory and high-performance computing. While complexity theory has always been a glamourous activity in theoretical computer science, it has only recently emerged as a focus of concerted activity in numerical circles (BLUM, CUCKER, SHUB and SMALE [1998], WERSCHULZ [1991]), occasionally leading to a measure of acrimony (PARLETT [1992], TRAUB and WOŹNIAKOWSKI [1992]). It is to be hoped that, sooner or later, computer scientists will find complexity issues involving real-number computations to be challenging, worthwhile and central to the understanding of theoretical computation. Likewise, the ongoing development of parallel computer architectures and the computational grid is likely to lead to considerably better numerical/computational interaction at the more practical, engineering-oriented end. The applied flank of numerical analysis is likely to be secured by the emergence of *scientific computing,* also known as *computational science:* the subject of scientific endeavour centred on computation and on a computational model as an alternative to experiment and expensive testing (WILSON [1989]). This is bound to lead to the emergence of a common agenda for scientists, engineers and numerical analysts.

Returning to the interaction with pure mathematics, we note that the latter was never as 'pure' and as detached from the 'real world' as its most fundamentalist adherents, or their fiercest critics, might choose to believe; cf. N. Young's careful deconstruction in YOUNG [1988]. It is difficult to imagine the invention of calculus, say, without the intertwined interests in navigation and astronomy. To move the time frame to the more recent past, imagine differential geometry, functional analysis and algebraic topology without the influence of relativity or quantum theory, or probability theory without statistics. We are in the midst of a new industrial revolution, despite the ending of the 'dot-com' boom. In place of mechanical contraptions machining physical goods, the 'new economy' is increasingly concerned with information, its derivation, transmission, storage and analysis. The computer is the information-processing engine of this emergence (the French word *ordinateur* is much more appropriate) and its influence on the themes and priorities of mathematics will be comparable to the role of physics in earlier centuries.

Note the absence of 'numerical analysis' in the last paragraph. This was deliberate. The joint border of mathematics and computing contains considerably more than numerical analysis alone (or, indeed, computer science) and grows daily. It is crucial to recognise that there is a two-way traffic across the border. On the one hand, pure mathematicians increasingly realise that many of their problems can be elucidated by computation: numerical analysis, numerical algebra, computational number theory, computational topology, computational dynamics, symbolic algebra and analysis.... Arguably, the day is nigh when every mathematical discipline will have its computational counterpart, a convenient means to play with ideas and investigate mathematical phenomena in a 'laboratory' setting. The idea is most certainly *not* to dispense with mathematical proof and rigour, but to use the computer to help with the necessary guesswork that leads to clear formulation and proof. On the other hand, seeing mathematical issues through the prism of computation inevitably leads to new *mathematical* problems, in need of traditional mathematical treatment. As we have already mentioned, such problems are often considerably more intricate than the mathematical phenomena that they approximate: discretizations in place of differential equations, iterative schemes in place of algebraic equations, maps in place of flows.

It is misleading to appropriate the name 'numerical analysis' for this activity. Firstly, we have already mentioned that the interaction of mathematics and computing is much wider and richer. Secondly, not all numerical analysts are concerned with mathematical issues: the design, programming and implementation of algorithms, i.e. their software engineering, are just as important and, arguably, more crucial for future applications than theorem proving. One should never confuse mathematics (or beauty, for that matter) with virtue. Thus, it makes sense to resurrect the old designation of *computational mathematics* to embrace the range of activities at the mathematics/computation interface. From this moment onward we will thus often abandon the dreaded words 'numerical analysis' altogether and explore the foundations of computational mathematics.

Mathematics is not some sort of an optional extra, bolted onto computational algorithms to allow us the satisfaction of sharing in the glory of Newton, Euler and Gauss. Mathematical understanding is ultimately crucial to the design of more powerful and useful algorithms. Indeed, part of the enduring profundity and beauty of mathematics is how often 'useless' mathematical concepts and constructs turn out to be very useful indeed. Hardy was right: mathematics reveals the structure and the underlying pattern of a problem. This is just as true with regard to mathematical computation as, say, in regard to algebraic geometry or analytic number theory.

2. The 'mathematics' in computational mathematics

The minimalist definition of 'the mathematics behind computational mathematics' is little more than linear algebra and Taylor expansions. The maximalist definition is that all mathematics is either relevant or potentially relevant to computation. The first definition is wrong but the second, while arguably correct, is neither useful nor informative. We need to refocus and explore a number of broad mathematical themes in greater detail, as they pertain to computation but always *from a mathematical standpoint*. This exercise, by necessity, is tentative, inexact, incomplete and coloured by our ignorance and prejudices. It lays no claims for any universal truth, being simply an attempt to foment debate.

2.1. Approximation theory

Suppose that we investigate a mathematical construct A, say, which, while probably amenable to some sort of mathematical analysis, is not amenable to computation. In the computational context it often makes sense to replace it by a different construct, A_ε, say, where ε denotes one or more built-in parameters. The virtue of A_ε is that it can be computed, but, to belabour the obvious, it is distinct from the original problem A. Thus some obvious pertinent questions follow: How closely does A_ε approximate A? Can we estimate, given ε, the error incurred by replacing A with A_ε? What is an appropriate measure for the error? How are the parameters and the error related?

These are classical approximation problems in numerical analysis and, whether in university courses or the research community, there is no simple dividing line between 'proper' numerical analysis and approximation theory. We all take it for granted that the basics of approximation theory should be acquired by any budding numerical analysis, and these fundamentals are usually taken to include interpolation, least-squares and Chebyshev approximation by polynomials, some orthogonal-polynomial theory, elements of the theory of splines, and possibly simple rational approximation. These topics are still important, but it is time to add to the list.

In the last two decades, approximation theory has undergone a revolution on several different fronts. In the long run, this revolution underscores, indeed deepens, its relevance to computational mathematics. The most significant developments are probably new approximating tools, nonlinear approximation and multivariate approximation.

The most familiar of emerging approximation tools are *box splines* (DE BOOR, HÖLLIG and RIEMENSCHNEIDER [1993]), *radial basis functions* (BUHMANN [2000]) and, perhaps the best known and understood, *wavelets* (DAUBECHIES [1992]). Each is based on a different premise and each confers a different advantage. It should also be noted that both box splines and radial basis functions are based on the fundamental work of I.J. Schoenberg.

The box spline M_X associated with the $d \times n$ matrix X with columns in $\mathbb{R}^d \setminus \{\mathbf{0}\}$ is the distribution defined by

$$\langle M_X, \phi \rangle := \int_{[0,1)^n} \phi(Xt)\,\mathrm{d}t,$$

where $\phi : \mathbb{R}^d \to \mathbb{R}$ can be any compactly supported smooth function. When $n \geqslant d$ and X is of full rank, then M_X can be identified with a function on the image of X. For example, when $X = (\,1 \quad 1\,)$, we obtain

$$\langle M_X, \phi \rangle = \int_{[0,1)^2} \phi(t_1 + t_2)\,\mathrm{d}t = \int_{\mathbb{R}} M_X(x)\phi(x)\,\mathrm{d}x,$$

where, slightly abusing notation, $M_X : \mathbb{R} \to \mathbb{R}$ also denotes the continuous function, supported by the interval $[0, 2]$, given by

$$M_X(x) = 1 - |1 - x|, \quad x \in [0, 2].$$

In this case, we have obtained a *univariate B-spline*, and, in fact, all B-splines arise in this way. However, when X is *not* a $1 \times n$ matrix, we obtain the greater generality of box splines. Such functions are piecewise polynomial, although this is not completely obvious from the geometric definition given here. Their mathematical properties are beautiful, but, until relatively recently, they were solutions looking for problems. However, their fundamental relevance to subdivision, a rapidly developing field of computer-aided geometric design, is now evident.

How did such objects arise? The characterization of B-splines as volumes of slices of polyhedra was discovered by CURRY and SCHOENBERG [1966], but their generalization, and the beautiful mathematics thereby generated, did not arise until de Boor's work in the late 1970s. In the 1980s, they became a central topic of interest in approximation theory. Micchelli, one of the most active mathematicians in this field, provided a fascinating historical survey in Chapter 4 of his lecture notes on computer-aided geometric design (CAGD) (MICCHELLI [1995]).

The development of radial basis functions has been less coherent than that of box splines. It is well known that *natural spline* interpolants provide univariate interpolants minimizing a certain integral, and this variational characterization was the origin of the term 'spline': the integral for cubic splines is an approximation to the bending energy of a thin rod, so that the 'draughtsman's spline', an archaic tool for drawing smooth curves before CAGD, is approximated by cubic spline interpolants. DUCHON [1976] modified this variational characterization to construct *thin plate splines* (although these functions had previously been studied in HARDER and DESMARAIS [1972], where they were termed 'surface splines'). Duchon found that the function $s:\mathbb{R}^2 \to \mathbb{R}$ that interpolates the data $s(z_j) = f_j$, $j = 1, 2, \ldots, n$, at non-collinear distinct points $z_1, \ldots, z_n$ in $\mathbb{R}^2$, and minimizes the integral

$$\int_{\mathbb{R}^2} \left(s_{xx}^2 + 2s_{xy}^2 + s_{yy}^2\right) \mathrm{d}x\, \mathrm{d}y$$

is given by

$$s(z) = \sum_{k=1}^{n} c_k \phi\left(\|z - z_k\|\right) + p(z), \qquad z \in \mathbb{R}^2,$$

where $\phi(r) = r^2 \log r$, for $r \geqslant 0$ ($\phi(0) = 0$), p is a linear polynomial, and the coefficients $c_1, \ldots, c_n$ are defined by the interpolation equations and the moment conditions

$$\sum_{k=1}^{n} c_k = \sum_{k=1}^{n} c_k z_k = 0.$$

Thus $s - p$ is a linear combination of translates of the radially symmetric function $\Phi(z) = \phi(\|z\|)$, for $z \in \mathbb{R}^2$, and Duchon's analysis established the nonsingularity of the underlying linear system. This construction can be generalized to any dimension, and the possibly surprising functional form of Φ becomes more plausible once the reader knows that Φ is a multiple of the fundamental function for the biharmonic operator Δ^2 in $\mathbb{R}^2$.

Shortly before Duchon's analysis, a geostatistician, Rolland Hardy had constructed an approximant with a similar functional form (R. HARDY [1971]). He took

$$s(z) = \sum_{k=1}^{n} c_k \phi\left(\|z - z_k\|\right), \qquad z \in \mathbb{R}^d,$$

but chose $\phi(r) = (r^2 + c^2)^{1/2}$, calling the resultant surfaces *multiquadrics*. Several workers found Hardy's multiquadrics to be useful interpolants, but no theoretical basis was then available; it was not even known if the linear system defining the interpolant was invertible. Nevertheless, their practical performance was too compelling for them to be abandoned, and a careful survey of Richard Franke found thin plate splines and multiquadrics to be excellent methods for scattered data interpolation (FRANKE [1982]). By this time, several variants had been suggested, so that Franke termed them 'global basis function type methods'. All of these methods used interpolants that were linear combinations of translates of a radially symmetric function, and the term 'radial basis function' seems to have entered mathematical conversation shortly thereafter, although we cannot pinpoint its genesis.

Franke had conjectured that the multiquadric interpolation matrix A, defined by

$$A_{jk} = \phi(\|x_j - x_k\|), \quad 1 \leqslant j, k \leqslant n,$$

for $\phi(r) = (r^2 + c^2)^{\pm 1/2}$, was nonsingular given distinct $x_1, \ldots, x_n \in \mathbb{R}^d$, whatever the dimension d of the ambient space. This conjecture was established by Charles Micchelli in his seminal paper (MICCHELLI [1986]), using mathematics originally developed by I.J. Schoenberg some forty years earlier, and this history is an excellent example of the path from pure mathematics to computational mathematics. In the 1930s, Fréchet, and others, had axiomatised metric spaces and were beginning to consider the following deep question: which metric spaces can be isometrically embedded in Hilbert spaces? This question was completely settled in a series of remarkable papers in SCHOENBERG [1935], SCHOENBERG [1937], SCHOENBERG [1938] and VON NEUMANN and SCHOENBERG [1941], and is too involved to present here in detail; a modern treatment may be found in BERG, CHRISTENSEN and RESSEL [1984]. To whet the reader's appetite, we remark that the interpolation matrix is positive definite, for any dimension d, if the function $f(t) := \phi(\sqrt{t})$ is *completely monotonic* (and nonconstant), that is, $(-1)^m f^{(m)}(t) \geqslant 0$, for every $t > 0$ and each nonnegative integer m (a new geometric proof of this fact may be found in BAXTER [2002]). This theory of positive definite functions on Hilbert space is equally fundamental to *learning theory*, where radial basis functions and data fitting are finding new applications.

One may ask what is so fundamental about box splines or radial basis functions? Our answer is based not on the specific definition of a box spline or of a radial basis function, but on the overall mathematical process underlying their construction. They are both a multivariate generalisation of the same familiar univariate construct, the spline function. Yet, they are not 'obvious' generalisations. After all, responding to a need for piecewise-polynomial approximation bases in a multivariate setting, in particular originating in the finite-element community, approximation theorists have been extending splines to $\mathbb{R}^m$, $m \geqslant 2$, for decades using tensor products. Yet, this procedure automatically restricts the underlying tessellation to parallelepipeds, and this imposes an unwelcome restriction on applications. An alternative, which has attracted impressive research effort, is an *ad hoc* construction of piecewise-smooth functions in more complicated geometries, one at a time. The twin approaches of box splines and of radial basis functions offer a multivariate generalisation of splines which allows an automatic generation of interpolating or approximating bases which are supported by fairly general polyhedra or, in the case of radial basis functions, which trade global support off for their ability to adjust to arbitrarily scattered data. The underlying trend is common to other foundational themes of computational mathematics: rather than extending existing constructs in a naive or *ad hoc* manner, dare to mix-and-match different mathematical concepts (in the present case, geometry and convexity theory, harmonic analysis, theory of distributions) to underly a more general computational construct.

2.2. *Harmonic analysis*

This is, of course, an enormous subject, whose *idée fixe* has been the construction of orthogonal decompositions of Hilbert spaces of square-integrable functions, but we should remark that there has again been a remarkable split between theoreticians and computational harmonic analysts. For example, the excellent books of KÖRNER [1988] and TERRAS [1999] describe both the Fast Fourier Transform algorithm and the Poisson summation formula, but neither observes that the Danielson–Lanczos lemma, which is the recursion at the heart of every FFT code, is *precisely* the Poisson summation formula relating Fourier coefficients of functions defined on $\ell^2(\mathbb{Z}_{2n})$ to those defined on $\ell^2(\mathbb{Z}_n)$. As another example, there are many deep theoretical works describing the Radon transform from the perspective of a pure mathematician studying the harmonic analysis of homogeneous spaces (HELGASON [1981]), but almost none mention its fundamental importance to computational tomography (NATTERER [1999]). Perhaps some *rapprochement* has occurred in the last decade with the emergence of wavelets and fast multipole methods, and it is to be hoped that this will flourish.

2.3. *Functional analysis*

For the traditionally minded, the phrase 'the mathematics of numerical analysis' is virtually synonymous with functional analysis. Even bearing in mind the observation that computational mathematics is substantially wider than just numerical analysis, there is no escaping the centrality of functional analysis in any description of the foundations of computing.

While rightfully scornful of the "just discretise everything in sight and throw it on the nearest computer" approach, which sadly prevails among many practitioners of scientific computing, we should remember that this used to be the prevailing attitude of mathematicians with an interest in solving partial differential equations. The magic wand that turned mathematical wishful thinking into a coherent theory was functional analysis, and it was wielded in the main by research groups inspired by two towering individuals, Richard Courant in New York and Jacques-Louis Lions in Paris. The traditional view of the discretization procedure is that continuous function values are approximated by discrete values at grid points. Although amenable to mathematical analysis, this approach rests on the obvious infelicity of comparing unlike with unlike. The alternative paradigm approximates an *infinite-dimensional* function space by a *finite-dimensional* subspace. Thus, both the exact solution of a differential equation and its approximation inhibit the same space and the discretization error can be measured in the natural topology of the ambient function space.

A natural bridge between the infinite-dimensional and its approximation is provided by one of two complementary approaches. The *Ritz method* replaces a boundary-value differential problem by a variational problem, thereby traversing the 'variational to differential' route, with a pedigree stretching to Newton, Euler and Lagrange, in the opposite direction. In a variational setting the original differential system reduces to global optimization in the underlying (infinite-dimensional) Hilbert space $\mathcal{H}$ and it is approximated by optimizing in a finite-dimensional subspace of $\mathcal{H}$. The norm $\|\cdot\|_{\mathcal{H}}$ provides the natural measuring rod to calculate the approximation error. The *Galerkin method* seeks a finite-dimensional approximation to the differential equation in the *weak* sense. Instead of searching for a function u such that $\mathcal{L}u = f$, where $\mathcal{L}$ is a differential operator and we impose suitable boundary conditions, we seek a function $u \in \mathcal{H}$ such that $\langle \mathcal{L}u - f, v\rangle_{\mathcal{H}} = 0$, $v \in \mathcal{H}_0$, where $\mathcal{H}$ is a Hilbert space of suitable functions (in practice, a *Sobolev space*) obeying the right boundary conditions, $\mathcal{H}_0$ is the same Hilbert space, except that boundary conditions are set to zero and $\langle\cdot,\cdot\rangle_{\mathcal{H}}$ is the underlying inner product. In a Galerkin method we seek a solution from a finite-dimensional subspace of $\mathcal{H}$.

All this is classical finite element theory, well understood and ably described in many monographs (CIARLET [1978], STRANG and FIX [1973]). Of course, this broad set of mathematical principles falls short of describing a viable computational approach. It needs to be supplemented with the practice of finite-element functions and their construction, the correct treatment of boundary conditions, and a plethora of mundane programming issues that often make the difference between a great mathematical idea of restricted practical importance and a great practical algorithm.

The role of functional analysis in the foundations of computational mathematics might thus be considered as the decisive, battle-winning weapon of a heroic, yet long past, age, the longbow of numerical analysis. This, however, is definitely false. Although we champion many new foundational themes and disciplines in this essay, functional analysis remains at the very heart of computational theory, and it is likely to stay there for a long time to come. The basic paradigm of functional analysis in the numerical setting, namely that approximation takes place in a finite-dimensional subspace of the infinite-dimensional space of all possible solutions, remains valid with regard to the many new methods for discretizing differential and integral operators: spectral and pseudospectral methods, boundary element and spectral element techniques, particle methods and, perhaps most intriguingly, multiresolution and multiscale techniques (DAHMEN [1997]). It is not an exaggeration to state that functional analysis provides the universal language and the rigour in the analysis of discretized partial differential equations.

An exciting recent development that brings together functional analysis, approximation theory and multiresolution techniques is *nonlinear approximation*. Most classical approximation problems, not least the problems from the previous subsection, are linear: we seek an approximating function from a linear space. As an example, consider the interpolation of univariate data by, say, cubic splines: we fix knots and interpolation points in the requisite interval and seek the unique member of the linear space of cubic splines with given knots that interpolates given data at the (fixed) interpolation points. This procedure and its error are well understood and, in particular, the practical solution of the problem can be reduced to the solution of a banded linear system of algebraic equations. However, suppose that we allow ourselves the freedom to vary the knots (and perhaps also the interpolation points) and are guided in our quest by the goal of minimising the approximation error in some norm. The problem ceases to be linear and it presents substantially more formidable challenge at both practical and analytic level.

The choice of optimal knots and the degree of improvement in the quality of the approximation depend very sensitively upon the choice of the underlying function space. Yet, substantial progress has recently been made in nonlinear approximation, in particular in connection with multiresolution methods and wavelet functions (DEVORE [1998], TEMLYAKOV [2002]). The underlying functional-analytic machinery is sophisticated, subtle and outside the scope of this brief essay. However, the importance of this development, like that of other important developments at the interface of mathematics and computation, is underscored by its impact on practical issues. 'Free choice' of knots is nothing else but adaptivity, a central issue in real-life computing to which we will return in the next section.

Another fairly recent development, particularly relevant as numerical problems increase in size, is the phenomenon of *concentration of measure*. This name refers to the increasing regularity seen in many structures as the dimension of the problem tends to infinity. One of the most striking ways to introduce the neophyte reader to this field is the following theorem of Lévy and Milman: a continuous function defined on the sphere $S^n = \{\boldsymbol{x} \in \mathbb{R}^{n+1}\colon \|\boldsymbol{x}\| = 1\}$ is, for large dimensions n, almost constant on almost all the sphere. More precisely, if our function f has median value M_f, and if A is that subset of points on the sphere at which f attains its median value, then

$$\mathbb{P}(\boldsymbol{x} \in A_\delta) \geqslant 1 - C\exp\bigl(-\delta^2 n/2\bigr),$$

where C is a (known) constant independent of n, $\mathbb{P}$ denotes normalized $(n-1)$-dimensional Lebesgue measure on the sphere, and

$$A_\delta := \bigl\{\boldsymbol{x} \in S^{n-1}\colon \rho(\boldsymbol{x}, A) \leqslant \delta\bigr\},$$

ρ denoting the *geodesic* metric on the sphere. This result seems to have been initially derived by Paul Lévy but "with a proof which has not been understood for a long time", according to MILMAN and SCHECHTMAN [1986]. This beautiful result deserves to be better known, so we shall now sketch its derivation, further details being available in MILMAN and SCHECHTMAN [1986]. We shall need the *isoperimetric inequality* on the sphere, stated in the following form: for each $\eta \in (0, 1)$ and $\delta > 0$, $\min\{\mathbb{P}(U_\delta)\colon\ U \subset S^n,\ \mathbb{P}(U) = \eta\}$ exists and is attained on a cap of suitable measure, that is, a set of the form $B(r) := \{\boldsymbol{x} \in S^n\colon\ \rho(\boldsymbol{x}, \boldsymbol{a}) \leqslant r\}$, where $\boldsymbol{a}$ can be any fixed point on the sphere. (To see the link with more familiar forms of the isoperimetric inequality, observe that $\mathbb{P}(U_\delta) \approx \mathbb{P}(U) + \ell(\partial U)\delta$, when the length $\ell(\partial U)$ of the boundary ∂U exists.) GARDNER [2002] has recently published a lucid derivation of this form of the isoperimetric inequality, based on the Brunn–Minkowski inequality. With the isoperimetric inequality to hand, we readily obtain the following crucial bound.

THEOREM 2.1. *If $U \subset S^n$ is any measurable set satisfying $\mathbb{P}(U) \geqslant 1/2$, then*

$$\mathbb{P}(U_\delta) \geqslant 1 - C_1 \mathrm{e}^{-\delta^2 n/2}, \tag{2.1}$$

for all sufficiently large n, where C_1 is an absolute constant whose value is close to $1/2$.

PROOF. The isoperimetric inequality tells us that it is sufficient to establish (2.1) when U is a cap. Thus we must show that

$$\mathbb{P}\left(B\left(\frac{\pi}{2}+\delta\right)\right) \geqslant 1 - C_1 \mathrm{e}^{-\delta^2 n/2}.$$

Now

$$\mathbb{P}\left(B\left(\frac{\pi}{2}+\delta\right)\right) = \frac{\int_{-\pi/2}^{\delta} \cos^n \theta \,\mathrm{d}\theta}{\int_{-\pi/2}^{-\pi/2} \cos^n \theta \,\mathrm{d}\theta} = \frac{\int_{-\pi\sqrt{n}/2}^{\delta\sqrt{n}} \cos^n (\tau/\sqrt{n})\,\mathrm{d}\tau}{\int_{-\pi\sqrt{n}/2}^{\pi\sqrt{n}/2} \cos^n (\tau/\sqrt{n})\,\mathrm{d}\tau},$$

using the substitution $\theta = \tau/\sqrt{n}$. If we now introduce the function

$$f_n(\tau) = \begin{cases} \cos^n(\tau/\sqrt{n}), & |\tau| < \pi\sqrt{n}/2, \\ 0, & |\tau| \geqslant \pi\sqrt{n}/2, \end{cases}$$

we observe that

$$0 \leqslant f_n(\tau) \leqslant f(\tau) := \mathrm{e}^{-\tau^2/2} \quad \text{for all } \tau \in \mathbb{R},$$

using the elementary inequality $\cos\sigma \leqslant \exp(-\sigma^2/2)$, for $|\sigma| < \pi/2$. Further, since $\lim_{n\to\infty} f_n(\tau) = f(\tau)$ at every point $\tau \in \mathbb{R}$, the dominated convergence theorem implies the relation

$$\lim_{n\to\infty} \int_{-\pi\sqrt{n}/2}^{\pi\sqrt{n}/2} \cos^n(\tau/\sqrt{n})\,\mathrm{d}\tau = \int_{\mathbb{R}} f(\tau)\,\mathrm{d}\tau = \sqrt{2\pi}.$$

Thus

$$1 - \mathbb{P}\left(B\left(\frac{\pi}{2}+\delta\right)\right) = \frac{\int_{\delta\sqrt{n}}^{\pi\sqrt{n}/2} \cos^n(\tau/\sqrt{n})\,\mathrm{d}\tau}{(1+\mathrm{o}(1))\sqrt{2\pi}} \leqslant \frac{\int_{\delta\sqrt{n}}^{\infty} f(\tau)\,\mathrm{d}\tau}{(1+\mathrm{o}(1))\sqrt{2\pi}},$$

and

$$\int_{\delta\sqrt{n}}^{\infty} f(\tau)\,\mathrm{d}\tau = \int_0^{\infty} \mathrm{e}^{-(\sigma+\delta\sqrt{n})^2/2}\,\mathrm{d}\sigma \leqslant \mathrm{e}^{-\delta^2 n/2} \int_0^{\infty} \mathrm{e}^{-\sigma^2/2}\,\mathrm{d}\sigma,$$

whence

$$1 - \mathbb{P}\left(B\left(\frac{\pi}{2}+\delta\right)\right) \leqslant \frac{\mathrm{e}^{-\delta^2 n/2}}{2(1+\mathrm{o}(1))},$$

as $n \to \infty$. □

Now let us define

$$A^+ := \{\boldsymbol{x} \in S^n \colon f(\boldsymbol{x}) \geqslant M_f\}$$

and

$$A^- := \{\boldsymbol{x} \in S^n \colon f(\boldsymbol{x}) \leqslant M_f\}.$$

By definition of the median, we have $\mathbb{P}(A^+) \geqslant 1/2$ and $\mathbb{P}(A^-) \geqslant 1/2$. Hence, applying our theorem, we deduce the inequalities

$$\mathbb{P}(A_\delta^\pm) \geqslant 1 - C_1 \mathrm{e}^{-\delta^2 n/2},$$

for all sufficiently large n. However, since $A = A^+ \cap A^-$ and $A_\delta = A_\delta^+ \cap A_\delta^-$, we deduce the relations

$$\begin{aligned}\mathbb{P}(A_\delta) &= \mathbb{P}(A_\delta^+) + \mathbb{P}(A_\delta^-) - \mathbb{P}(A_\delta^+ \cup A_\delta^-)\\ &\geqslant \mathbb{P}(A_\delta^+) + \mathbb{P}(A_\delta^-) - 1\\ &\geqslant 1 - 2C_1 \mathrm{e}^{-\delta^2 n/2}.\end{aligned}$$

Thus a continuous function defined on a large dimensional sphere is concentrated around its median value.

The concentration of measure phenomenon lies at the heart of much modern work on the geometry of Banach space, including Dvoretsky's famous theorem on almost Euclidean subspaces. For further work, we direct the reader to MILMAN and SCHECHTMAN [1986], and the delightful collection (LEVY [1997]).

The importance we see here is that many numerical problems *benefit* from concentration of measure: the probability that we obtain an unpleasant example is small for large n. Random graph theory contains many examples of such properties, but we suspect that concentration of measure is also responsible for the paucity of unpleasant numerical examples in many fields when the dimension is large. For example, Edelman has computed the probability distribution of the condition number of symmetric matrices, and his results exhibit concentration of measure for large n. To take another example, the *field of values* of a symmetric $n \times n$ matrix A is precisely the values taken by the continuous function $f(\boldsymbol{x}) := \boldsymbol{x}^\top A\boldsymbol{x}$, when $\boldsymbol{x} \in S^{n-1}$. Thus we can immediately apply the Lévy–Milman theorem to deduce that the field of values will concentrate for large n.

2.4. *Complexity theory*

Classical complexity theory is at the very heart of the interface between computer sciences and mathematics. Its framework is based on the 'distance' between a problem and algorithms for its solution. Each algorithm bears a *cost,* which might be conveniently expressed in terms of computer operations, length of input and sometimes other attributes.[1] Minimizing the cost is an important (and often overwhelming) consideration in the choice of an algorithm and in the search for new algorithms. The *complexity* of a problem is the lowest bound on the cost of any algorithm for its solution. Clearly, understanding complexity places the search for algorithms in a more structured and orderly framework. No wonder, thus, that classical complexity theory has attracted a

[1] We are assuming serial computer architecture. This is motivated not just by considerations of simplicity and economy of exposition, but also because complexity theory in parallel architectures is ill developed and, considering the sheer number of distinct and noncommensurable parallel architectures, arguably of little lasting importance.

great deal of intellectual effort, and that its main conundrum, the possible distinction (or otherwise) between the class P (problems which can be computed in polynomial time) and the class NP (problems whose possible solution can be checked in polynomial time) is considered one of the outstanding mathematical conjectures of the age. Yet, classical complexity theory is at odds with the paradigm of numerical computation. The starting point of complexity theory is the Turing machine and its framework is that of discrete operations on integer quantities. Even if a floating-point number is represented as a finite sequence of binary numbers, this provides little insight in trying to answer questions with regard to the cost and complexity of algorithms for problems in analysis.

Having been denied the unifying comfort of the Turing machine, 'real number complexity' has developed in a number of distinct directions. Perhaps the oldest is the ubiquitous counting of *flops* (floating point operations), familiar from undergraduate courses in numerical linear algebra. It is relatively straightforward to calculate the cost of finite algorithms (e.g., Gaussian elimination or fast Fourier transform) and, with more effort, to establish the cost of iterative procedures for a raft of linear-algebraic problems. Yet, it is considerably more difficult to establish the complexity of the underlying problem from simple flop-counting arguments, and the few successes of this approach (e.g., Winograd's result on the complexity of the discrete Fourier transform (WINOGRAD [1979])) are merely exceptions proving this rule. Moreover, counting flops becomes considerably less relevant once we endeavour to approximate *continuous* entities: differential and integral equations. The cost and complexity are equally relevant in that setting, but the question of how well a discrete system models a continuous one (to which we will return in a different setting in Section 2.6) is considerably more challenging than simply counting floating-point operations.

A more structured attempt to introduce complexity to real-number calculations is *information-based complexity*. The point of departure is typically a class of continuous problems, e.g., multivariate integrals or ordinary differential equations, defined very precisely in a functional-analytic formalism. Thus, the integrals might be over all functions that belong to a specific Sobolev space in a specific d-dimensional domain, ordinary differential equations might be restricted by size of the global Lipschitz constant and smoothness requirements in a specific interval, etc. An algorithm is a means of using imperfect (finite, discrete, perhaps contaminated by error) *information* in a structured setting to approximate the underlying continuous problem. Its cost is measured in terms of the information it uses, and its performance assessed in some norm over the entire class of 'acceptable' functions. Information-based complexity has elicited much controversy (PARLETT [1992], TRAUB and WOŹNIAKOWSKI [1992]) but it is safe to state that, concentrating our minds and helping to define the underlying subject matter and its methodology in precise mathematical terms, it has contributed a great deal to our understanding of complexity issues.

The latest – and most ambitious – attempt to fashion complexity for real-number calculations is due to Steve Smale and his circle of colleagues and collaborators. It is elaborated in heroic scope in BLUM, CUCKER, SHUB and SMALE [1998] and addressed by CHEUNG, CUCKER and YE [2003]. Recall that the organising focus for 'discrete complexity' is provided by the Turing machine but that the latter is inadequate for real-number computations. The central idea is to replace the Turing machine by a different formal construct, capable of basic operations on *real-number* inputs and producing real-number outputs. Further details are beyond the scope of this essay, but it suffices to mention that the entire edifice of 'Turing-machine complexity' can be extended, at great intellectual effort, to a substantially more elaborate framework, inclusive of questions (and partial answers) to issues of the type '$\mathrm{P} \neq \mathrm{NP}$?'.

The formal-machine approach to real-number complexity is already feeding back into numerical analysis 'proper', not least for the calculation of zeros of nonlinear functions (BLUM, CUCKER, SHUB and SMALE [1998], SMALE [1997]), but also affecting optimization (CHEUNG, CUCKER and YE [2003]). This, indeed, is the great virtue of complexity theory in its many manifestations: by focusing the mind, formalising sometimes vague activity in precise terms, quantifying the computational effort in an honest manner, it ultimately leads to a better understanding of existing algorithms and the frontiers of possible computation and, ultimately, to better algorithms.

2.5. *Probability theory and statistics*

The historical development of probability and statistics provides an interesting contrast with that of numerical analysis. Both fields were essentially despised for many years, but probability theory became respectable after Kolmogorov's measure-theoretic axiomatisation in 1933. However statistics has continued to be somewhat unfashionable, for reasons often similar to numerical analysis, and its relationship with probability theory has often faltered. In David Williams' memorable words in the preface to WILLIAMS [2001]: "Probability and Statistics used to be married; then they separated; then they got divorced; now they hardly see each other." There is no need for this breach to continue, and it is to be hoped that Williams' excellent text will help to introduce many mathematicians to statistics. Certainly, many of us were turned off by statistics courses concentrating on the mechanics of quantiles, medians and hypothesis testing, often omitting the beautiful mathematics underlying these topics. There is certainly an analogy to be made here with those numerical analysis courses which spend far too much time on floating point error analysis, whilst avoiding vastly more interesting (and important) topics. This is not to deny error analysis its place, merely to emphasize that it is just part of the picture, not its focal point.

It should be emphasized that there is really no clear boundary between numerical analysis and statistics. For example, given an approximation to a function known only through values corrupted by noise, how do we decide whether the fit is good? All numerical analysis texts cover least squares, but few cover the statistical theory of Chi-Squared tests. Similarly, many statistics texts cover least squares from the perspective of deriving minimum-variance unbiased estimators, but very few devote sufficient space (if any) to the singular value decomposition, or the fact that there are many reasons to *avoid* the least squares measure of error, given its particular sensitivity to outliers in the data, instead using the many excellent algorithms available for minimizing other norms of the error; see, for instance, WATSON [1998]. Further, both numerical analysis and statistics are evolving rapidly, driven by the importance of stochastic simulation. One particular common ground is that of mathematical finance, which often moves between partial differential equations and stochastic simulations in order to price options.

One topic which perfectly illustrates the deep connections between numerical analysis and statistics is that of *random matrices*. We recently rediscovered some properties of symmetric Gaussian matrices, which deserve to be better known. Furthermore, the analysis is a good example of numerical linear algebra that is unknown to most probabilists, as well as some probability theory forgotten, or never properly absorbed, by numerical analysts.

A *Gaussian random matrix* is simply a matrix whose elements are independent, normalized Gaussian random variables. We concern ourselves with the special case of *symmetric* Gaussian random matrices, for which only the elements on, or above, the diagonal are independent. Such matrices have been studied for decades, by physicists and mathematicians, and we refer the reader to Alan Edelman's excellent (as yet unpublished) book for their fascinating history and background (EDELMAN [20xx]). Here, we merely observe that their spectral properties display highly regular behaviour as the size of the matrices tends to infinity. Coming to this topic afresh via an initially unrelated problem in geometric integration, we began to consider spectral properties of such matrices, but were limited by the cost of the calculation. To generate the eigenvalues of a symmetric Gaussian random matrix, we simply generated $n(n+1)/2$ independent samples from a normalized Gaussian pseudo-random number generator and then computed the eigenvalues of the symmetric matrix, which requires $\mathcal{O}(n^3)$ floating point operations. This cubic cost in computation, and quadratic cost in storage, limit simulations to rather modest n and, displaying the cardinal mathematical virtue of impatience, we began to consider how this might be improved.

Now, given any vector $\boldsymbol{v} = (v_1, \ldots, v_n)^\top \in \mathbb{R}^n$, it is possible to construct a reflection $Q \in \mathbb{R}^{n\times n}$ for which

$$Q\boldsymbol{v} = \begin{pmatrix} v_1 \\ \sqrt{v_2^2 + \cdots + v_n^2} \\ 0 \\ \vdots \\ 0 \end{pmatrix},$$

and we observe that Q is a symmetric orthogonal matrix. In numerical analysis texts, it is traditional to refer to such orthogonal matrices as Householder reflections, in honour of A.S. Householder, a pioneer of numerical linear algebra. In fact, it is easy to check that

$$Q = I - 2\boldsymbol{u}\boldsymbol{u}^\top,$$

where $\boldsymbol{u} = \boldsymbol{w}/\|\boldsymbol{w}\|$ and $\boldsymbol{w}$ is defined by the equations

$$\boldsymbol{w} = \begin{pmatrix} 0 \\ v_2 + \|\boldsymbol{v}\| \\ v_3 \\ \vdots \\ v_n \end{pmatrix}.$$

Thus, given any Gaussian random matrix A and letting $\boldsymbol{v}$ be the first column of A, we obtain

$$A^{(1)} := QAQ = \begin{pmatrix} A_{1,1} & \sqrt{A_{1,2}^2 + \cdots + A_{1,n}^2} & 0 & \cdots & 0 \\ \sqrt{A_{1,2}^2 + \cdots + A_{1,n}^2} & & & & \\ 0 & & & & \\ \vdots & & & \hat{A} & \\ 0 & & & & \end{pmatrix}.$$

Of course, $A^{(1)}$ and A have identical eigenvalues because Q is orthogonal. Now let us recall that, if $\boldsymbol{X} \in \mathbb{R}^m$ is a normalized Gaussian random vector (i.e. its elements are independent normalized Gaussian random variables), then $Q\boldsymbol{X}$ is also a normalized Gaussian random vector. Further, the Euclidean norm $\|\boldsymbol{X}\|$ of a Gaussian random vector is said to have the χ_m distribution. Thus the elements $A_{1,2}^{(1)} = A_{2,1}^{(1)}$ have the χ_{n-1} distribution, whilst $\hat{A}$ is an $(n-1) \times (n-1)$ symmetric Gaussian random matrix. Recurring this construction, we obtain a symmetric, tridiagonal random matrix T, with the same spectrum as the original matrix A, whose elements have the following properties: (i) the random variables $T_{1,1}, \ldots, T_{n,n}$ and $T_{2,1}, T_{3,2}, \ldots, T_{n,n-1}$ are independent, (ii) the diagonal elements are normalized Gaussian random variables, and (iii) $T_{k+1,k}$ has the χ_{n-k} distribution, for $k = 1, 2, \ldots, n-1$. Now the square of a random variable with the χ_m distribution has, of course, the χ_m^2 distribution, and this is precisely a Gamma distribution (see WILLIAMS [2001]). Further, many papers have suggested efficient algorithms for the generation of pseudo-random numbers with the Gamma distribution (see, for example, WILLIAMS [2001]). Finally, the famous *QR algorithm* (GOLUB and VAN LOAN [1996]) enables the eigenvalues of a symmetric tridiagonal $n \times n$ matrix to be computed in $\mathcal{O}(n)$ operations. Thus we can now generate symmetric tridiagonal random matrices instead of random Gaussian matrices, the cost being reduced from cubic (in n) to linear.

We are unaware of an exposition of the above algorithm in the literature, hence our detailed treatment. STEWART [1980] and TROTTER [1984] seem to have been the first to observe these facts, the former being concerned with the generation of random orthogonal matrices, while the latter used the distribution of the random matrix T to derive Wigner's celebrated circle law. Here let us remark that the *expected value* $\mathbb{E}\,|T|$ of the random matrix $|T|$ is identically zero, except that the subdiagonal and superdiagonal elements form the sequence $1, \sqrt{2}, \ldots, \sqrt{n-2}, \sqrt{n-1}$. Thus, the principal minors of the Jacobi matrix $\mathbb{E}\,|T|$ obey the same three-term recurrence relation as monic Hermite polynomials, and this observation is a crucial step in Trotter's derivation of the circle law.

It is frustrating to rediscover a mathematical property, but a familiar feeling for us all. However, there might be similar insights awaiting us for other classes of random matrices. Furthermore, one of the most active areas of modern graph theory is that of *random graph theory*, where many recent papers study spectral properties of the adjacency matrix generated by a random graph. We suspect that there are some beautiful results to be discovered at the intersection of random graph theory, random matrix theory and numerical linear algebra.

2.6. *Nonlinear dynamical systems*

At the first level of approximation, each mathematical problem is one of a kind, separated from the universe of 'other' problems. This comforting state of affairs does not survive further deliberation: the 'space of mathematical problems' is continuous, not discrete. In particular, most meaningful problems in the realm of analysis are equipped with a multitude of parameters. Some parameters are 'explicit', other are conveniently buried in the formulation of the problem as initial and boundary values. Discretization methods for such problems depend on all these parameters, in addition to further parameters (e.g., the step size or the error tolerance) introduced by the numerical procedure. The theory of dynamical systems endeavours to elucidate the behaviour of time-dependent systems, whether continuous *flows* or discrete *maps,* when parameters vary. The main focus is on long-term dynamics and on abrupt qualitative changes (bifurcation points) in the underlying parameters that lead to a change in asymptotic behaviour.

Classical dynamical systems focus on a particular flow or a particular map. In a computational context, though, the situation is subtly different: we must consider two distinct dynamical systems in unison: the *flow* representing the differential system and the *map* corresponding to the discretization method. It is the similarity (or otherwise) of their asymptotic behaviour and their bifurcation patterns which is of interest in a computational setting.

Perhaps the most intriguing asymptotic phenomenon, which has penetrated public consciousness far beyond mathematical circles, is *chaos* – the capacity of some dynamical systems to display very sensitive behaviour in response to tiny changes in parameters. A celebrated case in point is that of the *Lorenz equations*, three coupled nonlinear ordinary differential equations originating in mathematical meteorology which exhibit $\mathcal{O}(1)$ variation in response to arbitrarily small variation in their initial conditions. It is important to bear in mind that chaos is not associated with total disorder. Typically (as is the case with Lorenz equations) the range of all possible states in a chaotic regime is restricted to a well-defined, lower-dimensional object, a *chaotic attractor.* It is this balancing act, chaos embedded within order and governed by precise mathematical laws, which makes the theory of nonlinear dynamical systems so enticing.

In the context of numerical methods, chaotic behaviour assumes added significance. It is perfectly possible for the underlying flow to be nonchaotic, while the discretized map exhibits chaotic behaviour. Thus, the numerical method may display behaviour which is qualitatively wrong. Although this can be avoided in principle by a judicious choice of discretization parameters, this is not always easy and sometimes impossible, because of hardware limitations on storage and speed. Therefore, it is essential to understand this phenomenon, which is elaborated by YAMAGUTI and MAEDA [2003].

'Numerical chaos' is just one possible way in which discretization can lead to the wrong conclusion with regard to the qualitative behaviour of the underlying differential system. With greater generality, the goal is to compare the entire menagerie of possible types of asymptotic behaviour (in other words, all global attractors) in flows and their discretization. A very great deal has been accomplished in that sphere. In particular, it is clear that not all discretization methods are alike: some are prone to exhibit incorrect asymptotic behaviour for even small time steps and space grids, while others are immune to this adverse behaviour (STUART and HUMPHRIES [1996]). The theory of nonlinear dynamical systems is, arguably, the most powerful tool at the disposal of a modern theoretical numerical analyst, attempting to explain long-time behaviour of discretization methods.

2.7. *Topology and algebraic geometry*

Mathematicians famously endeavour to reduce problems to other problems whose solutions are already known. The main concept of *continuation* (a.k.a. *homotopy*) methods is to travel in the opposite direction: commencing from a trivial problem, with known solution, deform it continuously to the problem whose solution we want to compute (ALLGOWER and GEORG [1990], KELLER [1987], LI [2003]).

To first order, this is an exercise in nonlinear dynamical systems: we have two problems, $\boldsymbol{f}$ and $\boldsymbol{g}$, say, for example, systems of algebraic or differential equations or eigenvalue problems. The solution of $\boldsymbol{f}$ is known and, seeking the solution of $\boldsymbol{g}$, we consider a new function, $\boldsymbol{\varphi}_\tau = (1-\tau)\boldsymbol{f} + \tau\boldsymbol{g}$, where the *continuation parameter* τ resides in $[0,1]$. Since $\boldsymbol{\varphi}_0 = \boldsymbol{f}$, we already know its solution and, like Theseus in the Minoan labyrinth, we grab the thread and pursue it, moving in small steps from $\boldsymbol{\varphi}_{\tau_k}$ to $\boldsymbol{\varphi}_{\tau_{k+1}}$, where $0 = \tau_0 < \tau_1 < \cdots < \tau_m = 1$, until we can reach the exit $\boldsymbol{g}$. Unfortunately, neither Ariadne's love nor bifurcation theory are sufficient to attain our goal. The most vexing issue is the number of solutions, and we recall that Theseus left Ariadne. Suppose, for example, that both $\boldsymbol{f}$ and $\boldsymbol{g}$ are systems of multivariate polynomial equations. In general, we do not know the number of solutions of $\boldsymbol{g}$, hence cannot choose $\boldsymbol{f}$ to match it and follow each solution along a different homotopy path. Moreover, there is no guarantee that a homotopy path from a solution of $\boldsymbol{f}$ will indeed take us to a solution of $\boldsymbol{g}$ (rather than, for example, meandering to infinity) or that two solutions of $\boldsymbol{f}$ will not merge at the same solution of $\boldsymbol{g}$.

The basic mathematical tool to investigate how the zeros, say, of $\boldsymbol{\varphi}_\tau$ change as τ varies from 0 to 1 is provided by *degree theory,* which explains the geometry of solutions in topological terms (ALLGOWER and GEORG [1990], KELLER [1987]). It allows the construction of algorithms that avoid unwelcome bifurcations and divergence.

In the important case of systems of multivariate polynomial equations we have at our disposal an even more powerful tool, namely *algebraic geometry.* At its very heart, algebraic geometry investigates *varieties,* sets of common zeros of polynomial systems, and the ideals of polynomials that possess such zeros. To illustrate the gap between the naive counting of zeros and even the simplest problems within the realm of homotopy methods, consider the problem of finding eigenvalues and normalised eigenvectors of a symmetric $n \times n$ matrix A. We can write it as a system of $n+1$ quadratic equations in $n+1$ variables,

$$g_1(\lambda, v_1, v_2, \ldots, v_n) = \sum_{k=1}^{n} A_{1,k} v_k - \lambda v_1,$$

$$\vdots$$

$$g_n(\lambda, v_1, v_2, \ldots, v_n) = \sum_{k=1}^{n} A_{n,k} v_k - \lambda v_n,$$

$$g_{n+1}(\lambda, v_1, v_2, \ldots, v_n) = \sum_{k=1}^{n} v_k^2 - 1.$$

According to the celebrated *Bézout theorem*, the upper bound on the number of all possible solutions of a multivariate polynomial system is the product of the degrees of all individual polynomials. In our case this bound is 2^{n+1}, wildly exceeding the sharp bound of $2n$ (there are n eigenvalues and, if $\boldsymbol{v}$ is a normalised eigenvector, so is $-\boldsymbol{v}$). The theory underlying homotopy algorithms must be powerful enough to know the correct number of solutions without help from linear algebra, since such help is available only in limited number of cases. Moreover, it must ensure that homotopy paths neither meander, nor merge, nor diverge. Modern techniques in algebraic geometry are equal to this challenge and provide a powerful theoretical underpinning for continuation methods (LI [2003]).

Homotopy is not the only concept at the intersection of topology and 'classical' computation. Another development of growing significance, pioneered by TONTI [1976] and recently embraced by an increasing number of professionals, is the use of *combinatorial topology* to embed laws of physics directly into the geometry of discretization space. Lest it be said that this is yet another (strange, beautiful and useless) flight of fancy of pure mathematicians, we should state at the outset that the origin of this set of ideas is in practical needs of physics (HESTENES [1984], TONTI [1976]), electromagnetism (HIPTMAIR [2002]) and geometric modelling (CHARD and SHAPIRO [2000]).

The work of CHARD and SHAPIRO [2000] is typical of this approach. Glossing over many technicalities and (essential) details, the construction of a 'topologically-aware' space discretization commences from an m-cell, a set homeomorphic to a closed unit m-ball in $\mathbb{R}^n$, $m \in \{0, 1, \ldots, n\}$. (The reader might think of an n-cell as an element in a finite-element tessellation and of the m-cells, $m < n$, as faces of higher-dimensional cells.) A *complex* is a collection $\mathcal{C}$ of cells, such that the boundary of each m-cell is a finite union of $(m-1)$-cells therein and a nonempty intersection of two cells in $\mathcal{C}$ is itself a cell in the complex. The *star* of a cell $X \in \mathcal{C}$, where $\mathcal{C}$ is a complex, is the set of all the adjacent cells of dimension greater than or equal to that of X. Chard and Shapiro propose constructing a collection of stars which automatically encodes for specific physical laws that admit topological interpretation. The most striking example is that of *balance*: a flow of physical quantity (energy, heat, mass) across a boundary of a domain is in perfect balance with storage, generation and destruction of that quantity inside. Another example is Stokes' theorem, which is automatically satisfied in the Chard–Shapiro 'starplex', in every cell, of any dimension. Once the physical model is formulated and discretized across this tessellation, it automatically obeys the laws in question. Rather than encoding physics in the software of the numerical method, it is already hardwired into the topology of the underlying space! Other applications of combinatorial topology to computation are similar. They can deal only with physical laws that can be expressed by topological concepts, but this is already powerful enough to allow for the incorporation of multiphysics into numerical methods for complicated partial differential equations (HIPTMAIR [2002]).

2.8. *Differential geometry*

A central reason for a widespread aversion to practical computing is that it abandons the nobility of mathematics, its quest for qualitative entities, its *geometry,* for mundane number crunching. Yet, there is no need to abandon geometry for the sake of computation and the two can coexist in peace and, indeed, in synergy. This is the main idea underlying the emerging discipline of *geometric integration* (BUDD and ISERLES [1999], BUDD and PIGGOTT [2003], HAIRER, LUBICH and WANNER [2002], MCLACHLAN and QUISPEL [2001]), whose language is differential geometry and whose aim is the discretization of differential equations whilst respecting their structural attributes. Many initial-value differential systems possess invariants, qualitative features that remain constant as the solution evolves in time:

- *Conservation laws,* i.e. functionals that stay constant along the solution trajectory. For example, the energy $H(\boldsymbol{p}(t), \boldsymbol{q}(t))$ of the *Hamiltonian* system

$$\boldsymbol{p} = -\frac{\partial H(\boldsymbol{p},\boldsymbol{q})}{\partial \boldsymbol{q}}, \qquad \boldsymbol{q}' = \frac{H(\boldsymbol{p},\boldsymbol{q})}{\partial \boldsymbol{p}}, \quad t \geqslant 0, \tag{2.2}$$

  remains constant along the solution trajectory. In other words, the exact solution of the system, which in principle is a trajectory in a $(2d)$-dimensional Euclidean phase space (where $\boldsymbol{p}, \boldsymbol{q} \in \mathbb{R}^d$), is restricted to the (usually nonlinear, differentiable) *manifold*

$$\mathcal{M} = \{(\boldsymbol{x}, \boldsymbol{y}) \in \mathbb{R}^d \times \mathbb{R}^d\colon\ H(\boldsymbol{x}, \boldsymbol{y}) = H(\boldsymbol{p}(0), \boldsymbol{q}(0))\}.$$

- *Symmetries:* transformations which, when applied to dependent and independent variables, produce another solution to the same differential system. Perhaps the most basic is the SO(3) symmetry, which is shared by all laws of mechanics (ARNOLD [1989]), since they must be independent of the frame of reference. Another is *self similarity,* addressed at some length in BUDD and PIGGOTT [2003]. Euclidean symmetry and self similarity might seem very different animals, yet they are examples of a wider class of symmetries generated by *Lie groups*, and can be phrased geometrically as acting on the vector field, rather than the configuration manifold, of the underlying equation. In other words, in the same manner that conservation laws are associated with a manifold $\mathcal{M}$, say, symmetries are associated with the *tangent bundle* $\mathrm{T}\mathcal{M}$.
- *Symplectic structures* occur in the context of Hamiltonian equations (2.2) – in fact, they characterize such equations (ARNOLD [1989]). There are several equivalent formulations of symplecticity and for our purpose probably the most illustrative is

$$\frac{\partial(\boldsymbol{p}(t), \boldsymbol{q}(t))}{\partial(\boldsymbol{p}(0), \boldsymbol{q}(0))}^{\top} \begin{pmatrix} O & I \\ -I & O \end{pmatrix} \frac{\partial(\boldsymbol{p}(t), \boldsymbol{q}(t))}{\partial(\boldsymbol{p}(0), \boldsymbol{q}(0))} = \begin{pmatrix} O & I \\ -I & O \end{pmatrix}, \quad t \geqslant 0.$$

  Brief examination confirms that the invariant in question evolves on the *cotangent bundle* $\mathrm{T}^*\mathcal{M}$ of some manifold $\mathcal{M}$: the set of all linear functionals acting on the tangent bundle $\mathrm{T}\mathcal{M}$.

We have mentioned three distinct types of invariants, and the list is by no means exhaustive. Two types of invariants have been illustrated by means of the Hamiltonian system

(2.2) and, indeed, it is important to emphasize that the same equation might possess different invariants, possibly of different types. The common denominator to all invariants is that they can be phrased in the language of differential geometry. Indeed, this appears to be the language of choice in geometric integration, which can be applied to a long list of other invariants, e.g. conservation of volume, angular momentum, potential vorticity and fluid circulation.

Geometric integration is a recent addition to the compendium of tools and ideas at the interface of computation and mathematics. Its current state of the art is reviewed in BUDD and PIGGOTT [2003], hence it will suffice to mention one example, for illustration, relevant to conservation laws of a special type. A *Lie group* is a manifold $\mathcal{G}$ endowed with a group structure, consistent with the underlying topology of the manifold. Useful examples comprise of the *orthogonal group* $\mathrm{O}(n)$ of $n \times n$ real orthogonal matrices and the *special linear group* $\mathrm{SL}(n)$ of $n \times n$ real matrices with unit determinant. The tangent space of $\mathcal{G}$ at its unit element has the structure of a *Lie algebra,* a linear space closed under commutation. For example, the Lie algebra corresponding to $\mathrm{O}(n)$ is $\mathfrak{so}(n)$, the linear space of $n \times n$ real skew-symmetric matrices, while $\mathfrak{sl}(n)$, the set of $n \times n$ real matrices with zero trace, is the Lie algebra of $\mathrm{SL}(n)$. The importance of Lie algebras in the present context is underscored by the observation that $X \in \mathfrak{g}$ implies $\mathrm{e}^X \in \mathcal{G}$, where $\mathcal{G}$ is a Lie group, $\mathfrak{g}$ 'its' Lie algebra and, in the case of matrix groups, e^X is the familiar matrix exponential of X.

A Lie group $\mathcal{G}$ is said to *act* on a manifold $\mathcal{M}$ if there exists a mapping $\lambda : \mathcal{G} \times \mathcal{M} \to \mathcal{M}$ which is consistent with the group multiplication: $\lambda(x_1, \lambda(x_2, y)) = \lambda(x_1 x_2, y)$ for all $x_1, x_2 \in \mathcal{G}$, $y \in \mathcal{M}$. A group action is *transitive* if, for every $y_1, y_2 \in \mathcal{M}$, there exists at least one $x \in \mathcal{G}$ such that $\lambda(x, y_1) = y_2$. A manifold endowed with a transitive group action is said to be *homogeneous.* Numerous manifolds arising in practical applications are homogeneous, e.g., $\mathrm{O}(n)$ acts transitively on the $(n-1)$-sphere $\mathrm{S}^{n-1} \subset \mathbb{R}^n$ by multiplication, while the *isospectral orbit*, consisting $\mathcal{I}(n, Y_0)$ of all $n \times n$ symmetric matrices which are similar to a specific matrix $Y_0 \in \mathrm{Sym}(n)$, can be represented via the orthogonal similarity action

$$\mathcal{I}(n, Y_0) = \left\{ Q Y_0 Q^\top : \ Q \in \mathrm{O}(n) \right\}.$$

Here $\mathrm{Sym}(n)$ is the space of $n \times n$ real symmetric matrices. As a matter of independent interest, we leave it to the reader to prove that it is also a homogeneous manifold (e.g., by using the *congruence action* $\lambda(X, Y) = XYX^\top$, where X is $n \times n$ and real, while $Y \in \mathrm{Sym}(n)$) and identify striking similarities between this and a familiar factorization procedure from numerical linear algebra.

Many ordinary differential equations of interest evolve on homogeneous manifolds. Thus, it is not difficult to verify that the solution of

$$\boldsymbol{y}' = A(t, \boldsymbol{y})\boldsymbol{y}, \quad t \geqslant 0, \qquad \boldsymbol{y}(0) = \boldsymbol{y}_0 \in \mathrm{S}^{n-1}, \tag{2.3}$$

where $A : \mathbb{R}_+ \times \mathrm{S}^{n-1} \to \mathfrak{so}(n)$, evolves on S^{n-1}, while the solution of the *isospectral flow*

$$Y' = \left[B(t, Y), Y \right], \quad t \geqslant 0, \qquad Y(0) = Y_0 \in \mathrm{Sym}(n),$$

where $B:\mathbb{R}_+ \times \mathcal{I}(n, Y_0) \to \mathfrak{so}(n)$ lives in $\mathcal{I}(n, Y_0)$. Yet, any attempt to discretize either equation with a standard numerical method whilst respecting its invariants (the Euclidean norm and the eigenvalues, respectively) is unlikely to be successful. Symplectic Runge–Kutta methods (for example, Gauss–Legendre schemes) will stay on a sphere, but no classical algorithm may respect isospectral structure for $n \geqslant 3$ (ISERLES, MUNTHE-KAAS, NØRSETT and ZANNA [2000]). The intuitive reason is that S^{n-1}, $\mathcal{I}(n, Y_0)$ and other homogeneous manifolds are nonlinear structures, while classical methods for differential equations – Runge–Kutta, multistep, Taylor expansions, etc. – do notoriously badly in adhering to nonlinear surfaces. This is a major justification for Lie-group solvers, the new breed of methods originating in geometric integration (ISERLES, MUNTHE-KAAS, NØRSETT and ZANNA [2000]). To give the simplest possible example, the standard Euler method, as applied to (2.3), produces the solution sequence $\boldsymbol{y}_{m+1} = [I + h_m A(t_m, \boldsymbol{y}_m)]\boldsymbol{y}_m$, $m \in \mathbb{Z}_+$, where $\boldsymbol{y}_m \approx \boldsymbol{y}(t_m)$ and $h_m = t_{m+1} - t_m$ is the step size. It is easy to check that, in general, $\boldsymbol{y}_{m+1} \notin \mathrm{S}^{n-1}$. Consider, however, the *Lie–Euler method*: Euler's scheme applied at the algebra level,

$$\boldsymbol{y}_{m+1} = \mathrm{e}^{h_m A(t_m, \boldsymbol{y}_m)} \boldsymbol{y}_m, \quad m \in \mathbb{Z}_+.$$

Since $h_m A(t_m, \boldsymbol{y}_m) \in \mathfrak{so}(n)$, (ii) is true that $\mathrm{e}^{h_m A(t_m, \boldsymbol{y}_m)} \in \mathrm{O}(n)$, whence $\boldsymbol{y}_m \in \mathrm{S}^{n-1}$. We refer to BUDD and PIGGOTT [2003], ISERLES, MUNTHE-KAAS, NØRSETT and ZANNA [2000] for a thorough discussion of Lie-group solvers.

Differential geometry is relevant to computation also in situations when the conservation of invariants, the essence of geometric integration, is not at issue. An important example is computer modelling and the manipulation of spatial images. How do we identify where one object ends and the other starts? How does an image move? How do we blend images? All these might be activities natural to the human (or animal) brain but it is far more challenging to teach a computer basic geometry. The very fuzziness of the above questions is not intrinsic, but reflects the sheer difficulty of formalising 'geometry in motion' in mathematical terms. As reported in SAPIRO [2003], important strides in this direction have been made in the last decade, combining differential-geometric principles with partial differential equations, variational principles and computational methods.

2.9. *Abstract algebra*

Zeros of polynomials again! Suppose thus that we wish to evaluate the zeros of a system of multivariate polynomial equations,

$$p_k(\boldsymbol{x}) = 0, \quad k = 1, 2, \ldots, m, \quad \boldsymbol{x} \in \mathbb{C}^n, \tag{2.4}$$

and, at the first instance, check whether this system has any zeros at all. This is one of the hallowed problems of mathematics, the 'master problem' of algebraic geometry, and immortalised by Hilbert's *Nullstellensatz*. Notwithstanding the success of purely numerical methods, upon which we have already commented in Section 2.7, they are accompanied by a multitude of perils. Some of them have to do with the notorious ill conditioning of polynomial zeros. Others originate in the fact that, although polynomials p_k often have 'nice' (e.g., integer) coefficients, numerical methods reduce everything to the lowest common denominator of floating-point arithmetic. Finally, numerical methods have a tendency to obscure: their final output, a long string of floating-point numbers, often hides interesting dependencies among the different polynomials and variables.

An alternative to numerics is symbolic computation, in our case a methodical conversion of $\{p_1, p_2, \dots, p_m\}$ into a *single* polynomial in a *single* variable, while retaining integer (or rational) coefficients whenever such coefficients prevail in the original problem. An early success of symbolic computation was an algorithm, due to BUCHBERGER [1970], which accomplishes this task employing methods of *commutative algebra*. Since most readers of this essay are likely to be numerical analysts, with little experience of what happens inside symbolic calculation packages, let us consider a simple example of a trivial version of the Buchberger algorithm, as applied to the polynomial system (2.4) with $n = m = 3$, $\boldsymbol{x} = (x, y, z)$ and

$$p_1 = 3x^2y - z - 2, \qquad p_2 = xyz + 1, \qquad p_3 = xz - 2xy + y^2.$$

We commence by imposing a *total ordering* on all terms of the form $x^\alpha y^\beta z^\gamma$. Among the many alternatives, we opt for the (far from optimal!) lexicographic ordering implied by $x \succ y \succ z$. Let $\mathrm{ht}(p)$, the *height* of the polynomial p, denote its largest term subject to our ordering, thus $\mathrm{ht}(p_1) = x^2y$, $\mathrm{ht}(p_2) = xyz$, $\mathrm{ht}(p_3) = xy$. We now set p_4 as equal to the linear combination (easily available from their height functions) which eliminates the largest terms in $\{p_1, p_2\}$,

$$p_4 = zp_1 - 3xp_2 = -z^2 - 2z - 3x.$$

Note that we can now express x in terms of the 'least' variable z by setting $p_4 = 0$, namely $x = -(2z + z^2)/3$. Next, we choose the pair $\{p_2, p_4\}$, eliminate the highest elements and obtain

$$p_5 = -3 + 2yz^2 + yz^3.$$

Although we can now recover y in terms of z, we desist, since we cannot (yet) express y as a *polynomial* in z. Instead, we turn our attention to the pair $\{p_3, p_5\}$, eliminate the highest terms and obtain $q = -6x + xz^4 + y^2z^3 + 4xyz^2$: a rather messy expression! However, subtracting multiples of p_3, p_4 and p_5, we can *reduce* q to a much simpler form, namely

$$p_6 = 9y + 12z + 5z^2 - 4z^4 - 4z^5 - z^6.$$

Thus, letting $p_6 = 0$, y is expressible as a polynomial in z. Finally, we shave off the highest terms in $\{p_2, p_6\}$, reduce using p_4, and derive

$$p_7 = -27 - 24z^3 - 24z^4 - 6z^5 + 8z^6 + 12z^7 + 6z^8 + z^9,$$

a univariate polynomial! The original system has been thus converted to the *Gröbner base* $\{p_4, p_6, p_7\}$.

On the face of it, we are dealing here with a variant of Gaussian elimination, eliminating polynomial terms by forming linear combinations with polynomial coefficients.

This hides the main issue: the Buchberger algorithm *always* terminates and provides a constructive means to compute a Gröbner base (BUCHBERGER and WINKLER [1998], COX, LITTLE and O'SHEA [1997], VON ZUR GATHEN AND GERHARD [1999]). Moreover, it can be generalised to *differential Gröbner bases,* dealing with differential (rather than polynomial) equations. In general, symbolic algebra is rapidly emerging as a major means of mathematical computation, capable of such diverse tasks as the verification of summation formulae (ZEILBERGER [1990]) or computing symmetries of differential equations (HUBERT [2000]). The theoretical framework rests upon commutative algebra, with substantial input from algebraic and differential geometry.

This review of 'pure mathematics influencing computational mathematics' would not be complete without mentioning a fascinating example of computational mathematics reciprocating in kind and enriching pure mathematics, specifically abstract algebra (and theoretical physics). Needless to say, computation helps in understanding mathematics by its very nature, by fleshing out numerical and visual data. Yet, here we wish to draw attention to a different pattern of influence: seeking mathematical structure to explain the behaviour of computational processes may lead to an insight into seemingly-unrelated domains of pure-mathematical research.

A pioneering effort to understand order conditions for Runge–Kutta methods led John Butcher, in the late 1960s, to a theory that illuminates the complicated expansions of Runge–Kutta maps via *rooted trees*. The complex and unpleasant procedure of deriving order conditions for high-stage methods thus reduces to a transparent exercise in recursion over rooted trees and this, indeed, is the method of choice in the modern numerical analysis of ordinary differential equations. In particular, endeavouring to explain the interaction of order conditions once several Runge–Kutta schemes are composed, Butcher endowed his expansion with an algebraic structure (BUTCHER [1972]). For almost thirty years (BUTCHER [1972]) was yet another a paper at the theoretical frontier of numerical analysis, only to be discovered and reinterpreted by algebraists in the late 1990s. It turns out that the *Butcher group* provides a superior representation of a *Hopf algebra*, with very important implications for both algebra and its applications in theoretical physics (BROUDER [1999], CONNES and KREIMER [1998]). It is gratifying to confirm that computational mathematicians can sometimes repay pure mathematics in kind.

2.10. ...and beyond

It is possible to extend the intellectual exercise of this section to most areas of pure mathematics, methodically progressing through the AMS (MOS) subject classification. Sometimes the connection is somewhat forced: most applications of *number theory* appear to be in random-number generation (ZAREMBA [1972]), say, and (to the best of our knowledge) the only significant way in which ergodic theory impacts upon computation is via the ergodic properties of some nonlinear dynamical systems. Sometimes the connection might appear mundane, at first glance, from the mathematical point of view: graph theory is often used in numerical analysis as a useful tool, e.g., in parallel computation, the numerical algebra of sparse matrices, expansions of Runge–Kutta methods, Lie-group and splitting algorithms in geometric integration, etc., but (arguably) these applications are not particularly profound from a graph-theoretical standpoint: they are all simple instances of planar graphs featuring in intricate analytic structures. Yet, one way or the other, our original contention stands: the mathematical foundations of computation are *all* of mathematics.

3. The 'computational' in computational mathematics

3.1. Adaptivity

Like any domain of intellectual endeavour, computational mathematics progresses from the simple to the complex. 'Simple' in this context means using an algorithm which is undifferentiated and does not respond dynamically to the computed solution. Thus, typically, the first program written by a student for the solution of differential equations might be a constant-step Runge–Kutta method: regardless of how the solution changes, whether it undergoes rapid oscillations, say, or stands still, whether the local error is huge or nil, the program ploughs on and spews out numbers which often bear little relevance to the underlying problem. Understandably, this is often also the *last* program that a disenchanted student writes....

Clearly, good computation requires the algorithm to respond to the data it is producing and change the allocation of computing resources (step size, size of the grid, number of iterations, even the discretization method itself) accordingly. The purpose of computation is not to produce a solution with least error but to produce reliably, robustly and *affordably* a solution which is within a user-specified tolerance. In particular, in time-dependent problems, a well-structured algorithm will typically employ two intermeshed mechanisms: a monitor of the error incurred locally during the computation and a means to respond to this error bound (or its estimate) by changing the algorithm's parameters (e.g., step size or space tessellation) or locally improving the accuracy of the solution.

The solution of time-dependent differential and integral equations is not the only area where adaptivity is conducive to good computing practice. An intriguing illustration of adaptivity in a different setting is *lossy data compression*, in a recent algorithm of Cohen, Dahmen, Daubechies and DeVore, using wavelet functions (COHEN, DAHMEN, DAUBECHIES and DEVORE [2001]). Instead of adapting a spatial or temporal structure, it is the expansion of an image (expressible as a long string of '0' and '1's) which is 'adapted', having been first arranged in a tree structure that lends itself to orderly extraction of the most significant bits.

All this emphasis on *adaptivity* is hardly 'foundational' and has been the focus of much effort in numerical analysis, scientific computing and software engineering. Having said this, adaptivity is linked to important foundational issues and, ideally, in place of its current niche of half-mathematics, half-art, should be brought into the fully mathematical domain.

It is premature to prophesize about the likely course of 'mathematical adaptivity', yet an increasing wealth of results implies at the potential of this approach. An important development is the emergence of *nonlinear approximation,* bringing together tools of functional and harmonic analysis and approximation theory to bear upon adaptivity in numerical methods for partial differential equations (DEVORE [1998], TEMLYAKOV

[2002]). This timely theory helps to elucidate the implementation of modern and highly flexible means of approximation, not least wavelet functions, to achieve the important goal of discretizing differential and integral equations. Another development that helps us to understand and fine-tune adaptivity is the introduction of monitor functions that reflect geometric and structural features of the underlying solution (BUDD and PIGGOTT [2003]). Thus Lie symmetries of the underlying differential equations can be used to adapt a numerical solution to deal successfully (and affordably) with discontinuities, blow ups and abrupt qualitative changes.

Adaptivity is not always free of conceptual problems. For example, 'classical' symplectic methods for Hamiltonian ordinary differential equations should be applied with constant (or virtually constant) step size, lest the benefits of symplecticity be voided (HAIRER, LUBICH and WANNER [2002]). At first glance, this represents a setback: we are forced to choose between the blessings of adaptivity and the benefits of symplecticity. However, as in the case of many other 'impossibility results', the main impact of this restriction is to motivate the development of algorithms, *Moser–Veselov integrators,* that, adopting an altogether-different approach, combine adaptivity and symplecticity (MARSDEN and WEST [2001]).

The above three instances of adaptivity, namely nonlinear approximation, geometric monitor functions and Moser–Veselov integrators, are distinct and probably unrelated. They hint at the tremendous potential scope of theoretical and foundational investigation into adaptivity.

3.2. Conditioning

It is a sobering thought that, even when a computational solution to a mathematical problem has been found, often following great intellectual and computational effort, its merit might be devalued by poor stability of the underlying algorithm. Most problems do not exist in some discrete universe but inhabit a continuum. Small changes in the statement of the problem, perturbations of a differential operator, of initial or boundary conditions or of other parameters of the problem, might lead to a significantly different *exact* solution. The situation is more complicated once we attempt to compute, because the numerical algorithm itself depends not only on all these parameters, but also on the intrinsic parameters of the algorithm, errors committed in the course of the solution, and the floating-point arithmetic of the underlying computer. This state of affairs is sometimes designated as 'stability' or 'well posedness' or 'conditioning' – purists may argue *ad nauseam* over the precise definitions of these concepts, but it is clear that, one way or the other, they play an instrumental role in computational mathematics. Naive complexity theory, say, makes very little sense if the 'optimal' algorithm is prone to instability or, worse, if the very problem that we endeavour to solve is ill conditioned in the first place. A more sophisticated concept of complexity is required and, indeed, has emerged in the last few years (CHEUNG, CUCKER and YE [2003]).

Stability is perhaps "the most misused and unstable concept in mathematics" (BELLMAN [1969]) and numerical analysis is probably the worst offender in this unwelcome proliferation. For example, in the numerical solution of ordinary differential equations we have A-stability, $A(\alpha)$-stability, A_0-stability, AN-stability, algebraic stability, B-stability, BN-stability..., and many remaining letters to afford every worker in the field the dubious benefit of yet another eponymous stability concept. The time has come perhaps to introduce order and understanding into the universe of stability, well posedness and conditioning. Although an all-encompassing stability theory belongs to the future, it is possible to make observations and venture ideas even at this early stage.

Time-dependent mathematical systems give rise to at least two broad stability concepts, which we might term *well posedness* (small variations in initial and boundary conditions and in internal parameters induce small variations in the solution in compact intervals) and *structural stability* (small variations do not induce qualitative changes in global dynamics). Both are inherited by discretization methods, although often terminology obscures this. Thus, for example, well posedness of discretization methods for partial differential equations of evolution is termed *stability,* a source of enduring confusion among students and some experienced researchers alike. Note that 'discrete well posedness' is not identical to 'continuous well posedness' since extra parameters (time step, size of the grid) are a consequence of discretization and they play an important role in the definition of numerical well posedness. Yet, in essence, it remains well posedness and should be distinguished from numerical structural stability, the province of computational dynamics.

Another dichotomy, extant in both computational analysis and computational algebra, is between traditional *forward* stability analysis and the approach of *backward error analysis* (a misnomer: in reality it refers to backward stability or conditioning analysis). Forward analysis asks the question that we have already posed time and again in this section: how does the behaviour vary when parameters are perturbed? An alternative approach, pioneered by WILKINSON [1960], is to investigate which perturbed problem is solved *exactly* by a computational algorithm. This is often the method of choice in numerical linear algebra, and it is gaining increasing prominence in the discretization of differential equations and in computational dynamics. Thus, for example, the theory of *shadowing* tells us that (as always, subject to a battery of side conditions) even imperfect modelling of chaotic systems produces exact orbits of the underlying system, with different parameter values (HAMMEL, YORKE and GREBOGI [1987]). This is indeed the reason why we can model chaotic attractors on a computer relatively successfully. Another powerful application of backward error analysis in the realm of differential equations is in the case of integrable Hamiltonian systems (HAIRER, LUBICH and WANNER [2002]). Since symplectic methods solve a nearby Hamiltonian problem (almost) exactly, their dynamics is identical to the original system (in particular, they evolve on invariant tori), Hamiltonian energy is conserved in an ergodic sense and, last but not least, numerical error accumulates much more slowly.

We do not delude ourselves that the preliminary and hesitant observations above are a basis for a unifying theory of stability and conditioning. Having said this, at least they seem to imply that it is possible to formulate meaningful statements within the realm of metastability. It is fair to assume that a better understanding of this issue might be one of the most useful consequences of the exploration of the foundations of computational mathematics.

3.3. Structure

Mathematics is not merely the science of number. It is the science of pattern and structure. Mathematical statements can be encoded in numbers (and the source file of this essay is merely a string of zeros and ones), but numbers as such are not an end in themselves. Once we wish to harness computation to *understand* mathematics, we are compelled to compute numerical results not as an end but as a means to reveal the underlying structure. This has a number of important implications for the future of computational algorithms.

Firstly, the impetus towards 'general' methods and 'general' software, which can cater to many different problems in a broad category, might be inimical to progress. Once we classify mathematical problems by their structural features, broad categories are much too coarse. We need to identify smaller sets of problems, each with its own common structural denominator and each deserving of separate computational treatment.

Secondly, the attitude of "don't think, the software will do it for you", comforting as it might be to some, will not do. There is no real dichotomy between mathematical and computational understanding. If you wish to compute, probably the best initial step is to learn the underlying mathematics, rather than rushing to a book of numerical recipes. Often when we air this point of view, some colleagues opine that it might create obstacles for scientists and engineers: this, we believe, is an unfair canard. In our experience, thoughtful scientists and engineers possess both the ability and the will to master necessary mathematics, often more so than the lazier members of the computational community.

Thirdly, 'structure' itself can assume different guises. It can refer to an attribute of a time-dependent system, in finite time or asymptotically; it can concern algebraic, analytic, geometric or stochastic features. Thus, it requires, on a case-by-case basis, different mathematical themes. In Section 2, we mentioned nonlinear dynamical systems and differential geometry, but clearly this does not exhaust the range of all mathematical subjects relevant to the retention of structure in mathematical computation.

Finally, the retention of structure can be for any of three reasons: mathematical, physical or computational. Sometimes we might wish to retain structure because of its mathematical significance. In other cases structure might possess physical interpretation which is essential to the modelling of the underlying phenomenon from science or engineering. In other cases the retention of structure leads to more accurate and affordable computational algorithms (HAIRER, LUBICH and WANNER [2002]). And, needless to say, there might be cases when retention of structure is of little or no importance.

Computation is not an activity where we can afford to 'paint by numbers', suspending our mathematical judgement and discernment. At its best, it requires us to be constantly on our mathematical guard and to deploy, as necessary, different weapons from the mathematical arsenal. The foundations of computational mathematics are the totality of mathematics and we believe that this is what makes our subject, once despised, so vibrant, exciting and full of promise. In a slightly different context, Goethe expressed this judgment beautifully in *Römische Elegien X*:

Wenn du mir sagst, du habest als Kind, Geliebte, den Menschen
  Nicht gefallen, und dich habe die Mutter verschmäht,
Bis du grösser geworden und still dich entwickelt, ich glaub es:
  Gerne denk ich mir dich als ein besonderes Kind.
Fehlet Bildung und Farbe doch auch der Blüte des Weinstocks,
Wenn die Beere, gereift, Menschen und Götter entzückt.[2]

Acknowledgements

We thank Raymond Brummelhuis (Birkbeck College, London), Felipe Cucker (City University, Hong Kong), Alan Edelman (MIT), Elizabeth Mansfield (University of Kent at Canterbury), Nilima Nigam (McGill University, Montreal) and Alex Scott (University College, London).

[2]In David Luke's translation (VON GOETHE [1997]):

When you were little, my darling, you tell me nobody liked you –
  Even your mother, you say, scorned you, until as the years
Passed, you quietly grew and matured; and I can believe it –
  It's rather pleasant to think you were a strange little child.
For though the flower of a vine may be still unformed and lack lustre,
In the ripe grape it yields nectar for gods and men.

References

ALLGOWER, E., GEORG, K. (1990). *Numerical Continuation Methods. An Introduction* (Springer-Verlag, Berlin).

ARNOLD, V. (1989). *Mathematical Methods of Classical Mechanics* (Springer-Verlag, Berlin).

BAXTER, B. (2002). Positive definite functions on Hilbert space, Tech. rep., Numerical Analysis Group, University of Cambridge.

BELLMAN, R. (1969). *Stability Theory of Differential Equations* (Dover, New York).

BERG, C., CHRISTENSEN, J.P.R., RESSEL, P. (1984). *Harmonic Analysis on Semigroups: Theory of Positive Definite and Related Functions*, Graduate Texts in Math. **100** (Springer, New York).

BLUM, L., CUCKER, F., SHUB, M., SMALE, S. (1998). *Complexity and Real Computation* (Springer, New York).

BROUDER, C. (1999). Runge–Kutta methods and renormalization. *Eur. Phys. J.* **12**, 521–534.

BUCHBERGER, B. (1970). Ein algorithmisches Kriterium für die Lösbarkeit eines algebraischen Gleichungssystems. *Aequationes Math.* **4**, 374–383.

BUCHBERGER, B., WINKLER, F. (eds.) (1998). *Gröbner Bases and Applications*. In: London Math. Soc. Lecture Note Ser. **251** (Cambridge University Press, Cambridge).

BUDD, C., ISERLES, A. (1999). Geometric integration: Numerical solution of differential equations on manifolds. *Philos. Trans. Royal Soc. A* **357**, 945–956.

BUDD, C., PIGGOTT, M. (2003). Geometric integration and its applications. In: Ciarlet, P.G., Cucker, F. (eds.), *Foundations of Computational Mathematics*. In: Handbook of Numerical Analysis **11** (North-Holland, Amsterdam), pp. 35–139.

BUHMANN, M. (2000). Radial basis functions. *Acta Numerica* **9**, 1–38.

BUTCHER, J. (1972). An algebraic theory of integration methods. *Math. Comp.* **26**, 79–106.

CHARD, J., SHAPIRO, V. (2000). A multivector data structure for differential forms and equations. *Math. Comput. Simulation* **54**, 33–64.

CHEUNG, D., CUCKER, F., YE, Y. (2003). Linear programming and condition numbers under the real number computation model. In: Ciarlet, P.G., Cucker, F. (eds.), *Foundations of Computational Mathematics*. In: Handbook of Numerical Analysis **11** (North-Holland, Amsterdam), pp. 141–207.

CIARLET, P.G. (1978). *The Finite Element Method for Elliptic Problems* (North-Holland, Amsterdam).

COHEN, A., DAHMEN, W., DAUBECHIES, I., DEVORE, R. (2001). Tree approximation and optimal encoding. *Appl. Comput. Harmon. Anal.* **11**, 192–226.

CONNES, A., KREIMER, D. (1998). Hopf algebras, renormalization and noncommutative geometry. *Comm. Math. Phys.* **199**, 203–242.

COX, D., LITTLE, J., O'SHEA, D. (1997). *Ideals, Varieties, and Algorithms. An Introduction to Computational Algebraic Geometry and Commutative Algebra* (Springer, New York).

CURRY, H., SCHOENBERG, I. (1966). Pólya frequency functions IV. The fundamental spline functions and their limits. *J. Anal. Math.* **17**, 71–107.

DAHMEN, W. (1997). Wavelet and multiscale methods for operator equations. *Acta Numerica* **6**, 55–228.

DAUBECHIES, I. (1992). *Ten Lectures on Wavelets* (SIAM, Philadelphia).

DE BOOR, C., HÖLLIG, K., RIEMENSCHNEIDER, S. (1993). *Box Splines* (Springer, New York).

DEVORE, R. (1998). Nonlinear approximation. *Acta Numerica* **7**, 51–150.

DUCHON, J. (1976). Interpolation des fonctions de deux variables suivant le principe de la flexion des plaques minces. *RAIRO Anal. Numer.* **10**, 5–12.

EDELMAN, A. (20xx). *Random matrices*, Unpublished course notes for MIT's class 18.325.

FRANKE, R. (1982). Scattered data interpolation: Tests of some methods. *Math. Comp.* **38**, 181–200.

GARDNER, R. (2002). The Brunn–Minkowski inequality. *Bull. Amer. Math. Soc.* **39**, 355–405.

GOLUB, G., VAN LOAN, C (1996). *Matrix Computations*, 3rd edn. (Johns Hopkins University Press, Baltimore).

HAIRER, E., LUBICH, C., WANNER, G. (2002). *Geometric Numerical Integration Structure-Preserving Algorithms for Ordinary Differential Equations* (Springer-Verlag, Berlin).

HAMMEL, S., YORKE, J., GREBOGI, C. (1987). Do numerical orbits of chaotic dynamical processes represent true orbits?. *J. Complexity* **3**, 136–145.

HARDER, R., DESMARAIS, R. (1972). Interpolation using surface splines. *J. Aircraft* **9**, 189–191.

HARDY, G.H. (1992). *A Mathematician's Apology* (Cambridge University Press, Cambridge).

HARDY, R. (1971). Multiquadric equations of tomography and other irregular surfaces. *J. Geophys. Res.* **76**, 1905–1915.

HELGASON, S. (1981). *Topics in Harmonic Analysis on Homogeneous Spaces* (Birkhäuser, Boston).

HESTENES, D. (1984). *Clifford Algebra to Geometric Calculus. A Unified Language for Mathematics and Physics* (D. Reidel, Dordrecht).

HIPTMAIR, R. (2002). Finite elements in computational electromagnetism. *Acta Numerica* **11**, 237–339.

HODGES, A. (1983). *Alan Turing, The Enigma* (Burnet Books, London).

HUBERT, E. (2000). Factorization-free decomposition algorithms in differential algebra. *J. Symbolic Comput.* **29**, 641–662.

ISERLES, A., MUNTHE-KAAS, H.Z., NØRSETT, S.P., ZANNA, A. (2000). Lie-group methods. *Acta Numerica* **9**, 215–365.

KELLER, H. (1987). *Lectures on Numerical Methods in Bifurcation Problems*, Tata Institute of Fundamental Research Lectures on Mathematics and Physics **79** (Springer-Verlag, Berlin).

KÖRNER, T. (1988). *Fourier Analysis* (Cambridge University Press, Cambridge).

LEVY, S. (ed.) (1997). *Flavors of Modern Geometry* (Cambridge University Press, Cambridge).

LI, T. (2003). Numerical solution of polynomial systems by homotopy continuation methods. In: Ciarlet, P.G., Cucker, F. (eds.), *Foundations of Computational Mathematics*. In: Handbook of Numerical Analysis **11** (North-Holland, Amsterdam), pp. 209–303.

MARSDEN, J., WEST, M. (2001). Discrete mechanics and variational integrators. *Acta Numerica* **10**, 357–514.

MCLACHLAN, R., QUISPEL, G. (2001). Six lectures on the geometric integration of ODEs. In: DeVore, R., Iserles, A., Süli, E. (eds.) *Foundations of Computational Mathematics, Oxford 1999* (Cambridge University Press, Cambridge), pp. 155–210.

MICCHELLI, C. (1986). Interpolation of scattered data: distance matrices and conditionally positive functions. *Constr. Approx.* **2**, 11–22.

MICCHELLI, C. (1995). *Mathematical Aspects of Geometric Modeling* (SIAM, Philadelphia).

MILMAN, V., SCHECHTMAN, G. (1986). *Asymptotic Theory of Finite-Dimensional Normed Spaces*, Lecture Notes in Math. **1200** (Springer-Verlag, Berlin).

NATTERER, F. (1999). Numerical methods in tomography. *Acta Numerica* **8**, 107–142.

PARLETT, B. (1992). Some basic information on information-based complexity theory. *Bull. Amer. Math. Soc.* **26**, 3–27.

SAPIRO, G. (2003). Introduction to partial differential equations and variational formulations in image processing. In: Ciarlet, P.G., Cucker, F. (eds.), *Foundations of Computational Mathematics*. In: Handbook of Numerical Analysis **11** (North-Holland, Amsterdam), pp. 383–461.

SCHOENBERG, I. (1935). Remarks to Maurice Fréchet's article: 'Sur la definition d'une classe d'espace distanciés vectoriellement applicable sur l'espace d'Hilbert. *Ann. of Math.* **36**, 724–732.

SCHOENBERG, I. (1937). On certain metric spaces arising from Euclidean space by a change of metric and their embedding in Hilbert space. *Ann. of Math.* **38**, 787–793.

SCHOENBERG, I. (1938). Metric spaces and completely monotone functions. *Ann. of Math.* **39**, 811–841.

SMALE, S. (1990). Some remarks on the foundations of numerical analysis. *SIAM Rev.* **32**, 211–220.

SMALE, S. (1997). Complexity theory and numerical analysis. *Acta Numerica* **6**, 523–551.

STEWART, G. (1980). The efficient generation of random orthogonal matrices with an application to condition estimators. *SIAM J. Numer. Anal.* **17**, 403–409.

STRANG, G., FIX, G. (1973). *An Analysis of the Finite Element Method* (Prentice-Hall, Englewood Cliffs, NJ).

STUART, A., HUMPHRIES, A. (1996). *Numerical Analysis of Dynamical Systems* (Cambridge University Press, Cambridge).

TEMLYAKOV, V. (2002). Nonlinear methods of approximation. *Found. Comp. Math.* **3**, 33–107.

TERRAS, A. (1999). *Fourier Analysis on Finite Groups and Applications*, London Mathematical Society Student Texts **4** (Cambridge University Press, Cambridge).

TONTI, E. (1976). Sulla struttura formale delle teorie fisiche. *Rend. Sem. Mat. Fis. Milano* **46**, 163–257 (in Italian).

TRAUB, J., WOŹNIAKOWSKI, H. (1992). Perspectives on information-based complexity. *Bull. Amer. Math. Soc.* **26**, 29–52.

TROTTER, H. (1984). Eigenvalue distributions of large Hermitian matrices; Wigner's semi-circle law and a theorem of Kac, Murdock and Szegő. *Adv. Math.* **54**, 67–82.

VON GOETHE, J. (1997). *Erotic Poems* (Oxford University Press) (transl. by D. Luke).

VON NEUMANN, J., SCHOENBERG, I. (1941). Fourier integrals and metric geometry. *Trans. Amer. Math. Soc.* **50**, 226–251.

VON ZUR GATHEN, J., GERHARD, J. (1999). *Modern Computer Algebra* (Cambridge University Press, New York).

WATSON, G. (1998). Choice of norms for data fitting and function approximation. *Acta Numerica* **7**, 337–377.

WERSCHULZ, A. (1991). *The Computational Complexity of Differential and Integral Equations. An Information-Based Approach* (Oxford University Press, New York).

WILKINSON, J. (1960). Error analysis of floating-point computation. *Numer. Math.* **2**, 319–340.

WILLIAMS, D. (2001). *Weighing the Odds* (Cambridge University Press, Cambridge).

WILSON, K. (1989). Grand challenges to computational science. *Future Gen. Comput. Sys.* **5**, 171–189.

WINOGRAD, S. (1979). On the multiplicative complexity of the discrete Fourier transform. *Adv. in Math.* **32**, 83–117.

YAMAGUTI, M., MAEDA, Y. (2003). Chaos in finite difference schemes. In: Ciarlet, P.G., Cucker, F. (eds.), *Foundations of Computational Mathematics*. In: Handbook of Numerical Analysis **11** (North-Holland, Amsterdam), pp. 305–381.

YOUNG, N. (1988). *An Introduction to Hilbert Space* (Cambridge University Press, Cambridge).

ZAREMBA, S. (ed.) (1972). *Applications of Number Theory to Numerical Analysis* (Academic Press, New York).

ZEILBERGER, D. (1990). A fast algorithm for proving terminating hypergeometric series identities. *Discrete Math.* **80**, 207–211.

Geometric Integration and its Applications

C.J. Budd

Department of Mathematical Sciences, University of Bath, Claverton Down, Bath, BA2 7AY, UK
e-mail: cjb@maths.bath.ac.uk

M.D. Piggott

Applied Modelling and Computation, Earth Science and Engineering, Imperial College of Science, Technology and Medicine, London SW7 2BP, UK
e-mail: m.d.piggott@imperial.ac.uk

Abstract

This review aims to give an introduction to the relatively new area of numerical analysis called geometric integration. This is an overall title applied to a series of numerical methods that aim to preserve the qualitative (and geometrical) features of a differential equation when it is discretised. As the importance of different qualitative features is a rather subjective judgement, a discussion of the qualitative theory of differential equations in general is given first. The article then develops both the underlying theory of geometrical integration methods and then illustrates their effectiveness on a wide ranging set of examples, looking at both ordinary and partial differential equations. Methods are developed for symplectic ODEs and PDEs, differential equations with symmetry, differential equations with conservation laws and problems which develop singularities and sharp interfaces. It is shown that there are clear links between geometrically based methods and adaptive methods (where adaptivity is applied both in space and in time). A case study is given at the end of this article on the application of geometrical integration methods to problems arising in meteorology. From this we can see the advantages (and the disadvantages) of the geometric approach.

Foundations of Computational Mathematics
Special Volume (F. Cucker, Guest Editor) of
HANDBOOK OF NUMERICAL ANALYSIS, VOL. XI
P.G. Ciarlet (Editor)

1. Introduction

The modern study of natural phenomena described by both ordinary and partial differential equations usually requires a significant application of computational effort. To understand the design and operation of such computer algorithms, numerical analysis is essential. A huge amount of effort over the past fifty years (and earlier) has thus been applied to the research of numerical methods for differential equations. This research has led to many ingenious algorithms and associated codes for the computation of solutions to such differential equations. Most of these algorithms are based upon the natural technique of discretising the equation in such a way as to keep the local truncation errors associated with the discretisation as small as possible, to ensure that these methods are stable so that the local errors do not grow, and in adaptive methods, to use meshes which constrain such errors to lie within specified tolerance levels. The resulting discrete systems are then solved with carefully designed linear and nonlinear solvers often dealing with very large numbers of unknowns. When coupled with effective (a-priori and a-posteriori) error control strategies these methods can often lead to very accurate solutions of fairly general classes of differential equations, provided that the times for integration are not long and the solution remains reasonably smooth with bounded derivatives.

However, methods based on the analysis of local truncation errors do not necessarily respect, or even take into account, the qualitative and global features of the problem or equation. It can be argued that in some situations these global structures tell us more about the underlying problem than the local information given by the expression of the problem in terms of differentials. The recent growth of geometric integration has, in contrast, led to the development of numerical methods which systematically incorporate qualitative information of the underlying problem into their structure. Such methods are sometimes existing algorithms (such as the Gauss–Legendre Runge–Kutta methods or the Störmer–Verlet leapfrog method) that turn out to have excellent qualitative properties, developments of existing methods (such as adaptive procedures or Moser–Veselov integrators) for which qualitative information can be included in a natural way, or are entirely new methods, such as Lie group integrators which use special techniques to incorporate qualitative structure. As a side benefit to the incorporation of qualitative features, the resulting numerical schemes are often more efficient than other schemes and are also easier to analyse. This is because we may exploit the qualitative theory of the underlying differential equations (such as its Hamiltonian structure or group invariance) by using the powerful backward error analysis methods developed by Hairer, Reich and their co-workers (HAIRER and LUBICH [1997], HAIRER and LUBICH [2000], REICH [1999]).

As it is often unclear a-priori what the correct qualitative features of a problem are (and it is highly unlikely that one numerical method can preserve all of them – although Gauss–Legendre methods have a jolly good try at doing this), the development of geometric integration algorithms to solve scientific problems necessarily involves a dialog between numerical analysts, applied mathematicians, engineers and scientists.

Some of the above comments may be exemplified by the following diagram (Fig. 1.1). Given a differential equation which we seek to approximate using a numerical method, the left-hand side of the diagram represents the use of classical methods for approx-

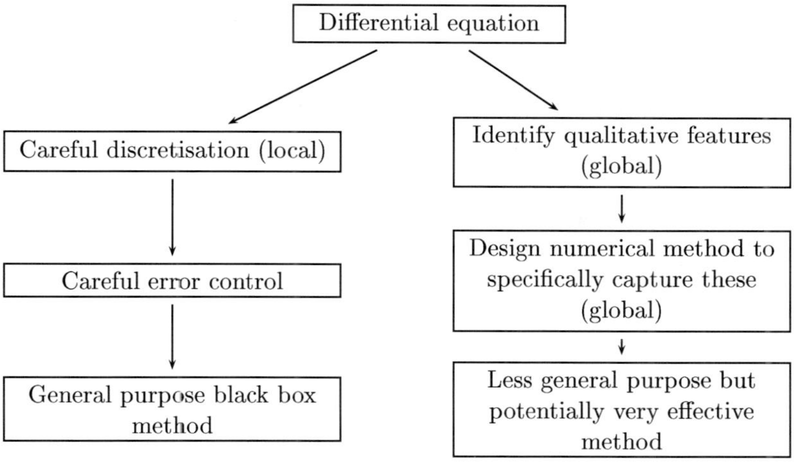

FIG. 1.1.

imating the equation, whereas the right-hand side displays the geometric integration philosophy.

As a simple illustration, consider the harmonic oscillator problem

$$\mathrm{d}u/\mathrm{d}t = v, \qquad \mathrm{d}v/\mathrm{d}t = -u.$$

Qualitatively, this system has the property that its solutions are all periodic and bounded and that $u^2 + v^2$ is a conserved quantity of the evolution. A forward Euler discretisation of this system with time step h gives

$$U_{n+1} = U_n + hV_n, \qquad V_{n+1} = V_n - hU_n.$$

It is easy to see that

$$U_{n+1}^2 + V_{n+1}^2 = (1 + h^2)(U_n^2 + V_n^2).$$

Thus, whilst over any finite time interval, the correct solution is obtained in the limit of $h \to 0$, the numerical method has lost all three qualitative features listed above. A geometrically based method instead aims to capture these.

The aim of this review is to introduce and explain, some ideas, methods, and techniques used in the field of geometric integration, basing our discussion around several carefully chosen examples. We will include some proofs of important results, but mostly will refer the reader to the literature for fuller explanations. We further aim to discuss how some of the mentioned ideas could be used in the very challenging problems arising in the fields of meteorology and numerical weather prediction. This discussion is necessarily incomplete and further details of geometric integration can be found in the recent reviews and discussions listed in the following references: BUDD and ISERLES [1999], MCLACHLAN and QUISPEL [2001], HAIRER, LUBICH and WANNER [2002], BUDD and PIGGOTT [2001], SANZ-SERNA [1997].

This review is organised as follows. In Section 2 we present an, obviously not exhaustive, description of some of the qualitative properties of both ordinary and partial differential equations that we have in mind.

In Sections 3–6 we look at methods for ordinary differential equations. In Section 3 we consider symplectic methods for Hamiltonian problems (including splitting and Runge–Kutta methods), looking at applications of these methods to a variety of problems in mechanics, and analysing their error through the use of backward error analysis. In Section 4 we consider methods that retain conservation laws of various forms. In Section 5 we look at a variety of methods that conserve symmetries of ordinary differential equations. These will include Lie group methods based upon Magnus series and/or approximations of the matrix exponential, methods for preserving reversing symmetries and time-adaptive methods for ordinary differential equations with scaling invariance symmetries. In particular developing methods for problems which allow the construction of numerical discretisations with uniform in time error estimates for certain problems. In Section 6 we make a comparison of various geometrically based methods for ordinary differential equations by comparing their performance for the (Hamiltonian) Kepler problem.

In Sections 7–12 we look at geometric integration methods in the context of partial differential equations. In Section 7 we study Poisson bracket preserving methods as applied to Hamiltonian problems (such as the nonlinear wave equation and the nonlinear Schrödinger equation). In Section 8 we look at methods which are derived from a Lagrangian formulation of a partial differential equation. In Section 9 we review the work of DORODNITSYN [1993a] which extends symmetry based methods for ordinary differential equations to the construction of moving mesh methods for partial differential equations which preserve all of the symmetries of these equations. These are extended in Section 10 to methods which use a discrete form of Noether's theorem to ensure the conservation of quantities related to the invariance of the Lagrangian to continuous transformation groups. In Section 11 we continue to look at adaptive methods for partial differential equations, by developing in detail methods for scale invariant partial differential equations, which include the cases of the nonlinear diffusion equation and the nonlinear Schrödinger equation. We study numerical methods (based on moving mesh partial differential equations) which have proved to be especially effective in integrating these in the presence of interfaces and singularities. In Section 12 we focus our attention on some model problems arising in meteorology for which geometric based ideas may be beneficial. In particular looking at the calculation of weather fronts and making ensemble forecasts. We briefly discuss what possible impacts geometric integration could have on their computational analysis.

Finally, in Section 13 we present some concluding remarks, looking in particular at the advantages and disadvantages of geometric integration and the likely future developments in this new field.

2. Qualitative properties

In this section we shall briefly look at some of the qualitative and global features of a system described by a differential equation which we have in mind when we talk about geometric integration. The discussions of such qualitative features will, out of necessity (and the vast complexity of possible behaviours of solutions of differential equations), be very brief and introductory. We shall wherever possible include references to material where much further information may be found.

2.1. A partial listing

There are many possible qualitative features which may be present in systems modelled by ordinary or partial differential equations. We shall not attempt to give a complete listing, however below we give a partial listing which covers a wide variety of possibilities and mention how the different qualitative properties may be linked to one another.

(1) *Geometrical structure*. Deep properties of the phase space in which the system is defined give enormous insight into the overall properties of its solutions. Perhaps the most important of these arise in Hamiltonian problems.

(2) *Conservation laws*. Underlying many systems are conservation laws. These may include the conservation of total quantities (usually integrals over the domain in which the system evolves) such as mass, energy and momentum, or instead quantities which are conserved along particle trajectories and flows such as fluid density or potential vorticity. The loss of energy in a system describing planetary motion will inevitably lead to the planet so modelled spiralling into the sun, which is clearly incorrect qualitatively. Similarly it is widely accepted (CULLEN, SALMOND and SMOLARKIEWICZ [2000]) that in large scale modelling of the oceans and atmosphere it is essential to conserve potential vorticity to retain the overall qualitative dynamics of the solution. The above are conservation laws associated directly with the manifold on which the solution evolves or the tangent bundle of this manifold, however, for Hamiltonian problems we may also have the conservation of symplectic structures in phase space which evolve on the cotangent bundle and volume preservation in divergence-free systems which are conservation laws associated with the phase space in which the solution is posed. As is common with many such properties, there are deep relationships between them. For example, the Hamiltonian of the solution is conserved in autonomous Hamiltonian systems and a wide class of Casimir functions are conserved along trajectories (MCLACHLAN [1994]).

(3) *Symmetries*. Many systems are invariant under the actions of symmetries such as Lie group, scaling and involution symmetries. Such symmetries may or may not be retained in the underlying solution but there may exist solutions (self-similar solutions) which do not change when the symmetry group acts. The possible symmetries may include the following

- *Galilean symmetries* such as translations, reflexions and rotations. One of the key problems in computer vision is to recognise an object which may be a translation or rotation of a known pattern. One way to do this is to associate invariants to the object (such as curvature) which do not change under the action of a Galilean symmetry. This is naturally achieved in the moving frames methods studied by OLVER [2001] within the context of the geometric integration approach. The study of the motion of rigid bodies in three-dimensional space (such as a satellite or a robot arm) or the buckling of a cylindrical shell are dominated by the fact that such systems are invariant under Galilean symmetries (MARSDEN and RATIU [1999]).

- *Reversal symmetries.* The solar system is an example of a system which is invariant under a time reversal, and more generally many physical systems are invariant under symmetries ρ satisfying the identity $\rho^2 = Id$. We can also ask whether a numerical method with time step h is *symmetric*, so that the inverse of taking such a time step is the same as replacing h by $-h$. (The forward Euler method does not have this property, in contrast to the trapezoidal rule which does.)
- *Scaling symmetries.* Many physical problems have the property that they are invariant under rescalings in either time or space. This partly reflects the fact that the laws of physics should not depend upon the units in which they are measured or indeed should not have an *intrinsic* length scale (BARENBLATT [1996]). An example of such a scaling law is Newton's law of gravitation which is invariant under a rescaling in time and space. This invariance leads directly to Kepler's third law linking the period of a planet on an elliptical periodic orbit to the length of the major axis of the ellipse – determining this law does not involve solving the differential equation. A numerical method invariant under the same scaling law will exhibit a similar relation between (discrete) period and scale.
- *Lie group symmetries.* These are deeper symmetries than those described above, often involving the invariance of the system to a (nonlinear) Lie group of transformations. An important example (which arises naturally in mechanics) is the invariance of a system to the action of the rotation group $SO(3)$. An excellent discussion of such symmetries is given in OLVER [1986]. The review article (ISERLES, MUNTHE-KAAS, NØRSETT and ZANNA [2000]) describes the numerical approach to computing solutions of ordinary differential equations with such symmetries.

(4) *Asymptotic behaviours.* The original development of numerical methods for differential equations centred on the study of minimising errors in the calculation of solutions over a fixed time T. However, when studying the *dynamics* of a system numerically and to gain insight into its long term behaviour we are often interested in the alternative question of taking a fixed method and applying it for an undefined number of time steps. Note that in a problem with widely varying time-scales (such as molecular dynamics where we try to model the behaviour of chemicals over seconds whilst molecular interactions occur over microseconds) the study of long term behaviour (measured in terms of the smallest time-scale) is unavoidable. Widely differing time-scales are natural in partial differential equations. For such studies we need to concentrate on the ability of a numerical method to preserve structurally stable features of the solution such as invariant sets, invariant curves and orbit statistics. A very thorough review of these issues and the way that numerical methods are used to study dynamical systems is given in the monograph of STUART and HUMPHRIES [1996] and fundamental contributions to the study of the preservation of invariant curves have

been made by BEYN [1987] (see also the work of STOFFER and NIPP [1991] in this context). An excellent example of such an investigation is the long term study of the solar system made by SUSSMAN and WISDOM in which numerical methods (special purpose splitting methods) are used to investigate whether the solar system has chaotic behaviour. As the errors of many 'conventional' methods accumulate exponentially with time, accurate qualitative investigations over long time using these methods is not possible, even if the methods are of high order and have very small local errors. Thus it is essential to use methods for which there is some control over the long term growth of errors, even if the local error made by such methods may appear to be very large in comparison to others. A hint of how this is possible is the fact that the differential system under study may evolve in time so that over a long time its dynamics in some sense simplifies. For example, it may ultimately evolve so that its dynamics is restricted to a low dimensional attractor (although its dynamics on this attractor may well be chaotic). Complex structures starting from arbitrary initial data may simplify into regular patterns (GRINDROD [1991]). Alternatively, the equation may have solutions which form singularities in finite time such as interfaces (singularities in curvature) weather fronts (derivative singularities) or combustion in which the solution itself becomes singular at a single point. All of these features can be incorporated into the design of a numerical scheme and should be reproduced by a good numerical method.

(5) *Orderings in the solutions.* The differential equation may possess some form of maximum principle which leads to a preservation of the solution orderings. For example, given two sets of initial data $u_0(x)$ and $v_0(x)$ for a partial differential equation, the solutions may respect the ordering that if $u_0(x) < v_0(x)$ for all x, then $u(x,t) < v(x,t)$ for all x and t. The linear heat equation $u_t = u_{xx}$ has this property as do many other parabolic equations. Maximum principles can play a significant role in understanding the *spatial* symmetries of the solution (FRAENKEL [2000]). A related concept is that of solution convexity. Indeed, the preservation of the convexity of a pressure function across a front is an important feature in some areas of numerical weather prediction (CULLEN, NORBURY and PURSER [1991]).

It is important to realise that these global properties may be closely linked to one another. For example, if the differential equation is derived from a variational principle linked to a Lagrangian function then, via Noether's theorem (OLVER [1986]), each continuous symmetry of the Lagrangian leads directly to a conservation law for the underlying equation. This has a beautiful application to numerical analysis. If a numerical method is also based upon a Lagrangian and this Lagrangian has symmetries then the numerical method automatically has a discrete conservation law associated with this symmetry (DORODNITSYN [1998]). We will explore this relation in the section on partial differential equations.

Symmetry when coupled with solution orderings frequently leads to an understanding of the asymptotic behaviour of the equation. In particular, self-similar solutions (which are invariant under the action of a scaling group) can be used to bound the actual solution from above and below. The solution behaviour is then constrained to follow that of the self-similar solution (SAMARSKII, GALAKTIONOV, KURDYUMOV and MIKHAILOV [1995]). Singularities in the equation often have more *local* symmetry than the general solution of the equation because the effects of boundary and initial conditions are less important (BARENBLATT [1996]). Conversely solutions can have *hidden* symmetries (GOLUBITSKY, SCHAEFFER and STEWART [1988]) not evident in the actual equation.

Some natural questions to ask of a numerical method which attempts to emulate some or all of these properties are as follows.

What is the benefit (if any) of designing numerical methods which take into account qualitative properties of the underlying solution? For systems which possess more than one important qualitative property, how much of this structure can we hope to preserve? Which qualitative properties turn out to be more important, or beneficial, from a computational viewpoint?

2.2. *Why preserve structure?*

There are several reasons why it is worthwhile to preserve qualitative structure. Firstly, many of the above properties can be found in systems which occur naturally in applications. For example, large scale molecular or stellar dynamics can be described by Hamiltonian systems with many conservation laws. Mechanical systems evolve under rotational constraints, as do many of the problems of fluid mechanics. Partial differential equations possessing scaling symmetries and self-similarity arise in fluid and gas dynamics, combustion, nonlinear diffusion and mathematical biology. Partial differential equations with a Hamiltonian structure are important in the study of solitons (such as the KdV and nonlinear Schrödinger equation) and the semi-geostrophic equations in meteorology also have a Hamiltonian structure. The list could be virtually endless.

In designing our numerical method to preserve some structure we hope to see some improvements in our computations. For a start we will be studying a discrete dynamical system which has the same properties as the continuous one, and thus can be thought of as being in some sense close to the underlying problem in that stability properties, orbits and long-time behaviour may be common to both systems. Geometric structures often (using backward error analysis and exploiting the geometric structure of the discretisations) make it easier to estimate errors, and in fact local and global errors may well be smaller for no extra computational expense. Geometric integration methods designed to capture specific qualitative properties may also preserve additional properties of the solution *for free*. For example, symplectic methods for Hamiltonian problems have excellent energy conservation properties and can conserve angular momentum or other invariants (which may not even be known in advance).

In conclusion, geometric integration methods (including Lie group methods, symplectic integrators, splitting methods, certain adaptive methods, etc.) can often 'go where other methods cannot'. They have had success in the accurate computation of singularities, long-term integration of the solar system, analysis of highly oscillatory systems (quantum physics for example). The list of application areas keeps on growing, see for example the many applications described by LEIMKUHLER [1999].

3. Hamiltonian ordinary differential equations and symplectic integration

Probably the first significant area where geometric ideas were used was in the (symplectic) integration of Hamiltonian ordinary differential equations. This is natural as Hamiltonian systems (often incorporating constraints) have very important applications in mechanics, celestial and molecular dynamics and optics, and their analysis, since Hamilton's original papers, has always centred on the geometrical structure of the equations. We will start our investigation of geometric integration methods by focusing on such problems (later on we will investigate Hamiltonian partial differential equations). A detailed description of methods for Hamiltonian problems is given in SANZ-SERNA and CALVO [1994], see also STUART and HUMPHRIES [1996]. The numerical methods for solving such problems are usually called *symplectic* and include certain Runge–Kutta methods, splitting methods and methods based upon discretising the Lagrangian. We structure this discussion by first describing Hamiltonian systems, we then consider methods for such systems, and finally look at an analysis of the error of these methods using the backward error technique.

3.1. The theory of Hamiltonian methods

For classic introductions to this material see ARNOLD [1978] and GOLDSTEIN [1980]. Consider initially a mechanical system with generalised coordinates $\mathbf{q} \in \mathbb{R}^d$ and Lagrangian $L = T - V$, where $T \equiv T(\mathbf{q}, \dot{\mathbf{q}})$ represents the kinetic energy of the system and $V \equiv V(\mathbf{q})$ its potential energy. The dynamics of such a system can be studied in terms of the calculus of variations by considering the action function constructed by integrating L along a curve $\mathbf{q}(t)$ and then computing variations of the action while holding the end points of the curve $\mathbf{q}(t)$ fixed (although it is sometimes advantageous not to impose this condition). See details of this derivation in MARSDEN and WEST [2001] together with a detailed discussion of the numerical methods derived by discretising the action functional. We shall return to this topic in Section 8. It can be shown easily that this procedure leads to the following Euler–Lagrange equations describing the motion

$$\frac{\mathrm{d}}{\mathrm{d}t}\left(\frac{\partial L}{\partial \dot{\mathbf{q}}}\right) - \frac{\partial L}{\partial \mathbf{q}} = 0. \tag{3.1}$$

Hamilton recognised that these equations could be put into a form which allowed a more geometrical analysis. In particular he introduced the coordinates

$$\mathbf{p} = \frac{\partial L}{\partial \dot{\mathbf{q}}} \in \mathbb{R}^d,$$

which are the conjugate generalized momenta for the system. He further defined the Hamiltonian via a Legendre transformation as

$$H(\mathbf{p}, \mathbf{q}) = \mathbf{p}^{\mathrm{T}}\dot{\mathbf{q}} - L(\mathbf{q}, \dot{\mathbf{q}})$$

and showed that (3.1) is equivalent to the following system of $2d$ first-order equations, called Hamilton's equations

$$\dot{\mathbf{p}} = -\frac{\partial H}{\partial \mathbf{q}}, \qquad \dot{\mathbf{q}} = \frac{\partial H}{\partial \mathbf{p}}. \tag{3.2}$$

This is called the *canonical form* for a Hamiltonian system. Note that for our considered mechanical system $H \equiv T + V$ and thus the Hamiltonian represents the total energy present.

More generally, if a system of ordinary differential is defined in terms of $\mathbf{u} \in \mathbb{R}^{2d}$ where $\mathbf{u} = (\mathbf{p}, \mathbf{q})^{\mathrm{T}}$ with $\mathbf{p}, \mathbf{q} \in \mathbb{R}^d$ such that

$$\dot{\mathbf{u}} = \mathbf{f}(\mathbf{u}) \tag{3.3}$$

then this system is canonically Hamiltonian if

$$\mathbf{f}(\mathbf{u}) = J^{-1}\nabla H, \tag{3.4}$$

where $H = H(\mathbf{p}, \mathbf{q})$ is the Hamiltonian function, ∇ is the operator

$$\left(\frac{\partial}{\partial p_1}, \frac{\partial}{\partial p_2}, \ldots, \frac{\partial}{\partial p_d}, \frac{\partial}{\partial q_1}, \ldots, \frac{\partial}{\partial q_d}\right),$$

and J is the skew-symmetric matrix

$$J = \begin{pmatrix} 0 & I_d \\ -I_d & 0 \end{pmatrix}. \tag{3.5}$$

Here I_d is the identity matrix of dimension d.

In this case $\mathbf{f}$ is called a Hamiltonian vector field. It is easy to see that if $\mathbf{f}_1$ and $\mathbf{f}_2$ are Hamiltonian vector fields then so is the vector field $\mathbf{f}_1 + \mathbf{f}_2$. We exploit this simple property in our analysis of splitting methods which follows shortly.

In addition to the conservation of the Hamiltonian a key feature of a Hamiltonian system is the *symplecticity* of its flow. The solution of the system (3.3) induces a transformation ψ on the phase space $\mathbb{R}^{2d}$ with associated Jacobian ψ'. Such a map is said to be *symplectic* if

$$\psi'^{\mathrm{T}} J \psi' = J \tag{3.6}$$

with J defined as above.

Symplectic maps have the very useful property that they combine to give other symplectic maps, as stated in the following.

LEMMA 3.1. *If ψ and φ are symplectic maps then so is the map $\psi \circ \varphi$.*

PROOF. We can see this as follows. If ψ and φ are both symplectic maps then

$$(\psi'\varphi')^{\mathrm{T}} J (\psi'\varphi') = \varphi'^{\mathrm{T}} \psi'^{\mathrm{T}} J \psi' \varphi' = \varphi'^{\mathrm{T}} J \varphi' = J.$$ □

Symplecticity has the following, important, geometrical interpretation. If M is any 2-dimensional manifold in $\mathbb{R}^{2d}$, we can define $\Omega(M)$ to be the integral of the sum over the oriented areas, of its projections onto the (p_i, q_i) plane (so that if $d = 1$ this is the area of M). If ψ is a symplectic map, then $\Omega(M)$ is conserved throughout the evolution. In particular in one-dimension ($d = 1$) areas are conserved. A key result of Poincaré (1899) relating symplectic flows to Hamiltonian flows is the following.

THEOREM 3.1. *The flow $\psi(t)$ induced by a Hamiltonian function H via the differential equation* (3.4) *is symplectic.*

PROOF. If ψ' is the Jacobian matrix of the flow induced by the ordinary differential equation (3.4), then it follows from the definition of the Hamiltonian system that

$$\frac{\mathrm{d}\psi'}{\mathrm{d}t} = J^{-1}H''\psi', \quad \text{with } H'' = \begin{pmatrix} H_{pp} & H_{pq} \\ H_{pq} & H_{qq} \end{pmatrix}.$$

Hence

$$\frac{\mathrm{d}}{\mathrm{d}t}\left(\psi'^{\mathrm{T}} J \psi'\right) = \psi'^{\mathrm{T}} H'' J^{-T} J \psi' + \psi'^{\mathrm{T}} H'' \psi' = 0.$$

However, when $t = 0$ we have $\psi' = I$ and hence $(\psi'^{\mathrm{T}} J \psi') = J$. The result then follows. □

It can further be shown (HAIRER, LUBICH and WANNER [2002]) that symplecticity holds iff a flow is Hamiltonian.

The symplecticity property is much stronger than simple preservation of $2d$-dimensional volume, since Hamiltonian systems preserve phase-space volume (Liouville's theorem) but it is possible to find volume preserving systems which are not Hamiltonian. From a perspective of dynamics, symplecticity plays a central role. In particular it means that behaviour of Hamiltonian systems is recurrent with solutions from any point in $\mathbb{R}^{2d}$ returning arbitrarily closely to their starting point. Furthermore (unlike dissipative systems) the dynamics of a Hamiltonian system cannot evolve onto a low dimensional attractor. Typical Hamiltonian dynamics is described by the celebrated KAM theorem (ARNOLD [1978]) which describes how integrable and near integrable Hamiltonian systems (whose solutions are generically periodic or are confined to a torus) perturb (under periodic Hamiltonian perturbations) to tori (of highly irrational period) surrounded by regions of chaotic behaviour. As stated by SANZ-SERNA [1997] 'many properties that general systems only possess under exceptional circumstances appear generically in Hamiltonian systems'. Further details may be found in ARNOLD [1978], MARSDEN and RATIU [1999], OLVER [1986].

3.2. *Symplectic numerical methods*

3.2.1. *Outline*

Suppose now that a numerical one-step method of constant step size h is applied to approximate the solutions $\mathbf{u}(t)$ of (3.3) at time $t = nh$ by the vector $\mathbf{U}_n \in \mathbb{R}^{2d}$. We will assume for the moment that h is constant. This numerical scheme will induce a discrete flow mapping Ψ_h on $\mathbb{R}^{2d}$ which will be an approximation to the continuous flow map $\psi(h)$. We define the map Ψ_h to be symplectic if it also satisfies the identity (3.6). A natural geometric property that we would require of Ψ_h is that it should be symplectic if ψ is. We will now consider the construction of numerical methods with this property. We will call any numerical scheme which induces a symplectic map a *symplectic numerical method*. As we will see, such methods retain many of the other features (in particular the ergodic properties) of their continuous counterparts.

If the time step h is not constant, such as in an adaptive method, then particular care has to be taken with this definition and much of the advantage of using a symplectic scheme is lost (SANZ-SERNA and CALVO [1994]) unless the equation defining h at each time step also preserves the geometric structure in some way. We will consider this issue further in Section 6.

The history of symplectic methods is interesting. Early work in the mathematics community which specifically aimed to construct symplectic methods for ordinary differential equations is given in DE VOGELAERE [1956], RUTH [1983] and KANG [1985]. These early constructions were rather involved and much simpler derivations have followed, in particular see the work of SANZ-SERNA and CALVO [1994]. However, symplectic integrators themselves have been used for much longer than this. A lot of well established and very effective numerical methods have been successful precisely because they are symplectic even though this fact was not recognised when they were first constructed. Three examples of this serendipitous behaviour are Gauss–Legendre methods, related collocation methods and the Störmer–Verlet (leapfrog) methods of molecular dynamics. In the literature quoted above symplectic methods are generally constructed using one of four different methods

(1) Generating functions,
(2) Runge–Kutta methods,
(3) Splitting methods,
(4) Variational methods.

Of these four methods, method (1) tends to lead to methods which are cumbersome and difficult to use, however see CHANNEL and SCOVEL [1990] for details and examples. We consider method (4) in Section 8 as it has a powerful application to partial differential equations. In this section we focus on methods (3) and (4).

Why bother? Whilst such methods preserve symplecticity and hence many of the qualitative features of the Hamiltonian problem they do not (in general) conserve the Hamiltonian itself (see GE and MARSDEN [1988]) unless an adaptive step h is used (KANE, MARSDEN and ORTIZ [1999]). (However they can remain exponentially (in h) close to H for exponentially long times.) However, they usually have more favourable error growth properties. Invariant sets and orbit statistics converge even if the orbit is not followed with pointwise accuracy (MACKAY [1992], MCLACHLAN and ATELA [1992], CANDY and ROZMUS [1991]). It is important also to observe that Hamiltonian systems arise naturally in partial differential equations (for example, the nonlinear Schrödinger equation) for which the associated differential equations (say obtained through a semi-discretisation) are usually stiff. A conventional stiff solver such as a BDF method may introduce artificial dissipation into higher order modes, producing quite false qualitative behaviour. To resolve the energy transfer into the higher modes and to retain the correct dynamics of these modes a symplectic solver is essential.

It is worth saying that in computations small round-off errors introduce non-symplectic perturbations to Hamiltonian problems which can be an important factor in long term integrations. Some effort has been made to use integer arithmetic and lattice maps to remove the effects of round-off errors in such calculations, see the work of EARN and TREMAINE [1992].

EXAMPLE. To motivate the rest of this section we consider a simple, but important, example of a symplectic method. This is the second-order implicit, (symmetric) mid-point Runge–Kutta method, which for the problem (3.3) takes the form

$$\mathbf{U}_{n+1} = \mathbf{U}_n + h\mathbf{f}\big((\mathbf{U}_n + \mathbf{U}_{n+1})/2\big) = \mathbf{U}_n + hJ^{-1}\nabla H\big((\mathbf{U}_n + \mathbf{U}_{n+1})/2\big). \tag{3.7}$$

A proof that this method is indeed symplectic is instructive.

PROOF. For a sufficiently small time step h the implicit mid-point rule defines a diffeomorphism $\mathbf{U}_{n+1} = \Psi_h(\mathbf{U}_n)$. Differentiating (3.7) then gives

$$\Psi_h' = \frac{\partial \mathbf{U}_{n+1}}{\partial \mathbf{U}_n} = I + \frac{h}{2}J^{-1}H''\big((\mathbf{U}_n + \mathbf{U}_{n+1})/2\big)\left(I + \frac{\partial \mathbf{U}_{n+1}}{\partial \mathbf{U}_n}\right).$$

Hence, the Jacobian matrix of the transformation $\mathbf{U}_n \to \mathbf{U}_{n+1}$ is given by

$$\begin{aligned}\Psi_h'(\mathbf{U}_n) = {} & \left[I - \frac{h}{2}J^{-1}H''\big((\mathbf{U}_n + \mathbf{U}_{n+1})/2\big)\right]^{-1} \\ & \times \left[I + \frac{h}{2}J^{-1}H''\big((\mathbf{U}_n + \mathbf{U}_{n+1})/2\big)\right].\end{aligned} \tag{3.8}$$

□

NOTE. This is a Cayley transformation. We shall return to this in the section on Lie group methods.

We have already mentioned that (3.6) is the condition we need to verify for a map to be symplectic, after substituting (3.8) into (3.6) we need to check that

$$\begin{aligned}&\left[I + \frac{h}{2}J^{-1}H''\big((\mathbf{U}_n + \mathbf{U}_{n+1})/2\big)\right]J\left[I + \frac{h}{2}J^{-1}H''\big((\mathbf{U}_n + \mathbf{U}_{n+1})/2\big)\right]^{\mathrm{T}} \\ &\quad = \left[I - \frac{h}{2}J^{-1}H''\big((\mathbf{U}_n + \mathbf{U}_{n+1})/2\big)\right]J\left[I - \frac{h}{2}J^{-1}H''\big((\mathbf{U}_n + \mathbf{U}_{n+1})/2\big)\right]^{\mathrm{T}}.\end{aligned}$$

The result now follows immediately from the symmetry of the Hessian matrix and the fact that $J^{\mathrm{T}} = -J$.

NOTE. The above proof is independent of the specific form of the operator J, requiring only that J be skew-symmetric. Precisely similar arguments to the above can also be used to show that mid-point rule preserves the Poisson structure for systems with constant Poisson structure. This is very useful in studies of partial differential equations (see later). If J depends on U then can be shown to be *Almost Poisson*, that is preserves the Poisson structure up to second order, see AUSTIN, KRISHNAPRASAD and WANG [1993]. The same idea is developed in *pseudo-symplectic methods* (AUBRY and CHARTIER [1998]).

As we shall see presently, a powerful way of understanding and analysing such schemes is through the use of modified equation analysis. In particular, if $\mathbf{u}(t)$ is the solution of the equation and $\mathbf{U}_n$ the discrete solution, then modified equation analysis aims to find a differential equation *close* to the original so that the solution $\hat{\mathbf{u}}(t)$ of this lies much closer to $\mathbf{U}_n$ than the true solution $\mathbf{u}$ does. Generally we construct $\hat{\mathbf{u}}$ such that $\mathbf{U}_n - \hat{\mathbf{u}}(t_n) = \mathcal{O}(h^{N+1})$ with $N > r$, where r is the order of the numerical method. Suppose that the modified equation is

$$\frac{\mathrm{d}\hat{\mathbf{u}}}{\mathrm{d}t} = \hat{\mathbf{f}}(\hat{\mathbf{u}}).$$

HAIRER, LUBICH and WANNER [2002] (see also MURUA and SANZ-SERNA [1999]) gives a construction of $\hat{\mathbf{f}}$ in terms of a B-series so that truncating a series of the form

$$\mathbf{f}_0(\hat{\mathbf{u}}) + h\mathbf{f}_1(\hat{\mathbf{u}}) + h^2\mathbf{f}_2(\hat{\mathbf{u}}) + \cdots \tag{3.9}$$

at order h^N gives an appropriate $\hat{\mathbf{f}}$ of order N as defined above. We give more details of this construction presently. Ideally the modified flow should have the same qualitative features as both the underlying equation and the discretisation. In this case the properties of the numerical *map* can be understood in terms of a smooth *flow* and thus analysed using methods from the theory of dynamical systems. This is discussed in STUART and HUMPHRIES [1996] and has also been the motivation behind a series of meetings on *The dynamics of numerics and the numerics of dynamics*, see BROOMHEAD and ISERLES [1992].

Presently we will use this procedure to analyse the performance of certain symplectic methods. For the moment we observe, that a numerical method is symplectic if and only if each of the modified equations obtained by truncating (3.9) generates a Hamiltonian system. In particular the modified equation will have a modified Hamiltonian $H^0 + hH^1 + \cdots$. Ignoring for the moment the (exponentially small) difference between the modified flow and the discrete solution this leads to the important observation that

> a symplectic discretisation of a Hamiltonian problem is a Hamiltonian perturbation of the original.

The importance of this observation cannot be over-emphasised. It implies that we can use the theory of perturbed Hamiltonian systems (in particular KAM theory) to analyse the resulting discretisation and it places a VERY strong constraint on the possible dynamics of this discrete system. It has been observed by Sanz-Serna that this procedure is not applicable to variable step methods as in this case the modified equation analysis breaks down. Whilst this is true in general, it can be overcome by a careful selection of h that preserves the Hamiltonian structure, and such choices may give improved methods (LEIMKUHLER [1999], HAIRER [1997], REICH [1999]). However, there are limits to this form of analysis. The modified equations typically lead to asymptotic series that may break down for large step sizes, causing a break up of invariant curves. Furthermore there may be exponential effects beyond all orders (such as arise in chaotic flows) which will simply not be detected by this approach.

3.2.2. *Symplectic Runge–Kutta methods*

It is fortunate that an important (accurate and A-stable) class of Runge–Kutta methods, namely the Gauss–Legendre methods (also known as Kuntzmann and Butcher methods), also turn out to be symplectic methods. The implicit mid-point rule considered earlier is a member of this class. Whilst this gives a very useful class of methods applicable to a wide class of general Hamiltonian problems, the resulting methods are highly implicit and for specific problems splitting methods, which may be explicit, may be preferable. To explain these Runge–Kutta methods, in this section we review the results in SANZ-SERNA and CALVO [1994]. Consider a standard s-stage Runge–Kutta method with Butcher tableau

$$\begin{array}{c|c} c & A \\ \hline & b^{\mathrm{T}} \end{array}$$

where $c, b \in \mathbb{R}^s$ and $A \equiv (a_{i,j}) \in \mathbb{R}^{s\times s}$. The following result (LASAGNI [1988], SANZ-SERNA [1988], SURIS [1989], see also SANZ-SERNA and CALVO [1994]) gives a complete characterisation of all symplectic Runge–Kutta methods

THEOREM. *A necessary and sufficient condition for a Runge–Kutta method to be symplectic is that for all* $1 \leqslant i, j \leqslant s$

$$b_i a_{ij} + b_j a_{ji} - b_i b_j = 0. \tag{3.10}$$

Note that the necessity of condition (3.10) requires that the method have no redundant stages, i.e. the method be irreducible, see SANZ-SERNA and CALVO [1994]. We can immediately see that if the matrix A was lower triangular then (3.10) would imply that $b_i = 0$, $\forall i$, this obviously contradicts the consistency condition $\sum b_i = 1$, and therefore we can conclude that symplectic Runge–Kutta methods must be implicit. This is not an ideal state of affairs, however we shall see presently that in a number of very important situations methods which are both symplectic and explicit may be found.

The proof of this result is a consequence of the fact that any method satisfying (3.10) automatically preserves any *quadratic* invariant of the solution of the equation to which it is applied (a point we return to in Section 4). In other words, any invariant of the form

$$u^{\mathrm{T}} A u$$

for an appropriate matrix A. The symplecticity condition is a quadratic invariant of the variational equation

$$\psi' = J^{-1} H''(\mathbf{u}) \psi$$

associated with the Hamiltonian equation. Thus applying the Runge–Kutta method to this derived equation gives the desired result, see BOCHEV and SCOVEL [1994].

All Gauss–Legendre methods have this property (and the implicit mid-point rule is simply the lowest order method of this form). For example, the second- (implicit mid-point) and fourth-order Gauss–Legendre methods have the form

$$\begin{array}{c|c} 1/2 & 1/2 \\ \hline & 1 \end{array} \qquad \begin{array}{c|cc} (3-\sqrt{3})/6 & 1/4 & (3-2\sqrt{3})/12 \\ (3+\sqrt{3})/6 & (3+2\sqrt{3})/12 & 1/4 \\ \hline & 1/2 & 1/2 \end{array}$$

and condition (3.10) can be seen to be satisfied for both methods. The fact that all Gauss–Legendre are symplectic implies that we can derive symplectic methods of arbitrarily high order.

The natural partitioning present in the Hamiltonian problem (3.2) suggests that we may think about using different numerical methods for different components of the problem. We shall now consider *partitioned Runge–Kutta* methods, where for our problem at hand we integrate the p components of (3.2) with the Runge–Kutta method given by the first Butcher tableau below, and the q components of (3.2) with the second tableau below.

$$\begin{array}{c|c} c & A \\ \hline & b^{\mathrm{T}} \end{array} \qquad \begin{array}{c|c} \tilde{c} & \tilde{A} \\ \hline & \tilde{b}^{\mathrm{T}} \end{array}$$

As we have done for symplectic Runge–Kutta methods we may now classify symplectic partitioned Runge–Kutta methods with the following result.

THEOREM. *A necessary and sufficient condition for a partitioned Runge–Kutta method to be symplectic is that for all* $1 \leqslant i, j \leqslant s$

$$b_i = \tilde{b}_i, \quad b_i\tilde{a}_{ij} + \tilde{b}_j a_{ji} - b_i\tilde{b}_j = 0. \tag{3.11}$$

The same comment as above about the need for irreducibility of the methods applies here, for more details see ABIA and SANZ-SERNA [1993]. For the special case of problems where the Hamiltonian takes the separable form $H(p,q) = T(p) + V(q)$, only the second part of (3.11) is required for symplecticity.

An example of such a method is given by Ruth's third-order method (RUTH [1983]) – one of the first symplectic methods appearing in the literature, it has the tableau

$$\begin{array}{c|ccc} & 7/24 & 0 & 0 \\ c & 7/24 & 3/4 & 0 \\ & 7/24 & 3/4 & -1/24 \\ \hline & 7/24 & 3/4 & -1/24 \end{array} \qquad \begin{array}{c|ccc} & 0 & 0 & 0 \\ \tilde{c} & 2/3 & 0 & 0 \\ & 2/3 & -2/3 & 0 \\ \hline & 2/3 & -2/3 & 1 \end{array}$$

For problems with a separable Hamiltonian (e.g., in celestial mechanics) the functional T often takes the special form $T(p) = p^{\mathrm{T}}M^{-1}p/2$, with M a constant symmetric invertible matrix (see the examples later). In this case Hamilton's equations (3.2) have the equivalent second order form

$$\ddot{q} = -M^{-1}V_q.$$

The fact that this special form for T arises often can now be seen to be a consequence of Newton's second law. The form that Hamilton's equations now take suggests the use of *Runge–Kutta–Nyström* methods, see HAIRER, NØRSETT and WANNER [1993]. The tableau here is of the form

$$\begin{array}{c|c} c & A \\ \hline & b^{\mathrm{T}} \\ \hline & B^{\mathrm{T}} \end{array}$$

which corresponds to the method given by

$$Q_i = q_n + c_i h M^{-1} p_n - h^2 \sum_{j=1}^{s} a_{ij} M^{-1} V_q(Q_j),$$

$$p_{n+1} = p_n - h \sum_{i=1}^{s} B_i V_q(Q_i),$$

$$q_{n+1} = q_n + h M^{-1} p_n - h^2 \sum_{i=1}^{s} b_i M^{-1} V_q(Q_i).$$

We now state a result giving conditions for a Runge–Kutta–Nyström method to be symplectic, see OKUNBOR and SKEEL [1992].

THEOREM. *The s-stage Runge–Kutta–Nyström given by the above tableau is symplectic if for all* $1 \leqslant i, j \leqslant s$

$$b_i = B_i(1 - c_i), \qquad B_i(b_j - a_{ij}) = B_j(b_i - a_{ji}). \tag{3.12}$$

3.2.3. Splitting and composition methods

These methods exploit natural decompositions of the problem, and particular of the Hamiltonian and have been used with very great success in studies of the solar system and of molecular dynamics (LEIMKUHLER [1999], SCHLIER and SEITER [1998]). Much recent work in this field is due to Yoshida and McLachlan. The main idea behind splitting methods is to decompose the discrete flow Ψ_h as a composition of simpler flows

$$\Psi_h = \Psi_{1,h} \circ \Psi_{2,h} \circ \Psi_{3,h} \cdots,$$

where each of the sub-flows is chosen such that each represents a simpler integration (perhaps even explicit) of the original. A geometrical perspective on this approach is to find useful geometrical properties of each of the individual operations which are preserved under combination. Symplecticity is just such a property, but we often seek to preserve reversibility and other structures.

Suppose that a differential equation takes the form

$$\frac{\mathrm{d}\mathbf{u}}{\mathrm{d}t} = \mathbf{f} = \mathbf{f}_1 + \mathbf{f}_2.$$

Here the functions $\mathbf{f}_1$ and $\mathbf{f}_2$ may well represent different physical processes in which case there is a natural decomposition of the problem (say into terms related to kinetic and potential energy).

The most direct form of splitting methods decompose this equation into the two problems

$$\frac{\mathrm{d}\mathbf{u}_1}{\mathrm{d}t} = \mathbf{f}_1 \quad \text{and} \quad \frac{\mathrm{d}\mathbf{u}_2}{\mathrm{d}t} = \mathbf{f}_2$$

chosen such that these two problems can be integrated *exactly in closed form* to give explicitly computable flows $\Psi_1(t)$ and $\Psi_2(t)$. We denote by $\Psi_{i,h}$ the result of applying the corresponding continuous flows $\psi_i(t)$ over a time h. A simple (first-order) splitting is then given by the *Lie–Trotter formula*

$$\Psi_h = \Psi_{1,h} \circ \Psi_{2,h}. \tag{3.13}$$

Suppose that the original problem has Hamiltonian $H = H_1 + H_2$, then as observed earlier this is the composition of two problems with respective Hamiltonians H_1 and H_2. The differential equation corresponding to each Hamiltonian leads to an evolutionary map $\psi_i(t)$ of the form described above. By definition, as ψ_i is the exact solution of a Hamiltonian system, it must follow that each operator $\Psi_{i,h}$ is a symplectic map. Hence, by Lemma 3.1, the combined map must also be symplectic. Thus operator splitting in this context automatically generates a symplectic map.

An interesting example of such a procedure in the context of a partial differential equation, are the so-called *equilibrium models* for calculating advective and reactive chemical flows of the form

$$u_t + au_x = -g(u, v), \qquad v_t = g(u, v).$$

To apply a splitting method to this system, an advective step without reaction is determined exactly by using the method of characteristics applied to the total concentration $c = u + v$. This step is then followed by a reactive step without advection. These models are widely used in the chemical industry and a numerical analysis of splitting methods for advection problems is given in LE VEQUE and YEE [1990]. Indeed, procedures of this form are widely used in any process where there are clearly two interacting components and each component is treated separately, see, for example, SANZ-SERNA and PORTILLO [1996].

The Lie–Trotter splitting (3.13) introduces local errors proportional to h^2 at each step and a more accurate decomposition is the *Strang splitting* given by

$$\Psi_h = \Psi_{1,h/2} \circ \Psi_{2,h} \circ \Psi_{1,h/2}. \tag{3.14}$$

This splitting method has a local error proportional to h^3.

A proof of these error estimates obtained by using the Baker–Campbell–Hausdorff formula will be given presently.

EXAMPLE 3.1. Suppose that a Hamiltonian system has a Hamiltonian which can be expressed as a combination of a kinetic energy and a potential energy term as follows

$$H(\mathbf{u}) = H_1(\mathbf{u}) + H_2(\mathbf{u}) \equiv T(\mathbf{p}) + V(\mathbf{q})$$

so that

$$\frac{\mathrm{d}\mathbf{p}}{\mathrm{d}t} = -H_{2,q}(\mathbf{q}), \qquad \frac{\mathrm{d}\mathbf{q}}{\mathrm{d}t} = H_{1,p}(\mathbf{p}).$$

This splitting of H is usually referred to as a separable or P-Q splitting. We immediately have that

$$\Psi_{1,h} = I - hH_{2,q} \quad \text{and} \quad \Psi_{2,h} = I + hH_{1,p},$$

where I represents the identity mapping.

Applying the Lie–Trotter formula directly to this splitting gives the *symplectic Euler method* (SE)

$$\mathbf{p}_{n+1} = \mathbf{p}_n - hH_{2,q}(\mathbf{q}_n), \quad \text{and} \quad \mathbf{q}_{n+1} = \mathbf{q}_n + hH_{1,p}(\mathbf{p}_{n+1}). \tag{3.15}$$

A more sophisticated splitting method for this problem based upon the Strang splitting is

$$\mathbf{p}_{n+1/2} = \mathbf{p}_n - \frac{h}{2}H_{2,q}(\mathbf{q}_n), \qquad \mathbf{q}_{n+1} = \mathbf{q}_n + hH_{1,p}(\mathbf{p}_{n+1/2}),$$
$$\mathbf{p}_{n+1} = \mathbf{p}_{n+1/2} - \frac{h}{2}H_{2,q}(\mathbf{q}_{n+1}).$$

For systems with such separable Hamiltonians we may combine subsequent Strang splittings (as described above) to give (after relabelling) the celebrated *Störmer–Verlet method* (SV)

$$\mathbf{q}^{n+1/2} = \mathbf{q}^{n-1/2} + hT_p(\mathbf{p}^n),$$
$$\mathbf{p}^{n+1} = \mathbf{p}^n - hV_q(\mathbf{q}^{n+1/2}).$$

The same method is obtained by apply the two-stage Lobatto IIIA-B Runge–Kutta pair to this separable case (see HAIRER, LUBICH and WANNER [2002]).

A form of this method appeared in the molecular dynamics literature (VERLET [1967]) many years before anyone realised that its remarkable success in that field was due to the fact that it was in fact a very efficient symplectic method.

EXAMPLE 3.2 (*Yoshida splittings*). The Strang splitting has the desirable property that it is symmetric so that $\Psi_h^{-1} = \Psi_{-h}$. By combining symmetric splittings, YOSHIDA [1990] derived a series of remarkable high order methods. Given a symmetric second-order *base method* $\Psi_h^{(2)}$, for example, in YOSHIDA [1990], Yoshida considers the case where the base method is given by the Strang splitting (3.14). Yoshida then constructs the method

$$\Psi_h^{(4)} = \Psi_{x_1 h}^{(2)} \circ \Psi_{x_0 h}^{(2)} \circ \Psi_{x_1 h}^{(2)}.$$

He proved that $\Psi_h^{(4)}$ is indeed a fourth-order symmetric method if the weights are chosen as

$$x_0 = -\frac{2^{1/3}}{2 - 2^{1/3}}, \qquad x_1 = \frac{1}{2 - 2^{1/3}}.$$

If the problem being considered is Hamiltonian and the second-order method is symplectic then our newly constructed fourth-order method will also be symplectic. This procedure can be extended and generalized as follows. Given a symmetric integrator $\Psi_h^{(2n)}$ of order $2n$, for example, when $n = 1$ we are simply restating the above construction and when $n = 2$ we may take the method constructed above. The method given by the composition

$$\Psi_h^{(2n+2)} = \Psi_{x_1 h}^{(2n)} \circ \Psi_{x_0 h}^{(2n)} \circ \Psi_{x_1 h}^{(2n)}$$

will be a symmetric method of order $2n + 2$ if the weights are chosen as

$$x_0 = -\frac{2^{1/(2n+1)}}{2 - 2^{1/(2n+1)}}, \qquad x_1 = \frac{1}{2 - 2^{1/(2n+1)}}.$$

NOTE. When considering partial differential equations MCLACHLAN [1994] uses alternative splittings of the Hamiltonian into linear and nonlinear components. We return to this decomposition presently.

3.3. *Examples of applications of symplectic integrators*

Before looking at the error analysis of these various symplectic methods we consider their application to four examples, the harmonic oscillator, the pendulum, molecular and stellar dynamics.

3.3.1. *Example 1. The Harmonic oscillator*

This well studied problem has a separable Hamiltonian of the form

$$H(p, q) = \frac{p^2}{2} + \frac{q^2}{2},$$

and has solutions which are circles in the (p, q) phase space. The associated differential equations are

$$\frac{\mathrm{d}q}{\mathrm{d}t} = p, \qquad \frac{\mathrm{d}p}{\mathrm{d}t} = -q.$$

Consider now the closed curve

$$\Gamma \equiv p^2 + q^2 = C^2,$$

the action of the solution operator of the differential equation is to map this curve into itself (conserving the area πC^2 of the enclosed region). The standard *forward Euler method* applied to this system gives the scheme

$$p_{n+1} = p_n - hq_n, \qquad q_{n+1} = q_n + hp_n,$$

so that Ψ_h is the operator given by

$$\Psi_h \mathbf{v} = \begin{pmatrix} 1 & -h \\ h & 1 \end{pmatrix} \mathbf{v}, \quad \text{with } \det(\Psi_h) = 1 + h^2.$$

It is easy to see that in this case, Γ evolves through the action of the discrete map Ψ_h to the new circle given by

$$p_{n+1}^2 + q_{n+1}^2 = C^2(1 + h^2)$$

and the area enclosed within the discrete evolution of Γ has increased by a factor of $1 + h^2$. Periodic orbits are not preserved by the forward Euler method – indeed all such discrete orbits spiral to infinity. Similarly, a discretisation using the backward Euler method leads to trajectories that spiral towards the origin.

Consider now the *symplectic Euler method* applied to this example. This gives rise to the discrete map

$$p_{n+1} = p_n - hq_n, \qquad q_{n+1} = q_n + h(p_n - hq_n) = (1 - h^2)q_n + hp_n.$$

The discrete evolutionary operator is then simply the matrix

$$\Psi_h \begin{pmatrix} p \\ q \end{pmatrix} = \begin{pmatrix} 1 & -h \\ h & 1 - h^2 \end{pmatrix} \begin{pmatrix} p \\ q \end{pmatrix}$$

which can easily be checked to be symplectic. For example, $\det(\Psi_h) = 1$. The circle Γ is now mapped to the ellipse

$$(1 - h^2 + h^4) p_{n+1}^2 - 2h^3 p_{n+1} q_{n+1} + (1 + h^2) q_{n+1}^2 = C^2$$

which has the same enclosed area. The symplectic Euler map is not symmetric in time so that

$$\psi_h^{-1} \neq \psi_{-h}.$$

Observe that the symmetry of the circle has been destroyed through the application of this mapping.

It is also easy to see that if

$$A = \begin{pmatrix} 1 & -\frac{h}{2} \\ -\frac{h}{2} & 1 \end{pmatrix},$$

then

$$\Psi_h^{\mathrm{T}} A \Psi_h = A.$$

Consequently, invariant curves of the map Ψ_h are given by the solutions of

$$\widehat{H}(p_n, q_n) \equiv p_n^2 + q_n^2 - hp_n q_n = C^2,$$

which are ellipses with major and minor axes of lengths $C(1 + h/2)$ and $C(1 - h/2)$, respectively. Observe that the modified Hamiltonian function $\widehat{H}$ is an $\mathcal{O}(h)$ perturbation of the original Hamiltonian. Thus the symplectic Euler method preserves the qualitative features of bounded solutions and periodic orbits lost by both the forward and backward Euler methods.

In Fig. 3.1 we compare the original periodic orbit with the perturbed periodic orbit in the case of $h = 0.25$, with starting values $p = 0, q = 1$.

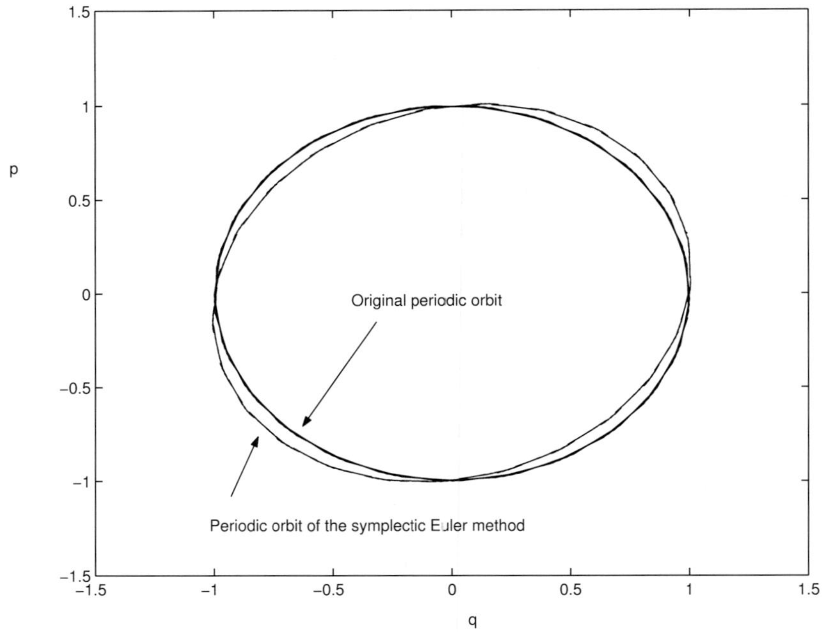

FIG. 3.1. Periodic orbit of the symplectic Euler method compared with the original periodic orbit.

Next consider the *Störmer–Verlet* method. This gives the discrete map

$$p_{n+1/2} = p_n - hq_n/2, \qquad q_{n+1} = q_n + h(p_n - hq_n/2),$$
$$p_{n+1} = p_{n+1/2} - \frac{h}{2}\left(q_n + h\left(p_n - \frac{h}{2}q_n\right)\right).$$

The discrete evolutionary operator ψ_h is now the symplectic matrix

$$\Psi_h \mathbf{v} = \begin{pmatrix} 1 - h^2/2 & -h + h^3/4 \\ h & 1 - h^2/2 \end{pmatrix} \mathbf{v}.$$

The curve Γ is to order h^2 mapped to a circle of the same radius. The Störmer–Verlet method preserves the symmetry of the circle to this order, and indeed is also symmetric in time so that $\psi^{-1} = \psi_{-h}$.

Finally we consider the implicit mid-point rule. For this we have

$$C\begin{pmatrix} p_{n+1} \\ q_{n+1} \end{pmatrix} = C^{\mathrm{T}}\begin{pmatrix} p_n \\ q_n \end{pmatrix},$$

where

$$C = \begin{pmatrix} 1 & \frac{h}{2} \\ -\frac{h}{2} & 1 \end{pmatrix}.$$

Observe that $CC^{\mathrm{T}} = (1 + h^2/4)I$. The evolution operator is then simply

$$\Psi_h = C^{-1}C^{\mathrm{T}}.$$

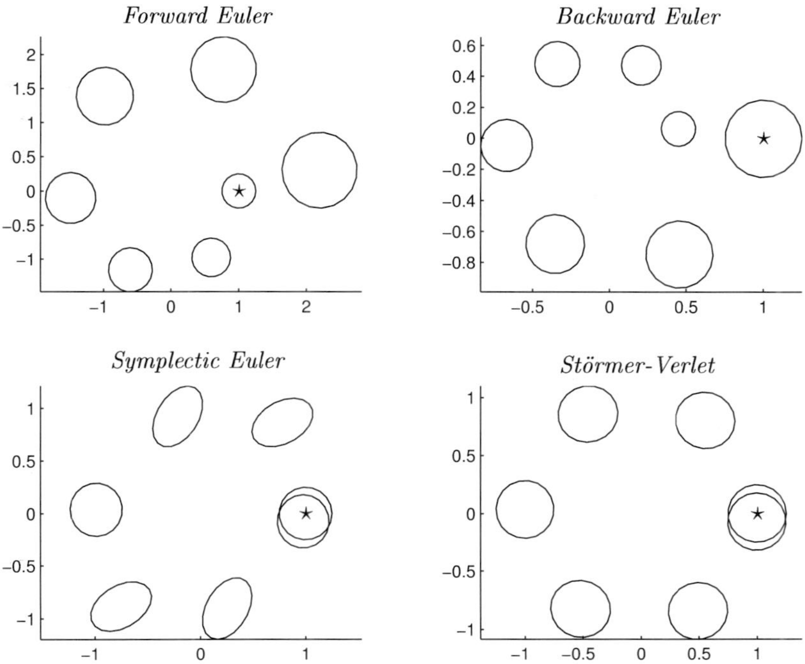

FIG. 3.2. Dynamics of four numerical methods applied to the harmonic oscillator problem. Forward and Backward Euler both have time step $2\pi/24$ and symplectic Euler and Störmer–Verlet have time step $2\pi/12$. The $\star$ is placed in the centre of the circle of initial conditions.

Observe further that if $\mathbf{U}_n = (p_n, q_n)^{\mathrm{T}}$ then

$$\left(1 + h^2/4\right)\mathbf{U}_{n+1}^{\mathrm{T}}\mathbf{U}_{n+1} = \mathbf{U}_{n+1}^{\mathrm{T}}C^{\mathrm{T}}C\mathbf{U}_{n+1} = \mathbf{U}_n^{\mathrm{T}}CC^{\mathrm{T}}\mathbf{U}_n = \left(1 + h^2/4\right)\mathbf{U}_n^{\mathrm{T}}\mathbf{U}_n.$$

Hence the quadratic invariant $\mathbf{U}_n^{\mathrm{T}}\mathbf{U}_n$ is exactly preserved by this method. This feature is shared by all symplectic Runge–Kutta methods.

Fig. 3.2 demonstrates the effect of applying some of these methods to the set Γ.

3.3.2. Example 2. The pendulum

The simple pendulum is a Hamiltonian system, the solution of which by symplectic methods is discussed in YOSHIDA [1993], HAIRER, NØRSETT and WANNER [1993]. For small values of p and q it reduces to the Harmonic oscillator and it has periodic orbits which are close to circles. For larger values the periodic orbits evolve towards heteroclinic connexions. For this problem we have the Hamiltonian

$$H(p, q) = \tfrac{1}{2}p^2 + \cos(q),$$

and associated differential equations

$$\frac{\mathrm{d}q}{\mathrm{d}t} = p, \qquad \frac{\mathrm{d}p}{\mathrm{d}t} = -\sin(q).$$

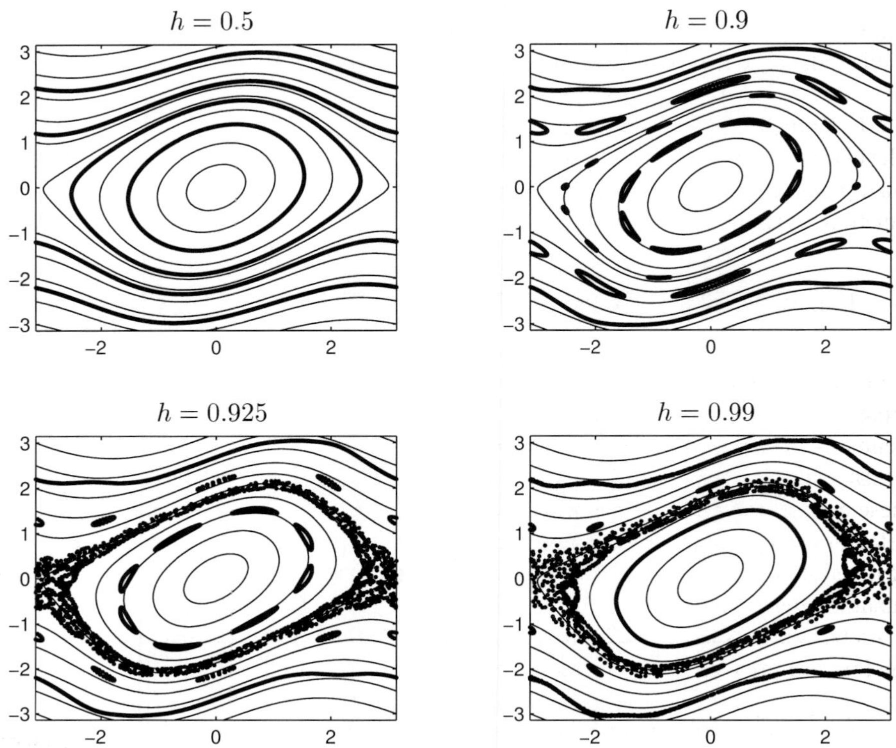

FIG. 3.3. Dynamics of the symplectic Euler method applied to the pendulum problem.

As above we apply the symplectic Euler method to the system to give the discrete system

$$p_{n+1} = p_n - h \sin q_n, \qquad q_{n+1} = q_n + h p_{n+1}.$$

In Fig. 3.3 we consider applying this method for a variety of values of h. For small values much of the original structure of the phase space is preserved. Observe the persistence of the periodic orbits, now distorted into ellipses as in the last example. For larger values of h this structure starts to break down with periodic orbits breaking up into invariant tori. For still larger values of h chaotic behaviour is observed. All of these figures are very similar to those of the 'standard map' in Hamiltonian dynamics (LICHTENBERG and LIEBERMAN [1983]) with dynamics determined by the results of the KAM theorem.

In this case the modified equation analysis presented in the next section predicts that to leading order the symplectic Euler method has the same dynamics as a Hamiltonian system with the modified Hamiltonian

$$\begin{aligned}\widetilde{H} = {} & \frac{p^2}{2} - \cos q - \frac{h}{2} p \sin(q) + \frac{h^2}{12}\bigl(\sin^2(p) + p^2 \cos(q)\bigr) \\ & - \frac{h^3}{12} p \cos(q) \sin(q) + \mathcal{O}\bigl(h^4\bigr),\end{aligned}$$

and it is level sets of this that are also plotted in Fig. 3.3. Observe that for small p and q this reduces to the modified Hamiltonian of the harmonic oscillator.

3.3.3. *Example 3. The many body problem*

The classical equations for a system of $N > 1$ particles (or heavenly bodies) interacting via a potential force V can be written in Hamiltonian form (3.2), with the Hamiltonian function given by

$$H(\mathbf{p}, \mathbf{q}) = \frac{1}{2}\sum_{i=1}^{N} m_i^{-1}\mathbf{p}_i^{\mathrm{T}}\mathbf{p}_i + \sum_{i<j} V_{i,j}\left(\|\mathbf{q}_i - \mathbf{q}_j\|\right). \tag{3.16}$$

Here the m_i represent masses of objects with positions and momenta $q_i, p_i \in \mathbb{R}^3$. $V(r)$ represents some potential function, for example, r^{-1} for gravitational problems and the Lennard-Jones potential $r^{-12} - r^{-6}$ for problems in molecular dynamics (ALLEN and TILDESLEY [1987]). Here $\|\cdot\|$ is the Euclidean distance. We would like to integrate these equations for long times where many near collisions (and hence large forces and velocities) may occur, whilst conserving the energy H and the total angular momentum of the system $L = \sum_{i=1}^{N} p_i \times q_i$. In practice L is relatively easy to conserve as it is a quadratic invariant.

As before we shall consider here three numerical schemes. The simplest possible is the forward (explicit) Euler method (FE)

$$\begin{aligned}\mathbf{p}^{n+1} &= \mathbf{p}^n - hH_q(\mathbf{p}^n, \mathbf{q}^n),\\ \mathbf{q}^{n+1} &= \mathbf{q}^n + hH_p(\mathbf{p}^n, \mathbf{q}^n).\end{aligned}$$

The Hamiltonian is again separable, so we may use the symplectic Euler method (SE)

$$\begin{aligned}\mathbf{p}^{n+1} &= \mathbf{p}^{n+1} - hH_q(\mathbf{p}^n, \mathbf{q}^n),\\ \mathbf{q}^{n+1} &= \mathbf{q}^n + hH_p(\mathbf{p}^{n+1}, \mathbf{q}^n)\end{aligned}$$

and the Störmer–Verlet method.

3.3.4. *Example 4. Stellar dynamics and the Kepler problem*

The Kepler (or two-body) problem, describing the motion under gravity of a planet around large stellar mass is a special case of Example 7.3 which may be written as a Hamiltonian system with

$$H(p_1, p_2, q_1, q_2) = \frac{1}{2}\left(p_1^2 + p_2^2\right) - \frac{a}{\sqrt{q_1^2 + q_2^2}}, \tag{3.17}$$

so that $V(r) = r^{-1}$ in (3.16). The exact dynamics of this system exactly preserve H which represents total energy, as well as the angular momentum given by

$$L = q_1 p_2 - q_2 p_1.$$

In addition, the problem has rotational, time-reversal and scaling symmetries, which we shall consider presently. In Fig. 3.4 we show some trajectories computed with the three

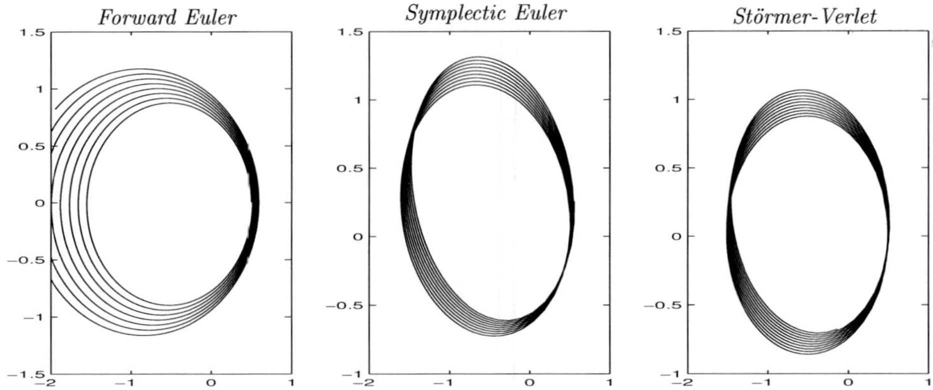

FIG. 3.4. Kepler trajectories with eccentricity of 0.5 computed with the forward Euler ($h = 0.002$), symplectic Euler ($h = 0.1$) and Störmer–Verlet ($h = 0.1$) methods.

methods introduced above. For the initial data used here the exact solution is an ellipse of eccentricity $e = 0.5$ with the origin at one focus. Notice that the forward Euler trajectory spirals outwards (in a similar manner to behaviour observed for the Harmonic oscillator) and so does not accurately reproduce the periodic solution to this problem. The symplectic Euler does better in this respect, the numerical solution lies much closer to an ellipse, however it exhibits clockwise precession (the ellipse rotates) about the origin.

In Fig. 3.5 we consider both the growth in the trajectory error (computed using the Euclidean norm) and the conservation (or lack of it) of Hamiltonian (or energy) for our methods. The forward Euler method acts to increase the energy of the system leading to a monotonic growth in the Hamiltonian. In contrast the Hamiltonian whilst not constant for the symplectic methods exhibits a bounded error. The symplectic methods have *linear* trajectory error growth as opposed to the *quadratic* growth observed in the non-symplectic methods. The various peaks in these graphs correspond to close approaches between the planet and the star.

These results are summarized in the following table, given in HAIRER, LUBICH and WANNER [2002]. Note that both the symplectic Euler and Störmer–Verlet methods preserve the quadratic invariant of the angular momentum exactly.

Method	Global error	Error in H	Error in L
FE	$\mathcal{O}(t^2h)$	$\mathcal{O}(th)$	$\mathcal{O}(th)$
SE	$\mathcal{O}(th)$	$\mathcal{O}(h)$	0
SV	$\mathcal{O}(th^2)$	$\mathcal{O}(h^2)$	0

See HAIRER, LUBICH and WANNER [2002], SANZ-SERNA and CALVO [1994] for similar experiments and discussions, as well as proofs and explanations of the apparent superiority of symplectic over non-symplectic methods.

More sophisticated methods for the N-body problem have been developed in the astrophysics literature. These methods have largely been based upon splitting the problem into a collection of independent two-body problems plus a potential term accounting for mutual planetary interactions, as well as careful integration of certain parts of close

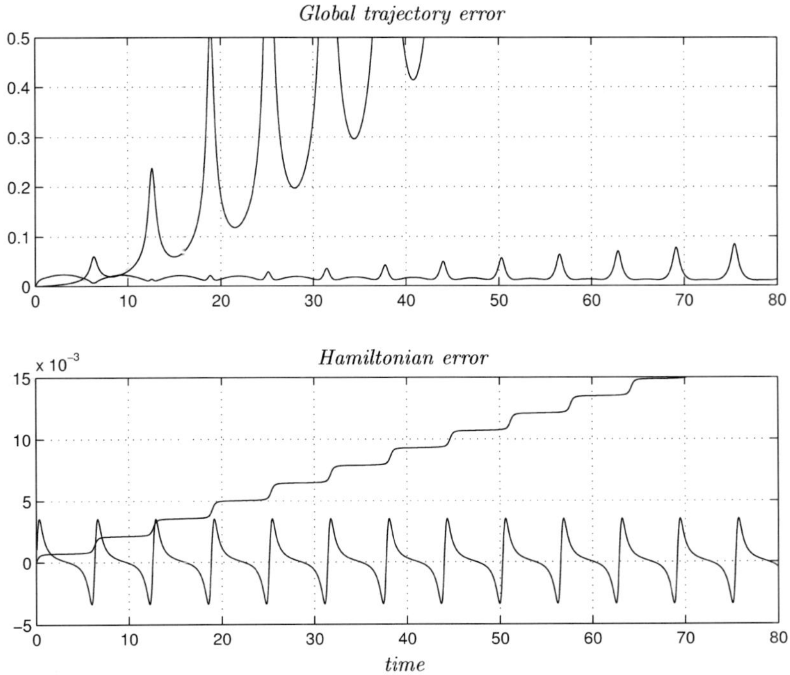

FIG. 3.5. Global trajectory error measured with Euclidean norm in four-dimensional phase space, and error in Hamiltonian for the Kepler problem with eccentricity $e = 0.5$. Methods shown are the forward Euler ($h = 0.0001$) lying in general above symplectic Euler ($h = 0.005$).

approach orbits. They have been used to compute the evolution of the solar system for many millions of years. In particular SUSSMAN and WISDOM [1992] computed the evolution of the nine planets for 100 million years using a time step of 7.2 days. They found that the solar system as a whole is chaotic with a Lyapunov exponent of 4 million years, and in particular they verified earlier work by showing that the motion of Pluto is chaotic with a Lyapunov exponent of 10 to 20 million years. Another calculation is presented by WISDOM and HOLMAN [1991], they calculated the evolution of the outer five planets for 1.1 billion years with a time step of 1 year and produced similar results. A possible disagreement between the chaos exhibited in numerical simulations and classical stability theories for the solar system is discussed and resolved by MURRAY and HOLMAN [1999].

Note that for problems with close approaches, and hence large forces and velocities, some form of adaptivity often needs to be employed and we discuss this later in the section on temporal adaptivity and singularities.

3.4. *Error analysis – a brief overview*

We now give a brief overview of the methods used to analyse the behaviour of symplectic and splitting methods. These ideas are covered in much greater detail elsewhere and for the most part we refer the reader to this literature.

3.4.1. The backward error analysis of symplectic methods

As described earlier, backward error analysis (or modified equation analysis) is the procedure for analysing a numerical method by finding a modified differential equation which has a flow which approximates the discrete flow better than the original equation. The modified equations can be constructed by using a simple iterative procedure outlined in HAIRER and LUBICH [2000]. Many deep theorems related to backward error analysis are given in REICH [1999]. This is a very rich field of analysis which we only touch on here, for a review and a discussion of its applicability to various problems see GRIFFITHS and SANZ-SERNA [1986].

We now give a more detailed derivation of the standard backward error formula. Suppose that $\mathbf{u}' = \mathbf{f}(\mathbf{u})$ is the original equation with flow $\psi(t)$ and that $\hat{\mathbf{u}}' = \hat{\mathbf{f}}(\hat{\mathbf{u}})$ is the modified equation. We will suppose that for small h we can develop an asymptotic series for $\hat{\mathbf{f}}$ of the form

$$\hat{\mathbf{f}}(\mathbf{u}) = \mathbf{f}(\mathbf{u}) + h\mathbf{f}_1(\mathbf{u}) + h^2\mathbf{f}_2(\mathbf{u}) + \cdots.$$

Assume that at some initial time $\mathbf{u} = \hat{\mathbf{u}}$. Using a simple Taylor series we can expand the solution of the modified ordinary differential equation so that

$$\hat{\mathbf{u}}(t+h) = \mathbf{u} + h\big(\mathbf{f} + h\mathbf{f}_2 + h^2\mathbf{f}_3 + \cdots\big) + \frac{h^2}{2}\big(\mathbf{f}' + h\mathbf{f}' + \cdots\big)(\mathbf{f} + \cdots) + \cdots.$$

We want to compare this solution with the action of the discrete operator Ψ_h. In general, for a consistent scheme, there will be functions $\mathbf{d}_k(h)$ so that

$$\Psi_h\mathbf{u} = \mathbf{u} + h\mathbf{f}(\mathbf{u}) + h^2\mathbf{d}_2(\mathbf{u}) + h^3\mathbf{d}_3(\mathbf{u}) + \cdots.$$

We can now equate the above two expressions. Taking terms at orders h^2 and h^3 gives

$$\mathbf{f}_2 = \mathbf{d}_2 - \tfrac{1}{2}\mathbf{f}'\mathbf{f} \quad \text{and} \quad \mathbf{f}_3 = \mathbf{d}_3 - \tfrac{1}{6}\big(\mathbf{f}''(\mathbf{f},\mathbf{f})(\mathbf{u}) + \mathbf{f}'\mathbf{f}'\mathbf{f}(\mathbf{u})\big) - \tfrac{1}{2}\big(\mathbf{f}'\mathbf{f}_2(\mathbf{u}) + \mathbf{f}_2'\mathbf{f}(\mathbf{u})\big).$$

This procedure can be continued to all orders and a Maple program to do this is described in HAIRER and LUBICH [2000]. In general it gives an asymptotic rather than a convergent series, which diverges if all terms are summed, but which can be optimally truncated to give errors which are (potentially exponentially) small in h.

An example given in HAIRER, LUBICH and WANNER [2002], GRIFFITHS and SANZ-SERNA [1986] is the blow-up or nonlinear heat equation. We consider this here as an example of backward error analysis and return to it later for a fuller discussion of the application of scale invariant adaptive methods. The blow-up equation is given by

$$u' = u^2, \qquad u(0) = 1. \tag{3.18}$$

Significantly, this equation has singular solution $u(t) = 1/(1-t)$ which becomes infinite as $t \to 1$.

When approximated by the Forward Euler method this problem has a discrete solution which is bounded for all time and does not in any way reproduce the singular behaviour. This will be true of any explicit method when applied to this problem. (In our later discussion on scale invariant methods we will show how this property can be recovered.) In this case we have $\mathbf{d}_2 = \mathbf{d}_3 = \cdots = 0$. Applying backward error analysis, the modified equation for (3.18) under this discretisation then takes the form

$$\hat{f}(u) = u^2 - hu^3 + (3/2)h^2u^4 - (8/3)h^3u^5 + \cdots.$$

The modified system does not admit singular solutions and is qualitatively much closer to the solution of the Forward Euler method than the original.

A key feature underlying the analytical approach of backward error analysis is the question of whether qualitative features of the underlying equation are reflected in the modified equation.

HAIRER, LUBICH and WANNER [2002] uses an inductive argument to show that if the underlying equation is Hamiltonian and the method is symplectic then the modified system constructed as above is also Hamiltonian. In particular there are smooth functions H_j such that

$$\mathbf{f}_j(\mathbf{u}) = J^{-1}\nabla H_j(\mathbf{u}).$$

For the Harmonic oscillator problem considered earlier, we have already seen that to leading order we have

$$H(\mathbf{u}) = \tfrac{1}{2}(p^2 + q^2) \quad \text{and} \quad H_1(\mathbf{u}) = -\tfrac{1}{2}pq.$$

It has also been shown that if the underlying system is reversible in time and a symmetric method (e.g., Störmer–Verlet) is used to solve it, then the resulting modified system is also reversible (HAIRER and STOFFER [1997]). In addition REICH [1999] proves that if f lies in a Lie algebra $\mathfrak{g}$ and the numerical method defines a discrete flow remaining within the associated Lie group G (see the section on Lie group methods later), then the $\hat{f}$ associated with the modified system also lies in $\mathfrak{g}$.

REICH [1999] and many others have presented conditions for the error between the modified equation, the discrete solution and the underlying differential equation. This error depends upon the truncation point of the series used to generate the modified equations. This is, in general, an *asymptotic* series which diverges if continued as an expansion with an infinite number of terms, but which has superconvergent properties if truncated optimally. (See similar results for the study of asymptotic series in general which also demonstrate exponential convergence when optimally truncated (CHAPMAN, KING and ADAMS [1998]).) In particular, it can be shown (BENETTIN and GIORGILLI [1994]) that an optimally truncated modified equation has an exponentially small error such that if ψ_h is the discrete flow and $\varphi_{N,h}$ the flow of the modified equation when truncated at the point N then there is a γ which depends on the method, a constant h^* and an optimal $N(h)$ so that

$$\|\Psi_h - \varphi_{N,h}\| < h\gamma\,\mathrm{e}^{-h^*/h}.$$

If the underlying system is Hamiltonian, this result extends to the theorem that the discrete Hamiltonian closely approximates the (constant) Hamiltonian of the modified problem over exponentially long time intervals. Indeed, in this case

$$\widehat{H}(y_n) = \widehat{H}(y_0) + \mathcal{O}(\mathrm{e}^{-h^*/2h}),$$

so that although the actual Hamiltonian is not conserved a modified form of it is, to a very good approximation, conserved over very long time intervals. This explains the results described in the calculations presented for the Kepler problem.

The pendulum equation discussed in the examples above has Hamiltonian

$$H = \frac{p^2}{2} - \cos(q).$$

When using the implicit mid-point rule the modified Hamiltonian (HAIRER, LUBICH and WANNER [2002]) is given by

$$\widehat{H} = \frac{p^2}{2} - \cos(q) + \frac{h^2}{48}\bigl(\cos(2q) - 2p^2\cos(q)\bigr).$$

For *sufficiently small* values of h the modified Hamiltonian $\widehat{H}$ is extremely well conserved. As we have seen from the earlier experiments, this agreement breaks down (with a consequent loss of integrability) when h takes larger values.

3.4.2. *An analysis of splitting methods*

The basic tool for analysis of splitting methods is the celebrated Baker–Campbell–Hausdorff (BCH) formula (VARADARAJAN [1974]) which allows an analysis of splitting methods to be made in terms of the (lack of) commutativity of the operators associated with the splitting.

To motivate this procedure, consider the system

$$\frac{d\mathbf{u}}{dt} = \mathbf{f}(\mathbf{u}) = \mathbf{f}_1 + \mathbf{f}_2$$

such that the two systems $d\mathbf{u}_i/dt = \mathbf{f}_i$ have respective flows ψ_t^i (which may be calculated exactly or approximately using a discrete (high order) method). Associated with each such flow is the *Lie derivative* D_i defined by its action on the function F so that

$$\frac{d}{dt}F\bigl(\mathbf{u}_i(t)\bigr) = (D_i F)\bigl(\mathbf{u}_i(t)\bigr) = F'(\mathbf{u}_i)f_i(\mathbf{u}_i),$$

where $\mathbf{u}_i(t)$ is the solution of the associated differential equation. Then if $d\mathbf{u}_i/dt = \mathbf{f}_i(\mathbf{u}_i)$ we have $d\mathbf{u}_i/dt = D_i\mathbf{u}_i(t)$, $d^2\mathbf{u}_i/dt^2 = D_i^2\mathbf{u}_i(t)$, etc. As a consequence we may express the flow φ_t^i as an exponential of the form

$$\varphi_t^i\mathbf{u}_0 = \exp(tD_i)\mathbf{u}_0 = \sum_n \frac{t^n D_i^n}{n!}\mathbf{u}_0,$$

and compose two flows in the manner

$$\bigl(\varphi_t^1 \circ \varphi_s^2\bigr)u_0 = \exp(tD_1)\exp(sD_2)\mathbf{u}_0.$$

Thus we see how a flow can be thought of in terms of the exponential of an operator. Now suppose that X and Y are general operators with associated exponentials

$$\exp(hX) = I + hX + \frac{h^2}{2}X^2 + \cdots, \qquad \exp(hY) = I + hY + \frac{h^2}{2}Y^2 + \cdots,$$

then the combined exponential (applying one operator and then the next) is given by

$$\exp(hX)\exp(hY)$$

whereas the simultaneous action of the two is given by

$$\exp(hX + hY).$$

If X and Y commute then these two operations are the same, however in general they are different, and the BCH formula gives a precise description of this difference.

THEOREM. *For the operators hX and hY defined above we have*

$$\exp(hX)\exp(hY) = \exp(hZ),$$

where

$$hZ = hX + hY + \frac{h^2}{2}[X, Y] + \frac{h^3}{12}\big([X, X, Y] + [Y, Y, X]\big) + \cdots, \tag{3.19}$$

where $[X, Y]$ is the commutator of the operators X and Y given by

$$[X, Y] = XY - YX, \quad [X, X, Y] = \big[X, [X, Y]\big], \quad \textit{etc.}$$

PROOF. See VARADARAJAN [1974]. □

We compare this with the operation of evolving under the action of $hX + hY$ given by $\exp(hX + hY)$. If hX and hY are discrete operators applied over a time period h then the principle error introduced by operator splitting is given by the order h^2 contribution $h^2[X, Y]/2$.

SANZ-SERNA [1997] considers analysing splitting methods using the BCH formula as follows

(1) Constructing modified systems corresponding to the discrete operators in the splitting (this is trivial for separable problems).
(2) Writing these flows in terms of exponentials X and Y as above.
(3) Using BCH to construct the modified equation for the splitting method.

A detailed analysis of this procedure is described in MURUA and SANZ-SERNA [1999].

EXAMPLE. Consider the use of the symplectic Euler method when applied to the Harmonic oscillator problem

$$\frac{du}{dt} = v, \qquad \frac{dv}{dt} = -u$$

so that

$$\frac{d\mathbf{u}}{dt} = \begin{pmatrix} 0 & 1 \\ -1 & 0 \end{pmatrix}\mathbf{u} \equiv A\mathbf{u}.$$

The continuous flow map ψ_h is then the rotation matrix

$$\psi_h \mathbf{u} = \exp(hA) = \begin{pmatrix} \cos(h) & \sin(h) \\ -\sin(h) & \cos(h) \end{pmatrix}\mathbf{u}.$$

The symplectic Euler method decomposes the matrix A into the two matrices A_1 and A_2 where

$$A_1 = \begin{pmatrix} 0 & 1 \\ 0 & 0 \end{pmatrix} \quad \text{and} \quad A_1 = \begin{pmatrix} 0 & 0 \\ -1 & 0 \end{pmatrix}.$$

Thus

$$\exp(hA_1) = \begin{pmatrix} 1 & h \\ 0 & 1 \end{pmatrix} \quad \text{and} \quad \exp(hA_2) = \begin{pmatrix} 1 & 0 \\ -h & 1 \end{pmatrix}$$

giving

$$\begin{aligned} \exp(hA_1)\exp(hA_2) &= \begin{pmatrix} 1-h^2 & h \\ -h & 1 \end{pmatrix} \\ &= \begin{pmatrix} \cos(h) & \sin(h) \\ -\sin(h) & \cos(h) \end{pmatrix} + \begin{pmatrix} -h^2/2 & 0 \\ 0 & h^2/2 \end{pmatrix} + \mathcal{O}(h^3). \end{aligned}$$

Observe that

$$\frac{h^2}{2}[A_1, A_2] = \begin{pmatrix} -h^2/2 & 0 \\ 0 & h^2/2 \end{pmatrix}$$

so that the above analysis is fully consistent with this result.

In contrast we may apply the same analysis to the *Strang splitting*. This gives

$$\begin{aligned} &\exp(hX/2)\exp(hY)\exp(hX/2) \\ &\quad = \exp(hX/2)\exp\bigl(hY + hX/2 + h^2[Y,X]/4 + \cdots\bigr) \\ &\quad = \exp\bigl(hY + hX + h^2[Y,X]/4 + h^2[X,Y]/4 + h^2[X,X]/8 + \cdots\bigr). \end{aligned}$$

The identities $[Y,X] + [X,Y] = 0$ and $[X,X] = 0$ imply that the dominant error to the Strang splitting is zero – accounting for the improved accuracy of this method. Indeed the next term in this construction is given by

$$\frac{h^3}{12}\left([Y,Y,X] - \frac{1}{2}[X,X,Y]\right),$$

demonstrating the improved accuracy of this method.

Returning to our example, it is easy to see that

$$\begin{aligned} &\exp(hA_1/2)\exp(hA_2)\exp(hA_1/2) \\ &\quad = \begin{pmatrix} 1-h^2/2 & h-h^3/4 \\ -h & 1-h^2/2 \end{pmatrix} = \exp(hA_1 + hA_2) + \mathcal{O}(h^3). \end{aligned}$$

This same procedure can be applied to general splitting decompositions of the form (McLachlan [1995])

$$\varphi^2_{b_m h} \circ \varphi^1_{a_m h} \circ \varphi^2_{b_{m-1} h} \circ \cdots \circ \varphi^1_{a_1 h}.$$

Careful choices of the coefficients $a_1, \ldots, b_m$ lead to higher order methods. A particular example of which is the Yoshida method described earlier which combines time-symmetric methods (such as the Störmer–Verlet method) with both *positive* and *negative* time steps to derive higher order methods. More recently BLANES and MOAN [2002] have constructed a series of higher order methods which are considerably more efficient than the Yoshida splittings and which remove the above error to leading and higher orders.

4. Conservation laws

4.1. Outline

The backward error analysis of Hamiltonian systems described in the last section has shown that a modified form of the Hamiltonian of a problem is conserved to within an exponentially small error by a symplectic method. As conservation laws are universal in physical applications, we can ask the question of how we should design a numerical method to capture them. Whilst the Hamiltonian (and many other quantities) are conserved in a Hamiltonian system, many other problems (which are not themselves Hamiltonian) have invariants. In this section we consider how well these can be conserved. Conservation laws have had a major impact in the development of schemes in (for example) fluid dynamics with the Arakawa scheme (ARAKAWA [1966]) for conserving energy and enstrophy (in partial differential equations) In another important study DE FRUTOS and SANZ-SERNA [1997] give an example of a conserving second-order method applied to the KdV equation outperforming a nonconserving third-order method, despite the fact that the higher order method has smaller local truncation errors. S. LI [1995] (in reference to numerical discretisations of the nonlinear Klein–Gordon equations) state that *in some areas, the ability to preserve some invariant properties of the original differential equation is a criterion to judge the success of a numerical simulation.*

We can think of a conservation law associated with the solutions of a differential equation as follows. Suppose that the underlying differential equation is $\mathrm{d}\mathbf{u}/\mathrm{d}t = \mathbf{f}(\mathbf{u})$, then a conservation law is given by the constraint $I(\mathbf{u}(t)) = \text{const}$ so that the solution evolves on the constraint manifold $\mathcal{M}$ where

$$\mathcal{M} = \left\{\mathbf{x} \in \mathbb{R}^d\colon\ I(\mathbf{x}) = I(\mathbf{u}_0)\right\}.$$

We distinguish these examples, where the manifold $\mathcal{M}$ arises as part of the solution (and is a natural constraint), from differential algebraic equations in which the differential equation system is underdetermined and the manifold $\mathcal{M}$ acts as an enforced constraint. First the bad news. Even in symplectic methods we cannot (with non-adaptive time steps) exactly conserve the underlying Hamiltonian, or indeed, in general anything other than *quadratic* invariants (GE and MARSDEN [1988]). Thus we need to look further than the symplectic methods outlined in the last section. Various approaches have been developed including moving frames (see OLVER [2001] as well as the references therein), discrete gradients, projection and methods based on Noether's theorem.

4.2. Serendipitous and enforced conservation laws

Now the good news. Many useful invariants can be preserved, either by general methods or by purpose designed methods.

A *linear* invariant takes the form

$$I = A\mathbf{u}$$

and a *quadratic* invariant the form

$$I = \mathbf{u}^{\mathrm{T}} C \mathbf{u},$$

where C is a symmetric or skew-symmetric square matrix. Angular momentum in the Kepler problem and in molecular dynamics is an example of a quadratic invariant. We will see the usefulness of the preservation of quadratic invariants when we consider the performance of a variety of methods, including the implicit mid-point rule, to the problem of rigid body motion.

The following theorem gives a broad classification of methods which conserve a wide class of such invariants.

THEOREM.
(i) *All Runge–Kutta and multi-step methods preserve linear invariants,*
(ii) *Runge–Kutta methods satisfying the condition* (3.10) *preserve quadratic invariants,*
(iii) *In a splitting method, if Ψ_h^i are methods that conserve quadratic invariants, then so does the composition of these methods.*

PROOF. We now present a proof of (ii) following SANZ-SERNA and CALVO [1994]. From the general structure of a Runge–Kutta method it is possible to write

$$\mathbf{u}_{n+1}^{\mathrm{T}} C \mathbf{u}_{n+1} = \mathbf{u}_n^{\mathrm{T}} C \mathbf{u}_n + 2h \sum_{i=1}^{s} b_i \mathbf{k}_i^{\mathrm{T}} C \mathbf{Y}_i + h^2 \sum_{i,j=1}^{s} (b_i b_j - b_i a_{ij} - b_j a_{ji}) \mathbf{k}_i^{\mathrm{T}} C \mathbf{k}_j, \tag{4.1}$$

where $\mathbf{k}_i = \mathbf{f}(\mathbf{Y}_i)$ and $\mathbf{Y}_i = \mathbf{u}_n + h \sum_j a_{ij} \mathbf{k}_j$. However, from

$$0 = \frac{\mathrm{d}}{\mathrm{d}t}\left(\mathbf{u}^{\mathrm{T}} C \mathbf{u}\right) = \mathbf{u}^{\mathrm{T}} C \mathbf{f}(\mathbf{u}),$$

and the symmetry or skew-symmetry of C, we see that the first sum in (4.1) vanishes. The result now follows since the second sum in (4.1) vanishes provided that the symplecticity condition (3.10) holds.

The other two parts of this theorem have similarly simple proofs. □

We can extend this theorem to say that for a partitioned systems with quadratic invariant $I(\mathbf{p}, \mathbf{q}) = \mathbf{p}^{\mathrm{T}} S \mathbf{q}$, S a constant square matrix, I is conserved if the system is integrated by a symplectic partitioned Runge–Kutta method, i.e. a method satisfying (3.11), see SANZ-SERNA and CALVO [1994] for additional details.

This very nice property of Runge–Kutta methods cannot be easily extended to other classes of ordinary differential equation solvers. Indeed, quadratic invariants are in general NOT preserved by linear multi-step methods. For example, the trapezoidal rule

$$u_{n+1} = u_n + \frac{h}{2}\big(f(u_n) + f(u_{n+1})\big)$$

(which is equivalent to the implicit mid-point rule for linear problems but differs for nonlinear problems) possesses many nice properties (such as symmetry) but does not preserve quadratic invariants (and hence is not symplectic). Although this may seem to limit the potential usefulness of this latter method it is shown in SANZ-SERNA and CALVO [1994] that the trapezoidal rule can be considered to be a symplectic map when applied to a system equivalent to the original but after a reversible change of coordinates, methods which satisfy this are called conjugate symplectic in SANZ-SERNA and CALVO [1994]. Thus many of the nice features of symplectic maps are inherited by this method.

A more general equation may have deeper invariants. An important class of such are isospectral flows of the form

$$\frac{\mathrm{d}}{\mathrm{d}t}u = [A, u] \equiv Au - uA,$$

where u and A are now matrices. Such flows have the entire spectrum of u as an invariant, provided that the matrix A is skew symmetric, and arise naturally in the computation of the Lyapunov exponents of a system. They also arise in the study of Lie group solvers which will be considered in the next section.

The theorem cannot be extended to more complicated invariants. For example, polynomial invariants of degree three or greater are in general not conserved, it is shown in HAIRER, LUBICH and WANNER [2002], for example, that no Runge–Kutta method can conserve all such invariants.

Faced with this problem, an obvious way to conserve a particular invariant is to simply enforce it during the evolution by continually projecting the solution onto the constraint manifold. An example of this procedure might be an integration of the many body problem with conservation of energy and angular momentum enforced. In another example we can consider a matrix equation in which we require that the determinant of the matrix at each stage of the evolution should be unity. This can be enforced by dividing the discrete solution obtained at each stage by its determinant scaled appropriately. HAIRER [2000] discusses such a procedure in which a standard numerical method is used to advance the solution by one step and the resulting calculated value is then projected (say using the Euclidean norm) on to the constraint manifold. This procedure can be systematised by using Lagrange multipliers.

Whilst this procedure may seem natural it has two significant disadvantages. Firstly, the projection procedure inevitably reduces the information present in the calculation. Secondly, it is far from clear that enforcing a natural constraint in any way is reproducing the correct geometrical features of the problem. It is quite possible that the dynamics on the manifold itself is completely wrong. Hairer makes this clear in HAIRER, LUBICH and WANNER [2002] in which he applies this procedure to the Kepler problem using both the forward Euler and the symplectic Euler methods. He finds that good qualitative agreement is obtained only when *all* of the invariants of the problem are maintained, and indeed if only one (for example, energy) is maintained then the overall performance of the method can in fact be degraded.

4.3. Discrete gradient methods

We now briefly discuss a more general approach for constructing methods which preserve conservation laws. These methods are termed *discrete gradient* methods and are based on the observation that an ODE problem $\dot{u} = f(u)$ with the invariant or first integral $I(u)$ may be written in the equivalent skew-gradient form

$$\frac{\mathrm{d}u}{\mathrm{d}t} = S(u)\nabla I(u), \qquad S^{\mathrm{T}} = -S. \tag{4.2}$$

Given a vector field f and first integral I there will in general be no uniqueness in the choice of S. One particular choice is given by $S = (f(\nabla I)^{\mathrm{T}} - (\nabla I)f^{\mathrm{T}})|\nabla I|^{-2}$. However, some problems may naturally be posed in such a way that an S is given immediately, for example, in Hamiltonian or Poisson systems. In general the method is then constructed by discretising (4.2) and forming the discrete skew gradient system, or discrete gradient method

$$\frac{u^{n+1} - u^n}{\Delta t} = \widetilde{S}\big(u^{n+1}, u^n, \Delta t\big)\widetilde{\nabla} I\big(u^n, u^{n+1}\big), \tag{4.3}$$

where $\widetilde{S}$ is a consistent ($\widetilde{S}(u, u, 0) = S(u)$) skew-symmetric matrix and $\widetilde{\nabla} I$ is defined to be a discrete gradient of I, that is

$$I(u') - I(u) = \widetilde{\nabla} I(u', u).(u' - u), \quad \text{and} \quad \widetilde{\nabla} I(u, u) = \nabla I(u).$$

The discretisation given by (4.3) can now be proven to conserve I

$$\begin{aligned} I\big(u^{n+1}\big) - I\big(u^n\big) &= \widetilde{\nabla} I\big(u^{n+1}, u^n\big).\big(u^{n+1} - u^n\big) \\ &= \Delta t \widetilde{\nabla} I\big(u^{n+1}, u^n\big)^{\mathrm{T}} \widetilde{S}\big(u^{n+1}, u^n, \Delta t\big)\widetilde{\nabla} I\big(u^{n+1}, u^n\big) \\ &= 0. \end{aligned}$$

There are many possible ways to choose the discrete gradient and the skew-symmetric matrix $\widetilde{S}$, see McLachlan, Quispel and Robodoux [1999], one possible choice of each is given by $\widetilde{S}(u, u', \tau) = S((u + u')/2)$ and

$$\widetilde{\nabla} I(u, u')_i = \frac{I(u^{(i)}) - I(u^{(i-1)})}{u'_i - u_i},$$

where $u^{(i)} = (u'_1, \ldots, u'_i, u_{i+1}, \ldots, u_m)$, called the coordinate increment discrete gradient in McLachlan, Quispel and Robodoux [1999] and attributed to Itoh and Abe [1988].

The above construction can be extended to derive methods which preserve more than one first integral, and also to cases where I is a Lyapunov exponent, see McLachlan and Quispel [2001]. Once a symmetric integral conserving method has been constructed using this technique (the method may be made symmetric by choosing $\widetilde{S}$

and $\widetilde{\nabla} I$ such that $\widetilde{S}(x, x', \tau) = \widetilde{S}(x', x, -\tau)$ and $\widetilde{\nabla} I(x, x') = \widetilde{\nabla} I(x', x)$), the ideas of Yoshida as discussed earlier may be used to construct higher-order methods.

5. Symmetry group methods

Whilst geometric methods were originally developed to solve Hamiltonian problems a very significant development of geometrical ideas has been in the computation of the solutions of problems with symmetries. Here we can distinguish between problems which have an underlying symmetry (such as a rotational symmetry) which the solutions may or may not share, or problems in which the solutions evolve on manifolds which are invariant under the action of a symmetry group. This situation becomes rather more complex for partial differential equations and we return to these later. In this section we consider progressively the development of methods for problems with Lie group, reversing and scaling symmetries. Here we start with a very general set of ideas and then specialise to a case where adaptivity and symmetry work together.

5.1. Lie group methods

This section summarises many of the ideas which are discussed in the very general survey article by ISERLES, MUNTHE-KAAS, NØRSETT and ZANNA [2000].

In the previous section we looked at problems in which the solutions were constrained to lie on a manifold $\mathcal{M}$. The task of discretising a system consistently with this manifold invariant is greatly assisted when $\mathcal{M}$ is *homogeneous*, namely when it is subjected to a transitive group action of a Lie group G with identity e and we can think of points on the manifold as being coupled by the group action. In other words there is a function $\lambda : G \times \mathcal{M} \to \mathcal{M}$ such that for every $g_1, g_2 \in G$ and $u \in \mathcal{M}$

$$\lambda\big(g_1, \lambda(g_2, u)\big) = \lambda(g_1 g_2, u), \quad \text{with } \lambda(e, u) = u$$

and for all $u_1, u_2 \in \mathcal{M}$ there is a g such that $\lambda(g, u_1) = u_2$.

Invariance of $\mathcal{M}$ under the action of such a Lie group G is a very strong special case of the invariants discussed in the last section. Homogeneous manifolds include Lie groups themselves, spheres, tori, Stiefel and Grassmann manifolds. An important example of such a problem is the evolution of an orthogonal matrix. Substantial work has been done by Iserles, Munthe-Kaas, Nørsett, Zanna and their co-workers (ISERLES [1999], ISERLES and NORSETT [1999], ISERLES, MUNTHE-KAAS, NØRSETT and ZANNA [2000], MUNTHE-KAAS [1998], MUNTHE-KAAS and OWREN [1999], MUNTHE-KAAS and ZANNA [1997], ZANNA [1999]) on the construction of efficient numerical schemes which ensure that solutions of appropriate differential equations evolve on such manifolds.

The most direct examples of homogeneous manifolds occur when $\mathcal{M}$ is equal to the Lie group G. Although this appears to be a special case, MUNTHE-KAAS and ZANNA [1997] have shown that, provided a numerical method can be constructed so that solutions of the differential equation evolve on a Lie-group, then these methods can be extended in a straightforward manner to every homogeneous space acted on by that group. Thus we can consider this simpler case only, and the remainder of this section will be on this topic.

Accordingly, suppose that G is a Lie group, we will consider the following differential equation.

$$\mathbf{u}' = A(t, \mathbf{u})\mathbf{u}, \quad t \geqslant 0, \qquad \mathbf{u}(t_0) = \mathbf{u}_0 \in G, \tag{5.1}$$

where $A : \mathbb{R}^+ \times G \to \mathfrak{g}$, and $\mathfrak{g}$ is the *Lie algebra* of G. Typically G will be a matrix Lie group and $\mathbf{u}$ a matrix. The motivation for considering this differential equation is the following lemma.

LEMMA. *If* $\mathfrak{g}$ *is the Lie algebra of the group* G *then the solutions* $\mathbf{u}$ *of the differential equation* (5.1) *are constrained to evolve on the Lie group* G.

Some examples of such matrix Lie groups and their corresponding Lie algebras are as follows

Lie group	Lie algebra
$SL_n = \{u\colon \det(u) = 1\}$	$\mathfrak{sl}_n = \{A\colon \operatorname{trace}(A) = 0\}$
$O_n = \{u\colon u^{\mathrm{T}}u = I\}$	$\mathfrak{so}_n = \{A\colon A^{\mathrm{T}} + A = 0\}$
$SO_n = \{u\colon u^{\mathrm{T}}u = I, \det(u) = 1\}$	$\mathfrak{so}_n = \{A\colon A^{\mathrm{T}} + A = 0\}$
$Sp_n = \{u\colon u^{\mathrm{T}}Ju = J\}$	$\mathfrak{sp}_n = \{A\colon JA + A^{\mathrm{T}}J = 0\}$
$GL_n = \{u\colon \det(u) \neq 0\}$	$\mathfrak{gl}_n = \text{all matrices}$

Observe, for example, that whilst the Lie group of orthogonal matrices forms a nonlinear manifold, the Lie algebra of skew symmetric matrices comprises a linear vector space. This is true for all of the other examples. Numerical solvers for the differential equation (5.1) tend to work on elements of the Lie algebra of G rather than the elements of G itself precisely because nonlinear constraints on G become linear constraints on $\mathfrak{g}$. Thus it is important to have methods which allow you to move from the Lie algebra to the Lie group. The most important such method is the *exponential map* which acts as a map between the Lie algebra and the corresponding Lie group. For $A \in \mathfrak{g}$ it is defined by

$$\exp(A) = \sum_{n=0}^{\infty} \frac{A^n}{n!}.$$

Calculating, or approximating, the exponential map plays an important role in most Lie group solvers as it allows us to transfer calculations from the (nonlinear) Lie group on to the (linear) Lie algebra. However, it is not an easy map to evaluate. MATLAB employs several alternative methods (empm, ..., expm3) for computing the exponential function, but finding the matrix exponential will always be an expensive process. Numerical methods have been used to find easy to compute approximations to it , for a review of many such methods see MOLER and VAN LOAN [1978], and for a geometric perspective see CELLEDONI and ISERLES [2000]. For certain cases it can be computed using a finite sum. For example, if $A \in \mathfrak{so}_3$ then Rodrigues formula (MARSDEN and RATIU

[1999]) gives

$$\exp(A) = I + \frac{\sin(\alpha)}{\alpha}A + \frac{1}{2}\left(\frac{\sin(\alpha/2)}{\alpha/2}A^2\right),$$

where

$$A = \begin{pmatrix} 0 & -\omega_3 & \omega_2 \\ \omega_3 & 0 & -\omega_1 \\ -\omega_2 & \omega_1 & 0 \end{pmatrix} \quad \text{and} \quad \alpha = \sqrt{\omega_1^2 + \omega_2^2 + \omega_3^2}.$$

It is worth observing that for certain Lie groups we may identify other maps between the Lie algebra and the Lie group which are much easier to compute than the exponential map. A notable example of such is the *Cayley map* defined by

$$\operatorname{cay}(A) = (I - A/2)^{-1}(I + A/2)$$

which requires one matrix inversion rather than an infinite sum.

LEMMA. *If* $A \in \mathfrak{so}_n$ *then* $\operatorname{cay}(A) \in O_n$.

PROOF. Let $O = (I - A/2)^{-1}(I + A/2)$ then

$$\begin{aligned} O^{\mathrm{T}} &= \left(I + A^{\mathrm{T}}/2\right)\left(I - A^{\mathrm{T}}/2\right)^{-1} = (I - A/2)(I + A/2)^{-1} \\ &= (I + A/2)^{-1}(I - A/2) = O^{-1}. \end{aligned}$$

□

This result may be extended to show that the Cayley map also works for the symplectic group, i.e. it maps $\mathfrak{sp}_n$ into Sp_n.

Most Lie-group solvers follow a set pattern: Let us assume that the underlying group is finite-dimensional. By Ado's theorem, it then follows that the Lie algebra $\mathfrak{g}$ is isomorphic to a matrix algebra (a subset of $\mathfrak{gl}_m(R)$, the Lie algebra of $m \times m$ real matrices). For simplicity's sake (and with moderate loss of generality) we can thus stipulate that $\mathfrak{g}$ is a matrix Lie algebra. Eq. (5.1) is *pushed* to the underlying Lie algebra $\mathfrak{g}$, solved there and the solution is *pulled back* to G with the exponential map. (In the restricted class of problems described above this step can be replaced by an application of the Cayley map.) This operation may take place repeatedly in the course of a single time step and is displayed in the following diagram,

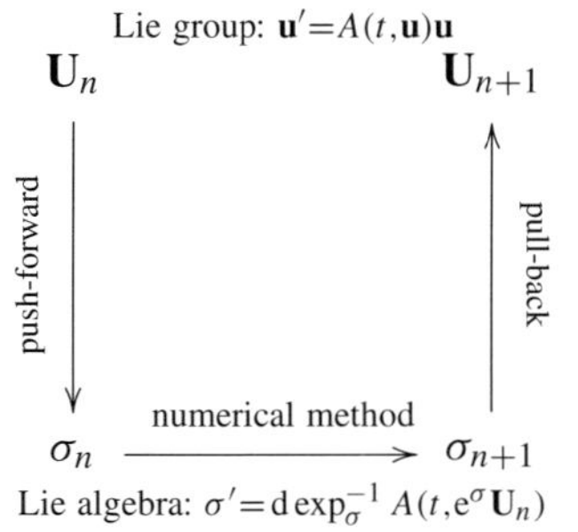

Here $A : \mathbb{R}^+ \times G \to \mathfrak{g}$ and $\sigma(t)$ evolves on the Lie algebra $\mathfrak{g}$. The numerical method is devised to solve a differential equation for σ rather than for u. The pull-back for matrix

Lie algebras is the standard matrix exponential (or Cayley map),

$$\mathbf{U}_{n+1} = \mathrm{e}^{\sigma_{n+1}}\mathbf{U}_n.$$

In contrast, the push-forward projects the equation onto the Lie algebra. For the exponential map it is given by the *dexpinv equation*

$$\sigma' = \sum_{l=0}^{\infty} \frac{B_l}{l!} ad_\sigma^l\, A\big(t, \mathrm{e}^\sigma \mathbf{U}_n\big), \quad t \geqslant t_0, \quad \sigma(t_n) = 0,$$

where $\{B_l\}_{l=0}^\infty$ are *Bernoulli numbers*, defined by

$$\frac{x}{\mathrm{e}^x - 1} = \sum_{k=0}^{\infty} B_k \frac{x^k}{k!}$$

while the *adjoint operator* ad_x in a Lie algebra is an iterated commutator,

$$ad_x^l y = \overbrace{[x, [x, \ldots, [x, y] \cdots]]}^{l \text{ times}}, \quad l \geqslant 0.$$

For a proof of this result, see HAIRER, LUBICH and WANNER [2002]. A similar expression to the dexpinv equation (the dcayinv equation) can be derived for the Cayley transform, see ISERLES [1999].

We can now use a numerical scheme to integrate the dexpinv equation. Here the constraint on this solver is that all computed solution should lie in the Lie algebra. However, as the Lie algebra is a linear vector space this is easy: as long as the numerical method employs just linear-space operations and commutators, it is bound to produce $\sigma_{n+1} \in \mathfrak{g}$. An important example of such methods are the *Runge–Kutta/Munthe-Kaas* (RKMK) schemes, which in the present formalism apply a (typically, explicit) Runge–Kutta method to the dexpinv equation, appropriately truncated (MUNTHE-KAAS [1998]). For example, a classical third-order Runge–Kutta method, applied to the Lie-group equation (5.1), reads

$$\begin{aligned}
\mathbf{k}_1 &= A(t_n, \mathbf{U}_n)\mathbf{U}_n, \\
\mathbf{k}_2 &= A\left(t_n + \frac{1}{2}h, \mathbf{U}_n + \frac{1}{2}h\mathbf{k}_1\right)\left(\mathbf{U}_n + \frac{1}{2}h\mathbf{k}_1\right), \\
\mathbf{k}_3 &= A(t_{n+1}, \mathbf{U}_n - h\mathbf{k}_1 + 2h\mathbf{k}_2)(\mathbf{U}_n - h\mathbf{k}_1 + 2h\mathbf{k}_2), \\
\mathbf{v} &= h\left(\frac{1}{6}\mathbf{k}_1 + \frac{2}{3}\mathbf{k}_2 + \frac{1}{6}\mathbf{k}_3\right), \\
\mathbf{U}_{n+1} &= \mathbf{U}_n + \mathbf{v}.
\end{aligned}$$

In general this cannot be expected to keep $\{\mathbf{U}_n\}_{n=0}^\infty$ in G. However, as soon as we change the configuration space from G to $\mathfrak{g}$, solve there and pull back, the ensuing RKMK scheme,

$$\begin{aligned}
\mathbf{k}_1 &= A(t_n, \mathbf{U}_n), \\
\mathbf{k}_2 &= A\left(t_n + \frac{1}{2}h, \mathrm{e}^{\frac{1}{2}h\mathbf{k}_1}\mathbf{U}_n\right),
\end{aligned}$$

$$\mathbf{k}_3 = A\big(t_{n+1}, \mathrm{e}^{-h\mathbf{k}_1+2h\mathbf{k}_2}\mathbf{U}_n\big),$$
$$\mathbf{v} = h\left(\frac{1}{6}\mathbf{k}_1 + \frac{2}{3}\mathbf{k}_2 + \frac{1}{6}\mathbf{k}_3\right),$$
$$\mathbf{U}_{n+1} = \exp\left(\mathbf{v} + \frac{h}{6}[\mathbf{v}, \mathbf{k}_1]\right)\mathbf{U}_n,$$

respects the Lie-group structure.

A different example of a class Lie-group solvers can be obtained from a direct manipulation of the dexpinv equation, in particular when the Lie-group equation is linear,

$$\mathbf{u}' = A(t)\mathbf{u}.$$

Perhaps the most powerful approach is to exploit the *Magnus expansion*. This is an explicit solution of the dexpinv equation for the linear problem above and which has the following form

$$\begin{aligned}\sigma(t) = {} & \int_0^t A(\xi)\,\mathrm{d}\xi - \frac{1}{2}\int_0^t\int_0^{\xi_1}\big[A(\xi_2), A(\xi_1)\big]\,\mathrm{d}\xi_2\,\mathrm{d}\xi_1 \\ & + \frac{1}{4}\int_0^t\int_0^{\xi_1}\int_0^{\xi_2}\big[\big[A(\xi_3), A(\xi_2)\big], A(\xi_1)\big]\,\mathrm{d}\xi_3\,\mathrm{d}\xi_2\,\mathrm{d}\xi_1 \\ & + \frac{1}{12}\int_0^t\int_0^{\xi_1}\int_0^{\xi_1}\big[A(\xi_3), \big[A(\xi_2), A(\xi_1)\big]\big]\,\mathrm{d}\xi_3\,\mathrm{d}\xi_2\,\mathrm{d}\xi_1 + \cdots, \end{aligned} \tag{5.2}$$

see MAGNUS [1954]. The analysis and numerical implementation of Magnus expansions is a far-from-trivial task and it is investigated in depth, using techniques from graph theory and quadrature, in ISERLES and NORSETT [1999]. Significantly, the Magnus expansion can be truncated at any point to give an approximation to σ which is comprised of linear operations and commutators, and thus must lie in the Lie algebra $\mathfrak{g}$. This is a key to the effective use of these methods. When applied to the problem $\mathbf{u}' = A(t)\mathbf{u}$ a typical method based upon the Magnus series (ISERLES and NORSETT [1999], ZANNA [1999]) replaces $A(t)$ locally by an interpolation polynomial $\hat{A}$ and then solves the equation $\mathbf{u}' = \hat{A}\mathbf{u}$ locally on the interval $[t_n, t_n + h]$ by using a truncated form of (5.2) and an appropriate quadrature method. Careful exploitation of the way in which the commutators are calculated can significantly reduce the computation labour in this calculation. ZANNA [1999] has also extended this method to certain classes of nonlinear equations.

EXAMPLE. If an interpolation polynomial of degree one is used with a two point Gaussian quadrature and the Magnus expansion is truncated at the fourth term then the following order four method is obtained (HAIRER, LUBICH and WANNER [2002]) (where we make the assumption that the exponential is evaluated exactly)

$$U_{n+1} = \exp\left(\frac{h}{2}(A_1 + A_2) + \frac{\sqrt{3}h^2}{12}[A_2, A_1]\right)U_n, \tag{5.3}$$

where $A_{1,2} = A(t_n + x_{1,2}h)$ with x_i the Gauss points $1/2 \pm \sqrt{3}/6$.

An alternative to using the Magnus expansion is the *Fer expansion*. In this we write the solution of the Lie-group equation $\mathbf{u}' = A(t)\mathbf{u}$ in the form

$$\mathbf{u}(t) = \exp\left[\int_0^t A(\xi)\,\mathrm{d}\xi\right]\mathbf{z}(t), \quad t \geqslant 0,$$

where $\mathbf{z}$ obeys the linear equation

$$\mathbf{z}' = \left[\sum_{l=1}^{\infty} \frac{(-1)^l}{(l+1)!} ad^l_{\int_0^t A(\xi)\,\mathrm{d}\xi} A(t)\right]\mathbf{z}, \quad t \geqslant 0, \quad \mathbf{z}(t_0) = \mathbf{u}(t_0).$$

This procedure can be iterated, ultimately producing the solution $\mathbf{u}$ as an infinite product of exponentials.

An important feature of RK/Munthe-Kaas, Magnus and Fer methods is that their implementation involves repeated computation of commutators. This activity, which represents the lion's share of computational expense, can be simplified a very great deal by exploiting linear dependencies among commutators. The analysis of this phenomenon is amenable to techniques from the theory of Lie algebras and constitutes the theme of MUNTHE-KAAS and OWREN [1999]. See also the work of BLANES, CASAS and ROS [2002] in the derivation of *optimal* Lie group methods which evaluate a minimum number of commutators.

5.1.1. Example 1. Evolution on the sphere

We now consider an example of a differential equation evolving on the sphere S_2. We use an example given in ENGØ, MARTHINSEN and MUNTHE-KAAS [1999], where it is used to demonstrate how to construct solutions to problems on manifolds with the matlab toolbox DiffMan. The problem is given by

$$\frac{\mathrm{d}\mathbf{u}}{\mathrm{d}t} = A(t)\mathbf{u}(t), \quad A(t) = \begin{pmatrix} 0 & t & -\cos(t)/2 \\ -t & 0 & t/2 \\ \cos(t)/2 & -t/2 & 0 \end{pmatrix}, \quad \mathbf{u}(0) = \begin{pmatrix} 0 \\ 0 \\ 1 \end{pmatrix}. \tag{5.4}$$

The matrix A is obviously a map from $\mathbb{R}$ into $\mathfrak{so}(3)$ and thus the quadratic quantity $I(\mathbf{u}) = \mathbf{u}^{\mathrm{T}}\mathbf{u}$ is an invariant of system (5.4), i.e.

$$u_1^2(t) + u_2^2(t) + u_3^2(t) = u_1^2(0) + u_2^2(0) + u_3^2(0), \quad t \geqslant 0,$$

and the solution evolves on the two-sphere S^2. In the case that u was a 3×3 matrix we would be in the situation of u evolving on the Lie group O_2.

Fig. 5.1 demonstrates the behaviour of four numerical methods applied to problem (5.4). As we have seen in previous examples the forward Euler method exhibits a growth in I and the solution curve consequently leaves the surface of the sphere. The RK method exhibits a monotonic decrease in I from 1 to less than 0.8, thus the numerical solution again leaves the surface of the sphere, however this time heading towards the origin. As expected from the previous section the RKMK and Magnus methods both exactly conserve I and hence evolve on the desired solution manifold.

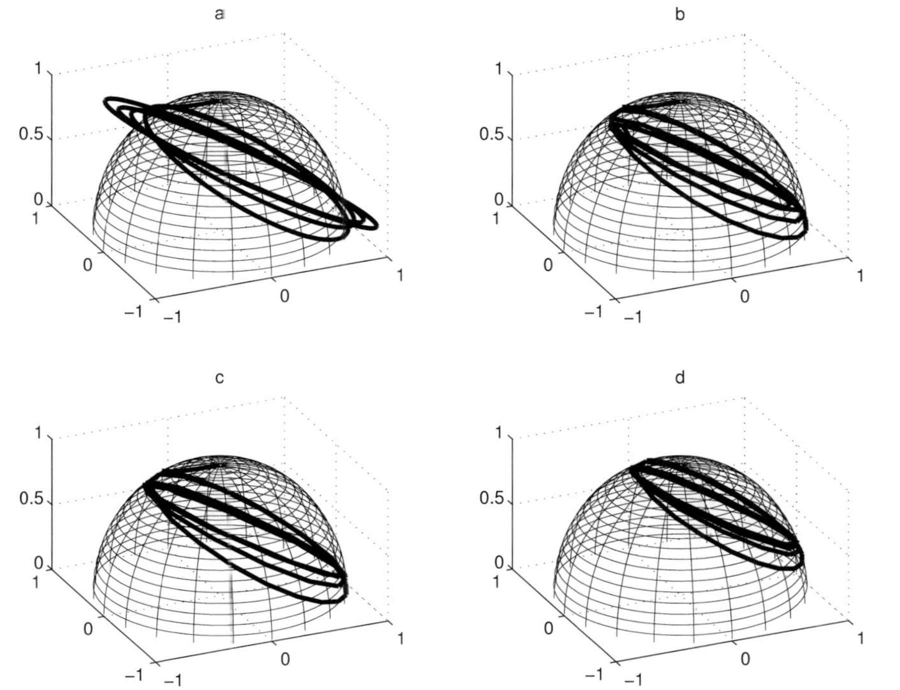

FIG. 5.1. The dynamics of four methods applied to problem (5.4). Forward Euler (a) is shown for 1400 steps of length 0.005. Classical RK (b), RKMK (c) and Magnus (d) methods are all shown for 70 steps of length 0.1.

5.1.2. Example 2. Rigid body motion

For our second example we consider the more subtle, nonlinear, physically based problem of a rigid body with centre of mass fixed at the origin. The body moves according to the Euler equations

$$
\begin{aligned}
\dot{u}_1 &= (I_2 - I_3)u_2u_3/I_2I_3,\\
\dot{u}_2 &= (I_3 - I_1)u_3u_1/I_3I_1,\\
\dot{u}_3 &= (I_1 - I_2)u_1u_2/I_1I_2
\end{aligned}
$$

where (I_1, I_2, I_3) are the moments of inertia about coordinate axes and (u_1, u_2, u_3) the body angular momenta. We may write this in our Lie group form

$$\dot{\mathbf{u}} = A(t, \mathbf{u})\mathbf{u},$$

where

$$A(t, \mathbf{u}) = \begin{pmatrix} 0 & I_3^{-1}u_3 & -I_2^{-1}u_2 \\ -I_3^{-1}u_3 & 0 & I_1^{-1}u_1 \\ I_2^{-1}u_2 & -I_1^{-1}u_1 & 0 \end{pmatrix}$$

is a map from $\mathbb{R}^3$ into $\mathfrak{so}(3)$ as in the previous example.

This problem has the additional structure of being able to be written in the non-canonical Hamiltonian form

$$\dot{\mathbf{u}} = \{\mathbf{u}, H\}, \tag{5.5}$$

where the conserved Hamiltonian function and the bracket operation on functions of $\mathbf{u}$ are given by

$$H(\mathbf{u}) = \frac{1}{2}\left(\frac{u_1^2}{I_1} + \frac{u_2^2}{I_2} + \frac{u_3^2}{I_3}\right), \qquad \{F, G\}(\mathbf{u}) = -\mathbf{u} \cdot (\nabla F \times \nabla G). \tag{5.6}$$

(See MARSDEN and RATIU [1999], OLVER [1986] for further details of the bracket operation, including the properties it must satisfy.) Following these operations through we can see that our rigid body system may be written

$$\dot{\mathbf{u}} = \mathbf{u} \times \nabla H(\mathbf{u}) \equiv J(\mathbf{u})\nabla H(\mathbf{u}). \tag{5.7}$$

Here $J(\mathbf{u})$ is the skew-symmetric matrix

$$J(\mathbf{u}) = \begin{pmatrix} 0 & -u_3 & u_2 \\ u_3 & 0 & -u_1 \\ -u_2 & u_1 & 0 \end{pmatrix}. \tag{5.8}$$

Since functions of $\mathbf{u}$ also evolve in the same manner as (5.5), we can obtain the additional conserved quantity of the system

$$S(\mathbf{u}) = u_1^2 + u_2^2 - u_3^2,$$

by observing that

$$\frac{\mathrm{d}}{\mathrm{d}t} S(\mathbf{u}) = \{S, H\}(\mathbf{u}) = -\mathbf{u} \cdot (\mathbf{u} \times \nabla H),$$

which is obviously zero. Notice that we can say further that $\{S, F\} = 0$ for all functions F, and so conservation of S follows from the form of the bracket operation and not the Hamiltonian, a quantity of this type is known as a *Casimir invariant* and such invariants arise very naturally in many Hamiltonian problems. We will consider these in more detail in the section on partial differential equations.

The conservation of S tells us that the motion of our system evolves on the sphere, and conservation of H that motion also evolves on an ellipsoid. An ideal numerical scheme will reproduce both of these invariants.

Fig. 5.2 shows the behaviour of some numerical methods applied to this problem with $I_1 = 7/8$, $I_2 = 5/8$ and $I_3 = 1/4$, and initial condition chosen such that $\mathbf{u}^{\mathrm{T}}\mathbf{u} = 1$. As we have come to expect now the forward Euler method exhibits a growth in both S and H and can be seen to leave the surface of the sphere. The RK method exhibits a decrease in both S and H and also leaves the sphere. The RKMK method (applied to the problem written in the form $\dot{\mathbf{u}} = A(t, \mathbf{u})\mathbf{u}$ rather than the Hamiltonian form (5.7)) preserves S and so the numerical solution lies on the sphere, although H is not conserved and hence the solution is not a closed curve. In the case of the implicit mid-point rule both S and H are conserved (since they are both quadratic invariants we know this from an earlier

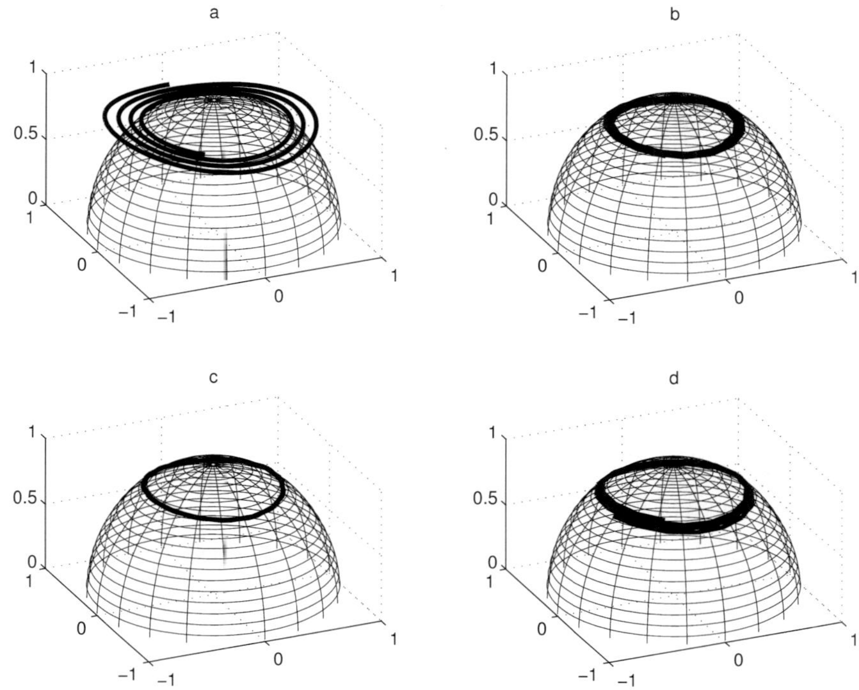

FIG. 5.2. The dynamics of four methods applied to the rigid body problem. Forward Euler (a) is shown for 500 steps of length 0.02. Classical RK (b), RKMK (c) and implicit mid-point (d) methods are all shown for 50 steps of length 0.2.

discussion), and the solution can be seen to be a closed curve lying on the surface of the sphere – exactly the ideal scenario we mentioned earlier.

The special form that J takes, and hence also the bracket operation, means that this problem is of *Lie–Poisson* type. That is we may write our skew-symmetric matrix (5.8) as

$$J_{ij}(\mathbf{u}) = \sum_{k=1}^{3} c_{ij}^{k} u_k,$$

where c_{ij}^k are structure constants for a Lie algebra, in this case $\mathfrak{so}(3)$. See MARSDEN and RATIU [1999], OLVER [1986] for further details. For a splitting based numerical method designed to preserve this structure see MCLACHLAN [1993], BARTH and LEIMKUHLER [1996], and for methods designed to preserve the Casimirs and Hamiltonians of such system see ENGØ and FALTINSEN [2001]. A recent review and comparison of various methods for this system may be found in BUSS [2000].

5.2. *Symmetries and reversing symmetries*

A manifold invariant under the action of a Lie group has a very special structure. However, a symmetry of a system need not satisfy all of the conditions of a Lie group of symmetries as outlined above. In particular, we may consider a general set of symmetries $\mathbf{h}$ from a phase space to itself such that if the underlying equation is

$$\frac{\mathrm{d}\mathbf{u}}{\mathrm{d}t} = \mathbf{f}(\mathbf{u})$$

and $\mathbf{v} = \mathbf{h}(\mathbf{u})$ is a map from the phase space to itself, then the new system will take the form

$$\frac{\mathrm{d}\mathbf{v}}{\mathrm{d}t} = \mathbf{g}(\mathbf{v}).$$

This system has $\mathbf{h}$ as a *symmetry* if $\mathbf{f} = \mathbf{g}$ and a *reversal symmetry* if $\mathbf{f} = -\mathbf{g}$ (so that the arrow of time is reversed in this case). A special case of symmetries, namely scaling symmetries, is considered in the next section. Symmetries relate the vector field at the two points $\mathbf{u}$ and $\mathbf{h}(\mathbf{u})$ and simplify and organise phase space.

The construction of methods which preserve symmetries and reversal symmetries is described in detail in McLachlan, Quispel and Turner [1998], McLachlan and Quispel [2001] and we follow their treatment here. They are important because many systems have symmetries and these help to reduce the complexity of the flow. Hamiltonian problems typically have reversing symmetries and this structure can be rather more easily preserved under a (variable time step) discretisation than symplecticity. Leimkuhler [1999] recommends the use of reversible adaptive methods when calculating the energy exchange in close approach orbits in the Kepler problem and in molecular dynamics and constructs some very effective methods with this property.

A special case occurs when $\mathbf{h}$ is a linear involution such that $\mathbf{h}^2 = Id$. In mechanics this arises with systems (p,q) with the action $\mathbf{h}(p,q) = (-p,q)$. The system is then reversible if

$$\mathbf{f}\big(\mathbf{h}(\mathbf{u})\big) = -\mathbf{h}\big(\mathbf{f}(\mathbf{u})\big).$$

The action of this map reverses the arrow of time. In this case if the flow map induced by the differential equation is ψ_t then

$$\mathbf{h}^{-1} \circ \psi_t \circ \mathbf{h} = \psi_t^{-1} = \psi_{-t}.$$

It is highly desirable that such systems should be integrated by numerical methods which preserve this reversible nature. In particular, if Ψ_h is the map on phase space induced by the numerical method then $\mathbf{h}$-reversibility is equivalent to time reversibility

$$\mathbf{h}^{-1} \circ \Psi_h \circ \mathbf{h} = \Psi_h^{-1}.$$

For a self-adjoint method we have further that

$$\Psi_h^{-1} = \Psi_{-h}.$$

In particular, a method such as the forward Euler method is not reversible, whereas the implicit mid-point rule does have this property.

These methods are studied in detail in Stoffer [1988], Stoffer [1995] where it is shown that they have many of the nice features of symplectic methods (for example, a reversible version of the KAM theorem exists, see Lamb and Roberts [1998], and this can be used to justify some results similar to those obtained for symplectic methods applied to Hamiltonian problems), but the desirable property that variable time steps may also be used. These methods are extended in HOLDER, LEIMKUHLER and REICH [2001], where a fully explicit time-reversible, variable time step Störmer–Verlet method is derived. See also the papers by HUANG and LEIMKUHLER [1997], CIRILLI, HAIRER and LEIMKUHLER [1999], BOND and LEIMKUHLER [1998] where such methods are analysed in detail.

5.3. *Scaling symmetries, temporal adaptivity and singularity capturing*

5.3.1. *Outline*

We shall consider now a special class of ordinary differential equation problems which are invariant under scaling symmetries which are a subset of linear symmetries. For such problems the Lie group is diagonalisable and all operations commute. Thus it may seem that this is a very restricted class of problems. However, it encompasses a wide class of differential equations that arise very naturally in theoretical physics, especially in considering systems which have no intrinsic length or time scales. We call such problems, and the methods used to solve them *scale invariant*. A natural approach to studying such problems is the use of *adaptive methods*. These have had a bit of a bad press in the geometric integration literature following the discovery by SANZ-SERNA and CALVO [1994] that symplectic methods with variable time steps often do not perform as well as similar methods with fixed time steps. This is because the variable time step methods could not, in general be analysed by using the backward error analysis methods described in Section 3.4.1. However, we demonstrate in this section that the additional structure given by scaling invariance when coupled with a geometrically designed adaptive method allows us to construct adaptive schemes with the remarkable property that they have uniform errors for all time (i.e. no error growth with time) when used to approximate self-similar solutions of the ordinary differential equations.

Suppose that we consider the differential equation system

$$\frac{\mathrm{d}\mathbf{u}}{\mathrm{d}t} = \mathbf{f}(\mathbf{u}), \quad \mathbf{u} = (u_1, u_2, \ldots, u_N), \quad \mathbf{f} = (f_1, f_2, \ldots, f_N).$$

A *scaling symmetry* of such an equation is an invariance under the linear transformation

$$t \to \lambda^{\alpha_0} t, \qquad u_i \to \lambda^{\alpha_i} u_i, \quad i = 1, \ldots, N,$$

where $\lambda > 0$ is an arbitrary scalar. The condition for this is that for all $\lambda > 0$

$$f_i\big(\ldots, \lambda^{\alpha_j} u_j, \ldots\big) = \lambda^{\alpha_i - \alpha_0} f_i(\ldots, u_j, \ldots).$$

This is a slight generalisation of previous sections as we are now allowing time to transform as well.

Note that such a definition may easily be extended to differential algebraic equations. This is especially useful when studying semi-discretisations of scale invariant partial differential equations.

The transformation above may be defined in terms of the vector $\boldsymbol{\alpha} = (\alpha_0, \alpha_1, \ldots, \alpha_N)$ and a system of differential equations may be invariant under the action of many such scaling groups. The vectors $\boldsymbol{\alpha}$ describing this set of invariances then form a (commutative) vector space.

Scaling symmetries arise very naturally in ordinary (and indeed partial) differential equations where they express an invariance of the equation to the units in which it is expressed. It is for this reason that many of the equations of mathematical physics have *polynomial* nonlinearities (BARENBLATT [1996]) as it is precisely these which are covariant under the action of such symmetries. For the purposes of numerical discretisations, scale invariant problems have a delicious structure. Consider the two operations of scaling and discretisation acting on a system of differential equations. If the discretisation is a multi-step, a Runge–Kutta or a Taylor series method then *the operations of scaling and discretisation commute*. The only other transformation with this properly is an affine transformation. Technically, such methods are *covariant* with respect to *linear* changes of variables, a result established in MCLACHLAN and QUISPEL [2001]. It is certainly not a property enjoyed by the action of a general Lie group. It is this observation which allows us the possibility of an easy derivation of scale invariant discretisations which admit discrete self-similar solutions.

To motivate this discussion we consider two examples of scale invariant equations. The first, which we looked at earlier in the discussion of backward error analysis is the blow-up equation

$$\frac{\mathrm{d}u}{\mathrm{d}t} = u^2 \tag{5.9}$$

which is invariant under the scaling

$$t \to \lambda t, \quad u \to \lambda^{-1} u.$$

The second example is the (two-dimensional) Kepler problem

$$\dot{x} = u, \quad \dot{u} = -ax/\sqrt{x^2+y^2}, \quad \dot{y} = v, \quad \dot{v} = -ay/\sqrt{x^2+y^2}.$$

This system is left unchanged following the rescaling

$$t \to \lambda t, \quad (x, y) \to \lambda^{2/3}(x, y), \quad (u, v) \to \lambda^{-1/3}(u, v),$$

for any arbitrary positive constant λ. We saw in a previous section that this problem can also be written in canonical Hamiltonian form (3.2), with Hamiltonian

$$H(x, y, u, v) = \frac{u^2}{2} + \frac{v^2}{2} - \frac{a}{\sqrt{x^2+y^2}}.$$

It is then immediate that the Hamiltonian scales in the manner

$$H\bigl(\lambda^{2/3}x, \lambda^{2/3}y, \lambda^{-1/3}u, \lambda^{-1/3}v\bigr) = \lambda^{-2/3}H(x, y, u, v)$$

from which we may deduce the 'conservation law'

$$\frac{2}{3}xH_x + \frac{2}{3}yH_y - \frac{1}{3}uH_u - \frac{1}{3}vH_v = -\frac{2}{3}H.$$

The scaling invariance of the Kepler problem transforms one solution to another, and when applied to periodic orbits immediately gives us Kepler's third law relating the cube of the semi-major axis of a periodic orbit to the square of its period. We observe also that the Kepler problem is invariant under the action of the rotation group O_2, and hence scaling and more general Lie groups may act together here.

5.3.2. Self-similar solutions

Whilst not all solutions of scale invariant systems of ODEs are themselves scale invariant, those that are play a special role in the overall theory. They are often attractors and therefore describe the long term asymptotics of more general solutions. A self-similar solution has the (invariant) form given by

$$u_i(\lambda^{\alpha_0} t) = \lambda^{\alpha_i} u_i(t).$$

It is easy to show that all such solutions take the form

$$u_i = t^{\alpha_i/\alpha_0} v_i,$$

where the *constants* v_i satisfy the *algebraic* equation

$$\alpha_i v_i = \alpha_0 f_i(\ldots, v_j, \ldots).$$

Self-similar solutions often describe singularities in the system. For the blow-up equation (5.9), the self-similar solutions are in fact all possible solutions, up to time translation, and these have the form

$$u(t) = \frac{1}{T - t},$$

where T is arbitrary. A singularity described by a self-similar solution can also occur in Kepler's problem in a finite time T provided that the angular momentum of the system is zero. This is called *gravitational collapse* and occurs when a particle falls into the sun. An example of a solution with this property is the following self-similar solution.

$$\begin{aligned} x &= \left(\frac{9}{2}\right)^{1/3} (T - t)^{2/3}, \\ u &= -\left(\frac{2}{3}\right)\left(\frac{9}{2}\right)^{1/3} (T - t)^{-1/3}, \\ y &= v = 0. \end{aligned} \tag{5.10}$$

5.3.3. Numerical methods for scale invariant problems

We immediately observe that conventional numerical methods, including symplectic ones, fail to reproduce the gravitational collapse, or indeed any singular in finite time phenomenon, if a fixed time step is used. An explicit method such as forward Euler will always give a bounded solution, and an implicit method may not have algebraic equations soluble as real numbers for all time steps. Some form of adaptivity therefore needs to be used for this problem, or indeed a large number of problems which have the common feature of developing singularities in a finite time. Thus we are strongly motivated to use an adaptive approach for this problem despite the difficulties of applying backward error analysis outlined in SANZ-SERNA and CALVO [1994].

More generally, we can further say that adaptive methods fall naturally within the geometric integration framework as fixed mesh methods impose constraints on the solution process whereas adaptivity allows the solution and method to evolve together. What we mean here is that if we reparameterize time it is possible for us to derive numerical methods which are themselves invariant under the scaling transformation. We now follow this through for the general case and by looking at the examples of the blow-up and gravitational collapse problems, examine what benefits we achieve.

To describe the adaptive approach we introduce a map which describes a rescaling of the time variable in terms of a new computational or fictive variable τ, given by

$$\frac{\mathrm{d}t}{\mathrm{d}\tau} = g(u).$$

We impose the constraint that $g(\lambda^{\alpha_i} u_i) = \lambda^{\alpha_0} g(u)$. For the blow-up problem a suitable choice is $g(u) = 1/u$ and for the Kepler problem in one-dimension a suitable choice is $g(x, u) = x^{3/2}$.

The motivation for introducing this new times variable is that the system describing u_i and t in terms of τ is then invariant under the scaling transformation with τ *fixed*, so that if $(t(\tau), u_i(\tau))$ is a solution, then so is $(\lambda^{\alpha_0} t(\tau), \lambda^{\alpha_i} u_i(\tau))$. Therefore, a numerical discretisation of our new system with fixed computational time step $\Delta\tau$ is also invariant, in that if $(t_n, u_{i,n})$ are discrete approximations to $t(n\Delta\tau), u_i(n\Delta\tau))$, then whenever $(t_n, u_{i,n})$ is a solution of our set of discrete equations induced by the discretisation then $(\lambda^{\alpha_0} t_n, \lambda^{\alpha_i} u_{i,n})$ is also a solution.

The rescaling in time is often called a Sundman transformation (LEIMKUHLER [1999]) and has the following direct link with adaptivity. If we take equal steps $\Delta\tau$ in the fictive time variable, then this will correspond to real time steps $\Delta t_n = t_{n+1} - t_n$. To leading order we then have

$$\Delta t_n = g(u)\Delta\tau$$

and hence we have a simple means of adapting the real time step.

Rescaling the whole system then gives

$$\frac{\mathrm{d}u_i}{\mathrm{d}\tau} = h_i(u), \qquad \frac{\mathrm{d}t}{\mathrm{d}\tau} = g(u), \tag{5.11}$$

where $h_i \equiv g f_i$. Observe that this transformation has had the effect of changing a problem with a variable step size into one with a constant step size $\Delta\tau$. Thus we may now use the full power of the analytical techniques developed for fixed time steps on these problems. STOFFER and NIPP [1991] used precisely this procedure for analysing the existence of invariant curves in variable time step discretisations of ODEs.

For the gravitational collapse problem with $a = 1$ an application of this procedure gives

$$\frac{\mathrm{d}x}{\mathrm{d}\tau} = x^{3/2} u, \qquad \frac{\mathrm{d}u}{\mathrm{d}\tau} = -\frac{1}{x^{1/2}}, \qquad \frac{\mathrm{d}t}{\mathrm{d}\tau} = x^{3/2}. \tag{5.12}$$

Observe that whereas this system is scale invariant it has lost its symplectic structure. A device to recover this structure (the Poincaré transform) is discussed presently. The choice of power 3/2 in the function g is also arrived at in BOND and LEIMKUHLER

[1998] by equalising the amount of fictive time required for both strong and weak collision events, this property has some similarity with Kepler's third law which our scaling invariant method automatically inherits (see later) and therefore the common choice of $3/2$ should be unsurprising.

The secret behind the use of the scale invariant adaptive methods is a careful discretisation of the extended system (5.11), in which all equations in this system are discretised to the same order of accuracy.

For example, consider a forward Euler discretisation of (5.12) with constant step size $\Delta\tau$.

$$\begin{aligned} x_{n+1} - x_n &= x_n^{3/2} u_n \Delta\tau \\ u_{n+1} - u_n &= -x_n^{-1/2} \Delta\tau \\ t_{n+1} - t_n &= x_n^{3/2} \Delta\tau. \end{aligned} \tag{5.13}$$

The following properties of linear multi-step discretisations of the scaled system (5.11) are listed below.

(1) Linear multi-step or Runge–Kutta discretisations of our transformed system (for the above example (5.12)) have relative local truncation errors which are independent of scale.
(2) Global properties of the continuous solution which are derived from scaling invariance may be inherited by the numerical scheme, one example being Kepler's third law.
(3) Any continuous self-similar solutions of the problem are uniformly (in time) approximated by discrete self-similar solutions admitted by the numerical method. These discrete solutions also inherit the stability of the continuous ones.

The importance of points (1) and (3) can be seen from the following. Suppose we are computing a solution to a problem in which a singularity is forming, and that the formation is progressively described through the action of the scaling group. Our adaptive numerical method will continue to compute with no overall loss in relative accuracy.

To elaborate on the first part of point (3) we firstly define a *discrete self-similar solution* of the discretised system. This is a discrete solution which has the property that it is invariant under the same changes of scale as the original. In particular a discrete self-similar solution $(t_n, u_{i,n})$ must take the form

$$u_{i,n} = z^{\alpha_i n} V_i, \qquad t_n = z^{\alpha_0 n} T,$$

for appropriate V_i, T and z satisfying a suitable algebraic equation. We now state the connexion between the discrete and self-similar solutions via the following theorem.

THEOREM. *Let $u_i(t)$ be a self-similar solution of our ordinary differential equation, then for sufficiently small $\Delta\tau$ there is a discrete self-similar solution $(t_n, u_{i,n})$ of the discrete scheme approximating our rescaled system such that for **all** n*

$$u_i(t_n) = u_{i,n}\big(1 + \mathcal{O}(\Delta\tau^p)\big),$$

where p is the order of the discretisation and the constant implied in the $\mathcal{O}(\Delta\tau^p)$ term does not depend on n.

PROOF. See BUDD, LEIMKUHLER and PIGGOTT [2001]. □

The most interesting physical self-similar solutions are those which act as attractors since they determine asymptotic behaviour of more general solutions. The stability result mentioned in point 3 ensures that we are not destroying this property in our numerical method.

Crucial to the success of this method is the calculation of the function $g(u)$. This can be done a-priori by finding solutions to the functional equation

$$g(\ldots, \lambda^{\alpha_i} u_i, \ldots) = \lambda^{\alpha_0} g(\ldots, u_i, \ldots),$$

by differentiating with respect to λ, setting $\lambda = 1$ and then solving the resulting linear hyperbolic equation. (This has many possible solutions, all of which are in principle appropriate functions g.) This procedure can be automated in Maple. Alternatively (and maybe preferably) g can be estimated a-posteriori. If the underlying system is scale invariant, then indeed error estimates obtained using the Milne device do lead to equations which are first order discretisations of the equation $\mathrm{d}t/\mathrm{d}\tau = g$ with scale invariant functions g, however, we should note that the higher order accuracy in the approximation of self-similar solutions is lost in the use of such methods. Research is ongoing in the application and analysis of a-posteriori scale invariant methods.

5.3.4. Examples

Now we see how this theorem applies to our two examples. Firstly consider the blow-up equation

$$\frac{\mathrm{d}u}{\mathrm{d}t} = u^2.$$

Taking $g(u) = 1/u$ we end up with the following rescaled equation

$$\frac{\mathrm{d}u}{\mathrm{d}\tau} = u, \qquad \frac{\mathrm{d}t}{\mathrm{d}\tau} = u^{-1}.$$

Now consider an implicit mid-point rule discretisation of this system, this gives the discrete system

$$u_{n+1} - u_n = \frac{\Delta\tau}{2}(u_n + u_{n+1}), \qquad t_{n+1} - t_n = \frac{2\Delta\tau}{u_n + u_{n+1}}.$$

Now seek a discrete self-similar solution in the form of

$$u_n = z^{-n} V, \qquad t_n = z^n.$$

Substituting gives

$$\begin{aligned}
&\left(z^{-(n+1)} - z^{-n}\right) V = \frac{\Delta\tau}{2} V\left(z^{-(n+1)} - z^{-n}\right),\\
&z^{n+1} - z^n = \frac{2\Delta\tau}{V(z^{-(n+1)} + z^{-n})}
\end{aligned}$$

so that, dividing by factors of z^n we have

$$(1-z)=\frac{\Delta\tau}{2}(1+z), \quad \text{and} \quad z-1=\frac{2z\Delta\tau}{V(1+z)},$$

and hence

$$z=\frac{1-\Delta\tau/2}{1+\Delta\tau/2} \quad \text{and} \quad V=-\big(1-(\Delta\tau/2)^2\big).$$

Thus

$$t_n=\left(\frac{1-\Delta\tau/2}{1+\Delta\tau/2}\right)^n, \qquad u_n=-\big(1-(\Delta\tau/2)^2\big)\left(\frac{1-\Delta\tau/2}{1+\Delta\tau/2}\right)^{-n},$$

so that

$$u_n=-\big(1-(\Delta\tau/2)^2\big)t_n^{-1}.$$

Now the true self-similar solution is $u=-1/t$ so that we see we have *uniformly* approximated this to an accuracy of $\Delta\tau^2$.

We now look at some results from an implementation of the forward Euler discretisation to solve the gravitational collapse problem with initial conditions $x=1$ and $u=0$, at (without loss of generality) $t=1$. The true solution for these initial conditions is not self-similar, but it evolves towards a true self-similar solution as the collapse time T is approached.

In Fig. 5.3 we plot t_n and x_n both as functions of τ. Observe that t_n tends towards the constant value of $T_{\Delta\tau}$ whilst x_n tends to zero. In Fig. 5.4 we present a plot of x_n as a function of t_n in this case. Observe the singular nature of collapse of the solution.

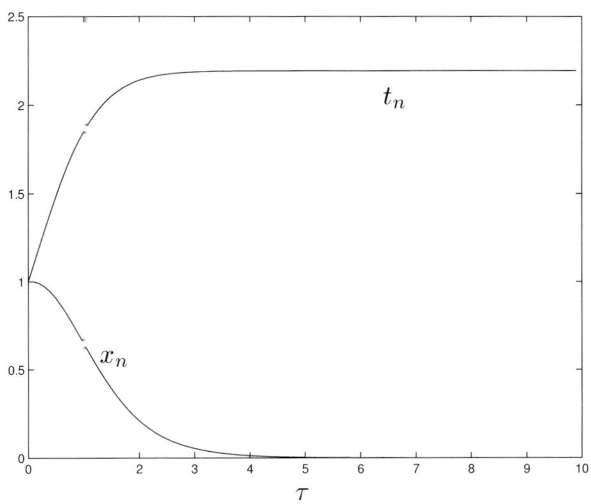

FIG. 5.3. Convergence properties of t_n and x_n as functions of τ.

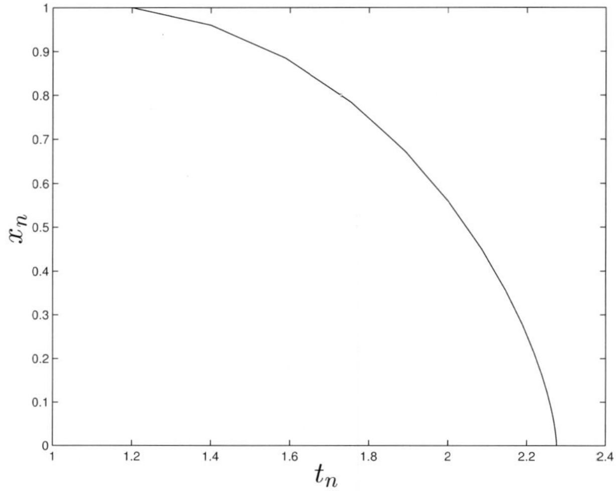

FIG. 5.4. The collapse of x_n as $t_n \to T_{\Delta\tau}$.

Now compare these results with the theoretical predictions. Our transformed system (5.12) has the collapsing self-similar solution (from (5.10))

$$x = R\,\mathrm{e}^{2\mu\tau/3}, \qquad u = -v\,\mathrm{e}^{-\mu\tau/3}, \qquad t = T - \mathrm{e}^{\mu\tau}, \tag{5.14}$$

where

$$\mu = -\left(\frac{9}{2}\right)^{1/2}.$$

Observe that as $\tau \to \infty$ we have $x \to 0$, $|u| \to \infty$ and $t \to T$, and that $t = T - 1$ at $\tau = 0$. Similarly our numerical scheme (5.13) admits a discrete collapsing self-similar solution of the form

$$x_n = \widehat{X} z^{2n/3}, \qquad u_n = -\widehat{V} z^{-n/3}, \qquad t_n = T_{\Delta\tau} - z^n.$$

Comparing with (5.14) we see that z is an analogue of $\exp(\mu\Delta\tau)$ if $n\Delta\tau = \tau$. Here $|z| < 1$ so that $x_n \to 0$, $|u_n| \to \infty$ and $t_n \to T_{\Delta\tau}$ as $n \to \infty$, (cf. CHEN [1986]). $T_{\Delta\tau}$ is a discrete collapse time which need not necessarily coincide with the true collapse time T, however as $\Delta\tau \to 0$ we have $T_{\Delta\tau} \to T$. Substituting into our forward Euler discretisation we find the values of the constants (see Table 5.1) notice that in agreement with the theory we have

$$\widehat{X} = X\bigl(1 + \mathcal{O}(\Delta\tau)\bigr), \quad \widehat{V} = V\bigl(1 + \mathcal{O}(\Delta\tau)\bigr), \quad z = \mathrm{e}^{\mu\Delta\tau}\bigl(1 + \mathcal{O}(\Delta\tau^2)\bigr),$$

where from (5.10) we have

$$X = 1.65096, \quad V = 1.100642,$$

and $\exp(\mu\Delta\tau)$ takes the values 0.65425, 0.80886, and 0.89937 for the three choices of $\Delta\tau$ above. See BUDD, LEIMKUHLER and PIGGOTT [2001] for further results and examples.

TABLE 5.1

$\Delta\tau$	$\widehat{X}$	$\widehat{V}$	z
0.2	1.46704	1.04761	0.64461
0.1	1.55633	1.07440	0.80584
0.05	1.60298	1.08760	0.89852

6. A comparison of methods

To conclude this discussion on ordinary differential equations we now look at a comparison of different geometrically based methods (for example, methods designed around the symplectic structure or around the scaling structure) applied to the same problem, and we also look at some nongeometrical methods. In keeping with our earlier discussion we look at the Kepler two-body problem, which exhibits both a Hamiltonian structure as well as various symmetries, and study periodic solutions with high eccentricity where errors might be expected to be introduced during the close approaches.

We gave the Hamiltonian for Kepler's problem in Eq. (3.17), Hamilton's corresponding equations are

$$\dot{p}_i = -\frac{q_i}{(q_1^2+q_2^2)^{3/2}}, \qquad \dot{q}_i = p_i, \quad i = 1, 2. \tag{6.1}$$

We may think of q representing the position and p the velocity of a heavenly body moving (in a two-dimensional pane) around the sun positioned at the origin of our coordinate system. In our idealized state considered here both objects have the same mass. Throughout this section we take the initial conditions for this problem to be

$$q_1(0) = 1-e, \quad q_2(0) = 0, \quad p_1(0) = 0, \quad p_2(0) = \sqrt{\frac{1+e}{1-e}}$$

the exact solution is given by a conic section. We consider here the case of an ellipse (periodic solutions of period 2π), i.e. we consider eccentricities between zero and one, $0 \leqslant e < 1$. Along with the Hamiltonian which represents the total energy of the system, the angular momentum of the system given by $L(\mathbf{q},\mathbf{p}) = q_1p_2 - q_2p_1$ is also a conserved quantity.

For this problem large forces are experienced close to the origin, i.e. when the planet passes close by the other (near-collision), whereas away from the origin the effect of the force is more sedate. From the standpoint of equidistributing the local truncation error of a numerical method between two adjacent time steps at any point on the orbit, it is often considered desirable to use a method which incorporates an adaptive time stepping strategy. In SANZ-SERNA and CALVO [1994] symplectic methods with standard adaptive time stepping strategies are tested and disappointingly they are shown to behave in a non-symplectic way. Thus although an individual step of an individual orbit may be symplectic, the overall map on the whole of phase space may not be symplectic. Moreover, there is no obvious shadowing property, in that backward error analysis is hard to apply and the solution of the numerical method is not closely approximated by the solution of a nearby Hamiltonian system. As a consequence, the symplectic method with a (badly chosen) variable time step exhibits a quadratic rather than linear error growth for large times. For other discussions regarding this matter see SKEEL and GEAR [1992], STOFFER [1995].

To overcome this problem we may consider other ways of varying the time step, and not concentrate exclusively on symplectic integrators. An obvious method for performing temporal adaptivity is through the Sundman transform we discussed in the previous section for preserving scaling symmetries. However, as we shall now see there is a problem with this approach for Hamiltonian problems (especially if our aim is to approximate invariant curves such as periodic orbits rather than singular behaviour). Kepler's problem is scaling invariant under the transformation

$$t \to \lambda t, \quad (q_1, q_2) \to \lambda^{2/3}(q_1, q_2), \qquad (p_1, p_2) \to \lambda^{-1/3}(p_1, p_2),$$

for any arbitrary positive constant λ. Suppose that we perform the Sundman transform with the function

$$g = \left(q_1^2 + q_2^2\right)^{3/4}.$$

This gives the new, scale invariant, system

$$\frac{\mathrm{d}p_i}{\mathrm{d}\tau} = -q_i\left(q_1^2 + q_2^2\right)^{-3/4}, \qquad \frac{\mathrm{d}q_i}{\mathrm{d}\tau} = p_i\left(q_1^2 + q_2^2\right)^{3/4}, \quad i = 1, 2. \tag{6.2}$$

However in general this procedure fails to preserve any Hamiltonian structure present in a problem. Since the function g has been chosen in a special way any numerical method applied to (6.2) will preserve the scaling invariance of the problem, as in the previous section. There would appear no particular reason to use a symplectic method on the transformed system. The question arises as to whether there is any way to construct a method which manages to preserve both the Hamiltonian and scaling invariance properties for problems such as Kepler's problem which possess them both.

REICH [1999] and HAIRER [1997] combine the use of symplectic methods with adaptive time stepping through the use of the *Poincaré transformation*, see also LEIMKUHLER [1999]. Suppose that the original Hamiltonian H is time independent. Now, with $e = H(\mathbf{p}_0, \mathbf{q}_0)$ introduce a modified Hamiltonian $\widehat{H}$ defined by

$$\widehat{H}(\mathbf{p}, \mathbf{q}, t, e) = g(\mathbf{p}, \mathbf{q})\left\{H(\mathbf{p}, \mathbf{q}) - e\right\}, \tag{6.3}$$

the Hamiltonian system corresponding to $\widehat{H}$ is given by

$$\begin{aligned}
\frac{\mathrm{d}\mathbf{p}}{\mathrm{d}\tau} &= -g\nabla_q H - \{H - e\}\nabla_q g, \\
\frac{\mathrm{d}\mathbf{q}}{\mathrm{d}\tau} &= g\nabla_p H + \{H - e\}\nabla_p g, \\
\frac{\mathrm{d}t}{\mathrm{d}\tau} &= g, \qquad \frac{\mathrm{d}e}{\mathrm{d}\tau} = 0.
\end{aligned} \tag{6.4}$$

Here $(\mathbf{p}, t)^{\mathrm{T}}$ and $(\mathbf{q}, e)^{\mathrm{T}}$ are now conjugate variables in the extended phase space $\mathbb{R}^{2d} \times \mathbb{R}^2$. Along the *exact* solution of the problem $H(\mathbf{p}, \mathbf{q}) = e$, and thus the first

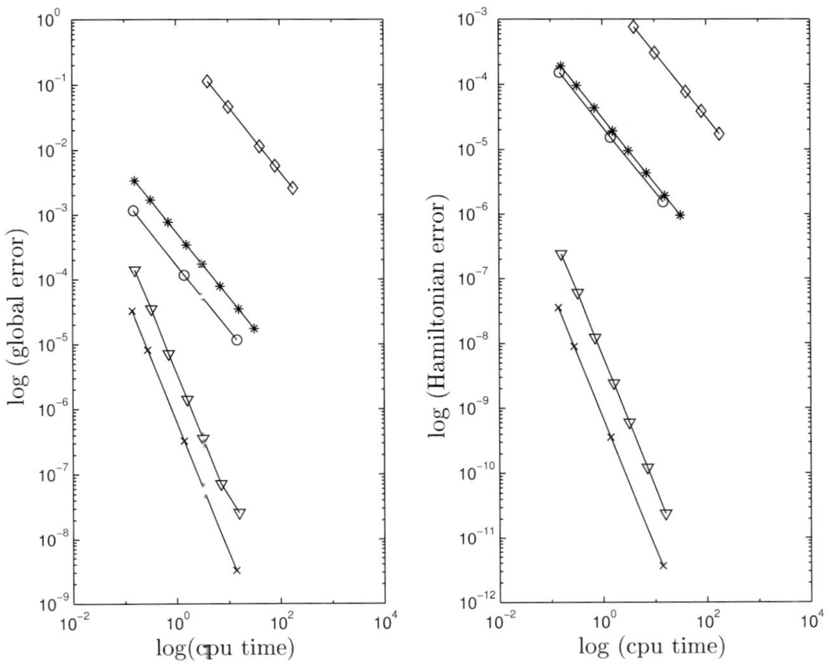

FIG. 6.1. Kepler problem with eccentricity of 0.5 for 10 orbits. Symplectic Euler ($*$), Störmer–Verlet (∇), Sundman Symplectic Euler ($\diamond$), Poincaré Symplectic Euler ($\circ$), adaptive Verlet Sundman ($\times$).

two equations of (6.4) simply reduce in this case to a system transformed by using the Sundman transformation described earlier. We may thus think of the last terms in (6.4) as perturbations of the Sundman transformed system which make the system Hamiltonian. We can obviously now apply a symplectic method with fixed time step $\Delta\tau$ to the transformed system and the favourable properties of such methods should follow. Notice that the function g is performing exactly the same rôle that it did in the Sundman transform method of performing time reparameterization. The scale invariant choices for g derived above therefore also gives a Poincaré transformed system which is scale invariant without the need to scale τ.

For simplicity we discretise this system using the symplectic Euler method to achieve a first-order symplectic method. The Poincaré transformation has the disadvantage that it destroys the separability property of the Hamiltonian and therefore the symplectic Euler method applied to this problem is implicit. We use Newton iteration to solve the nonlinear equations, however it is possible in this case to simply solve a quadratic equation, see HAIRER [1997].

For comparison we form a time-reversible, second order, angular momentum conserving method by applying the second-order Lobatto IIIa–b pair to the Sundman transformed system, the method is usually termed the adaptive Verlet method (HUANG and LEIMKUHLER [1997], see also HOLDER, LEIMKUHLER and REICH [2001], LEIMKUHLER [1999]). An explanation for the reciprocal choice of time step update is given in CIRILLI, HAIRER and LEIMKUHLER [1999]. For the Kepler problem de-

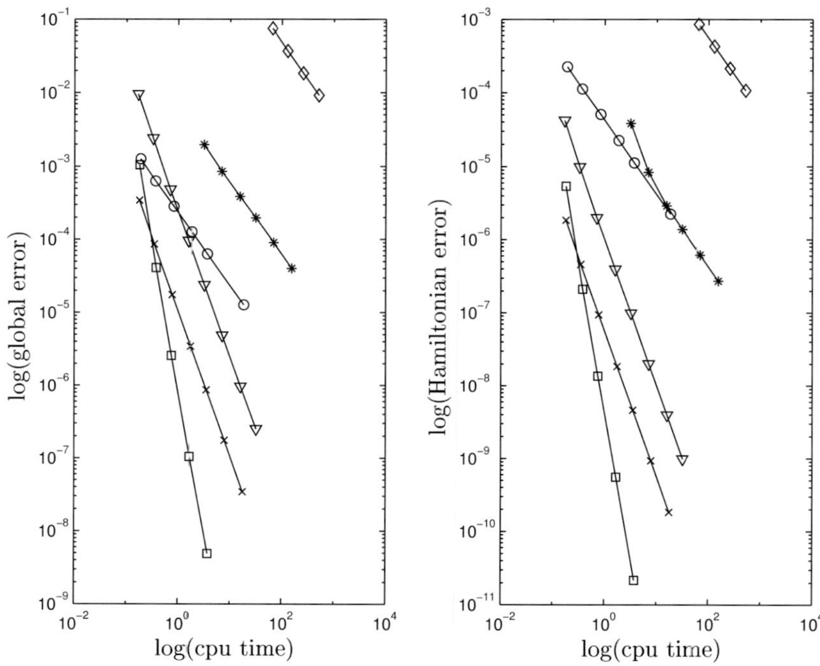

FIG. 6.2. Kepler problem with eccentricity of 0.9 for 10 orbits. Symplectic Euler ($*$), Störmer–Verlet (∇), Sundman Symplectic Euler ($\diamond$), Poincaré Symplectic Euler ($\circ$), adaptive Verlet Sundman ($\times$), Ruth's third-order method ($\Box$).

scribed earlier, with function g depending only on $\mathbf{q}$ this scheme can be written as (where $r = \sqrt{q_1^2 + q_2^2}$)

$$
\begin{aligned}
\mathbf{q}_{n+1/2} &= \mathbf{q}_n + \frac{\Delta\tau}{2\rho_n}\mathbf{p}_n, \\
\rho_{n+1} &= \frac{2}{g(\mathbf{q}_{n+1/2})} - \rho_n, \\
\mathbf{p}_{n+1} &= \mathbf{p}_n - \frac{\Delta\tau}{2}\left\{\frac{1}{\rho_n} + \frac{1}{\rho_{n+1}}\right\}\frac{\mathbf{q}_{n+1/2}}{r_{n+1/2}^3}, \\
\mathbf{q}_{n+1} &= \mathbf{q}_{n+1/2} + \frac{\Delta\tau}{2\rho_{n+1}}\mathbf{p}_{n+1}, \\
t_{n+1} &= t_n + \frac{\Delta\tau}{2}\left\{\frac{1}{\rho_n} + \frac{1}{\rho_{n+1}}\right\}.
\end{aligned}
$$

In Fig. 6.1 we show the results of applying the SE and SV methods to the untransformed Kepler problem as well as SE and adaptive Verlet applied to the Sundman transformed system and SE applied to the Poincaré transformed system. For this experiment we only integrate for 10 orbits of eccentricity 0.5 – a fairly simple problem. Straight away we see the correct order for the methods with both the fixed and variable step size formulations.

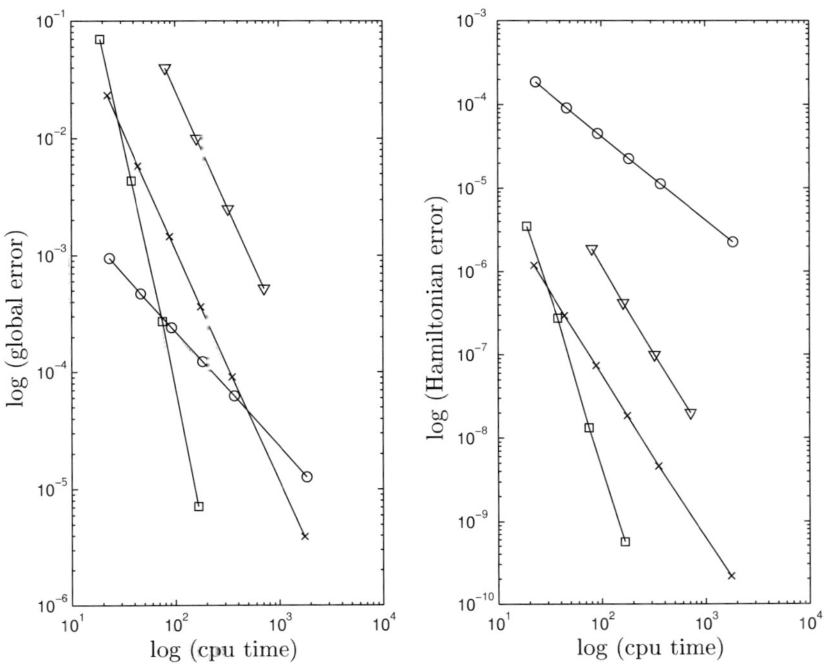

FIG. 6.3. Kepler problem with eccentricity of 0.9 for 1000 orbits. Störmer–Verlet (▽), Poincaré Symplectic Euler (○), adaptive Verlet Sundman (×), Ruth's third order method (□).

The adaptive methods which preserve either the symplectic or reversible properties can be seen to have better performance even for this problem where adaptivity is not vital for efficiency. However the symplectic Euler method applied to the Sundman transformed system demonstrates a definite reduction in performance, presumably due to the loss of both symplecticity and reversibility.

We perform a similar experiment in Fig. 6.2, this time for a higher eccentricity of 0.9, where adaptive methods should come into their own. We see the desired result of the adaptive methods designed to preserve symplecticity or reversibility performing well in comparison to their fixed step size counterparts. Again, the symplectic Euler method applied to the Sundman transformed system is seen to perform poorly. Ruth's third order method is also included for comparison purposes.

In Fig. 6.3 we again perform computations for the problem with eccentricity of 0.9, but now integrate over the much longer time scale of 1000 orbits. Again we see the improvements the use of adaptive time stepping affords, although this example clearly demonstrates that for high accuracy the use of high-order methods appears to be beneficial.

Finally in Fig. 6.4 we demonstrate the desirable linear error growth property of the methods which preserve symplecticity or reversibility applied to this problem. We also include the classical third-order Runge–Kutta method mentioned in a previous section, it can be seen to exhibit quadratic error growth demonstrating the fact that for long-

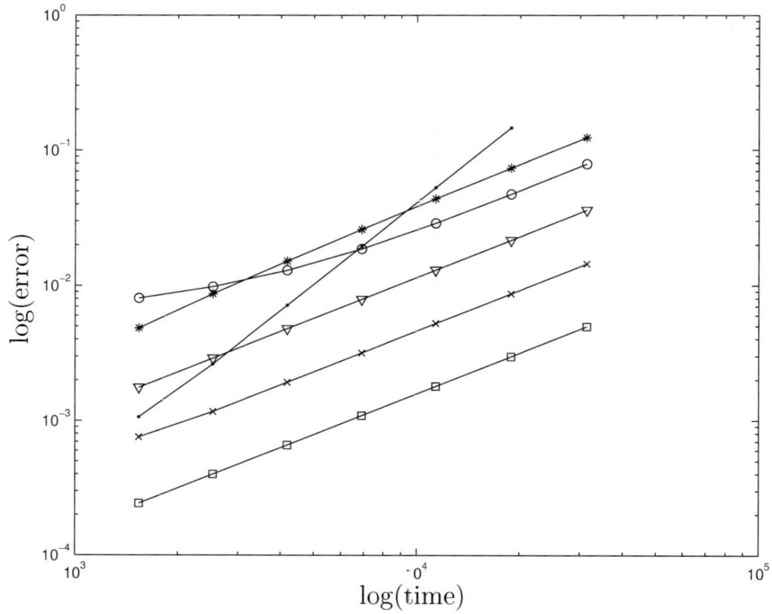

FIG. 6.4. Linear error growth of methods applied to Kepler's problem with eccentricity 0.5. Symplectic Euler (⋆), Störmer–Verlet (▽), adaptive Verlet Sundman (×), Poincaré Symplectic Euler (∘), Ruth's third order method (□). For comparison third-order classical Runge–Kutta (·) is shown, clearly exhibiting quadratic error growth.

term simulations a geometric integrator is to be preferred in many situations. Note that no effort has been made to choose fictive time steps in a consistent manner here, and therefore no conclusions regarding the relative accuracies of these methods should be inferred.

Similar experiments are carried out by CALVO, LÓPEZ-MARCOS and SANZ-SERNA [1998]. They come to the similar conclusion that for Hamiltonian problems a code based on high-order Gauss formulae with Poincaré transformations may outperform standard software.

7. Partial differential equations

7.1. Overview

There has been far less treatment of partial differential equations in the geometric integration literature than that of ordinary differential equations. Indeed the excellent accounts of geometric integration by HAIRER, LUBICH and WANNER [2002], MCLACHLAN and QUISPEL [2001] and by SANZ-SERNA [1997] do not mention PDEs at all. For certain types of PDE (e.g., strongly hyperbolic) this is understandable as they represent significantly more complex systems than ODEs and when discretised as systems of ODEs are far more complex to work with and are usually stiff – containing widely different timescales. On the other hand, it is precisely geometrically based methods which are well placed to deal with them as the spatial variation implicit in such problems adds a very strong geometric structure to the ODE systems which we can subsequently exploit. An important example of this is an extension of the symmetry group methods described in Section 5.3 to include the spatial variable so that we strongly couple spatial and temporal structures. Here adaptivity plays a central role in all of our calculations – indeed adaptivity finds a natural geometrical setting. We start our discussion however by looking at Hamiltonian PDEs and then at PDEs with a Lagrangian structure.

7.2. Symplectic methods for Hamiltonian PDEs

Many partial differential equations can be put into a Hamiltonian formulation and admitted symmetries of this formulation can lead, via Noether's theorem, to conservation laws. There are several approaches aimed at exploiting this structure in discretisations of the equations, all of which can loosely be labelled geometrical. One is based upon using the Hamiltonian formulation of the PDEs on multi-symplectic structures, generalising the classical Hamiltonian structure by assigning a distinct symplectic operator for each space direction and time (BRIDGES [1997], BRIDGES and REICH [2001], REICH [2000]). Variational integrators (MARSDEN and WEST [2001], see also the next section) generalise Veselov-type discretisations of PDEs in variational form. In this section we will look at methods for discretising the Hamiltonian PDEs using a Poisson bracket approach and/or various splitting methods. In Sections 9 and 10 we look at the use of semi-discretisations (in space) of the equations and discrete analogues of Noether's theorem.

7.2.1. Basic theory

Here we summarise some of the ideas presented in the monograph by OLVER [1986].

Many differential equations of the form

$$\mathbf{u}_t = \mathbf{F}(x, \mathbf{u}, \mathbf{u}_x, \mathbf{u}_{xx}, \ldots)$$

can be represented in the Hamiltonian form

$$\mathbf{u}_t = \mathcal{D}\left(\frac{\delta\mathcal{H}}{\delta\mathbf{u}}\right), \qquad \mathcal{H}[\mathbf{u}] = \int H(x, \mathbf{u}, \mathbf{u}_x, \ldots)\,\mathrm{d}x.$$

Typically the domain of integration in the above is the real line or the circle (corresponding to periodic boundary conditions for the PDE). In this expression $\mathcal{H}$ is a functional map from the function space in which $\mathbf{u}$ is defined to the real line. The variational (or functional) derivative $\delta\mathcal{H}/\delta\mathbf{u}$ is the function defined via the Gateaux derivative

$$\left(\frac{\mathrm{d}}{\mathrm{d}\varepsilon}\mathcal{H}[\mathbf{u} + \varepsilon\mathbf{v}]\right)_{\varepsilon=0} = \int \frac{\delta\mathcal{H}}{\delta\mathbf{u}} v\,\mathrm{d}x,$$

so that if u is a scalar function and $H \equiv H(u, u_x)$ then

$$\frac{\delta\mathcal{H}}{\delta u} = \frac{\partial H}{\partial u} - \frac{\partial}{\partial x}\left(\frac{\partial H}{\partial u_x}\right).$$

The operator $\mathcal{D}$ is *Hamiltonian* if given the functionals $\mathcal{P}[\mathbf{u}]$ and $\mathcal{L}[\mathbf{v}]$ it generates a Poisson bracket $\{\cdot,\cdot\}$ given by

$$\{\mathcal{P},\mathcal{L}\} = \int \frac{\delta\mathcal{P}}{\delta\mathbf{u}}\cdot\mathcal{D}\frac{\delta\mathcal{L}}{\delta\mathbf{u}}\,\mathrm{d}x.$$

Where the Poisson bracket must satisfy the skew symmetry condition

$$\{\mathcal{P},\mathcal{L}\} = -\{\mathcal{L},\mathcal{P}\}$$

and the Jacobi identity

$$\bigl\{\{\mathcal{P},\mathcal{L}\},\mathcal{R}\bigr\} + \bigl\{\{\mathcal{L},\mathcal{R}\},\mathcal{P}\bigr\} + \bigl\{\{\mathcal{R},\mathcal{P}\},\mathcal{L}\bigr\} = 0,$$

for all functionals $\mathcal{P}$, $\mathcal{L}$ and $\mathcal{R}$.

The Hamiltonian partial differential equation describing the time evolution of the system is then given by

$$\mathbf{u}_t = \{\mathbf{u},\mathcal{H}\}.$$

Such Hamiltonian systems have invariants (Casimirs) which are functionals $\mathcal{C}$ which satisfy

$$\{\mathcal{C},\mathcal{F}\} = 0 \quad \text{for all functionals } \mathcal{F},$$

and so these invariants are not associated with properties of the Hamiltonian, they arise through any degenerate nature of the Poisson structure. In particular, Casimirs are conserved during the evolution. The Hamiltonian functional $\mathcal{H}$ is conserved when it is not explicitly dependent on time.

7.2.2. *Examples*

We now give three examples of such PDEs which we will refer to throughout this section.

EXAMPLE 7.1 (*The nonlinear wave equation*). In this case $\mathbf{u} = (p,q)^{\mathrm{T}}$, $\mathcal{D}$ is the *canonical* Poisson operator given by

$$\mathcal{D} = \begin{bmatrix} 0 & -1 \\ 1 & 0 \end{bmatrix} \quad \text{and} \quad H = \frac{p^2}{2} + \frac{q_x^2}{2} + V(q).$$

Thus, the partial differential equation is then

$$\begin{pmatrix} p_t \\ q_t \end{pmatrix} = \begin{pmatrix} -\delta\mathcal{H}/\delta q \\ \delta\mathcal{H}/\delta p \end{pmatrix} = \begin{pmatrix} q_{xx} - V'(q) \\ p \end{pmatrix}.$$

EXAMPLE 7.2 (*The KdV equation*). This integrable equation is used to describe the motion of water waves in long channels. It has soliton solutions which can be determined by using the inverse scattering transformation. It is described in detail in Drazin and Johnson's monograph on solitons (DRAZIN and JOHNSON [1989]). In this case

$$\mathcal{D} = \frac{\partial}{\partial x} \quad \text{and} \quad H = -\frac{u^3}{6} + \frac{u_x^2}{2}.$$

The associated PDE is then

$$u_t = \frac{\partial}{\partial x}\left(\frac{\delta \mathcal{H}}{\delta u}\right) = \frac{\partial}{\partial x}\left(-\frac{u^2}{2} - u_{xx}\right)$$

so that

$$u_t + uu_x + u_{xxx} = 0.$$

EXAMPLE 7.3 (*The one-dimensional nonlinear Schrödinger equation*). This is an important example of an integrable PDE which models the modulational instability of water waves and is also used to describe the dynamics of lasers and of nonlinear optics. In a later section we will look at cases of determining the singular solutions for this equation posed in three dimensions which have a rather different behaviour from the soliton solutions which occur in one dimension.

In the one-dimensional case we take $u = (\psi, \bar{\psi})^{\mathrm{T}}$ where ψ and $\bar{\psi}$ are treated as independent variables, we take $\mathcal{D}$ to be i times the canonical Poisson operator and

$$H(\psi, \bar{\psi}) = \psi_x \bar{\psi}_x - \tfrac{1}{2}\psi^2\bar{\psi}^2.$$

The corresponding partial differential equation is then given by

$$\mathrm{i}\psi_t = \frac{\delta \mathcal{H}}{\delta \bar{\psi}} = -\psi_{xx} - \psi|\psi|^2.$$

This system has many conserved quantities including $\mathcal{H}$, $\int |u|^2\,\mathrm{d}x$ and

$$\int |\psi_{xx}|^2 + 2|\psi|^6 - 6|\psi_x|^2|\psi|^2 - \left((|\psi|^2)_x\right)^2 \mathrm{d}x.$$

Hamiltonian wave equations of the form described in the examples above have been studied in great detail by many authors. An excellent survey of results and methods is given by MCLACHLAN [1994]. The basic approach suggested by these authors is to discretise $\mathcal{H}$ and $\mathcal{D}$ separately to obtain a symplectic set of ordinary differential equations. In this paper McLachlan recommends that (pseudo) spectral methods should be used whenever possible to discretise the equations in space, in particular differential operators such as $\mathcal{D}$ (in the KdV example) as finite differences introduce excessive dispersion into the system. Indeed, it is considered essential that the higher modes typically present in Hamiltonian PDEs should be correctly resolved.

The resulting system of ordinary differential equations can then be integrated using a symplectic method, typically based upon one of the splitting methods described in Section 3. Two types of splitting can be considered, both based upon a natural decomposition of the Hamiltonian. If $\mathbf{u} = (p, q)^{\mathrm{T}}$ and $\mathcal{H} = T(p) + V(q)$ then a $P - Q$ splitting can be used exactly as before, with explicit time integration at each stage.

An alternative, potentially more stable, and certainly recommended method is to use an $L - N$ splitting. In this it is assumed that $\mathcal{H}$ can be decomposed as

$$H(\mathbf{u}) = L(\mathbf{u}) + N(\mathbf{u}),$$

where L corresponds to the linear dynamics of the partial differential equation associated with the higher order derivatives and N is typically a nonlinear operator associated with lower order derivatives. In this case the time flow of the vector field corresponding to L is typically integrated by using a Fourier transform.

It is significant that Casimirs, which are invariants of the flow, are in general conserved by symplectic splitting methods. This is a natural consequence of the results described in Section 4.

We now describe two examples of the use of these methods taken from MCLACHLAN [1994].

EXAMPLE 7.4 (*The nonlinear wave equation*). As described above this is given by

$$\frac{\partial q}{\mathrm{d}t} = p, \qquad \frac{\partial p}{\mathrm{d}t} = q_{xx} - V'(q),$$

and we consider this with periodic boundary conditions on the interval $[0, 2\pi]$. A convenient way of discretising the Hamiltonian formulation of this equation is to use a spectral method. Accordingly, applying a full spectral decomposition in space gives

$$p(x,t) = \sum_m p_m(t)\,\mathrm{e}^{\mathrm{i}mx}, \qquad q(x,t) = \sum_m q_m(t)\,\mathrm{e}^{\mathrm{i}mx}.$$

Substituting into the partial differential equation leads to the following system of ordinary differential equations

$$\dot{q}_m = p_m, \qquad \dot{p}_m = m^2 q_m - \frac{1}{2\pi}\int_0^{2\pi} V'\Big(\sum q_k\,\mathrm{e}^{\mathrm{i}kx}\Big)\,\mathrm{e}^{-\mathrm{i}mx}\,\mathrm{d}x.$$

This system is canonically Hamiltonian, with conjugate variables q_m and p_{-m} and Hamiltonian

$$H = \frac{1}{2}p_0^2 + \sum_{m\neq 0}\big(p_m p_{-m} + m^2 q_m q_{-m}\big) + \frac{1}{2\pi}\int_0^{2\pi} V\Big(\sum q_m\,\mathrm{e}^{\mathrm{i}mx}\Big)\,\mathrm{d}x.$$

Note that the function H above is simply $\mathcal{H}/2\pi$ where $\mathcal{H}$ is defined as in the example and the operator $\mathcal{D}$ is the same in both cases. McLachlan suggests that this system should be discretised by truncating p, q (and hence H) to a finite number of modes and evaluating the integral inside H by using a trapezoidal quadrature rule. This leads to the Hamiltonian system

$$\dot{q}_m = p_m, \qquad \dot{p}_m = -m^2 q_m - \big(FV'(F^{-1}q)\big)_m,$$

where F is the discrete Fourier transform used to map between Fourier space and the solution space. Difficulties arise in this discretisation due to aliasing of the nonlinear terms, which can be reduced by padding the solution with zeros.

Both the $P-Q$ and the $L-N$ splittings may be used on this system. For the $P-Q$ splitting we can use an explicit Störmer–Verlet method exactly as before. For the $L-N$ splitting the problem is decomposed as

$$L\colon\ \dot{q}_m = p_m, \quad \dot{p}_m = -m^2 q_m$$

and

$$N\colon \dot{q}_m = 0, \quad \dot{p}_m = -\big(FV'(F^{-1}q)\big)_m.$$

In this splitting, the linear system L is the harmonic oscillator and can be integrated explicitly. Similarly the nonlinear system N has a trivial integral. It is shown in MCLACHLAN [1994] that for weakly nonlinear problems the $L - N$ splitting has both a smaller truncation error and no linear stability limit, making them generally preferable to $P - Q$ splittings.

EXAMPLE 7.5 (*The nonlinear Schrödinger equation*). We consider again the one-dimensional NLS equation

$$\mathrm{i}\psi_t + \psi_{xx} + \psi|\psi|^2 = 0, \qquad \psi(0,t) = \psi(L,t).$$

This partial differential equation is integrable and can be solved by using the inverse scattering transform (DRAZIN and JOHNSON [1989]). It admits plane wave solutions $\psi = \exp(2\mathrm{i}|a|^2 t)a$ and in the symmetric case with $\psi(x,t) = \psi(L-x,t)$ it also admits homoclinic (localised wave-packet type) orbits connecting the periodic orbits which have fine spatial structure. It is possible to devise discretisations of the equation which are themselves integrable (ABLOWITZ and HERBST [1990]) and the latter authors recommend this approach when resolving the structure of the homoclinic solutions. If instead a standard Runge–Kutta–Merson method in time is used together with a central difference discretisation in space then there is a rapid flow of energy into the high frequencies due to poor resolution of the spatial structure, and temporal chaos in the time-series (a phenomenon not present in the underlying system). In both cases the spatial mesh introduces a large perturbation from integrability.

Instead a symplectic (integrable) method based on the $L - N$ splitting can be used in the form

$$L\colon \mathrm{i}\psi_t + \psi_{xx} = 0, \qquad N\colon \mathrm{i}\psi_t + \psi|\psi|^2 = 0.$$

In MCLACHLAN [1994] a spectral decomposition to integrate the linear evolution equation L in Fourier space is used rather than the discrete Fourier transform, the nonlinear equation N is integrated in real space. In both cases the time integration can be performed using a symplectic method such as the Störmer–Verlet method.

8. Lagrangian and action based methods for PDEs

Variational integrators for a variety of problems, including Hamiltonian ones, can be constructed from a discrete version of Lagrangian mechanics, based on a discrete variational principle. They have the advantage that much of the structure of Lagrangian systems can be replicated on a discrete level, with variational proofs (such as Noether's theorem) extending from the continuous to the discrete setting. When applied to PDEs they can give rise to a natural multi-symplectic setting. For ODEs they lead to many symplectic rules such as the Störmer–Verlet method, the Newmark method in structural dynamics and the Moser–Veselov method for integrating the equations of motion for a rigid body considered earlier. A very complete account of these methods is given in the review by MARSDEN and WEST [2001].

To give an example of the application of Lagrangian based methods, the methods described in the previous section can be extended to a wider class of partial differential equations satisfying Dirichlet, Neumann or periodic boundary conditions that are derived as variational derivatives of a suitable energy function $\mathcal{G}(u, u_x)$ in the manner

$$u_t = \left(\frac{\partial}{\partial x}\right)^{\alpha} \frac{\delta \mathcal{G}}{\delta u}, \tag{8.1}$$

where we define an analogous function $\mathcal{G}$ to the Hamiltonian $\mathcal{H}$ by

$$\mathcal{G}(u) = \int G(u)\,\mathrm{d}x$$

and assume that u satisfies the following boundary condition

$$\left[\frac{\partial G}{\partial u_x}\frac{\partial u}{\partial t}\right]_0^L = 0,$$

which is true for problems with Dirichlet, natural or periodic boundary conditions.

This class of partial differential equations includes some of the Hamiltonian examples described earlier, for example, the first order wave equation

$$u_t = u_x, \quad \text{where } G(u) = \frac{u^2}{2}, \ \alpha = 1,$$

and the KdV equation

$$u_t = \frac{\partial}{\partial x}\left(\frac{u^2}{2} + u_{xx}\right), \quad \text{where } G = \frac{u^3}{6} - \frac{u_x^2}{2}, \ \alpha = 1.$$

In particular, if α is *odd* then

$$\frac{\partial \mathcal{G}}{\partial t} = 0. \tag{8.2}$$

However, this class also includes other dissipative (non Hamiltonian) problems such as the nonlinear diffusion equation

$$u_t = (u^n)_{xx}, \quad \text{where } G = -\frac{u_x^{n+1}}{n+1}, \ \alpha = 0,$$

and the Cahn–Hilliard equation

$$u_t = \left(pu + ru^3 + qu_{xx}\right)_{xx},$$

where

$$G = \left(p\frac{u^2}{2} + r\frac{u^4}{4} - q\frac{u_x^2}{2}\right), \quad p < 0,\ q < 0,\ r > 0, \quad \alpha = 2.$$

Indeed, when α is even we have

$$(-1)^{\alpha/2+1}\frac{\partial \mathcal{G}}{\partial t} \leqslant 0. \tag{8.3}$$

This broader class of problems is considered by FURIHATA [1999] and we consider his derivation of a discretisation for this class of problems here. In particular Furihata considers schemes which preserve the conditions (8.2), (8.3) together with certain other conserved quantities. For a very complete discussion of integrators based upon action integrals, see the review article of MARSDEN and WEST [2001].

To derive a geometrically based method, Furihata considers that the integral $\mathcal{G}$ be replaced by a discrete trapezium rule sum $\mathcal{G}_d$ of a discrete form G_d of G in which u_x is replaced by a suitable finite difference equivalent. Note that this differs from the methods considered by McLachlan. Following this procedure we obtain

$$J_d(U) = T\sum_{k=0}^{N} G_d(U)_k \Delta x,$$

where the operator $T\sum_{k=0}^{N} f_k$ refers to the Trapezium rule sum

$$\left(\frac{1}{2}f_0 + \sum_{k=1}^{N-1} f_k + \frac{1}{2}f_N\right).$$

The operation of integration by parts (and its repeated applications) can then be replaced under this discretisation by the operation of summation by parts (and its repeated applications). This is a crucial observation as it allows many of the properties of the variational derivative to be directly inherited by the discretisation. In particular the summation by parts identity

$$T\sum_{k=0}^{N} f_k\left(\delta_k^+ g_k\right)\Delta x + T\sum_{k=0}^{N}\left(\delta_k^- f_k\right)g_k \Delta x = \left[\frac{f_k(s_k^+ g_k) + (s_k^- f_k)g_k}{2}\right]_{k=0}^{N},$$

where, for the sequence $f_1, f_2, \ldots, f_k, \ldots$ we have operators

$$s_k^+ f_k = f_{k+1}, \quad s_k^- f_k = f_{k-1}, \quad \delta_k^+ f_k = \frac{f_{k+1} - f_k}{\Delta x}, \quad \delta_k^- f_k = \frac{f_k - f_{k-1}}{\Delta x}.$$

By doing this the discrete variation in G_d can be calculated to satisfy the identity

$$\mathcal{G}_d(U) - \mathcal{G}_d(V) = T\sum_{k=0}^{N} \frac{\delta \mathcal{G}_d}{\delta(U, V)_k}(U_k - V_k)\Delta x,$$

where the variational derivative is replaced by a discrete analogue $\delta\mathcal{G}_g/\delta(U, V)_k$. The derivation of the expression for the variational derivative is a key result given in FURIHATA [1999] and can be expressed most easily for discrete functionals of the form

$$G_d(U)_k = \sum_{l=1}^{m} f_l(U_k)g_l^+\left(\delta_k^+ U_k\right)g_l^-\left(\delta_k^- U_k\right),$$

where f, g^+, g^- are arbitrary functions. By repeated application of summation by parts it can be shown that the variational derivative of $\mathcal{G}_d$ has (for this formulation) the following discrete form

$$\frac{\delta \mathcal{G}_d}{\delta(U,V)_k} = \sum_{l=1}^{m} \left(\frac{\mathrm{d} f_l}{\mathrm{d}(U_k, V_k)} \frac{g_l^+(\delta_k^+ U_k) g_l^-(\delta_k^- U_k) + g_l^+(\delta_k^+ V_k) g_l^-(\delta_k^- V_k)}{2} \right.$$
$$\left. - \delta_k^+ W_l^-(U,V)_k - \delta_k^- W_l^+(U,V)_k \right). \tag{8.4}$$

Here the functions W above correspond to the derivative of G with respect to u_x so that

$$W_l^+(U,V)_k = \left(\frac{f_l(U_k) + f_l(V_k)}{2} \right) \left(\frac{g_l^-(\delta_k^- U_k) + g_l^-(\delta_k^- V_k)}{2} \right) \frac{\mathrm{d} g_l^+}{\mathrm{d}(\delta_k^+ U_k, \delta_k^+ V_k)},$$

$W_l^-(U,V)$ is defined similarly with all signs reversed and

$$\frac{\mathrm{d} f}{\mathrm{d}(a,b)} = \frac{f(a) - f(b)}{a - b} \quad \text{if } a \neq b,$$

and

$$\frac{\mathrm{d} f}{\mathrm{d}(a,b)} = \frac{\mathrm{d} f}{\mathrm{d} a} \quad \text{if } a = b.$$

The expression (8.4) gives a discrete representation of the variational derivative when finite difference methods are applied. The differential equation can then be discretised in space by replacing the right hand side of the original equation (8.1) by the discrete variational derivative (8.4) and a suitable time discretisation then used for the left hand side. In its simplest form this leads to the discrete evolutionary equation

$$\frac{U_k^{(n+1)} - U_k^{(n)}}{\Delta t} = \delta_k^{\langle \alpha \rangle} \frac{\delta \mathcal{G}_d}{\delta(U^{(n+1)}, U^{(n)})_k}, \tag{8.5}$$

where

$$\delta_k^{\langle 1 \rangle} = \frac{s_k^+ - s_k^-}{2\Delta x}, \quad \delta_k^{\langle 2 \rangle} = \frac{s_k^+ - 2 + s_k^-}{\Delta x^2}, \quad \delta_k^{\langle 2m+1 \rangle} = \delta_k^{\langle 1 \rangle} \delta_k^{\langle 2m \rangle}.$$

It is shown in FURIHATA [1999] that if α is odd then both $\mathcal{G}$ and $\mathcal{G}_d$ are conserved during the evolution along with the total discrete mass $T \sum U_k \Delta x$ if the continuous mass is also conserved.

In contrast, if α is *even* then both $\mathcal{G}$ and $\mathcal{G}_d$ decrease along solution trajectories and hence both the underlying problem and its discretisation are dissipative.

EXAMPLE. Following FURIHATA [1999] we consider the KdV equation as given by

$$u_t = \frac{\partial}{\partial x}\left(\frac{u^2}{2} + u_{xx} \right), \qquad G(u, u_x) = \frac{1}{6}u^3 - \frac{1}{2}u_x^2,$$

together with periodic boundary conditions

$$u(x + L, t) = u(x, t).$$

The KdV equation conserves both $\mathcal{G}$ and the total mass of the solution and it is integrable by means of the inverse scattering transformation, with soliton solutions. Many numerical methods to integrate the KdV equation have been considered, some of which share many of the properties of the splitting methods for the nonlinear wave equation described in the last section. A selection of these methods are described in DE FRUTOS and SANZ-SERNA [1997], GRIEG and MORRIS [1976], GODA [1975], HERMAN and KNICKERBOCKER [1993], P.W. LI [1995], PEN-YU and SANZ-SERNA [1981], SANZ-SERNA [1982], TAHA and ABLOWITZ [1984], ZABUSKY and KRUSKAL [1965].

To construct a numerical method for integrating the KdV equation which is directly based on the action integral, we start with a 'natural' discretisation G_d of G given by

$$G_d(U)_k = \frac{1}{6}(U_k)^3 - \frac{1}{2}\frac{(\delta_k^+ U_k)^2 + (\delta_k^- U_k)^2}{2}.$$

From this it immediately follows that

$$\frac{\delta \mathcal{G}_d}{\delta(U^{(n+1)}, U^{(n)})_k} = \frac{1}{2}\frac{(U_k^{(n+1)})^2 + U_k^{(n+1)}U_k^{(n)} + (U_k^{(n)})^2}{3} + \delta_k^{\langle 2\rangle}\frac{U_k^{(n+1)} + U_k^{(n)}}{2}.$$

Following the procedure described above then leads to the following implicit scheme

$$\frac{U_k^{(n+1)} - U_k^{(n)}}{\Delta t} = \delta_k^{\langle 1\rangle}\left(\frac{1}{2}\frac{(U_k^{(n+1)})^2 + U_k^{(n+1)}U_k^{(n)}) + (U_k^{(n)})^2}{3} + \delta_k^{\langle 2\rangle}\frac{U_k^{(n+1)} + U_k^{(n)}}{2}\right). \tag{8.6}$$

This scheme conserves both the integrals of G and of u during the evolution, although it is not symplectic. In FURIHATA [1999] it is claimed that (8.6) is unconditionally stable although no proof is given.

In Fig. 8.1 we present a calculation of the evolution of the KdV system comprising two colliding solitons which is obtained using the above method in which we take interval of $[0, 5]$ a discrete step size $\Delta x = 5/100$ and a discrete time step of $\Delta t = 1/2000$. The implicit equations in (8.6) were solved at each time step using the Powell-hybrid solver SNSQE.

This rather brief survey has not included a detailed discussion about suitable treatments of the various types of boundary conditions encountered in the partial differential equations described by such actions. For more details on this topic see FURIHATA [1999] and the references therein. Further application of the method described in FURIHATA [1999] to a series of nonlinear wave equations and also to the Cahn–Hilliard equation are given in FURIHATA [2001a], FURIHATA [2001b], MATSUO and FURIHATA [2001], MATSUO, SUGIHARA, FURIHATA and MORI [2001]. In MARSDEN and WEST [2001] variationally based methods for many other PDE problems are discussed, including the development of circulation preserving integrators for fluid systems.

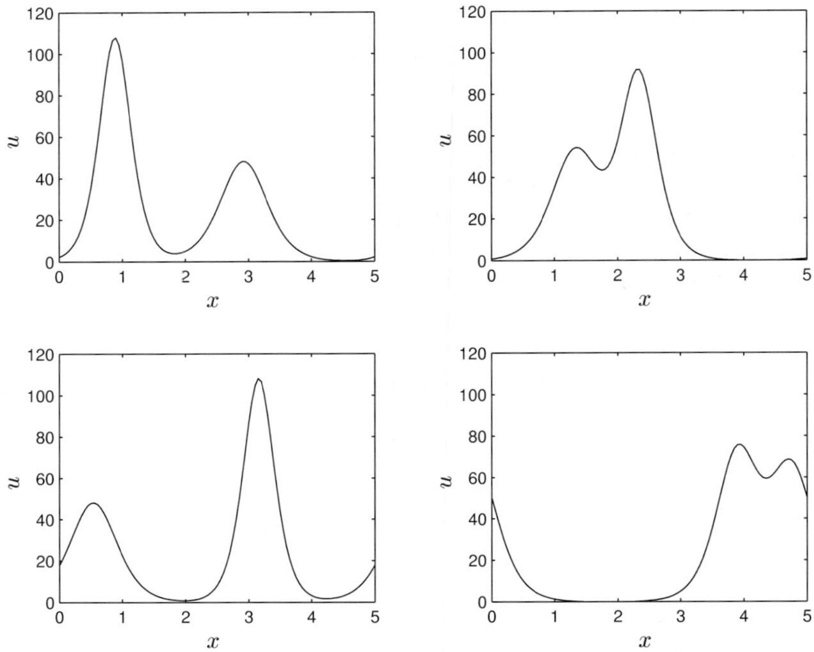

FIG. 8.1. Interaction of two solitons obtained using the scheme (8.6).

9. Methods for partial differential equations that preserve all symmetries

In this section we review some of the work of Dorodnitsyn and his co-workers in studying methods, fully invariant under the action of all symmetries. Whereas the two previous sections have considered discretisations with a *constant* spatial mesh, we now relax this condition to allow for meshes which vary in both space and time.

As discussed earlier, symmetries are fundamental features of the differential equations of mathematical physics and if possible should be retained when the equations are discretised on either a fixed or an adaptive mesh.

In a very similar manner to that discussed earlier, we can consider the action of Lie point symmetries on a partial differential equation. Dorodnitsyn lists the following reasons for considering the action of such symmetries:

- The group action transforms one set of solutions into another.
- There exists a standard procedure (OLVER [1986]) for obtaining (differential) invariants for a symmetry group. This can allow a reduction of the PDE to an ODE under certain circumstances (for example, we can study self-similar solutions).
- The action of the symmetry can reduce the order of the equation.
- The invariance of the PDE is a necessary condition for the application of Noether's theorem on variational problems to obtain conservation laws. We return to this point later in this section.

- Lie point transformations have a clear geometrical interpretation and one can construct the orbits of a group in a finite dimensional space of independent and dependent variables.

Dorodnitsyn and his group have concentrated on constructing adaptive mesh methods that inherit all symmetries of the system. We describe these here and in the next section the use of them in finding discrete analogues of Noether's theorem. In the section after that we look at the special case of scaling symmetries.

Suppose that the partial differential equation in which we consider u to be a function of x and t which is to be discretised is given by

$$F(x,t,u_t,u,u_x,u_{xx})=0. \tag{9.1}$$

In the section on ordinary differential equations we studied the action of Lie groups on a manifold. Similarly we can consider a Lie group to act on this partial differential equation noting that the action of the group is now to couple the temporal and the spatial structures. The action of the Lie point transformations on then can be considered by looking at the one-parameter transformation groups given by

$$X=\xi^t\frac{\partial}{\partial t}+\xi^x\frac{\partial}{\partial x}+\eta\frac{\partial}{\partial u}+\cdots.$$

Here the functions ξ^x, ξ^t and η are considered to be functions of t, x and u.

The case of constant functions ξ^x, ξ^t and η corresponds to an invariance of the partial differential equation to translations in the appropriate variables, for example, any autonomous differential equation is invariant under the translation $t\to t+T$ generated by the action $X=\partial_t$ and, for example, the wave equation is invariant under spatial translations of the form $x\to x+L$ generated by the action $X=\partial_x$. (Solutions invariant under the two actions are simply travelling waves.)

The case of linear functions

$$\xi^t=\alpha t,\quad \xi^x=\beta x.\quad \text{and}\quad \eta=\gamma u,$$

corresponds to scaling symmetries of the form

$$t\to\lambda^\alpha t,\quad x\to\lambda^\beta x,\quad u\to\lambda^\gamma u,$$

which were considered earlier in the context of ODEs and presently in the context of PDEs.

In general, the group generated by X transforms a solution point (x,t,u,u_t,u_x,u_{xx}) to a new one and also transforms the associated partial differential equation.

Dorodnitsyn et al. consider discretising (9.1) on a mesh comprising a finite set of points in time and space $z^1, z^2, \dots$ to give a difference equation

$$F(z)=0. \tag{9.2}$$

Now, when discretising any partial differential equation using a finite difference or a finite element method it is essential to also define a mesh. This mesh should ideally also reflect the underlying symmetries of the differential equation. In particular, suppose that the mesh is defined by an equation of the form

$$\Omega(z,h)=0. \tag{9.3}$$

If we seek a discretisation of the differential equation which is invariant to the Lie group action then both the difference scheme (9.2) and the mesh equation (9.3) need to be invariant under the action of the group. In particular we require the two conditions

$$XF(z) = 0 \quad \text{and} \quad X\Omega(z, h) = 0 \tag{9.4}$$

to be satisfied iff $F = \Omega = 0$. If this can be done (and this is a significant challenge) for *all* the possible invariant group actions then the method is termed *fully invariant*. More usually it can only be achieved for a subset of the possible group actions.

Dorodnitsyn et al. specifically consider the use of uniform grids that are invariant under all group actions. We relax this condition in the next section, but consider their approach in more detail here. A uniform mesh is given when the left spatial step at any point in space equals the right spatial step so that

$$h^+ = h^-.$$

We note that a *uniform* mesh, in which the mesh spacing can evolve with time, should not be confused with a *constant* mesh which does not evolve with time.

In this case we can define D_{+h} and D_{-h} as right and left difference derivative operators with corresponding shift operators S_{+h} and S_{-h}. It is shown in DORODNITSYN [1991], see also AMES, ANDERSON, DORODNITSYN, FERAPONTOV, GAZIZOV, IBRAGIMOV and SVIRSHEVSKII [1994] that the exact representation of total differentiation D_x in the space of difference operators is then given by a prolonged form of the above given by

$$D^+ = \frac{\partial}{\partial x} + \widehat{D}_{+h}\frac{\partial}{\partial u} + \cdots, \qquad \widehat{D}_{+h} = \sum_{n=1}^{\infty} \frac{(-h)^{n-1}}{n} D^n_{+h}.$$

If the symmetry group action is given by the one-parameter transformation group generated by the operator

$$X = \xi^x \frac{\partial}{\partial x} + \eta \frac{\partial}{\partial u}$$

then the action of this must be prolongated on the discrete mesh to give

$$X = \xi^x \frac{\partial}{\partial x} + \eta \frac{\partial}{\partial u} + \cdots + \left[S_{+h}(\xi^x) - \xi^x\right]\frac{\partial}{\partial h^+} + \left[\xi^x - S_{-h}(\xi^x)\right]\frac{\partial}{\partial h^-}.$$

A *uniform* mesh then retains its invariance under this prolonged action if

$$S_{+h}(\xi^x) - 2\xi^x + S_{-h}(\xi^x) = 0 \quad \text{or} \quad D_{+h}D_{-h}(\xi^x) = 0.$$

Similarly it is uniform under the action of the operator η^t (in time) if

$$D_{+\tau}D_{-\tau}(\xi^t) = 0,$$

where $+\tau$ and $-\tau$ are temporal time steps. It is often desirable that orthogonality should also be preserved under the action of the group. It is shown in DORODNITSYN [1993a] that a necessary and sufficient condition for this is that

$$D_{+h}(\xi^t) + D_{+\tau}(\xi^x) = 0.$$

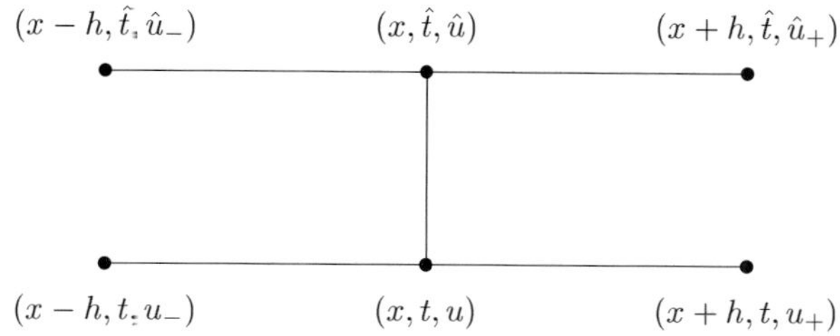

FIG. 9.1. Stencil for an orthogonal mesh.

A further simple condition, given by

$$D_{+h}D_{+\tau}(\xi^t)=0,$$

preserves the flatness of a layer of the grid in the time direction.

EXAMPLE 9.1 (*Nonlinear diffusion with a source*). We discuss here an example investigated by DORODNITSYN and KOZLOV [1997] (see also KOZLOV [2000]) with regard to a fully invariant discretisation. In the next section we look at the same problem using an equidistribution based method.

The nonlinear diffusion equation with a source describes the diffusion and reaction of a chemical in a porous medium. For example, the flow of pollutant materials through rock. It is given by the equation

$$u_t=(u^\sigma u_x)+au^n. \tag{9.5}$$

It admits two translation groups and one scaling group given by

$$X_1=\frac{\partial}{\partial t},\quad X_2=\frac{\partial}{\partial x},\quad X_3=2(n-1)t\frac{\partial}{\partial t}+(n-\sigma-1)x\frac{\partial}{\partial x}-2u\frac{\partial}{\partial u}.$$

This set satisfies the conditions for a uniform, orthogonal mesh. The operators for the stencil and the mesh are given in Fig. 9.1.

Corresponding to time translation, space translation and scaling we have seven difference invariants of the Lie algebra given by

$$\frac{\hat{u}}{u},\quad \frac{u_+}{u},\quad \frac{u_-}{u},\quad \frac{\hat{u}_+}{\hat{u}},\quad \frac{\hat{u}_-}{\hat{u}},\quad \frac{\tau^{(n-\sigma-1)/2(n-1)}}{h},\quad \tau u^{n-1}.$$

An example of an invariant implicit scheme is then given by

$$\frac{\hat{u}-u}{\tau}=\frac{1}{(\sigma+1)h^2}\left(\hat{u}_+^{\sigma+1}-2\hat{u}^{\sigma+1}+\hat{u}_-^{\sigma+1}\right)+a\hat{u}^n. \tag{9.6}$$

EXAMPLE 9.2 (*The semilinear heat equation*). A weakly reacting chemical system can be described by the equation

$$u_t=u_{xx}+\delta u\log(u).$$

This is invariant under the one-parameter groups generated by the operators

$$X_1 = \frac{\partial}{\partial t}, \quad X_2 = \frac{\partial}{\partial x}, \quad X_3 = 2\,\mathrm{e}^{\delta t}\frac{\partial}{\partial x} - \delta\,\mathrm{e}^{\delta t} x u \frac{\partial}{\partial u}.$$

Unlike the previous example an orthogonal mesh cannot be used to model this equation with an invariant scheme, although we can use one with flat time layers. The four differential invariants are given by

$$J^1 = \mathrm{d}t, \quad J^2 = \left(\frac{u_x}{u}\right)^2 - \frac{u_{xx}}{u},$$

$$J^3 = 2\frac{u_x}{u} + \frac{\mathrm{d}x}{\mathrm{d}t}, \quad J^4 = \frac{1}{u}\frac{\mathrm{d}u}{\mathrm{d}t} - \delta\log(u) + \frac{1}{4}\left(\frac{\mathrm{d}x}{\mathrm{d}t}\right)^2.$$

Using these we may reformulate the differential equation in a more suggestive manner. Setting $J^3 = 0$ gives a Lagrangian condition for the evolution of the mesh so that

$$\frac{\mathrm{d}x}{\mathrm{d}t} = -2\frac{u_x}{u}. \tag{9.7}$$

Similarly, setting $J^2 = J^4$ gives the following equation for the evolution of the solution:

$$\frac{\mathrm{d}u}{\mathrm{d}t} = u_{xx} + \delta u\log(u) - 2\frac{u_x^2}{u}.$$

Thus explicit conservation of the differential invariants of this system is achieved by both a motion of the mesh and an evolution of the solution. Observe that we now have to solve two equations, one for the solution and another for the mesh. This is exactly analogous to the extended system that we investigated when we looked at time adaptivity in Section 5.3. Following Kozlov we now examine the difference invariants for the systems defined using a similar (but non-orthogonal) stencil as for the nonlinear diffusion equation. The following are difference invariants

$$I^1 = \tau, \quad I^2 = h^+, \quad I^3 = h^-, \quad I_4 = \hat{h}^+, \quad I^5 = \hat{h}^-$$
$$I^6 = \big(\log(u)\big)_x - \big(\log(u)\big)_{\bar{x}}, \quad I^7 = \big(\log(\hat{u})\big)_x - \big(\log(\hat{u})\big)_{\bar{x}},$$

where

$$u_x = \frac{u_+ - u}{h} \quad \text{and} \quad u_{\bar{x}} = \frac{u - u_-}{h},$$

$$I^8 = \delta\Delta x + 2(\mathrm{e}^{\delta\tau} - 1)\left(\frac{h^-}{h^+ + h^-}\big(\log(u)\big)_x + \frac{h^+}{h^+ + h^-}\big(\log(u)\big)_{\bar{x}}\right),$$

$$I^9 = \delta\Delta x + 2(1 - \mathrm{e}^{-\delta\tau})\left(\frac{\hat{h}^-}{\hat{h}^+ + \hat{h}^-}\big(\log(\hat{u})\big)_x + \frac{\hat{h}^+}{\hat{h}^+ + \hat{h}^-}\big(\log(\hat{u})\big)_{\bar{x}}\right),$$

$$I^{10} = \delta(\Delta x)^2 + 4(1 - \mathrm{e}^{-\delta\tau})\big(\log(\hat{u}) - \mathrm{e}^{\delta\tau}\log(u)\big)$$

where $\Delta x = \hat{x} - x$.

Again, by exploiting these invariants we may construct fully invariant difference schemes (effectively discretisations of the above coupled systems) which evolve both the mesh and the solution on the mesh. In KOZLOV [2000] it is suggested that an explicit scheme can be constructed so that

$$I^8 = 0 \quad \text{and} \quad I^{10} = \frac{8}{\delta}\frac{(e^{\delta I^1} - 1)^2}{I^2 + I^3} I^6.$$

Thus the equation for the evolution of the mesh becomes

$$\delta \Delta x + 2(e^{\delta\tau} - 1)\left(\frac{h^-}{h^+ + h^-}\big(\log(u)\big)_x + \frac{h^+}{h^+ + h^-}\big(\log(u)\big)_{\bar{x}}\right) = 0, \tag{9.8}$$

which is a discretisation of (9.7). Similarly, the equation for the evolution of the solution then becomes

$$\begin{aligned} &\delta(\Delta x)^2 + 4(1 - e^{-\delta\tau})\big(\log(\hat{u}) - e^{\delta\tau}\log(u)\big) \\ &\quad = \frac{8}{\delta}\frac{(e^{\delta\tau} - 1)^2}{h^+ + h^-}\big[\big(\log(u)\big)_x - \big(\log(u)\big)_{\bar{x}}\big]. \end{aligned} \tag{9.9}$$

It is shown in KOZLOV [2000] that this discretisation admits an invariant solution which has a mesh evolving in the manner

$$x = x_0 \frac{e^{\delta t} + \alpha}{1 + \alpha}$$

and a solution which evolves in the manner

$$u(x,t) = \exp\left(e^{\delta t}\left(f(0) - \frac{e^{\delta\tau} - 1}{2}\sum_{j=1}^{n-1}\frac{e^{-\delta t_j}}{1 + \alpha e^{-\delta t_j}}\right) - \frac{\delta e^{\delta t}}{\alpha + e^{\delta t}}\frac{x^2}{4}\right).$$

Here $x = x_i^j$ and $t = t_j$ and f is a function determined from the initial conditions. For constant α this solution is invariant under the action of the combined operation $2\alpha X_2 + X_3$.

10. Noether's theorem for discrete schemes

A combination of the action based methods for partial differential equations with the symmetry preserving methods described in the last section leads to difference schemes that inherit a discrete form of Noether's theorem relating symmetries of a Lagrangian to conservation laws. An account of these methods is given in the articles by DORODNITSYN [1998], DORODNITSYN [1993b] and the thesis of KOZLOV [2000] and we follow the latter treatment here. See also the discussion of Lagrangian based integrators in MARSDEN and WEST [2001]. For convenience we have considered only the case of ordinary differential equations. The application of the same methods to partial differential equations follows very similar lines and is described in KOZLOV [2000].

Suppose that a differential equation is derived from the variation of a Lagrangian functional $\mathcal{L}$ with

$$\mathcal{L} = \int L(x, u, du/dx)\, dx.$$

This functional achieves its extremal values when $u(t)$ satisfies the Euler–Lagrange equations

$$\frac{\delta \mathcal{L}}{\delta u} = \frac{\partial L}{\partial u} - D\left(\frac{\partial L}{\partial u'}\right) = 0,$$

where D is the total derivative operator

$$D = \frac{\partial}{\partial x} + u' \frac{\partial}{\partial u} + \cdots.$$

If X is a Lie point symmetry given by

$$X = \xi \frac{\partial}{\partial x} + \eta \frac{\partial}{\partial u} \tag{10.1}$$

then X is an *infinitesimal divergence symmetry* if there is a function $B(x, u)$ such that

$$X(L) + LD(\xi) = D(B). \tag{10.2}$$

Note that this is a statement about symmetries of the Lagrangian, rather than of the underlying differential equation. Noether's theorem applied to this system then leads to a conservation law generated by X as follows

THEOREM (Noether). *If X is an infinitesimal divergence symmetry of the functional $\mathcal{L}$ then there is a first integral J of the Euler–Lagrange equations given by*

$$J \equiv N(L) - B = \text{const}, \quad \text{where } N = \xi + (\eta - \xi u') \frac{\partial}{\partial u'}. \tag{10.3}$$

PROOF. See OLVER [1986]. □

Now we seek a discrete version of this theorem. The Lagrangian functional can be discretised directly in a very similar manner to that described by Furihata. Suppose that the numerical method is constructed over a one-dimensional lattice given by

$$h_+ = \varphi(x, x_+, u, u_+),$$

where x_-, x and x_- are adjacent mesh points with $h_- = x - x_-$, $h_+ = x_+ - x$ and u_-, u, u_+ (and also ξ_-, ξ, ξ_+) are defined on these points. An appropriate discretisation of $\mathcal{L}$ then takes the form

$$\mathcal{L}_d = \sum_{\Omega_h} L_d(x, x_+, u, u_+) h_+.$$

The definition of an infinitesimal divergence symmetry can then be extended to the discrete case in the manner

$$\xi \frac{\partial L_d}{\partial x} + \xi^+ \frac{\partial L_d}{\partial x_+} + \eta \frac{\partial L_d}{\partial u} + \eta^+ \frac{\partial L_d}{\partial u_+} + L_d D_h(\xi) = D_h(B), \tag{10.4}$$

together with an invariance condition on the mesh given by

$$S_h(\xi) - \xi = X(\varphi). \tag{10.5}$$

Here the discrete operators D_h and S_h are defined by

$$S_h\big(f(x)\big) = f(x_+) \quad \text{and} \quad D_h(f) = \big(S_h(f) - f\big)/h_+,$$

and there is a function $B = B(x, u, x_+, u_+)$.

Finding conditions for the extremals of the discrete Lagrangian leads to conditions on both the solution and on the mesh. Corresponding to the Euler–Lagrange equations are two *quasiextremals* given by

$$\frac{\delta L_d}{\delta x} = h_+ \frac{\partial L_d}{\partial x} + h_- \frac{\partial L_d^-}{\partial x} + L_d^- - L_d \quad \text{and} \quad \frac{\delta L_d}{\delta u} = h_+ \frac{\partial L_d}{\partial u} + h_- \frac{\partial L_d^-}{\partial u}.$$

If the functional $\mathcal{L}_d$ is stationary along the orbits of the group generated by X and the complete group acting on the system is at least two-dimensional then both of the quasiextremals must vanish so that

$$\frac{\delta L_d}{\delta x} = 0 \quad \text{and} \quad \frac{\delta L_d}{\delta u} = 0. \tag{10.6}$$

This leads to a condition on both the solution and the mesh.

The discrete version of Noether's theorem is then given as follows

THEOREM. *Let the discrete Lagrangian $\mathcal{L}_d$ be divergence invariant under the action of a Lie group G of dimension $n \geqslant 2$. If the quasiextremals* (10.6) *vanish then symmetry X leads to the invariant*

$$J = h_- \eta \frac{\partial L_d}{\partial u} + h_- \eta \frac{\partial L_d}{\partial x} + \eta L_d^- - B.$$

PROOF. See DORODNITSYN [1993b]. □

EXAMPLE. The following example of an application of these methods to the Pinney equation, is given in KOZLOV [2000]. Consider the following second order ordinary differential equation

$$u'' = u^{-3}.$$

This is the Euler–Lagrange equation corresponding to the Lagrangian

$$L = u'^2 - \frac{1}{u^2}.$$

It is invariant under the action of a three-dimensional Lie group G generated by the operators

$$X_1 = \frac{\partial}{\partial x}, \quad X_2 = 2x\frac{\partial}{\partial x} + u\frac{\partial}{\partial u}, \quad X_3 = x^2\frac{\partial}{\partial x} + xu\frac{\partial}{\partial u}.$$

The action of the (prolongation of) the symmetry operators X_i leads to the invariances

$$X_1 L + LD(\xi_1) = 0, \qquad X_2 L + LD(\xi_2) = 0,$$
$$X_3 L + LD(\xi_3) = 2u'u = D(u^2).$$

Applying Noether's theorem then gives the following first integrals

$$J_1 = u'^2 + \frac{1}{u^2}, \qquad J_2 = 2\frac{x}{u^2} - 2(u - u'x)u', \qquad J_3 = \frac{x^2}{u^2} + (u - xu')^2.$$

A natural discretisation of L is given by

$$L_d = \left(\frac{u_+ - u}{h_+}\right)^2 - \frac{1}{uu_+}.$$

This admits the invariances

$$X_1 L_d + L_d D_h(\xi_1) = 0, \qquad X_2 L_d + L_d D_h(\xi_2) = 0,$$
$$X_3 L_d + L_d D_h(\xi_3) = D_h(u^2).$$

The corresponding quasiextremals (invariant under X_i) are given by

$$\frac{\delta L_d}{\delta u}\colon\ 2(u_x - u_{\bar{x}}) - \frac{h_+}{u^2 u_+} + \frac{h_-}{u^2 u_-} = 0,$$
$$\frac{\delta L_d}{\delta x}\colon\ (u_x)^2 + \frac{1}{uu_+} - (u_{\bar{x}})^2 - \frac{1}{uu_-} = 0.$$

Here

$$u_x \equiv \frac{u_+ - u}{h_+}, \qquad u_{\bar{x}} \equiv \frac{u_- u_-}{h_-}.$$

The discrete form of Noether's theorem then gives the following three conserved discrete integrals

$$J_1 = u_x^2 + \frac{1}{uu_+}, \qquad J_2 = \frac{x + x_+}{uu_+} - u_x\big((u + u_+) - (x + x_+)u_x\big),$$
$$J_3 = \frac{xx_+}{uu_+} + \left(\frac{u + u_+}{2} - \frac{x + x_+}{2}u_x\right)^2.$$

11. Spatial adaptivity, equidistribution and scaling invariance

11.1. Scaling invariance of partial differential equations

As we have demonstrated in Section 9, problems described by a partial differential equation often involve a complex interaction between temporal and spatial structures. As for ordinary differential equations this coupling can take many forms, however, as in our discussion on ordinary equations, a common feature of the equations of mathematical physics is that a frequently encountered group of transformations are *scaling transformations*. These generalise those we investigated for ordinary differential equations in that changes in the temporal scale and scale of the solution are also related to changes in the spatial scale. Thus partial differential equations often fall naturally into the class of problems (considered earlier in the context of ordinary differential equations) which are invariant under the action of a scaling transformation.

Although scaling symmetries are (perhaps) the most important example of differential invariants and play a key role in the analysis of partial differential equations, in comparison with the work on ordinary differential equations scaling symmetries have been less exploited in the development of numerical algorithms. There are several good reasons for this. Firstly, a partial differential equation is often acted upon independently by several symmetry groups (indeed sometimes by a continuum of such) and it is unlikely that one discretisation can preserve invariance under all of these, so some choice has to be made in advance. An example of this is given by considering the linear *heat equation*

$$u_t = u_{xx}. \tag{11.1}$$

This equation, in the absence of boundary conditions, is invariant under arbitrary translations in time t and space x and to any *scaling* transformation of the form

$$t \to \lambda t, \qquad x \to \lambda^{1/2} x, \qquad u \to \lambda^{\alpha} u, \quad \forall \alpha \text{ and } \lambda > 0, \tag{11.2}$$

where α is determined by other conditions such as boundary conditions or integral constraints. (This invariance can also be expressed very naturally in terms of the action of elements in the tangent bundle.) The classical self-similar point source solution is precisely the solution of the linear heat equation invariant under the action of (11.2) with constant first integral so that $\alpha = -1/2$. In contrast if we consider the nonlinear heat equation

$$u_t = u_{xx} + u^2, \tag{11.3}$$

then this is invariant under the scaling group (11.2) in the case of $\alpha = -1$ only. This invariance plays a key role in the understanding of singular behaviour of the solutions of (11.3) when the initial data is large.

A second reason for the lack of methods which exploit scaling invariance is that in a practical situation, a partial differential equation is defined for arbitrary initial and boundary conditions. Although the equation in the absence of these can be invariant under the action of a group, a general solution will not be so. Often the group invariance is asymptotic in the sense that it describes the *intermediate asymptotic behaviour* of the solutions after a sufficiently long period of evolution that the effects of boundary and initial conditions have become unimportant, and before the system has reached an equilibrium state (BARENBLATT [1996]). This behaviour is very evident for the problem (11.1), where solutions from an almost arbitrary initial condition evolve to become asymptotically invariant under the action of (11.2).

A third, and rather subtle reason, is that in many cases the precise nature of the action of a symmetry group on a partial differential equation can only be determined *after* the equation has been solved. For example, in many travelling wave problems, the solutions invariant under the group actions of translation in space and time are precisely travelling waves, the study of which reduces to that of an ordinary differential equation. However the wave speed is in general undetermined until *after* the equation has been solved – indeed determining it is a nonlinear eigenvalue problem. Thus it is difficult *a priori* to find a coordinate system travelling along with the solution in which the solution is invariant.

11.2. Spatial adaptivity

It is possible to capture much of this scaling behaviour using an adaptive method based upon geometric ideas. We now describe a general method for adapting the spatial mesh and then show how geometric ideas can naturally be incorporated into it to give a scale invariant method for partial differential equations which generalises our earlier scale invariant method for ordinary differential equations.

Consider a general partial differential equation discretised on a nonuniform mesh with spatial points $X_{i,j}$. In an analogous manner to the way in which we constructed an adaptive temporal mesh through a time reparameterization or transformation function, where the time t was described in terms of a differential equation in a fictive variable τ, we can think of the spatial mesh $X_{i,j}$ as being point values of a function $X(\xi, \tau)$ of a fictive spatial variable ξ such that X satisfies a (partial differential) equation in ξ. Here we will assume that this function has a high degree of regularity (i.e. we progress beyond thinking of a mesh as a piece-wise constant function of ξ). The study and implementation of adaptive methods then becomes the analysis of the coupled systems describing both the underlying discrete solution $U(\xi, \tau)$ and of the mesh $X(\xi, \tau)$. Invoking a new system to describe the mesh makes the whole discrete system rather larger and more complex and apparently harder to analyse. However, just as in the case of the ODE systems we looked at in Section 5.3, a close geometrical coupling of the equations for the mesh and for the solution actually simplifies their analysis. A widely used, (and from a geometrical point of view natural) procedure for generating such a mesh is that of *equidistribution* of an appropriate *monitor* function.

The approach we describe here to study adaptive discretisations of partial differential equations in one-dimension is based on the framework which is developed in HUANG, REN and RUSSELL [1994]. It is possible to use similar ideas in higher dimensions, see HUANG and RUSSELL [1999], but this is naturally a more complex process. Equidistribution can be loosely thought of as a process for changing the *density* of the mesh points in response to the solution (in contrast to Lagrangian methods which tend to change the mesh points themselves). Our discussion will be confined to the one-dimensional case.

We define the physical mesh, that is the mesh upon which the physical problem is posed, in terms of a mesh function $X(\xi, t)$ which maps the computational (fictive) coordinate $\xi \in [0, 1]$ onto a physical coordinate (which we assume without loss of generality to be in $[0, 1]$) such that

$$X\left(\frac{j}{N}, t\right) = X_j(t), \quad j = 0, \ldots, N.$$

Therefore our $N + 1$ mesh points $X_j(t)$, which are permitted to vary with time, are simply the image of a uniform *computational* mesh under a time dependent mesh function.

We now introduce a monitor function $M(t, x, u, u_x, u_{xx})$ which classically represents some measure of computational difficulty in the problem, and invoke the principle of equidistribution by defining our mesh function X through the relation

$$\int_0^X M \,\mathrm{d}x = \xi \int_0^1 M \,\mathrm{d}x. \tag{11.4}$$

Differentiating with respect to ξ we see that the mesh-density X_ξ satisfies the equation

$$X_\xi = \frac{1}{M}\int_0^1 M\,\mathrm{d}x. \tag{11.5}$$

so that the mesh density is inversely proportional to M. The equation above may be differentiated again to give the following partial differential equation for the mesh – referred to as MMPDE1 (moving mesh partial differential equation (3.1))

$$(MX_\xi)_\xi = 0.$$

This equation may then be solved for the mesh by discretising the function X appropriately (HUANG, REN and RUSSELL [1994]). In practice this can lead to an unstable system and many related partial differential equations have been derived to stabilise this process – with details given in HUANG, REN and RUSSELL [1994]. An example of such a stabilisation is MMPDE6 given by

$$\varepsilon X_{t\xi\xi} = -(MX_\xi)_\xi,$$

where ε is a small relaxation parameter. A further advantage of this approach is that it allows the use of an initially uniform mesh (although we note that starting with such a mesh can lead to a stiff system as the equilibrium manifold of an equidistributed mesh is approached). Ideally the stabilisation process should also inherit the symmetries of the original.

To solve the original partial differential equation, both the partial differential equation for the mesh function X and the partial differential equation for $u(x,t)$ are discretised. (It often pays to use a high order discretisation in both cases.) One of the mesh equations may then be coupled with the original PDE, giving a new system.

The new coupled system may or may not inherit the qualitative features of the original problem. The geometric integration viewpoint is to produce a mesh in which the mesh equation inherits as many qualitative features as possible. For the purposes of this section, we follow the analysis presented for ordinary differential equations and seek a mesh so that the new system has the same scaling invariances as the original. As the mesh is governed by the monitor function M this problem reduces to that of choosing M such that the coupled system is invariant with respect to the same scaling transformation group as the original equation. By doing this we ensure that all of the scaling symmetry structure of the underlying partial differential equation will be inherited by the resulting numerical method. It is possible to do this for a wide variety of problems with relatively simple choices of M leading to some elegant scaling invariant methods.

We shall now look at a specific example in which we can compare these ideas with the symplectic methods considered earlier.

11.3. Singularities in the nonlinear Schrödinger equation (NLS)

In Section 7 we looked at the integrable partial differential equation described by the one-dimensional nonlinear Schrödinger equation. In higher dimensions the radially symmetric solutions of the cubic nonlinear Schrödinger equation satisfy the partial differential equation

$$\mathrm{i}u_t + u_{xx} + \frac{d-1}{x}u_x + u|u|^2 = 0. \tag{11.6}$$

Here d is the dimension of the spatial variable and x is the distance from the origin. This is a natural extension of the one-dimensional nonlinear Schrödinger equation discussed earlier. However it has two very significant differences, firstly it is not an integrable equation if $d > 1$. Furthermore, if $d \geqslant 2$ then this problem has solutions for a wide variety of initial data which blow-up so that they develop singularities at the origin in a finite time T (in a manner similar to that of gravitational collapse) so that the solution tends to infinity in an increasingly narrow (in space) peak.

As remarked earlier, the nonlinear Schrödinger equation is a model for the modulational instability of water waves and plasma waves, and is important in studies of nonlinear optics where the refractive index of a material depends upon the intensity of a laser beam. In the latter case, blow-up corresponds to a self-focusing of the wave. A general discussion of the singularity formation of the NLS equation is given in SULEM and SULEM [1999].

The NLS in higher dimensions remains unitary, and can also be written in precisely the same Hamiltonian form as in one dimension. However in higher dimensions it is no longer integrable and has far fewer conserved quantities. Indeed, during the evolution the following are the only two invariant quantities

$$\int_0^\infty |u|^2 x^{d-1}\,\mathrm{d}x \quad \text{and} \quad \int_0^\infty \left(|u_x|^2 - \tfrac{1}{2}|u|^4\right) x^{d-1}\,\mathrm{d}x. \tag{11.7}$$

The integration strategy for the one-dimensional NLS described in Section 7 strongly exploited its Hamiltonian structure. It is much less clear in the context of higher dimensions whether this approach is the correct one as there is now a play-off between symplectic methods and scale invariant methods.

We now briefly discuss a numerical method to solve (11.6) when $d \geqslant 2$ which is based upon scale invariant geometrical ideas and which is very effective at computing the blow-up solutions. Details of this method are given in BUDD, CHEN and RUSSELL [1999]. It is based on preserving the scaling symmetries of the problem rather than either of the two invariant quantities (11.7) or indeed the overall Hamiltonian structure.

To apply this procedure we consider the scaling and translational groups under which NLS is invariant. In particular (11.6) is invariant under the action of either of the transformations the *scaling transformation*

$$t \to \lambda t, \qquad x \to \lambda^{1/2}x, \qquad u \to \lambda^{-1/2}u, \tag{11.8}$$

and *translations in phase* given by

$$u \to \mathrm{e}^{\mathrm{i}\varphi}u, \quad \varphi \in \mathbb{R}.$$

We seek a numerical method which inherits both the scaling and phase invariance symmetries, and achieve this through the use of adaptivity of both the temporal and spatial meshes.

The temporal and spatial adaptivity used for this example are given by solving the following equations

$$\frac{dt}{d\tau} = \frac{1}{|u(0,t)|^2},$$

coupled with MMPDE6 The choice of monitor function

$$M = |u|^2$$

results in a combined system for the mesh and the solution which is invariant under the full set of transformations involving changes both in scale and in phase.

The resulting coupled system is discretised in space using a collocation method. This leads to a system of ordinary differential equations which are scale invariant but do not (necessarily) inherit the Hamiltonian structure of the underlying partial differential equation. It is thus highly appropriate to discretise the resulting ordinary differential equation system using a scale invariant temporal adaptive method as described in Section 5.3, but less appropriate to use a symplectic solution method for the same system. (It is an unresolved question of how we may systematically include all the geometrical features of a partial differential equation into an adaptive method, indeed the equations for the mesh obtained above do not in general have a symplectic structure, even if the underlying partial differential equation does have this structure.) Accordingly, we do not use a symplectic method to integrate the ordinary differential equations but use instead a stable, scale invariant, spatially adaptive BDF discretisation (HAIRER, NØRSETT and WANNER [1993]). The resulting method has proved very effective at computing singular solutions using only a modest number ($N = 81$) of mesh points in the computational domain. The computations have revealed much new geometrical structure in the problem of blow-up in the NLS including new multi-bump blow-up structures (BUDD [2001]).

We now present the results of a computation using this method which evolves towards the singular blow-up self-similar solution which has a monotone profile.

Fig. 11.1 shows two solutions taken when $d = 3$ with initial data $u(0,x) = 6\sqrt{2}\exp(-x^2)$ if $0 \leqslant x \leqslant 5$, and $u(0,x) = 0$ if $x > 5$. In this case the estimated value of T is $T = 0.0343013614215$. These two solutions are taken close to the blow-up time when the amplitude of $|u|$ is around 10^5 and the peak has width around 10^{-5}. Observe that the resolution of the peak is very good, indicating that the mesh points are adapting correctly.

In Fig. 11.2 we present a plot of $X_i(t)|u(0,t)|$ as a function of $|\log(T-t)|$ for a range in which u varies from 100 to 500000. According to (11.8) the quantity ux is an invariant of the scaling transformation and is therefore also a constant for a self-similar solution. Observe that in the computations these values rapidly evolve towards constants, demonstrating that the mesh and the discrete solution are both evolving in a self-similar manner.

Thus, as in the case of applying a scale invariant temporal adaptive approach to the ordinary differential equations describing gravitational collapse, the application of a scale invariant spatial approach to adaptivity has led to an attracting self-similar solution being precisely recreated by the numerical solution and the mesh.

A more refined approach to integrate this PDE which combines both the scale invariance and the Hamiltonian structure is described in BUDD and DORODNITSYN [2001].

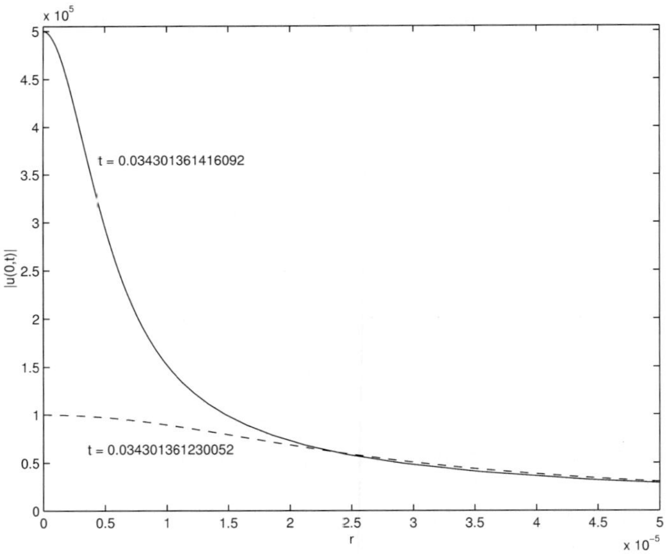

FIG. 11.1. The solution when $|u(0,t)| = 100000$ and 500000.

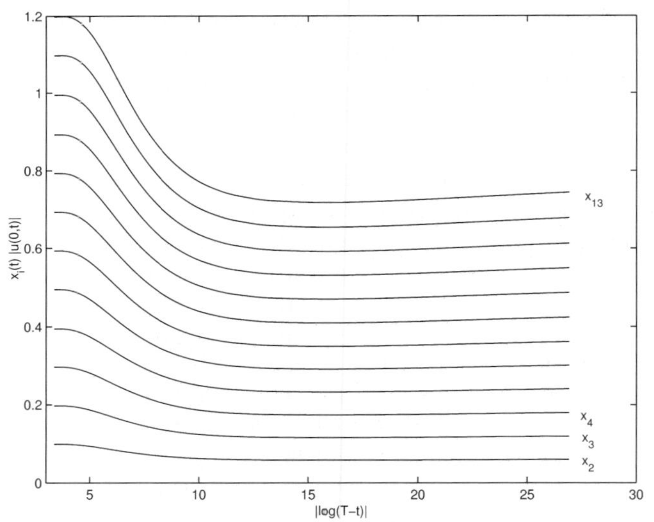

FIG. 11.2. The scaled mesh $X_i(t)|u(0,t)|$.

11.4. The nonlinear diffusion equation

In Section 9 we looked at the methods described by Dorodnitsyn to integrate the nonlinear diffusion equation. We now compare and contrast these with methods based on scale invariance and equidistribution. We consider here the special case of

$$u_t = (uu_x)_x. \tag{11.9}$$

This equation admits four continuous transformation groups, the two groups of translations in time and space and the two-dimensional vector space of scaling symmetry groups spanned by the operators

$$X_1 = t\frac{\partial}{\partial t} + \frac{1}{2}x\frac{\partial}{\partial x} \quad \text{and} \quad X_2 = t\frac{\partial}{\partial t} - u\frac{\partial}{\partial u}.$$

Eq. (11.9) admits a family of self-similar solutions of the form

$$u(x,t) = t^{\gamma} v(x/t^{\beta}) \tag{11.10}$$

for any values of β and γ which satisfy the algebraic condition

$$2\beta - \gamma = 1.$$

Without additional conditions any such solution is possible, however, if we impose the condition that $u(x,t)$ decays as $|x| \to \infty$ then a simple calculation shows that if the mass I of the solution is given by

$$I = \int u(x,t)\,\mathrm{d}x \tag{11.11}$$

then I is constant for all t. The only self-similar solution with this property has $\gamma = -1/3$ and $\beta = 1/3$. Reducing the partial differential equation down to an ordinary differential equation and solving this gives a one-parameter family of compactly supported self-similar solutions of the form

$$u(x,t) = t^{-1/3}(a - x^2/t^{2/3})_+.$$

These solutions were discovered independently by BARENBLATT [1996].

Some of these results can be found in the paper of BUDD, COLLINS, HUANG and RUSSELL [1999]. As described, the equation

$$u_t = (uu_x)_x = (u^2/2)_{xx},$$

has self-similar solutions of constant first integral (11.11). Both these, and the solution from general compactly supported initial data, are compactly supported and have interfaces at points s_- and s_+ which move at a finite speed. To discretise this equation we introduce and adaptive mesh $X(\xi,t)$ such that $X(0,t) = s_-$ and $X(1,t) = s_+$. To determine X we use a monitor function and a moving mesh partial differential equation. As the evolution of the porous medium equation is fairly gentle it is possible to use the mesh equation MMPDE1 without fear of instability in the mesh. This then allows a wide possible choice of scale invariant monitor functions of the form $M(u) = u^{\gamma}$. A convenient function to use is

$$M(u) = u$$

the choice of which is strongly motivated by the conservation law

$$\int_{s_-}^{s_+} u\,\mathrm{d}x = C.$$

Here C is a constant which we can take to equal 1 without loss of generality. Setting $M = u$ and $C = 1$ in the equidistribution principle gives

$$\int_{s_-}^{X} u \, \mathrm{d}x = \xi, \tag{11.12}$$

so that differentiation with respect to ξ we have

$$u X_\xi = 1 \tag{11.13}$$

as the equation for the mesh which we will discretise. Note that this is invariant under the group action $u \to \lambda^{-1/3} u$, $X \to \lambda^{1/3} X$. Now, differentiating (11.12) with respect to t gives

$$0 = X_t u + \int_{s_-}^{X} u_t \, \mathrm{d}x = X_t u + \int_{s_-}^{X} (u u_x)_x \, \mathrm{d}x = u(X_t + u_x).$$

Thus, for the continuous problem we also have that X satisfies the Lagrangian equation

$$X_t = -u_x. \tag{11.14}$$

Substituting (11.13) and (11.14) into the rescaled equation gives, after some manipulation, the following equation for u in the computational coordinates:

$$u(\xi, t)_t = \tfrac{1}{2} u^2 (u^2)_{\xi\xi}. \tag{11.15}$$

Eqs. (11.13), (11.14) and (11.15) have a set of self-similar solutions of the form

$$\hat{u}(\xi, t) = (t + C)^{-1/3} w(\xi), \qquad \widehat{X}(\xi, t) = (t - C)^{1/3} Y(\xi),$$

where C is arbitrary and the functions w and Y satisfy differential equations in ξ only. Now, consider semi-discretisations of (11.13) and (11.15) so that we introduce discrete approximations $U_i(t)$ and $X_i(t)$ to the continuous functions $u(\xi, t)$ and $X(\xi, t)$ over the computational mesh

$$\xi = i/N, \quad 1 \leqslant i \leqslant (N-1),$$

with $U_0(t) = U_N(t) = 0$. A simple centred semi-discretisation of (11.15) for $1 \leqslant i \leqslant (N-1)$ is given by

$$\frac{\mathrm{d}U_i}{\mathrm{d}t} = \frac{N^2}{2} U_i^2 \bigl(U_{i+1}^2 - 2U_i^2 + U_{i-1}^2\bigr). \tag{11.16}$$

To define the mesh X_i we discretise (11.13) to give the algebraic system

$$(X_{i+1} - X_i)(U_{i+1} + U_i) = \frac{2}{N}, \quad 0 \leqslant i \leqslant (N-1). \tag{11.17}$$

We observe that this procedure has the geometric property of automatically conserving the discrete mass

$$\sum_{i=0}^{N-1} (X_{i+1} - X_i)(U_{i+1} + U_i). \tag{11.18}$$

An additional equation is needed to close the set of equations for the unknowns X_i and we do this by insisting that (as in the true solution) the discrete centre of mass is conserved (without loss of generality at 0) so that

$$\sum_{i=0}^{N-1} (X_{i+1}^2 - X_i^2)(U_{i+1} + U_i) = 0. \tag{11.19}$$

Observe that Eq. (11.16) for the solution and Eqs. (11.17), (11.19) for the mesh have decoupled in this system. This makes it much easier to analyse. In particular (11.16) has two key geometrical features. *Firstly*, it is invariant under the group action

$$t \to \lambda t, \quad U_i \to \lambda^{-1/3} U_i.$$

Thus it admits a (semi) discrete self-similar solution of the form

$$\widehat{U}_i(t) = (t + C)^{-1/3} W_i, \tag{11.20}$$

where $W_i > 0$ satisfies an algebraic equation approximating the differential equation for $w(\xi)$. Observe, that for all C we have

$$t^{1/3} \widehat{U}_i \to W_i.$$

Secondly, the discretisation satisfies a maximum principle, so that if $U_i(t)$ and $V_i(t)$ are two solutions with $U_i(0) < V_i(0)$ for all i then $U_i(t) < V_i(t)$ for all i. (See BUDD and PIGGOTT [2001].)

The consequences of these two results are profound. Suppose that $U_i(0) > 0$ is a general set of initial data. By choosing values of C appropriately (say $C = t_0$ and $C = t_1$) we can write

$$t_0^{-1/3} W_i < U_i(0) < t_1^{-1/3} W_i, \quad 1 \leqslant i \leqslant N-1.$$

Consequently, applying the maximum principle to the self-similar solution we then have that for all t

$$(t + t_0)^{-1/3} W_i < U_i(t) < (t + t_1)^{-1/3} W_i, \quad 1 \leqslant i \leqslant N-1,$$

and hence we have the convergence result

$$t^{1/3} U_i(t) \to W_i, \tag{11.21}$$

showing that the solution U_i and hence the mesh X_i converges *globally* to the self-similar solution. Hence the numerical scheme is predicting precisely the correct asymptotic behaviour. This is precisely because the discretisation has same underlying geometry as the partial differential equation.

We illustrate this result by performing a computation in which the differential-algebraic equations are decoupled. The resulting ODEs are solved with a BDF method and a linear system is inverted to give the mesh. Fig. 11.3 shows a solution with arbitrary initial data being *squeezed* between two discrete self-similar solutions. The self-similar solutions shown here are those with C taking the values 0.9 and 2.3.

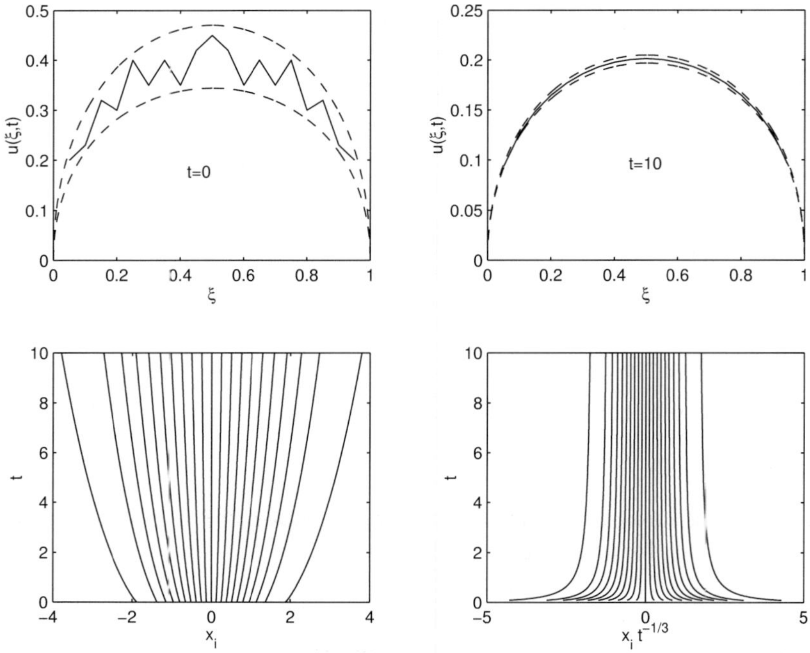

FIG. 11.3. Convergence of the solution in the computational domain and invariance of the computed mesh.

12. The Euler equations and some problems from meteorology

12.1. The Euler equation

Before we consider the problems with a purely meteorological flavour we shall now discuss the two-dimensional Euler equations describing inviscid incompressible fluid flow. This problem appears extensively in the mathematics literature as well as the meteorology literature where it is sometimes referred to as the baratropic vorticity equation. The Euler equation has the form

$$\dot{\omega} = \frac{\partial(\omega, \Psi)}{\partial(x, y)}, \tag{12.1}$$

where the right-hand side of (12.1) is simply the Jacobian determinant $\omega_x \Psi_y - \omega_y \Psi_x$, and the streamfunction Ψ ($\Psi_x = v$, $\Psi_y = -u$) is related to the vorticity $\omega = v_x - u_y$ through the relation

$$\omega = \nabla^2 \Psi.$$

This is another example of a problem which may be written in Hamiltonian form (see OLVER [1986], MARSDEN and RATIU [1999]), with Hamiltonian functional given by the energy

$$\mathcal{H} = \frac{1}{2}\int \mathrm{d}\mathbf{x}\,|\mathbf{u}|^2 = \frac{1}{2}\int \mathrm{d}\mathbf{x}\,|\nabla\Psi|^2 = -\frac{1}{2}\int \mathrm{d}\mathbf{x}\,\Psi\omega$$

and Poisson bracket given by

$$\{\mathcal{F},\mathcal{G}\} = -\int \mathrm{d}\mathbf{x}\,\frac{\delta\mathcal{F}}{\delta\rho(\mathbf{x})}\left(\frac{\partial(\omega(\mathbf{x}),\frac{\delta\mathcal{G}}{\delta\rho(\mathbf{x})})}{\partial(x,y)}\right) = \int \mathrm{d}\mathbf{x}\,\omega(\mathbf{x})\left(\frac{\partial(\frac{\delta\mathcal{F}}{\delta\rho(\mathbf{x})},\frac{\delta\mathcal{G}}{\delta\rho(\mathbf{x})})}{\partial(x,y)}\right),$$

and therefore, following the notation of Section 7.2.1, we have

$$\mathcal{D}\bullet = -\frac{\partial(\omega,\bullet)}{\partial(x,y)}.$$

12.1.1. The Arakawa Jacobian

With the correct assumptions on boundary conditions this problem can be shown to preserve the domain integrated vorticity, the enstrophy and the energy, given respectively by

$$\int \omega\,\mathrm{d}\mathbf{x},\quad \int \omega^2\,\mathrm{d}\mathbf{x},\quad \text{and}\quad \int |\nabla\Psi|^2\,\mathrm{d}\mathbf{x}. \tag{12.2}$$

Numerical methods for solving the Euler equations may be subject to nonlinear instabilities, in particular aliasing errors where there is a spurious transfer of energy from large spatial scales to unresolvable smaller spatial scales. Although finite element discretisations automatically obey discrete analogues of (12.2), the same is not in general true for finite difference spatial approximations. However, ARAKAWA [1966] constructed finite difference analogues of the Jacobian operator in (12.1) which do preserve discrete analogues of (12.2). This property of a method can be shown to prevent the nonlinear instabilities described above. Following the notation of DURRAN [1999] and expanding Ψ into a Fourier series

$$\Psi = \sum_{k,l}\Psi_{k,l},\qquad \Psi_{k,l} = a_{k,l}\,\mathrm{e}^{\mathrm{i}(kx+ly)},$$

we can now show that the enstrophy and energy in Fourier space are

$$\overline{\nabla\Psi\cdot\nabla\Psi} = -\overline{\Psi\nabla^2\Psi} = \sum_{k,l}\kappa^2\overline{\Psi_{k,l}^2}$$

and

$$\overline{\omega^2} = \overline{(\nabla^2\omega)^2} = \sum_{k,l}\kappa^4\overline{\Psi_{k,l}^2},$$

where overbar denotes domain integration and $\kappa = \sqrt{k^2+l^2}$ is the total wave number. We can therefore define a new invariant for this system – the average wave number

$$\kappa_{\mathrm{avg}} = \sqrt{\frac{\overline{\omega^2}}{\overline{\nabla\Psi\cdot\nabla\Psi}}}$$

The conservation of which demonstrates that there can be no net transfer of energy into small length scales. The Arakawa discretisation of the Jacobian prevents nonlinear instabilities by preserving discrete analogues of the enstrophy and energy and therefore also the discrete form of the average wave number. Arakawa's first second-order approximation to the Jacobian operator on a uniform grid ($\Delta x = \Delta y = d$) is given by

$$\left.\frac{\partial(\omega,\Psi)}{\partial(x,y)}\right|_{i,j} \approx \frac{1}{3}\big(J_{i,j}^{++}(\omega,\Psi) + J_{i,j}^{+\times}(\omega,\Psi) + J_{i,j}^{\times+}(\omega,\Psi)\big),$$

where

$$J_{i,j}^{++}(\omega,\Psi) = -\frac{1}{4d^2}\big[(\omega_{i+1,j} - \omega_{i-1,j})(\Psi_{i,j+1} - \Psi_{i,j-1}) \\ - (\omega_{i,j+1} - \omega_{i,j-1})(\Psi_{i+1,j} - \Psi_{i-1,j})\big],$$

and $J_{i,j}^{+\times}$, $J_{i,j}^{\times+}$ are given by similar expressions, see ARAKAWA [1966] for more details. A procedure for preserving arbitrary conserved integrals in PDEs is given by MCLACHLAN [Preprint].

12.1.2. Sine bracket type truncation

For references to this material see ARNOLD and KHESIN [1998], MCLACHLAN [1993], MCLACHLAN, SZUNYOGH and ZEITLIN [1997], ZEITLIN [1991]. We begin with our advection equation in the following form

$$\frac{\partial\omega}{\partial t} = \frac{\partial(\omega,\Psi)}{\partial(x,y)} \equiv \omega_x\Psi_y - \omega_y\Psi_x. \tag{12.3}$$

We assume (2π) periodic boundary conditions and therefore we evolve on a torus T^2. We now decompose our system into Fourier modes

$$\omega = \sum_{\mathbf{m}} \omega_{\mathbf{m}}\, \mathrm{e}^{\mathrm{i}(\mathbf{m},\mathbf{x})}$$

($(\mathbf{m},\mathbf{x}) = \mathbf{m}\cdot\mathbf{x}$) and consider the resulting system of infinitely many ODEs describing their evolution in time. Decomposing Ψ in a similar manner and using the above relation between ω and Ψ we arrive at

$$\Psi_{\mathbf{m}} = \frac{\omega_{\mathbf{m}}}{|\mathbf{m}|^2}.$$

If we substitute into (12.3) we get after some cancellations

$$\dot{\omega}_{\mathbf{m}}(t) = \sum_{\mathbf{n}\neq\mathbf{0}} \frac{\mathbf{m}\times\mathbf{n}}{|\mathbf{n}|^2}\omega_{\mathbf{m}+\mathbf{n}}\,\omega_{-\mathbf{n}}$$

where $\mathbf{m}\times\mathbf{n} = m_1n_2 - m_2n_1$, and for real ω we have that $\omega_{-\mathbf{n}} = \omega_{\mathbf{n}}^{\star}$.

This turns out to be Lie–Poisson with the following

$$H(\omega) = \frac{1}{2}\sum_{\mathbf{n}\neq\mathbf{0}} \frac{\omega_{\mathbf{n}}\omega_{-\mathbf{n}}}{|\mathbf{n}|^2}, \qquad \nabla H(\omega)_{\mathbf{k}} = \frac{\omega_{-\mathbf{k}}}{|\mathbf{k}|^2},$$

and Poisson structure (structure constants) defined by

$$J_{\mathbf{mn}}(\omega) = (\mathbf{m}\times\mathbf{n})\omega_{\mathbf{m}+\mathbf{n}} \equiv \sum_{\mathbf{k}} C_{\mathbf{mn}}^{\mathbf{k}}\omega_{\mathbf{k}}, \qquad C_{\mathbf{mn}}^{\mathbf{k}} = (\mathbf{m}\times\mathbf{n})\delta_{\mathbf{m}+\mathbf{n}-\mathbf{k},\mathbf{0}}.$$

So that finally we may write our system in the form

$$\dot{\omega}_{\mathbf{m}} = \sum_{\mathbf{k},\mathbf{l},\mathbf{n}} a^{\mathbf{nl}} C^{\mathbf{k}}_{\mathbf{mn}} \omega_{\mathbf{k}} \omega_{\mathbf{l}},$$

and the metric (inverse inertia tensor) is given by

$$a^{\mathbf{nl}} = \frac{1}{|\mathbf{n}|^2} \delta_{\mathbf{n}+\mathbf{l},\mathbf{0}}.$$

The finite-dimensional truncation of our bracket is now achieved by defining the new structure constants (N finite)

$$C^{\mathbf{k}}_{\mathbf{mn}} = \frac{N}{2\pi} \sin\left(\frac{2\pi}{N}(\mathbf{m} \times \mathbf{n})\right) \delta_{\mathbf{m}+\mathbf{n}-\mathbf{k},\mathbf{0}}$$

and so we have the finite-dimensional (Poisson) bracket

$$J_{\mathbf{mn}} = \frac{N}{2\pi} \sin\left(\frac{2\pi}{N}(\mathbf{m} \times \mathbf{n})\right) \omega_{\mathbf{m}+\mathbf{n}} \Big|_{\bmod N}.$$

Note that these structure constants are those for the algebra $\mathfrak{su}(N)$, and the consistency of this truncation relies on the fact that, in some sense, $SU(N) \to \mathrm{Diff}_{\mathrm{Vol}}(T^2)$ as $N \to \infty$.

We then reduce indices modulo N to the periodic lattice $-M \leqslant m_1, m_2 \leqslant M$ where $N = 2M + 1$. The Hamiltonian is truncated to a finite sum and we can now write down the Sine–Euler equations

$$\begin{aligned} \dot{\omega}_{\mathbf{m}} &= J_{\mathbf{mn}}(\omega) \nabla H_{\mathbf{n}}(\omega) \\ &= \sum_{\substack{n_1,n_2=-M \\ \mathbf{n}\neq\mathbf{0}}}^{M} \frac{N}{2\pi |\mathbf{n}|^2} \sin\left(\frac{2\pi}{N}(\mathbf{m} \times \mathbf{n})\right) \omega_{\mathbf{m}+\mathbf{n}} \omega_{-\mathbf{n}}. \end{aligned} \tag{12.4}$$

McLachlan [1993] constructs a Poisson integrator for (12.4) which is explicit, fast, and preserves analogues of $N - 1$ Casimir's to within round-off error. McLachlan, Szunyogh and Zeitlin [1997] were able to study baroclinic instability in a two-layer quasi-geostrophic type model using these ideas.

12.2. The semi-geostrophic equations, frontogenesis and prediction ensembles

To put this work into some context we will now consider how a variety of geometrical integration ideas have been used in combination to tackle certain problems which arise in the numerical solution of meteorological problems. These complex problems abound with geometrical features (such as the conservation of potential vorticity) and are very challenging numerically as they involve solving large systems of equations over long periods of time, often with the formation of singular structures. We now consider two meteorological problems for which a geometric integration based numerical approach to their solution appears to give some improvement over existing methods of solution.

The three-dimensional Boussinesq equations of semi-geostrophic theory describe the ideal, inviscid flow of a fluid on a plane with constant Coriolis force f. If u and v are the wind velocities then they may be written in the form (for many further details see HOSKINS [1975], CULLEN and PURSER [1984], CULLEN and PURSER [1987])

$$\left.\begin{aligned}\frac{Du_g}{Dt} - fv + \frac{\partial\varphi}{\partial x} &= \tfrac{Du_g}{Dt} - f(v - v_g) = 0,\\ \frac{Dv_g}{Dt} + fu + \frac{\partial\varphi}{\partial y} &= \tfrac{Dv_g}{Dt} + f(u - u_g) = 0,\end{aligned}\right\} \tag{12.5}$$

$$\frac{D\theta}{Dt} = 0, \tag{12.6}$$

$$\nabla_x \cdot \mathbf{u} = 0,$$

$$\nabla_x\varphi = \left(fv_g, -fu_g, g\frac{\theta}{\theta_0}\right).$$

Here (u_g, v_g) is the geostrophic wind, θ the potential temperature, and φ the geopotential. The energy integral E defined by

$$E = \int_D \left(\frac{1}{2}(u_g^2 + v_g^2) - \frac{g\theta z}{\theta_0}\right) \mathrm{d}x\,\mathrm{d}y\,\mathrm{d}z$$

is an invariant of this set of equations. This equation set is of interest for many reasons, one being that it allows the study of idealised atmospheric weather fronts. That is, just as for the nonlinear Schrödinger equation it admits solutions which form singularities in finite time (although the nature of these singularities is quite different).

It is usual to define a coordinate transformation from $(x, y, z)^{\mathrm{T}}$ coordinates to isentropic geostrophic momentum coordinates

$$\mathbf{X} \equiv (X, Y, Z)^{\mathrm{T}} = \left(x + \frac{v_g}{f}, y - \frac{u_g}{f}, \frac{g\theta}{f^2\theta_0}\right)^{\mathrm{T}}. \tag{12.7}$$

In a similar manner to the previous section we can think of $(X, Y, Z)^{\mathrm{T}}$ as being (fictive) computational coordinates introduced in earlier sections, (12.7) then describes the evolution of a spatial mesh. In terms of these new coordinates (12.5) and (12.6) transform to

$$\frac{D\mathbf{X}}{Dt} = \mathbf{u}_g \equiv (u_g, v_g, 0)^{\mathrm{T}}, \tag{12.8}$$

and hence the motion in these new coordinates is exactly geostrophic, as well as nondivergent $\nabla_X \cdot \mathbf{u}_g = 0$. A numerical method based on (12.8) will perform in an adaptive way – a Lagrangian form of mesh adaptivity where the mesh is moved at exactly the speed of the underlying velocity field.

The ratio of volume elements in dual space to that of physical space is given by

$$q = \frac{\partial(X, Y, Z)}{\partial(x, y, z)}.$$

This relates the scaling of the spatial mesh to the computational mesh in an analogous manner to the use of monitor functions described earlier. The expression q defines a consistent form of the Ertel potential vorticity (PV) in SG theory, satisfying

$$\frac{Dq}{Dt} = 0.$$

It is possible to write our coordinate transformation as $\mathbf{X} = \nabla_x P$ where

$$P(\mathbf{x}) = \frac{\varphi}{f^2} + \frac{1}{2}(x^2 + y^2).$$

Hence q, the PV, is equivalently the determinant of the Hessian of P with respect to the coordinates $\mathbf{x}$,

$$q = \det(\mathrm{Hess}_x(P)).$$

Also, $\mathbf{x} = \nabla_{\mathbf{X}} R$, where $P(\mathbf{x})$ and $R(\mathbf{X})$ are a pair of Legendre transforms related by

$$P + R = \mathbf{x} \cdot \mathbf{X}.$$

We can now introduce a stream function for the geostrophic velocities,

$$\Psi = f^2\left(\frac{1}{2}(X^2 + Y^2) - R(\mathbf{X})\right), \qquad (u_g, v_g) = \frac{1}{f}\left(-\frac{\partial \Psi}{\partial Y}, \frac{\partial \Psi}{\partial X}\right). \tag{12.9}$$

Defining ρ (the pseudo-density) to be

$$\rho \equiv q^{-1} = \det(\mathrm{Hess}_X(R)), \tag{12.10}$$

it can be shown that

$$\frac{D_X \rho}{Dt} \equiv \frac{\partial \rho}{\partial t} - \frac{1}{f}\frac{\partial(\rho, \Psi)}{\partial(X, Y)} = 0. \tag{12.11}$$

It is now possible to compute the evolution of this system using the following procedure,

(1) given an initial distribution of a pseudo-density solve the nonlinear elliptic equation of Monge–Ampère type (12.10) for R,
(2) using the streamfunction (12.9) compute a new velocity field (u_g, v_g),
(3) advect the pseudo-density distribution using (12.11) and return to start.

We thus have two distinct numerical problems to solve. The first being the computation of a solution to the Monge–Ampère equation (12.10). This is obviously linked to determining the coordinate transformation (12.7) in exactly the manner of the adaptive mesh methods described earlier, since for a given R we have

$$\mathbf{x} = \nabla_{\mathbf{X}} R,$$

and hence fits in well with our discussions of coordinate transformations and adaptivity earlier.

The second numerical challenge is that of solving the advection equation (12.11). We will now show that this also has a nice geometric structure very similar to the Euler equations considered in the previous subsection. We shall then return to the connection with two-dimensional adaptivity through a unifying geometrical framework.

12.2.1. Hamiltonian formulations

We now combine the theory described earlier for Hamiltonian partial differential equations with the meteorological equations and consider Hamiltonian formulations for this problem. ROULSTONE and NORBURY [1994] discuss two Hamiltonian formulations of the semi-geostrophic equations. The first, a canonical (infinite-dimensional extension of (3.2)) representation of the equations of motion (12.8), with Hamiltonian functional

$$\mathcal{H}[\mathbf{X}] = f \int \mathrm{d}\mathbf{a} \left(\frac{1}{2}\big(X^2(\mathbf{a}) + Y^2(\mathbf{a})\big) - R\big(\mathbf{X}(\mathbf{a})\big) \right),$$

where $\mathbf{a}$ is a Lagrangian particle labelling coordinate. The standard canonical Poisson bracket is given by

$$\{\mathcal{F}, \mathcal{G}\}_c = \int \mathrm{d}\mathbf{a} \left(\frac{\delta \mathcal{F}}{\delta X(\mathbf{a})} \frac{\delta \mathcal{G}}{\delta Y(\mathbf{a})} - \frac{\delta \mathcal{F}}{\delta Y \mathbf{a})} \frac{\delta \mathcal{G}}{\delta X(\mathbf{a})} \right).$$

As for the rigid body bracket (5.6), refer to MARSDEN and RATIU [1999], OLVER [1986], ROULSTONE and NORBURY [1994] for further details of this bracket operation. It is possible to prove conservation of PV (equivalently pseudo-density) along trajectories by demonstrating that

$$\{q, \mathcal{H}\}_c = 0.$$

Again following ROULSTONE and NORBURY [1994] we may write the advection equation (12.11) in Hamiltonian form. Using the previous Hamiltonian, this time evaluated in phase space solely as a functional of ρ. That is,

$$\mathcal{H}[\rho] = f \int \mathrm{d}\mathbf{X}\, \rho(\mathbf{X}) \left(\frac{1}{2}(X^2 + Y^2) - R(\mathbf{X}) \right),$$

$$\frac{\delta \mathcal{H}}{\delta \rho} = f^2 \left(\frac{1}{2}(X^2 + Y^2) - R \right) \equiv \Psi.$$

We are now in a position to write the equations of motion (12.11) as

$$\frac{\partial \rho(\mathbf{X})}{\partial t} = \big\{\rho(\mathbf{X}), \mathcal{H}\big\}, \tag{12.12}$$

where the noncanonical Poisson bracket (see ROULSTONE and NORBURY [1994], MARSDEN and RATIU [1999], OLVER [1986]) is given by

$$\{\mathcal{F}, \mathcal{G}\} = \int_\Gamma \mathrm{d}\mathbf{X}\, \frac{\delta \mathcal{F}}{\delta \rho(\mathbf{X})} \left(\frac{\partial\big(\rho(\mathbf{X}), \frac{\delta \mathcal{G}}{\delta \rho(\mathbf{X})}\big)}{\partial(X, Y)} \right).$$

Note that in these coordinates, and with this bracket, the pseudo-density becomes a Casimir invariant for the system, $\{\rho, \mathcal{F}\} = 0$ for all functionals $\mathcal{F}(\rho)$, (cf. the quantity S in the rigid body problem).

The geometric integration problem of numerically integrating these equations whilst preserving the pseudo-density or potential vorticity along the flow turns out to be intimately related to preserving the Hamiltonian (or Poisson bracket) structure of the problem. See ROULSTONE and NORBURY [1994] for more details, and also MCLACHLAN [1993], MCLACHLAN, SZUNYOGH and ZEITLIN [1997] for some applications to similar problems where explicit methods capturing the Hamiltonian structure and the Casimir invariants are derived.

12.2.2. Links with moving mesh theory

We now take a closer look at the coordinate transformation from physical to geostrophic or dual coordinates, we also choose to sometimes use the term computational coordinates for X since these are the variables in which computing will be carried out. Recall from earlier we had

$$\mathbf{X} \equiv (X, Y, Z) = \left(x + \frac{v_g}{f}, y - \frac{u_g}{f}, \frac{g\theta}{f^2\theta_0}\right)^{\mathrm{T}},$$

and

$$q = \frac{\partial(X, Y, Z)}{\partial(x, y, z)} = \det\big(\mathrm{Hess}_x(P)\big), \tag{12.13}$$

we shall now show some links with the theory of moving mesh partial differential equations.

It is possible to write the continuous form of the first one-dimensional moving mesh partial differential equation (MMPDE) in the form (cf. (11.5))

$$\frac{\partial x}{\partial \xi} = \frac{1}{M},$$

where M is our monitor function. Notice the similarity with (12.13) if we take $M = q$, and identify the computational coordinates X and ξ. In particular in one-dimension we simply have that $M = P_{xz}$.

In three-dimensions the HUANG and RUSSELL [1999] approach to construct coordinate transformations ($\boldsymbol{\xi} = \boldsymbol{\xi}(\mathbf{x}, t)$) is to minimize the following integral

$$I[\boldsymbol{\xi}] = \frac{1}{2}\int \sum_{i=1}^{3} (\nabla\xi^i)^{\mathrm{T}} G_i^{-1} \nabla\xi^i \,\mathrm{d}\mathbf{x},$$

where the G_i are monitor functions, three by three symmetric positive definite matrices (concentrates mesh points in regions where G_i is large). The Euler–Lagrange equations for which are

$$-\frac{\delta I}{\delta \xi^i} = \nabla \cdot \big(G_i^{-1} \nabla \xi^i\big) = 0, \quad i = 1, 2, 3.$$

But now notice what happens when we take our monitor functions to be equal $G_i \equiv G$ and

$$G = \mathrm{Hess}_x(P).$$

For a start the determinant of our monitor function is simply the potential vorticity, and one possible solution to the Euler–Lagrange equations is $\boldsymbol{\xi} = \nabla_x P$, since then (using the symmetry of the Hessian if necessary),

$$G^{-1}\nabla\xi^i = \mathbf{e}_i, \quad i = 1, 2, 3; \qquad \mathbf{e}_1 = (1, 0, 0)^{\mathrm{T}}, \quad \text{etc.}$$

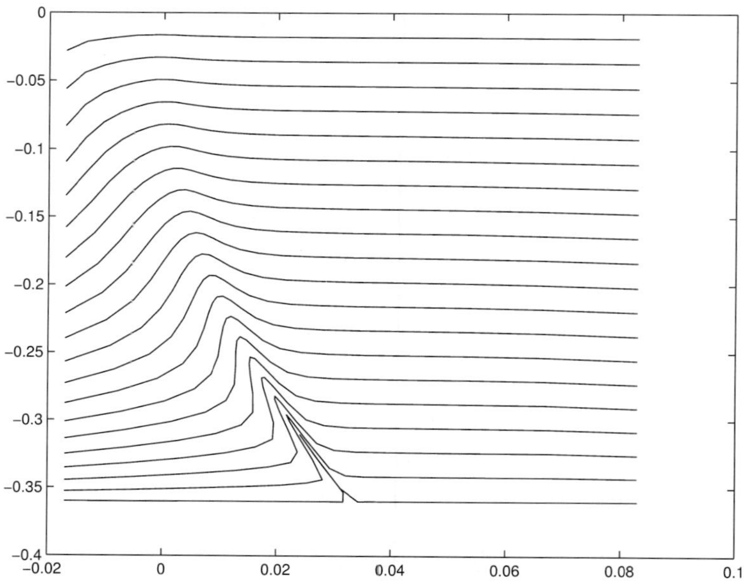

FIG. 12.1. The structure of the transformation obtained via the parabolic umbilic example.

We have thus shown a link between the usual analytical transformation found in the literature and the moving mesh adaptivity ideas discussed in previous sections. A key point to take on board from these is that the equations governing the mesh transformation should be solved to high order, i.e. a smooth mesh should be used. This contrasts with the piecewise constant mesh transformation discussed in CULLEN and PURSER [1984], as well as the geometric method (a numerical method based on the piecewise constant construction, see CHYNOWETH [1987]).

We hope to employ some of the geometric integration ideas discussed in this paper to the semi-geostrophic equations, with the aim of obtaining the superior results we observed for the nonlinear Schrödinger, and other equations.

We now show some results from a very early calculation based on the above results. Fig. 12.1 demonstrates the singular structure in the coordinate transformation (12.7) for an example discussed by CHYNOWETH, PORTER and SEWELL [1988] where the parabolic umbilic is used as a simple example of an atmospheric front. Fig. 12.2 shows a corresponding mesh which was computed using the potential vorticity as a monitor function.

12.2.3. Calculating Lyapunov exponents

Weather forecasting makes *prediction ensembles* to see how reliable a forecast is. This involves discretising the underlying partial differential equations in space to get a system of ordinary differential equations, then looking for the p largest modes of error growth of this system. A systematic way of finding such modes is to calculate Lyapunov exponents of the system.

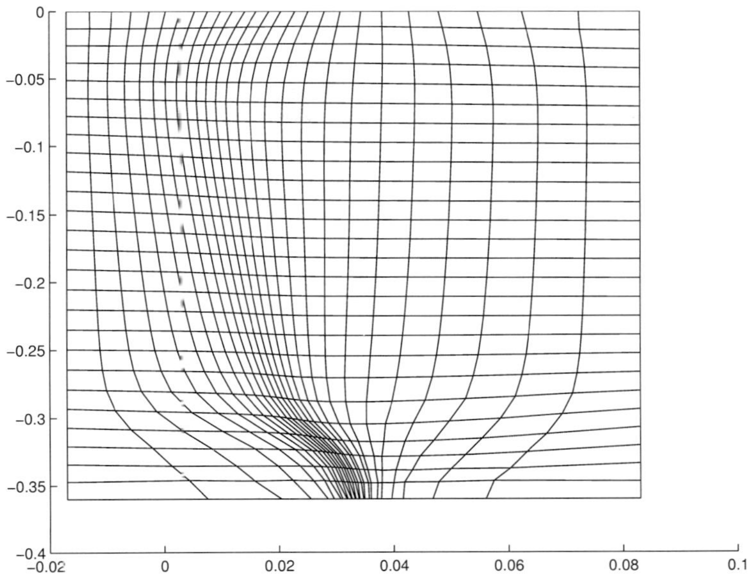

FIG. 12.2. The mesh computed for the parabolic umbilic example with potential vorticity as our monitor function.

The calculation of Lyapunov exponents is a natural candidate for a geometric integration based approach as it involves integrating matrix ordinary differential equations which require the preservation of structural properties of the matrices. An interesting algorithm to do this was proposed in DIECI, RUSSELL and VAN VLECK [1997] and involved integrating systems of equations defining orthogonal matrices. An alternative procedure (HAUSER [2000]) is to solve the system

$$\dot{Y} = A(t)Y, \quad Y \in M^{n\times p}$$

and to find a singular value decomposition (see TREFETHEN and BAU III [1997]) of Y using

$$Y = U(t)S(t)V(t),$$

where U is an $n \times p$ matrix with orthogonal columns, V is a $p \times p$ matrix with orthogonal columns and S is a diagonal $p \times p$ matrix of singular values. The growth rates of the dominant modes can then be calculated via the Lyapunov exponents

$$\lambda_i = \lim_{t\to\infty}\left[\frac{1}{t}\log S_{ii}(t)\right].$$

Essential to this calculation is determining the matrices U and V. An efficient way to proceed is to determine ordinary differential equations for U and V which typically take the form

$$\dot{U} = AU + U\big(H - U^{\mathrm{T}}AU\big), \tag{12.14}$$

where H is a derived $p \times p$ matrix and U the $n \times p$ matrix with orthogonal columns. Eq. (12.14) is a naturally arising evolutionary system in which keeping the columns of U orthogonal is important for the accuracy of the estimates of the Lyapunov exponents. Conventional integrators will not do this easily, however this structure is amenable to methods based on the Lie group invariance of such systems (ISERLES, MUNTHE-KAAS, NØRSETT and ZANNA [2000]). The geometric integration approach is to decompose U as a product of Householder reflections (TREFETHEN and BAU III [1997]) and solve the ordinary differential equations that define the reflections. The advantage of using such a procedure for (12.14) is that it has very efficient estimates (exponential convergence) of subspaces to spaces spanned by the fastest growing modes, although great care needs to be taken when singular values coalesce.

13. Conclusion

We have demonstrated in this review that the consideration of qualitative properties in ordinary and partial differential equations is important when designing numerical schemes and we have given some examples of general methods and techniques which may be employed to capture geometric features in discretisations of continuous problems. Traditionally numerical analysts have not necessarily followed this philosophy, they have tended to prefer the black box approach as described in the left hand side of the diagram in Section 1. This approach has led to excellent results in many situations and the numerical analyst has been able to avoid looking too far into the underlying structure of their problems. In this brief article on a subject which is growing and evolving at a fast rate we have hopefully shown using some theory as well as examples the possible advantages which the geometric integration approach to numerical analysis may yield. We have looked at methods applied to both ordinary and partial differential equation problems with Hamiltonian structure, singularity formation, and evolution on manifolds, as well as discussing the possibility of applying these ideas to some problems arising in meteorology and numerical weather prediction – although work on this area is ongoing. In many examples geometric integration has shown that a consideration of the underlying structure of a problem may well be worthwhile since this opens up the possibility of discretising the problem in such a way that the structure is not totally destroyed in the discrete system, this has led to robust methods which are both more accurate and stable for classes of problems. To follow the geometric integration philosophy numerical analysts will therefore need to be conversant with both current results and any new developments in the qualitative theory of differential equations.

Geometric integration is still a relatively new subject area and much work has yet to be done, and with every new development in the underlying theory of differential equations this task will grow. In this review we have shown that it is possible to incorporate qualitative features of many diverse types and for many different problems into numerical algorithms. However a study of the feasibility and practicality of such methods for a wide range of industrial sized real world problems is required. With our brief discussion of some possible meteorological applications we have shown that there is some hope in this direction, but also that there is the possibility of a real two way exchange of ideas between numerical analysis and application areas, where techniques used for specific problems may be generalised to a wider class of problems. Geometric integration also allows us the chance of thinking about old and popular methods in a new light, maybe to improve or generalise, or maybe to help explain behaviour and performance not done so before, or even to help us discover behaviour previously not noticed.

Acknowledgements

Much of this work was completed under the financial support of the EPSRC Computational Partial Differential Equations grant GR/M30975. We are especially grateful to Reinout Quispel, J.F. Williams and Sergio Blanes for reading and commenting on earlier versions of this article.

References

ABIA, L., SANZ-SERNA, J.M. (1993). Partitioned Runge–Kutta methods for separable Hamiltonian problems. *Math. Comput.* **60**, 617–634.

ABLOWITZ, M., HERBST, B.M. (1990). On homoclinic structure and numerically induced chaos for the nonlinear Schrödinger equation. *SIAM J. Appl. Math.* **50**, 339–351.

ALLEN, M.P., TILDESLEY, D.J. (1987). *Computer Simulations of Liquids* (Clarendon Press, Oxford).

AMES, W.F., ANDERSON, R.L., DORODNITSYN, V.A., FERAPONTOV, E.V., GAZIZOV, R.K., IBRAGIMOV, N.H., SVIRSHEVSKII, S.R. (1994). *CRC Handbook of Lie Group Analysis of Differential Equations, Vol. 1, Exact Solutions and Conservation Laws* (CRC Press).

ARAKAWA, A. (1966). Computational design for long-term numerical integration of the equations of fluid motion: two-dimensional incompressible flow. *J. Comput. Phys.* **1**, 119–143.

ARNOLD, V.I. (1978). *Mathematical Methods of Classical Mechanics* (Springer-Verlag, New York).

ARNOLD, V.I., KHESIN, B.A. (1998). *Topological Methods in Hydrodynamics* (Springer-Verlag, New York).

AUBRY, A., CHARTIER, P. (1998). Pseudo-symplectic Runge–Kutta methods. *BIT* **38**, 439–461.

AUSTIN, M.A., KRISHNAPRASAD, P.S., WANG, L.-S. (1993). Almost Poisson integration of rigid body systems. *J. Comput. Phys.* **107**, 105–117.

BARENBLATT, G.I. (1996). *Scaling, Self-Similarity and Intermediate Asymptotics* (Cambridge Univ. Press).

BARTH, E., LEIMKUHLER, B. (1996). Symplectic methods for conservative multibody systems. *Comm. Fields Inst.* **10**.

BENETTIN, G., GIORGILLI, A. (1994). On the Hamiltonian interpolation of near-to-the-identity symplectic mappings with applications to symplectic integration algorithms. *J. Stat. Phys.* **74**, 1117–1143.

BEYN, W.J. (1987). On invariant closed curves for one-step methods. *Numer. Math.* **51**, 103–122.

BLANES, S., CASAS, F., ROS, J. (2002). High order optimized geometric integrators for linear differential equations. *BIT* **42** (2), 262–284.

BLANES, S., MOAN, P.C. (2002). Practical symplectic partitioned Runge–Kutta and Runge–Kutta–Nyström methods. *J. Comput. Appl. Math.* **142**, 313–330.

BOCHEV, P.B., SCOVEL, C. (1994). On quadratic invariants and symplectic structure. *BIT* **34**, 337–345.

BOND, S.D., LEIMKUHLER, B.J. (1998). Time-transformations for reversible variable stepsize integration. *Numer. Algorithms* **19**, 55–71.

BRIDGES, T.J. (1997). Multi-symplectic structures and wave propagation. *Math. Proc. Cambridge Philos. Soc.* **121**, 147–190.

BRIDGES, T.J., REICH, S. (2001). Multi-symplectic integrators: numerical schemes for Hamiltonian PDEs that conserve symplecticity. *Phys. Lett. A* **284**, 184–193.

BROOMHEAD, D.S., ISERLES, A. (eds.) (1992). *The Dynamics of Numerics and the Numerics of Dynamics* (Clarendon Press, Oxford).

BUDD, C.J. (2001). Asymptotics of new blow-up self-similar solutions of the nonlinear Schrödinger equation. *SIAM J. Appl. Math.* **62**, 801–830.

BUDD, C.J., CHEN, S., RUSSELL, R.D. (1999). New self-similar solutions of the nonlinear Schrödinger equation with moving mesh computations. *J. Comput. Phys.* **152**, 756–789.

BUDD, C.J., COLLINS, G.J., HUANG, W.-Z., RUSSELL, R.D. (1999). Self-similar discrete solutions of the porous medium equation. *Philos. Trans. Roy. Soc. London A* **357**, 1047–1078.

BUDD, C.J., DORODNITSYN, V.A. (2001). Symmetry adapted moving mesh schemes for the nonlinear Schrödinger equation. *J. Phys. A* **34**, 10 387–10 400.

BUDD, C.J., ISERLES, A. (Eds.) (1999). Geometric integration: numerical solution of differential equations on manifolds. *Philos. Trans. Roy. Soc. London A* **357**, 943–1133.

BUDD, C.J., LEIMKUHLER, B., PIGGOTT, M.D. (2001). Scaling invariance and adaptivity. *Appl. Numer. Math.* **39**, 261–288.

BUDD, C.J., PIGGOTT, M.D. (2001). The geometric integration of scale invariant ordinary and partial differential equations. *J. Comput. Appl. Math.* **128**, 399–422.

BUSS, S.R. (2000). Accurate and efficient simulations of rigid-body rotations. *J. Comput. Phys.* **164**, 377–406.

CALVO, M.P., LÓPEZ-MARCOS, M.A., SANZ-SERNA, J.M. (1998). Variable step implementation of geometric integrators. *Appl. Numer. Math.* **28**, 1–16.

CANDY, J., ROZMUS, W. (1991). A symplectic integration algorithm for separable Hamiltonian functions. *J. Comput. Phys.* **92**, 230–256.

CELLEDONI, E., ISERLES, A. (2000). Approximating the exponential from a Lie algebra to a Lie group. *Math. Comp.* **69**, 1457–1480.

CHANNEL, P.J., SCOVEL, C. (1990). Symplectic integration of Hamiltonian systems. *Nonlinearity* **3**, 231–259.

CHAPMAN, S.J., KING, J.R., ADAMS, K.L. (1998). Exponential asymptotics and Stokes lines in nonlinear ordinary differential equations. *Proc. Roy. Soc. London A* **454**, 2733–2755.

CHEN, Y.G. (1986). Asymptotic behaviours of blowing-up solutions for finite difference analogue of $u_t = u_{xx} + u^{1+\alpha}$. *J. Fac. Sci. Univ. Tokyo Sect. IA Math.* **33**, 541–574.

CHYNOWETH, S. (1987). The semi-geostrophic equations and the Legendre transform. Ph.D thesis, University of Reading, UK.

CHYNOWETH, S., PORTER, D., SEWELL, M.J. (1988). The parabolic umbilic and atmospheric fronts. *Proc. Roy. Soc. London A* **419**, 337–362.

CIRILLI, S., HAIRER, E., LEIMKUHLER, B. (1999). Asymptotic error analysis of the adaptive Verlet method. *BIT* **39**, 25–33.

CULLEN, M.J.P., NORBURY, J., PURSER, R.J. (1991). Generalised Lagrangian solutions for atmospheric and oceanic flows. *SIAM J. Appl. Math.* **51**, 20–31.

CULLEN, M.J.P., PURSER, R.J. (1984). An extended Lagrangian theory of semi-geostrophic frontogenesis. *J. Atmos. Sci.* **41**, 1477–1497.

CULLEN, M.J.P., PURSER, R.J. (1987). Properties of the Lagrangian semi-geostrophic equations. *J. Atmos. Sci.* **46**, 2684–2697.

CULLEN, M., SALMOND, D., SMOLARKIEWICZ, P. (2000). Key numerical issues for the development of the ECMWF model. In: *Proceedings of ECMWF Workshop on Developments in Numerical Methods for Very High Resolution Global Models.*

DIECI, L., RUSSELL, R.D., VAN VLECK, E.S. (1997). On the computation of Lyapunov exponents for continuous dynamical systems. *SIAM J. Numer. Anal.* **34** (1), 402–423.

DORODNITSYN, V.A. (1991). Transformation groups in mesh spaces. *J. Sov. Math.* **55**, 1490–1517.

DORODNITSYN, V.A. (1993a). Finite-difference models exactly inheriting symmetry of original differential equations. In: Ibragimov, N.H , et al. (eds.), *Modern Group Analysis: Advanced Analytical and Computational Methods in Mathematical Physics* (Kluwer, Dordrecht), pp. 191–201.

DORODNITSYN, V.A. (1993b). Finite difference analog of the Noether theorem. *Dokl. Akad. Nauk* **328**, 678.

DORODNITSYN, V.A. (1998). Noether-type theorems for difference equations. IHES/M/98/27, Bures-sur-Yvette, France.

DORODNITSYN, V.A., KOZLOV, R. (1997). The whole set of symmetry preserving discrete versions of a heat transfer equation with a source. Preprint 4/1997, NTNU, Trondheim, Norway.

DRAZIN, P.G., JOHNSON, R.S. (1989). *Solitons: An Introduction* (Cambridge Univ. Press).

DURRAN, D. (1999). *Numerical Methods for Wave Equations in Geophysical Fluid Dynamics* (Springer-Verlag, New York).

EARN, D.J.D., TREMAINE, S. (1992). Exact numerical studies of Hamiltonian maps: Iterating without roundoff error. *Physica D* **56**, 1–22.

ENGØ, K., FALTINSEN, S. (2001). Numerical integration of Lie–Poisson systems while preserving coadjoint orbits and energy. *SIAM J. Numer. Anal.* **39**, 128–145.

ENGØ, K., MARTHINSEN, A., MUNTHE-KAAS, H.Z. (1999). DiffMan: an object oriented MATLAB toolbox for solving differential equations on manifolds. Technical Report No. 164, Dept. of Informatics, University of Bergen.

FRAENKEL, L.E. (2000). *An Introduction to Maximum Principles and Symmetry in Elliptic Problems* (Cambridge Univ. Press).

DE FRUTOS, J., SANZ-SERNA, J.M. (1997). Accuracy and conservation properties in numerical integration: the case of the Korteweg–de Vries equation. *Numer. Math.* **75**, 421–445.

FURIHATA, D. (1999). Finite difference schemes for $\partial u/\partial t = (\partial/\partial x)^{\alpha}\delta G/\delta u$ that inherit energy conservation or dissipation property. *J. Comput. Phys.* **156**, 181–205.

FURIHATA, D. (2001a). A stable and conservative finite difference scheme for the Cahn–Hilliard equation. *Numer. Math.* **87**, 675–699.

FURIHATA, D. (2001b). Finite-difference schemes for nonlinear wave equation that inherit energy conservation property. *J. Comput. Appl. Math.* **134**, 37–57.

GE, Z., MARSDEN, J.E. (1988). Lie–Poisson Hamilton–Jacobi theory and Lie–Poisson integrators. *Phys. Lett. A* **133**, 134–139.

GODA, K. (1975). On the instability of some finite difference schemes for the Korteweg–de Vries equation. *J. Phys. Soc. Japan* **39**, 229–236.

GOLDSTEIN, H. (1980). *Classical Mechanics*, 2nd edn. (Addison-Wesley, Reading, MA).

GOLUBITSKY, M., SCHAEFFER, STEWART, I. (1988). *Singularities and Groups in Bifurcation Theory* (Springer, Berlin).

GRIEG, I.S., MORRIS, J.L. (1976). A hopscotch method for the Korteweg–de Vries equation. *J. Comput. Phys.* **20**, 64–80.

GRIFFITHS, D.F., SANZ-SERNA, J.M. (1986). On the scope of the method of modified equations. *SIAM J. Sci. Stat. Comput.* **7**, 994–1008.

GRINDROD, P. (1991). *Patterns and Waves* (Oxford University Press, New York).

HAIRER, E. (1997). Variable time step integration with symplectic methods. *Appl. Numer. Math.* **25**, 219–227.

HAIRER, E. (2000). Geometric integration of ordinary differential equations on manifolds. *BIT*.

HAIRER, E., LUBICH, CH. (1997). The life-span of backward error analysis for numerical integrators. *Numer. Math.* **76**, 441–462.

HAIRER, E., LUBICH, CH. (2000). Asymptotic expansions and backward analysis for numerical integrators. In: de la Llave, R., Petzold, L.R., Lorenz, J. (eds.), *Dynamics of Algorithms*. In: IMA Volumes in Mathematics and its Applications **118**, pp. 91–106.

HAIRER, E., LUBICH, C., WANNER, G. (2002). *Geometric Numerical Integration*, Springer Series in Computational Mathematics **31** (Springer-Verlag, Heidelberg).

HAIRER, E., NØRSETT, S.P., WANNER, G. (1993). *Solving Ordinary Differential Equations. I*, 2nd edn., Springer Series in Computational Mathematics **8** (Springer-Verlag, Berlin).

HAIRER, E., STOFFER, D. (1997). Reversible long-term integration with variable stepsizes. *SIAM J. Sci. Comput.* **18**, 257–269.

HAUSER, R. (2000). Private communication.

HERMAN, R., KNICKERBOCKER, C. (1993). Numerically induced phase shift in the KdV soliton. *J. Comput. Phys.* **104**, 50–55.

HOLDER, T., LEIMKUHLER, B., REICH, S. (2001). Explicit variable step-size and time-reversible integration. *Appl. Numer. Math.* **39**, 367–377.

HOSKINS, B.J. (1975). The geostrophic momentum approximation and the semi-geostrophic equations. *J. Atmos. Sci.* **32**, 233–242.

HUANG, W., LEIMKUHLER, B. (1997). The adaptive Verlet method. *SIAM J. Sci. Comput.* **18**, 239–256.

HUANG, W., REN, Y., RUSSELL, R.D. (1994). Moving mesh partial differential equations (MMPDES) based on the equidistributional principle. *SIAM J. Numer. Anal.* **31**, 709–730.

HUANG, W., RUSSELL, R.D. (1999). Moving mesh strategy based on a gradient flow equation for two-dimensional problems. *SIAM J. Sci. Comput.* **20**, 998–1015.

ISERLES, A. (1999). On Cayley-transform methods for the discretization of Lie-group equations. Tech. Report 1999/NA4, DAMTP, University of Cambridge.

ISERLES, A., MUNTHE-KAAS, H.Z., NØRSETT, S.P., ZANNA, A. (2000). Lie-group methods. *Acta Numerica*, 215–365.

ISERLES, A., NORSETT, S. (1999). On the solution of linear differential equations in Lie groups. *Philos. Trans. Roy. Soc. London A* **357**, 983–1019.

ITOH, T., ABE, K. (1988). Hamiltonian-conserving discrete canonical equations based on variational difference quotients. *J. Comput. Phys.* **77**, 85–102.

KANG, F. (1985). On difference schemes and symplectic geometry. In: *Proc. 1984 Beijing Symp. Diff. Geometry and Diff. Equations* (Science Press, Beijing), pp. 42–58.

KANE, C., MARSDEN, J.E., ORTIZ, M. (1999). Symplectic-energy-momentum preserving variational integrators. *J. Math. Phys.* **40**, 3353–3371.

KOZLOV, R. (2000). Symmetry applications to difference and differential-difference equations. PhD Thesis, Institut for matematiske fag, NTNU Trondheim.

LAMB, J.S.W., ROBERTS, J.A.G. (1998). Time-reversal symmetry in dynamical systems: a survey. *Physica D* **112**, 1–39.

LASAGNI, F.M. (1988). Canonical Runge–Kutta methods. *Z. Angew. Math. Phys.* **39**, 952–953.

LEIMKUHLER, B. (1999). Reversible adaptive regularization: perturbed Kepler motion and classical atomic trajectories. *Philos. Trans. Roy. Soc. London A* **357**, 1101–1133.

LE VEQUE, R.J., YEE, H.C. (1990). A study of numerical methods for hyperbolic conservation laws with stiff source terms. *J. Comput. Phys.* **86**, 187–210.

LI, P.W. (1995). On the numerical study of the KdV equation by the semi-implicit and leap-frog methods. *Comput. Phys. Comm.* **88**, 121–127.

LI, S. (1995). Finite difference calculus invariant structure of a class of algorithms for the nonlinear Klein–Gordon equation. *SIAM J. Numer. Anal.* **32**, 1839–1875.

LICHTENBERG, A.J., LIEBERMAN, M.A. (1983). *Regular and Stochastic Motion* (Springer-Verlag, New York).

MACKAY, R.S. (1992). Some aspects of the dynamics and numerics of Hamiltonian systems. In: Broomhead, D., Iserles, A. (eds.), *The Dynamics of Numerics and the Numerics of Dynamics* (Clarendon Press, Oxford), pp. 137–193.

MAGNUS, W. (1954). On the exponential solution of differential equations for a linear operator. *Comm. Pure Appl. Math.* **VII**, 649–673.

MARSDEN, J.E., RATIU, T.S. (1999). *Introduction to Mechanics and Symmetry*, 2nd edn. (Springer-Verlag, Berlin).

MARSDEN, J.E., WEST, M. (2001). Discrete mechanics and variational integrators. *Acta Numerica*, 357–514.

MATSUO, T., FURIHATA, D. (2001). Dissipative and conservative finite difference schemes for complex-valued nonlinear partial differential equations. *J. Comput. Phys.* **171**, 425–447.

MATSUO, T., SUGIHARA, M., FURIHATA, D., MORI, M. (2001). Spatially accurate dissipative or conservative finite difference schemes derived by the discrete variational method. *Japan J. Industr. Appl. Math.*, to appear.

MCLACHLAN, R.I. (1993). Explicit Lie–Poisson integration and the Euler equations. *Phys. Rev. Lett.* **71**, 3043–3046.

MCLACHLAN, R.I. (1994). Symplectic integration of Hamiltonian wave equations. *Numer. Math.* **66**, 465–492.

MCLACHLAN, R.I. (1995). On the numerical integration of ordinary differential equations by symmetric composition methods. *SIAM J. Sci. Comp.* **16**, 151–168.

MCLACHLAN, R.I. (to appear). Spatial discretization of partial differential equations with integrals. Submitted to SIAM J. Numer. Anal.

MCLACHLAN, R.I., ATELA, P. (1992). The accuracy of symplectic integrators. *Nonlinearity* **5**, 541–562.

MCLACHLAN, R.I., QUISPEL, G.R.W. (2001). Six lectures on the geometric integration of ODEs. In: DeVore, R., Iserles, A., Süli, E. (eds.), *Foundations of Computational Mathematics*. In: London Math. Soc. Lecture Note Ser. **284**, pp. 155–210.

MCLACHLAN, R.I., QUISPEL, G.R.W., ROBODOUX, N. (1999). Geometric integration using discrete gradients. *Philos. Trans. Roy. Soc. London A* **357**, 1021–1045.

MCLACHLAN, R.I., QUISPEL, G.R.W., TURNER, G.S. (1998). Numerical integrators that preserve symmetries and reversing symmetries. *SIAM J. Numer. Anal.* **35**, 586–599.

McLachlan, R., Szunyogh, I., Zeitlin, V. (1997). Hamiltonian finite-dimensional models of baroclinic instability. *Phys. Lett. A* **229**, 299–305.

Moler, C., Van Loan, C. (1978). Nineteen dubious ways to compute the exponential of a matrix. *SIAM Rev.* **20**, 801–836.

Munthe-Kaas, H. (1998). Runge–Kutta methods on Lie groups. *BIT* **38**, 92–111.

Munthe-Kaas, H., Owren, B. (1999). Computations in a free Lie algebra. *Philos. Trans. Roy. Soc. London A* **357**, 957–981.

Munthe-Kaas, H., Zanna, A. (1997). Numerical integration of differential equations on homogeneous manifolds. In: Cucker, F., Shub, M. (eds.), *Foundations of Computational Mathematics* (Springer, Berlin), pp. 305–315.

Murray, N., Holman, M. (1999). The origin of chaos in the outer solar system. *Science* **283**, 1877–1881.

Murua, A., Sanz-Serna, J.M. (1999). Order conditions for numerical integrators obtained by composing simpler integrators. *Philos. Trans. Roy. Soc. London A* **357**, 1079–1100.

Okunbor, D., Skeel, R.D. (1992). An explicit Runge–Kutta–Nyström method is canonical if and only if its adjoint is explicit. *SIAM J. Numer. Anal.* **29** (2), 521–527.

Olver, P.J. (1986). *Applications of Lie Groups to Differential Equations* (Springer, New York).

Olver, P.J. (2001). Moving frames – in geometry, algebra, computer vision and numerical analysis. In: DeVore, R., Iserles, A., Süli, E. (eds.), *Foundations of Computational Mathematics*. In: London Math. Soc. Lecture Note Ser. **284**, pp. 267–297.

Pen-Yu, K., Sanz-Serna, J.M. (1981). Convergence of methods for the numerical solution of the Korteweg–de Vries equation. *IMA J. Numer. Anal.* **1**, 215–221.

Reich, S. (1999). Backward error analysis for numerical integrators. *SIAM J. Numer. Anal.* **36**, 1549–1570.

Reich, S. (2000). Multi-symplectic Runge–Kutta collocation methods for Hamiltonian wave equations. *J. Comput. Phys.* **157**, 473–499.

Roulstone, I., Norbury, J. (1994). A Hamiltonian structure with contact geometry for the semi-geostrophic equations. *J. Fluid Mech.* **272**, 211–233.

Ruth, R.D. (1983). A canonical integration technique. *IEEE Trans. Nucl. Sci.* **30**, 2669–2671.

Samarskii, A., Galaktionov, V., Kurdyumov, S., Mikhailov, A. (1995). *Blow-up in Quasilinear Parabolic Equations* (Walter de Gruyter, Berlin).

Sanz-Serna, J.M. (1982). An explicit finite-difference scheme with exact conservation properties. *J. Comput. Phys.* **47**, 199–210.

Sanz-Serna, J.M. (1988). Runge–Kutta schemes for Hamiltonian systems. *BIT* **28**, 877–883.

Sanz-Serna, J.M. (1997). Geometric Integration. In: Duff, I.S., Watson, G.A. (eds.), *The State of the Art in Numerical Analysis* (Clarendon, Oxford), pp. 121–143.

Sanz-Serna, J.M., Calvo, M.P. (1994). *Numerical Hamiltonian Problems* (Chapman and Hall, London).

Sanz-Serna, J.M., Portillo, A. (1996). Classical numerical integrators for wave-packet dynamics. *J. Chem. Phys.* **104**, 2349–2355.

Schlier, Ch., Seiter, A. (1998). Symplectic integration and classical trajectories: a case study. *J. Chem. Phys. A* **102** (102), 9399–9404.

Skeel, R.D., Gear, C.W. (1992). Does variable step size ruin a symplectic integrator?. *Physica D* **60**, 311–313.

Stoffer, D.M. (1988). Some geometric and numerical methods for perturbed integrable systems. Ph.D. Thesis, ETH, Zürich.

Stoffer, D.M. (1995). Variable steps for reversible integration methods. *Computing* **55**, 1–22.

Stoffer, D., Nipp, K. (1991). Invariant curves for variable step size integrators. *BIT* **31**, 169–180.

Stuart, A.M., Humphries, A.R. (1996). *Dynamical Systems and Numerical Analysis* (Cambridge Univ. Press).

Sulem, C., Sulem, P.-L. (1999). *The Nonlinear Schrödinger Equation*, Applied Mathematical Sciences **139** (Springer, Berlin).

Suris, Y.B. (1989). The canonicity of mappings generated by Runge–Kutta type methods when integrating the system $\ddot{x} = -\partial U/\partial x$. *USSR Comput. Math. Phys.* **29**, 138–144.

Sussman, G.J., Wisdom, J. (1992). Chaotic evolution of the solar system. *Science* **257**, 56–62.

Taha, T.R., Ablowitz, M.J. (1984). Analytical and numerical aspects of certain nonlinear evolution equations. III. Numerical, Korteweg–de Vries equation. *J. Comput. Phys.* **55**, 231–253.

TREFETHEN, L.N., BAU, D. III (1997). *Numerical Linear Algebra* (SIAM, Philadelphia, PA).

VARADARAJAN, V.S. (1974). *Lie Groups, Lie Algebras, and their Representations* (Prentice-Hall, Englewood Cliffs, NJ).

VERLET, L. (1967). Computer "experiments" on classical fluids. I. Thermodynamic properties of Lennard-Jones molecules. *Phys. Rev.* **159**, 98–103.

DE VOGELAERE, R. (1956). Methods of integration which preserve the contact transformation property of Hamiltonian equations. Tech. report No. 4, Dept. Math., Univ. Notre Dame.

WISDOM, J., HOLMAN, M. (1991). Symplectic maps for the N-body problem. *Astron. J.* **102**, 1528–1538.

YOSHIDA, H. (1990). Construction of higher order symplectic integrators. *Phys. Lett. A* **150**, 262–268.

YOSHIDA, H. (1993). Recent progress in the theory and application of symplectic integrators. *Celestial Mechanics and Dynamical Astronomy* **56**, 27–43.

ZABUSKY, N.J., KRUSKAL, M.D. (1965). Interaction of 'solitons' in a collisionless plasma and the recurrence of initial states. *Phys. Rev. Lett.* **15**, 240–243.

ZANNA, A. (1999). Collocation and relaxed collocation for the Fer and the Magnus expansions. *SIAM J. Numer. Anal.* **36**, 1145–1182.

ZEITLIN, V. (1991). Finite-mode analogues of 2D ideal hydrodynamics: Coadjoint orbits and local canonical structure. *Physica D* **49**, 353–362.

Linear Programming and Condition Numbers under the Real Number Computation Model

Dennis Cheung, Felipe Cucker

City University of Hong Kong, Department of Mathematics, 83 Tat Chee Avenue, Kowloon, Hong Kong
e-mail: macucker@sobolev.cityu.edu.hk

Yinyu Ye

The University of Iowa, College of Business Administration, Department of Management Sciences, S384 Pappajohn Building, Iowa City, IA 52242-1000, USA
e-mail: yyye@dollar.biz.uiowa.edu

1. Introduction

1.1. A few words on linear programming

Linear programming, hereafter LP, plays a distinguished role in complexity theory. In one sense it is a continuous optimization problem since the goal is to minimize a linear objective function over a convex polyhedron. But it is also a combinatorial problem involving selecting an extreme point among a finite set of possible vertices.

Linear programming is also widely applied. Businesses, large and small, use linear programming models to optimize communication systems, to schedule transportation networks, to control inventories, to plan investments, and to maximize productivity.

Foundations of Computational Mathematics
Special Volume (F. Cucker, Guest Editor) of
HANDBOOK OF NUMERICAL ANALYSIS, VOL. XI
P.G. Ciarlet (Editor)

A set of linear inequalities defines a polyhedron, properties of which have been studied by mathematicians for centuries. Ancient Chinese and Greeks calculated volumes of simple polyhedra in three-dimensional space. Fourier's fundamental research connecting optimization and inequalities dates back to the early 1800s. At the end of 19th century, Farkas and Minkowski began basic work on algebraic aspects of linear inequalities. In 1910 De La Vallée Poussin developed an algebraic technique for minimizing the infinity-norm of $b - Ax$ that can be viewed as a precursor of the simplex method. Beginning in the 1930s, such notable mathematicians as von Neumann, Kantorovich, and Koopmans studied mathematical economics based on linear inequalities. During World War II, it was observed that decisions involving the best movement of personnel and optimal allocation of resources could be posed and solved as linear programs. Linear programming began to assume its current popularity.

An optimal solution of a linear program always lies at a vertex of the feasible region, which itself is a polyhedron. Unfortunately, the number of vertices associated with a set of n inequalities in m variables can be exponential in the dimensions – in this case, up to $n!/m!(n-m)!$. Except for small values of m and n, this number is sufficiently large to prevent examining all possible vertices for searching an optimal vertex.

The simplex method, invented in the mid-1940s by George Dantzig, is a procedure for examining optimal candidate vertices in an intelligent fashion. It constructs a sequence of adjacent vertices with improving values of the objective function. Thus, the method travels along edges of the polyhedron until it hits an optimal vertex. Improved in various way in the intervening four decades, the simplex method continues to be the workhorse algorithm for solving linear programming problems.

Although it performs well in practice, the simplex method will examine every vertex when applied to certain linear programs. Klee and Minty in 1972 gave such an example. These examples confirm that, in the worst case, the simplex method needs an exponential number of iterations to find the optimal solution. As interest in complexity theory grew, many researchers believed that a good algorithm should be polynomial – i.e. broadly speaking, the running time required to compute the solution should be bounded above by a polynomial in the "size", or the total data length, of the problem. Thus, the simplex method is not a polynomial algorithm.[1]

In 1979, a new approach to linear programming, Khachijan's ellipsoid method, received dramatic and widespread coverage in the international press. Khachijan proved that the ellipsoid method, developed during the 1970s by other mathematicians, is a polynomial algorithm for linear programming under a certain computational model. It constructs a sequence of shrinking ellipsoids with two properties: the current ellipsoid always contains the optimal solution set, and each member of the sequence undergoes a guaranteed reduction in volume, so that the solution set is squeezed more tightly at each iteration.

The ellipsoid method was studied intensively by practitioners as well as theoreticians. Based on the expectation that a polynomial linear programming algorithm would be faster than the simplex method, it was a great disappointment that, in practice, the best implementations of the ellipsoid method were not even close to being competitive.

[1] We will be more precise about complexity notions such as "polynomial algorithm" in Section 1.2 below.

Thus, after the dust eventually settled, the prevalent view among linear programming researchers was that Khachijan had answered a major open question on the polynomiality of solving linear programs, but the simplex method remained the clear winner in practice.

This contradiction, the fact that an algorithm with the desirable theoretical property of polynomiality might nonetheless compare unfavorably with the (worst-case exponential) simplex method, set the stage for exciting new developments. It was no wonder, then, that the announcement by KARMARKAR [1984] of a new polynomial time algorithm, an interior-point method, with the potential to dramatically improve the practical effectiveness of the simplex method made front-page news in major newspapers and magazines throughout the world.

Interior-point algorithms are continuous iterative algorithms. Computational experience with sophisticated procedures suggests that the number of necessary iterations grows very slowly with the problem size. This provides the potential for dramatic improvements in computation effectiveness. One of the goals of this article is to provide some understanding on the complexity theoretic properties of interior-point algorithms.

1.2. A few words on complexity theory

Complexity theory is arguably the foundational stone of computer algorithms. The goal of the theory is twofold: to develop criteria for measuring the effectiveness of various algorithms (and thus, to be able to compare algorithms using these criteria), and to assess the inherent difficulty of various problems.

The term "complexity" refers to the amount of resources required by a computation. In this article we will focus on a particular resource namely, the computing time. In complexity theory, however, one is not interested on the execution time of a program implemented in a particular programming language, running on a particular computer over a particular input. There are too many contingent factors here. Instead, one would like to associate to an algorithm some more intrinsic measures of its time requirements.

Roughly speaking, to do so one needs to fix:

- a notion of *input size*,
- a set of *basic operations*, and
- a *cost* for each basic operation.

The last two allow one to define the *cost of a computation*. If x is any input, the cost $C(x)$ of the computation with input x is the addition of the costs of all the basic operations performed during this computation.

The intrinsic measures mentioned above will actually take the form of functions with domain $\mathbb{N}$ and range in $\mathbb{R}^+$. Let $\mathcal{A}$ be an algorithm and $\mathcal{I}_n$ be the set of all its inputs having size n. The *worst-case cost function* of $\mathcal{A}$ is the function $T^w_{\mathcal{A}}$ defined by

$$T^w_{\mathcal{A}}(n) = \sup_{x \in \mathcal{I}_n} C(x).$$

If $\mathcal{I}_n$ is endowed with a probability measure for each $n \in \mathbb{N}$ and $\mathbf{E}_n$ denotes expectation with respect to this measure then we may consider the *average-case cost function* $T^a_{\mathcal{A}}$

given by

$$T^a_{\mathcal{A}}(n) = \mathbf{E}_n\big(C(x)\big).$$

If $\mathcal{A}$ and $\mathcal{B}$ are two algorithms solving the same computational problem, a way of comparing their time requirements is by comparing $T^w_{\mathcal{A}}$ and $T^w_{\mathcal{B}}$ (or their corresponding T^a) for large values of n. This allows to compare the intrinsic time requirements of $\mathcal{A}$ and $\mathcal{B}$ and is independent of the contingencies mentioned above.

The use of T^a instead of T^w seems a more "realistic" choice and most of the times is. However, estimates for average complexity are more difficult to obtain than those for worst-case complexity. In addition, it is not clear which probability measure better reflects "reality".

Let us now discuss how the objects in the three items above are selected. The selection of a set of basic operations is generally easy. For the algorithms we will consider in this article, the obvious choice is the set $\{+, -, \times, /, \leqslant\}$ of the four arithmetic operations and the comparison. Selecting a notion of input size and a cost for the basic operations is more delicate and depends on the kind of data dealt with by the algorithm. Some kinds can be represented within a fixed amount of computer memory, some others require a variable amount depending on the data at hand.

Examples of the first are fixed-precision floating-point numbers. Any such number is stored in a fixed amount of memory (usually 32 or 64 bits). For this kind of data the size of an element is usually taken to be 1 and consequently to have *unit size*.

Examples of the second are integer numbers which require a number of bits approximately equal to the logarithm of their absolute value. This logarithm is usually referred to as the *bit size* of the integer. Similar ideas apply for rational numbers.

In this article we will only consider unit size and bit size.

Let A be some kind of data and $x = (x_1, \ldots, x_n) \in A^n$. If A is of the first kind above then we define $\text{size}(x) = n$. Else, we define $\text{size}(x) = \sum_{i=1}^n \text{bit-size}(x_i)$.

Similar considerations apply for the cost of arithmetic operations. The cost of operating two unit-size numbers is taken to be 1 and, as expected, is called *unit cost*. In the bit-size case, the cost of operating two numbers is the product of their bit-sizes (for multiplications and divisions) or its maximum (for additions, subtractions, and comparisons).

The consideration of integer or rational data with their associated bit size and bit cost for the arithmetic operations is usually referred to as the *Turing model of computation*. The consideration of idealized reals with unit size and unit cost is today referred as the *BSS model of computation* (from BLUM, SHUB and SMALE [1989]).

REMARK 1.1. The choice of one or the other model to analyze the complexity of an algorithm or a problem should depend, as we already remarked, of the kind of data the algorithm will deal with. If the chosen model is the BSS one, an additional issue that should be addressed is the behavior of the algorithm regarding round-off errors (we will return to this issue in Section 4). Note that the same problem can be considered in both models and complexity bounds for one model do not imply complexity bounds for the other. So, while comparing algorithms, one should begin by making clear the model of computation used to derive complexity bounds for these algorithms.

A basic concept related to both the Turing and the BSS models of computation is that of *polynomial time*. An algorithm $\mathcal{A}$ is said to be a polynomial time algorithm if $T^w_{\mathcal{A}}$ is bounded above by a polynomial. A problem can be solved in polynomial time if there is a polynomial time algorithm solving the problem. The notion of *average polynomial time* is defined similarly replacing $T^w_{\mathcal{A}}$ by $T^a_{\mathcal{A}}$.

The notion of polynomial time is usually taken as the formal counterpart of the more informal notion of efficiency. One not fully identify polynomial-time with efficiency since high degree polynomial bounds can hardly mean efficiency. Yet, many basic problems admitting polynomial-time algorithms can actually be efficiently solved. On the other hand, the notion of polynomial-time is extremely robust and it is the ground upon which complexity theory is built (cf. BALCÁZAR, DÍAZ and GABARRÓ [1988], PAPADIMITRIOU [1994], BLUM, CUCKER, SHUB and SMALE [1998]).

Most of this article will deal with the complexity of algorithms. Lower bounds for the complexity of computational problems is a subject we will barely touch. The interested reader may consult BLUM, CUCKER, SHUB and SMALE [1998], BÜRGISSER, CLAUSEN and SHOKROLLAHI [1996] for developments in this direction.

2. Decision and optimization problems

Intuitively, a computational problem $\mathcal{P}$ is a set of admissible instances together with, for each such instance $\mathcal{I}$, a solution set of possible outcomes. Thus, we describe the problem $\mathcal{P}$ by the set of instances $\mathcal{Z}_P$ (called the *data set*) and the family of *solution sets* $\{\mathcal{S}_\mathcal{I}\}_{\mathcal{I}\in\mathcal{Z}_P}$. Note that the solution set $\mathcal{S}_\mathcal{I}$ may be empty, i.e. instance $\mathcal{I}$ may have no solution.

We will no further develop this intuitive idea. Instead, in what follows, we list several decision and optimization problems specifying, for each of them, what $\mathcal{Z}_P$ is and what $\mathcal{S}_\mathcal{I}$ is for an instance $\mathcal{I}\in\mathcal{Z}_P$.

2.1. System of linear equations

Given $A\in\mathbb{R}^{m\times n}$ and $b\in\mathbb{R}^m$, the problem is to solve m linear equations for n unknowns:

$$Ax=b.$$

The data and solution sets are

$$\mathcal{Z}_P=\{A\in\mathbb{R}^{m\times n},\ b\in\mathbb{R}^m\}\quad\text{and}\quad \mathcal{S}_\mathcal{I}=\{x\in\mathbb{R}^n\colon\ Ax=b\}.$$

In this case $\mathcal{S}_\mathcal{I}$ is an affine set.

To highlight the analogy with the theories of linear inequalities and linear programming, we list several well-known results of linear algebra. The first theorem provides two basic representations of a linear subspace. If $A:\mathbb{R}^n\to\mathbb{R}^m$ is a linear map we denote its *nullspace* and *rowspace* by $\mathcal{N}(A)$ and $\mathcal{R}(A)$, respectively.

THEOREM 2.1. *Each linear subspace of $\mathbb{R}^n$ can be generated by finitely many vectors, and can also be written as the intersection of finitely many linear hyperplanes. That* *is, for each linear subspace of L of $\mathbb{R}^n$ there exist matrices A and C such that $L = \mathcal{N}(A) = \mathcal{R}(C)$.*

The following theorem was observed by Gauss. It is sometimes called the *fundamental theorem* of linear algebra.

THEOREM 2.2. *Let $A \in \mathbb{R}^{m\times n}$ and $b \in \mathbb{R}^m$. The system $\{x\colon\ Ax = b\}$ has a solution if and only if that $A^{\mathrm{T}}y = 0$ implies $b^{\mathrm{T}}y = 0$.*

A vector y, with $A^{\mathrm{T}}y = 0$ and $b^{\mathrm{T}}y = 1$, is called an *infeasibility certificate* for the system $\{x\colon\ Ax = b\}$.

EXAMPLE 2.1. Let $A = (1; -1)$ and $b = (1; 1)$. Then, $y = (1/2; 1/2)$ is an *infeasibility certificate* for $\{x\colon\ Ax = b\}$.

2.2. Linear least-squares problem

Given $B \in \mathbb{R}^{n\times m}$ and $c \in \mathbb{R}^n$, the system of equations $By = c$ may be over-determined or have no solution. Such a case usually occurs when the number n of equations is greater than the number m of variables. Then, the problem is to find $y \in \mathbb{R}^m$ or $s \in \mathcal{R}(B)$ such that $\|By - c\|$ or $\|s - c\|$ is minimized. We can write the problem in the following format:

$$\text{(LS)}\quad \begin{array}{ll} \text{minimize} & \|By - c\|^2 \\ \text{subject to} & y \in \mathbb{R}^m, \end{array}$$

or, equivalently,

$$\text{(LS)}\quad \begin{array}{ll} \text{minimize} & \|s - c\|^2 \\ \text{subject to} & s \in \mathcal{R}(B). \end{array}$$

In the former format, the term $\|By - c\|^2$ is called the *objective function* and y is called the *decision variable*. Since y can be any point in $\mathbb{R}^m$, we say this (optimization) problem is *unconstrained*. The data and solution sets are

$$\mathcal{Z}_P = \{B \in \mathbb{R}^{n\times m},\ c \in \mathbb{R}^n\}$$

and

$$\mathcal{S}_{\mathcal{I}} = \{y \in \mathbb{R}^m\colon\ \|By - c\|^2 \leqslant \|Bx - c\|^2 \text{ for every } x \in \mathbb{R}^m\}.$$

Let's now discuss how to find an y minimizing $\|By - c\|^2$. By abuse of notation let's denote also by B the linear map from $\mathbb{R}^m$ to $\mathbb{R}^n$ whose matrix is B. Let $\mathbf{w} \in \mathcal{R}(B)$ be the point whose distance to c is minimal. Then

$$S = \{y \in \mathbb{R}^m \mid By = \mathbf{w}\}$$

is an affine subspace of $\mathbb{R}^m$ of dimension $\dim(\mathcal{N}(B))$. In particular, the least squares problem has a unique solution $y \in \mathbb{R}^m$ if and only if B is injective. If this is the case, let us denote by B_* the bijection $B_*: \mathbb{R}^m \to \mathcal{R}(B)$ given by $B_*(y) = B(y)$. The next result is immediate.

PROPOSITION 2.1. *Let $B: \mathbb{R}^m \to \mathbb{R}^n$ be injective and $c \in \mathbb{R}^n$. Then the solution of the least squares problem is given by $y = B^\dagger c$ where $B^\dagger = B_*^{-1}\pi$. Here $\pi: \mathbb{R}^n \to \mathcal{R}(B)$ is the orthogonal projection onto $\mathcal{R}(B)$.*

Recall that the orthogonal complement to $\mathcal{R}(B)$ in $\mathbb{R}^n$ is $\mathcal{N}(B^{\mathrm{T}})$. Thus, for every $y \in \mathbb{R}^m$,

$$\begin{aligned} y = B^\dagger c \quad &\Leftrightarrow \quad By = \pi c \quad \Leftrightarrow \quad By - \pi c \in \mathcal{R}(B)^\perp \\ &\Leftrightarrow \quad B^{\mathrm{T}}(By - c) = 0 \quad \Leftrightarrow \quad y = \left(B^{\mathrm{T}}B\right)^{-1}B^{\mathrm{T}}c. \end{aligned}$$

The map $B^\dagger = (B^{\mathrm{T}}B)^{-1}B^{\mathrm{T}}$ is called the *Moore–Penrose inverse* of the injective map B. So, we have shown that $y = B^\dagger c$. In particular, y is a linear function of c.

The linear least-squares problem is a basic problem solved on each iteration of any interior-point algorithm.

2.3. *System of linear inequalities (linear conic systems)*

Given $A \in \mathbb{R}^{m\times n}$ and $b \in \mathbb{R}^m$, the problem is to find a point $x \in \mathbb{R}^n$ satisfying $Ax \leqslant b$ or to prove that no such point exists. The inequality problem includes other forms such as finding an x that satisfies the combination of linear equations $Ax = b$ and inequalities $x \geqslant 0$ (or prove that no such point exists). The data and solution sets of the latter are

$$\mathcal{Z}_P = \left\{A \in \mathbb{R}^{m\times n},\ b \in \mathbb{R}^m\right\} \quad \text{and} \quad \mathcal{S}_{\mathcal{I}} = \left\{x \in \mathbb{R}^n\colon\ Ax = b,\ x \geqslant 0\right\}.$$

Traditionally, a point in $\mathcal{S}_{\mathcal{I}}$ is called a *feasible solution*, and a strictly positive point in $\mathcal{S}_{\mathcal{I}}$ is called a *strictly feasible* or *interior feasible solution.*

The following results are Farkas' lemma and its variants.

THEOREM 2.3 (Farkas' lemma). *Let $A \in \mathbb{R}^{m\times n}$ and $b \in \mathbb{R}^m$. The system $\{x\colon\ Ax = b,\ x \geqslant 0\}$ has a feasible solution x if and only if, for all $y \in \mathbb{R}^m$, $A^{\mathrm{T}}y \leqslant 0$ implies $b^{\mathrm{T}}y \leqslant 0$.*

A vector y, with $A^{\mathrm{T}}y \leqslant 0$ and $b^{\mathrm{T}}y = 1$, is called a (*primal*) infeasibility certificate for the system $\{x\colon\ Ax = b,\ x \geqslant 0\}$. Geometrically, Farkas' lemma means that if a vector $b \in \mathbb{R}^m$ does not belong to the cone generated by $a_{.1}, \ldots, a_{.n}$, then there is a hyperplane separating b from $\operatorname{cone}(a_{.1}, \ldots, a_{.n})$.

THEOREM 2.4 (Farkas' lemma variant). *Let $A \in \mathbb{R}^{m\times n}$ and $c \in \mathbb{R}^n$. The system $\{y\colon\ A^{\mathrm{T}}y \leqslant c\}$ has a solution y if and only if, for all $x \in \mathbb{R}^n$, $Ax = 0$ and $x \geqslant 0$ imply $c^{\mathrm{T}}x \geqslant 0$.*

Again, a vector $x \geqslant 0$, with $Ax = 0$ and $c^{\mathrm{T}}x = -1$, is called a (*dual*) infeasibility certificate for the system $\{y\colon\ A^{\mathrm{T}}y \leqslant c\}$.

2.4. Linear programming (LP)

Given $A \in \mathbb{R}^{m \times n}$, $b \in \mathbb{R}^m$ and $c, l, u \in \mathbb{R}^n$, the linear programming (LP) problem is the following optimization problem:

$$\begin{array}{ll} \text{minimize} & c^{\mathrm{T}} x \\ \text{subject to} & Ax = b, \quad l \leqslant x \leqslant u, \end{array}$$

where some elements in l may be $-\infty$ meaning that the associated variables are unbounded from below, and some elements in u may be ∞ meaning that the associated variables are unbounded from above. If a variable is unbounded both from below and above, then it is called a "*free*" variable.

The standard form linear programming problem which we will use throughout this article is the following:

$$\begin{array}{lll} \text{(LP)} & \text{minimize} & c^{\mathrm{T}} x \\ & \text{subject to} & Ax = b, \quad x \geqslant 0. \end{array}$$

The linear function $c^{\mathrm{T}} x$ is called the *objective function*, and the components of x are called the *decision variables*. In this problem, $Ax = b$ and $x \geqslant 0$ enforce *constraints* on the selection of x. The set $\mathcal{F}_P = \{x\colon\ Ax = b,\ x \geqslant 0\}$ is called *feasible set* or *feasible region*. A point $x \in \mathcal{F}_P$ is called a *feasible point* (or a *feasible solution* or simply a *solution* of (LP)). A feasible point x^* is called an *optimal solution* if $c^{\mathrm{T}} x^* \leqslant c^{\mathrm{T}} x$ for all feasible points x. If there is a sequence $\{x^k\}$ such that x^k is feasible and $c^{\mathrm{T}} x^k \to -\infty$, then (LP) is said to be *unbounded*.

The data and solution sets for (LP), respectively, are

$$\mathcal{Z}_P = \{A \in \mathbb{R}^{m \times n},\ b \in \mathbb{R}^m,\ c \in \mathbb{R}^n\}$$

and

$$\mathcal{S}_{\mathcal{I}} = \{x \in \mathcal{F}_P\colon c^{\mathrm{T}} x \leqslant c^{\mathrm{T}} y \text{ for every } y \in \mathcal{F}_P\}.$$

Again, given $x \in \mathbb{R}^n$, to determine whether $x \in \mathcal{S}_{\mathcal{I}}$ is as difficult as the original problem. However, due to the duality theorem, we can simplify the representation of the solution set significantly.

To every instance of (LP) we associate another linear program, called the dual (LD), defined by:

$$\begin{array}{lll} \text{(LD)} & \text{maximize} & b^{\mathrm{T}} y \\ & \text{subject to} & A^{\mathrm{T}} y \leqslant c, \end{array}$$

where $y \in \mathbb{R}^m$. Denote by $\mathcal{F}_D$ the set of all $y \in \mathbb{R}^m$ that are feasible for the dual, i.e. $\mathcal{F}_D = \{y \in \mathbb{R}^m \mid A^{\mathrm{T}} y \leqslant c\}$.

The following theorems give us an important relation between the two problems.

THEOREM 2.5 (Weak duality theorem). *Let $\mathcal{F}_P$ and $\mathcal{F}_D$ be nonempty. Then, for all $x \in \mathcal{F}_P$ and $y \in \mathcal{F}_D$*

$$c^{\mathrm{T}} x \geqslant b^{\mathrm{T}} y.$$

This theorem shows that a feasible solution to either problem yields a bound on the value of the other problem. If x and y are solutions of the primal and the dual, respectively, we call $c^{\mathrm{T}}x - b^{\mathrm{T}}y$ the *duality gap* of x and y.

THEOREM 2.6 (Strong duality theorem). *Let $\mathcal{F}_P$ and $\mathcal{F}_D$ be nonempty. Then, x^* is optimal for* (LP) *if and only if the following conditions hold*:

(i) $x^* \in \mathcal{F}_P$;

(ii) *there is $y^* \in \mathcal{F}_D$ such that $c^{\mathrm{T}}x^* = b^{\mathrm{T}}y^*$.*

The following result follows easily from Theorems 2.5 and 2.6.

THEOREM 2.7 (LP duality theorem). *If* (LP) *and* (LD) *both have feasible solutions then both problems have optimal solutions and the optimal objective values of the objective functions are equal.*

If one of (LP) *or* (LD) *has no feasible solution, then the other is either unbounded or has no feasible solution. If one of* (LP) *or* (LD) *is unbounded then the other has no feasible solution.*

The above theorems show that if a pair of feasible solutions can be found to the primal and dual problems with equal objective values, then these are both optimal. The converse is also true; there is no "gap" (i.e. if optimal solutions x^* and y^* exist for both problems then $c^{\mathrm{T}}x^* = b^{\mathrm{T}}y^*$).

An equivalent (and much used) form of the dual problem is the following

$$\begin{array}{lll}\text{(LD)} & \text{maximize} & b^{\mathrm{T}}y \\ & \text{subject to} & A^{\mathrm{T}}y + s = c, \quad s \geqslant 0,\end{array}$$

where $y \in \mathbb{R}^m$ and $s \in \mathbb{R}^n$. The components of s are called *dual slacks*. Written this way, (LD) is a linear programming problem where y is a "free" vector.

Using the theorems above we can describe the solution set for both (LP) and (LD) with the system

$$\mathcal{S}_{\mathcal{I}} = \left\{ (x, y, s) \in \left(\mathbb{R}^n_+, \mathbb{R}^m, \mathbb{R}^n_+\right): \begin{array}{r} c^{\mathrm{T}}x - b^{\mathrm{T}}y = 0, \\ Ax = b, \\ -A^{\mathrm{T}}y - s = -c \end{array} \right\}, \tag{2.1}$$

which is a system of linear inequalities and equations. Now it is easy to verify whether or not a triple (x, y, s) is optimal.

For feasible x and (y, s), $x^{\mathrm{T}}s = x^{\mathrm{T}}(c - A^{\mathrm{T}}y) = c^{\mathrm{T}}x - b^{\mathrm{T}}y$ is also called the *complementarity gap*. Thus, (x, y, s) is optimal if and only if $x^{\mathrm{T}}s = 0$.

If $x^{\mathrm{T}}s = 0$ we say x and s are *complementary* to each other. Since both x and s are nonnegative, $x^{\mathrm{T}}s = 0$ implies that $x_j s_j = 0$ for all $j = 1, \ldots, n$. If, in addition, $x_j + s_j > 0$ for $j = 1, \ldots, n$ we say that x and s are *strictly complementary*. Note that the equality $c^{\mathrm{T}}x - b^{\mathrm{T}}y = x^{\mathrm{T}}s$ allows us to replace one equation (in $2n$ variables) plus nonnegativity into n equations (each of them in 2 variables) also plus nonnegativity.

Equations in (2.1) become

$$\begin{aligned} Xs &= 0, \\ Ax &= b, \\ -A^{\mathrm{T}}y - s &= -c, \end{aligned} \tag{2.2}$$

where we have used X to denote the diagonal matrix whose diagonal entries are the elements of x. We will use this notation repeatedly in this article. This system has in total $2n+m$ unknowns and $2n+m$ equations including n nonlinear equations.

The following theorem plays an important role in analyzing LP interior-point algorithms. It gives a unique partition of the LP variables in terms of complementarity.

THEOREM 2.8 (Strict complementarity theorem). *If* (LP) *and* (LD) *both have feasible solutions then there exists a pair of strictly complementary solutions* $x^* \geqslant 0$ *and* $s^* \geqslant 0$, *i.e. a pair* (x^*, s^*) *satisfying*

$$X^*s^* = 0 \quad \textit{and} \quad x^* + s^* > 0.$$

Moreover, the partition of $\{1, \ldots, n\}$ *given by the supports*

$$P^* = \{j\colon x_j^* > 0\} \quad \textit{and} \quad Z^* = \{j\colon s_j^* > 0\}$$

do not depend on the pair (x^*, s^*) *of strictly complementary solutions.*

Given (LP) or (LD), the pair of P^* and Z^* is called the (strict) *complementarity partition*.

If $I \subset \{1, \ldots, n\}$ then we denote by A_I the submatrix of A obtaining by removing all columns with index not in I. The vector x_I is defined similarly. The set

$$\{x \in \mathbb{R}^n\colon A_{P^*}x_{P^*} = b,\ x_{P^*} \geqslant 0,\ x_{Z^*} = 0\}$$

is called the *primal optimal face*, and the set

$$\{y \in \mathbb{R}^m\colon c_{Z^*} - A_{Z^*}^{\mathrm{T}}y \geqslant 0,\ c_{P^*} - A_{P^*}^{\mathrm{T}}y = 0\}$$

the *dual optimal face*.

Select m linearly independent columns, denoted by the index set B, from A. Then matrix A_B is nonsingular and we may uniquely solve

$$A_Bx_B = b$$

for the m-vector x_B. By setting the components of x corresponding to the remaining columns of A equal to zero, we obtain a solution x such that

$$Ax = b.$$

Then, x is said to be a (*primal*) *basic solution* to (LP) with respect to the *basis* A_B. The components of x_B are called *basic variables*. A dual vector y satisfying

$$A_B^{\mathrm{T}}y = c_B$$

is said to be the corresponding *dual basic solution*. If a basic solution x satisfies $x \geqslant 0$, then x is called a *basic feasible solution*. If the dual solution is also feasible, that is

$$s = c - A^{\mathrm{T}} y \geqslant 0,$$

then x is called an *optimal basic solution* and A_B an *optimal basis*. A basic feasible solution is a vertex on the boundary of the feasible region. An optimal basic solution is an optimal vertex of the feasible region.

If one or more components in x_B are zero, that basic solution x is said to be (*primal*) degenerate. Note that in a nondegenerate basic solution the basic variables and the basis can be immediately identified from the nonzero components of the basic solution. If all components, s_N, in the corresponding dual slack vector s, except for s_B, are nonzero, then y is said to be (*dual*) nondegenerate. If both primal and dual basic solutions are nondegenerate, A_B is called a *nondegenerate basis*.

THEOREM 2.9 (LP fundamental theorem). *Given* (LP) *and* (LD) *where A has full row rank m,*

(i) *if there is a feasible solution, there is a basic feasible solution;*
(ii) *if there is an optimal solution, there is an optimal basic solution.*

The above theorem reduces the task of solving a linear program to that of searching over basic feasible solutions. By expanding upon this result, the simplex method, a finite search procedure, is derived. The simplex method examines the basic feasible solutions (i.e. the extreme points of the feasible region) moving from one such point to an adjacent one. This is done in a way so that the value of the objective function is continuously decreased. Eventually, a minimizer is reached. In contrast, interior-point algorithms will move in the interior of the feasible region and reduce the value of the objective function, hoping to by-pass many extreme points on the boundary of the region.

2.5. *Semi-definite programming (SDP)*

Let $\mathcal{M}^n$ denote the set of real $n \times n$ symmetric matrices. Given $C \in \mathcal{M}^n$, $A_i \in \mathcal{M}^n$, $i = 1, 2, \ldots, m$, and $b \in \mathbb{R}^m$, the semi-definite programming problem is to find a matrix $X \in \mathcal{M}^n$ for the optimization problem:

$$\begin{aligned} &(\text{SDP}) \quad \inf C \bullet X \\ &\qquad\qquad \text{subject to} \quad A_i \bullet X = b_i, \ i = 1, 2, \ldots, m, \ X \succeq 0. \end{aligned}$$

Recall that the $\bullet$ operation is the matrix inner product

$$A \bullet B := \operatorname{tr} A^{\mathrm{T}} B.$$

The notation $X \succeq 0$ means that X is a positive semi-definite matrix, and $X \succ 0$ means that X is a positive definite matrix. If a point $X \succ 0$ and satisfies all equations in (SDP), it is called a (*primal*) strictly or interior feasible solution.

The data set of (SDP) is

$$\mathcal{Z}_P = \left\{ A_i \in \mathcal{M}^n, \ i = 1, \ldots, m, \ b \in \mathbb{R}^m, \ C \in \mathcal{M}^n \right\}.$$

The dual problem to (SDP) can be written as:

$$\begin{aligned} \text{(SDD)} \quad & \sup b^{\mathrm{T}} y \\ & \text{subject to} \quad \textstyle\sum_i^m y_i A_i + S = C, \quad S \succeq 0, \end{aligned}$$

which is analogous to the dual (LD) of LP. Here $y \in \mathbb{R}^m$ and $S \in \mathcal{M}^n$. If a point $(y, S \succ 0)$ satisfies all equations in (SDD), it is called a *dual interior feasible solution.*

EXAMPLE 2.2. Let $P(y \in \mathbb{R}^m) = -C + \sum_i^m y_i A_i$, where C and A_i, $i = 1, \ldots, m$, are given symmetric matrices. The problem of minimizing the max-eigenvalue of $P(y)$ can be cast as a (SDD) problem.

In positive semi-definite programming, we minimize a linear function of a matrix in the positive semi-definite matrix cone subject to affine constraints. In contrast to the positive orthant cone of linear programming, the positive semi-definite matrix cone is nonpolyhedral (or "nonlinear"), but convex. So positive semi-definite programs are convex optimization problems. Positive semi-definite programming unifies several standard problems, such as linear programming, quadratic programming, and convex quadratic minimization with convex quadratic constraints, and finds many applications in engineering, control, and combinatorial optimization.

We have several theorems analogous to Farkas' lemma.

THEOREM 2.10 (Farkas' lemma in SDP). *Let $A_i \in \mathcal{M}^n$, $i = 1, \ldots, m$, have rank m (i.e. $\sum_i^m y_i A_i = 0$ implies $y = 0$) and $b \in \mathbb{R}^m$. Then, there exists a symmetric matrix $X \succ 0$ with*

$$A_i \bullet X = b_i, \quad i = 1, \ldots, m,$$

if and only if $\sum_i^m y_i A_i \preceq 0$ and $\sum_i^m y_i A_i \neq 0$ implies $b^{\mathrm{T}} y < 0$.

Note the difference between the above theorem and Theorem 2.3.

THEOREM 2.11 (Weak duality theorem in SDP). *Let $\mathcal{F}_P$ and $\mathcal{F}_D$, the feasible sets for the primal and dual, be nonempty. Then,*

$$C \bullet X \geqslant b^{\mathrm{T}} y \quad \text{where } X \in \mathcal{F}_P,\ (y, S) \in \mathcal{F}_D.$$

The weak duality theorem is identical to that of (LP) and (LD).

COROLLARY 2.1 (Strong duality theorem in SDP). *Let $\mathcal{F}_P$ and $\mathcal{F}_D$ be nonempty and have an interior. Then, X is optimal for* (PS) *if and only if the following conditions hold*:

(i) $X \in \mathcal{F}_P$;
(ii) *there is* $(y, S) \in \mathcal{F}_D$;
(iii) $C \bullet X = b^{\mathrm{T}} y$ *or* $X \bullet S = 0$.

Again note the difference between the above theorem and the strong duality theorem for LP.

Two positive semi-definite matrices are complementary to each other, $X \bullet S = 0$, if and only if $XS = 0$. From the optimality conditions, the solution set for certain (SDP) and (SDD) is

$$\mathcal{S}_{\mathcal{I}} = \{X \in \mathcal{F}_P,\ (y, S) \in \mathcal{F}_D\colon C \bullet X - b^{\mathrm{T}} y = 0\},$$

or

$$\mathcal{S}_{\mathcal{I}} = \{X \in \mathcal{F}_P,\ (y, S) \in \mathcal{F}_D\colon XS = 0\},$$

which is a system of linear matrix inequalities and equations.

In general, we have

THEOREM 2.12 (SDP duality theorem). *If one of* (SDP) *or* (SDD) *has a strictly or interior feasible solution and its optimal value is finite, then the other is feasible and has the same optimal value. If one of* (SDP) *or* (SDD) *is unbounded then the other has no feasible solution.*

Note that a duality gap may exists if neither (SDP) nor (SDD) has a strictly feasible point. This is in contrast to (LP) and (LD) where no duality gap exists if both are feasible.

Although semi-definite programs are much more general than linear programs, they are not much harder to solve. It has turned out that most interior-point methods for LP have been generalized to solving semi-definite programs. As in LP, these algorithms possess polynomial worst-case complexity under certain computation models. They also perform well in practice. We will describe such extensions later.

3. Linear programming and complexity theory

We can now summarize some basic facts about algorithms for linear programming. In this section we consider the problems in Sections 2.3 and 2.4. The complexity bounds we quote are valid for both.

A preliminary remark, of the greatest importance, is that for all the algorithms we will mention in this section, we can consider both floating-point and rational data. Therefore, we shall evaluate their complexity in both the bit model and the unit cost model (as we described them in Section 1.2). Thus, besides the dimensions n and m of the matrix A, in the bit model (which assumes A, b and c to have rational entries) we will also consider the parameter

$$L = \sum_{i,j} \mathrm{size}(a_{ij}) + \sum_{i} \mathrm{size}(b_i) + \sum_{j} \mathrm{size}(c_j),$$

where, in this case, "size" means bit size. In this model, "polynomial time" means polynomial in n, m and L.

The first algorithm we mention is the simplex method. This is a direct method. As we remarked its worst-case complexity may be exponential and one may find instances requiring time $\Omega(2^{\min\{m,n\}})$ in both the unit cost model and the bit model.

On the other hand, in the unit cost model, results by BORGWARDT [1982] and SMALE [1983] later improved by HAIMOVICH [1983] show that, if the coefficients of A are i.i.d. random variables with standard Gaussian distribution then the number of vertices examined by the simplex method is, on the average, linear in $\min\{m,n\}$. Since the number of arithmetic operation performed at each such examined vertex is $\mathcal{O}(mn + (\min\{m,n\})^2)$ we obtain an average complexity of $\mathcal{O}(\mu(mn + \mu^2))$ where $\mu = \min\{m,n\}$.

The ellipsoid and interior-point methods radically differ in their conception from the simplex. They are iterative methods and, for both of them, the cost of each iteration is polynomial in n and m (and also in L in the bit model). But to bound the number of iterations they perform a new insight is required. In the bit model, this number of iterations can be bounded by a polynomial in m, n and L thus yielding the so talked about polynomial time algorithms for linear programming. The ellipsoid method has a complexity bound of $\mathcal{O}((\min\{m,n\})^2(mn + (\min\{m,n\})^2)L^2)$ and the interior-point method of

$$\mathcal{O}\big((\max\{m,n\})^{0.5}\big((\min\{m,n\})^2(\max\{m,n\})^{0.5} + \min\{m,n\}\max\{m,n\}\big)L^2\big).$$

What about their number of iterations in the unit cost model? It turns out that a bound depending only on m and n for this number is not yet available and in practice, it turns out that two instances with the same (unit) size may result in drastically different performances under these algorithms. This has lead during the past years to some research developments relating complexity of ellipsoid and interior-point algorithms with certain "condition" measures for linear programming (or linear conic systems) instances.

The goal of this article is to describe some of these recent developments and to suggest a few directions in which future progress might be made on the real number complexity issues of linear programming.

4. Condition measures

The notion of conditioning first appeared in relation with round-off analysis and eventually found a place in complexity analysis as well. We next describe some basics about finite precision and round-off errors, and use this setting to introduce conditioning in linear algebra (where it first appeared) and linear conic systems.

4.1. Finite precision and round-off errors

The dawn of digital computers brought the possibility of mechanically solving a plethora of mathematical problems. It also rekindled the interest for round-off analysis. The issue here is that, within the standard floating point arithmetic, all computations are carried in a subset $\mathbb{F} \subset \mathbb{R}$ instead of on the whole set of real numbers $\mathbb{R}$. A characteristic property of floating point arithmetic is the existence of a number $0 < u < 1$, the *round-off unit*, and a function $r : \mathbb{R} \to \mathbb{F}$, *the rounding function*, such that, for all $x \in \mathbb{R}$,

$|r(x) - x| \leqslant u|x|$. Arithmetic operations in $\mathbb{R}$ are then replaced by "rounded" versions in $\mathbb{F}$. The result of, for instance, multiplying $x, y \in \mathbb{F}$ is $r(xy) \in \mathbb{F}$.

During a computation these errors accumulate and the final result may be far away from what it should be. A prime concern when designing algorithms is thus to minimize the effects of this accumulation. Algorithms are consequently analyzed with this regard and compared between them in the same way they are compared regarding their running times. This practice, which today is done more or less systematically, was already present in Gauss' work.

> Since none of the numbers we take out from logarithmic or trigonometric tables admit of absolute precision, but are all to a certain extent approximate only, the results of all calculations performed by the aid of these numbers can only be approximately true. [...] It may happen, that in special cases the effect of the errors of the tables is so augmented that we may be obliged to reject a method, otherwise the best, and substitute another in its place.
>
> Carl Friedrich Gauss, *Theoria Motus*
> (cited in GOLDSTINE [1977], p. 258).

REMARK 4.1. Most of the work related to finite precision assumes that this precision is fixed (i.e. the round-off unit remain unchanged during the execution of the algorithm). This implies a fixed cost for each arithmetic operation and therefore, the use of the unit cost model described in Section 1.2. The total cost of a computation is, up to a constant, the number of arithmetic operations performed during that computation.

The unit cost model is not, however, a reasonable complexity model for variable precision algorithms. A more realistic assumption assigns cost $(\log u)^2$ to any multiplication or division between two floating-point numbers with round-off unit u, since this is roughly the number of elementary operations performed by the computer to multiply or divide these numbers. For an addition, subtraction or comparison the cost is $|\log u|$. The cost of the integer arithmetic necessary for computing variables' addresses and other quantities related with data management may be (and is customarily) ignored.

4.2. Linear systems

To study how errors accumulate during the execution of an algorithm it is convenient to first focus on a simplified situation namely, that in which errors occur only when reading the input. That is, an input $a = (a_1, \ldots, a_n) \in \mathbb{R}^n$ is rounded to $r(a) = (r(a_1), \ldots, r(a_n)) \in \mathbb{F}^n$ and then the algorithm is executed with infinite precision over the input $r(a)$.

Let's see how this is done for the problem of linear equation solving. Let A be an invertible $n \times n$ real matrix and $b \in \mathbb{R}^n$. We are interested in solving the system

$$Ax = b$$

and want to study how the solution x is affected by perturbations in the input (A, b).

Early work by TURING [1948] and VON NEUMANN and GOLDSTINE [1947] identified that the key quantity was

$$\kappa(A) = \|A\|\|A^{-1}\|,$$

where $\|A\|$ denotes the operator norm of A defined by

$$\|A\| = \max_{\|x\|=1} \|A(x)\|.$$

Here $\| \ \|$ denotes the Euclidean norm in $\mathbb{R}^n$ both as a domain and codomain of A. Turing called $\kappa(A)$ the *condition number* of A. A main result for $\kappa(A)$ states that, for an input error $(\Delta A, \Delta b)$ with ΔA small enough,

$$\frac{\|\Delta x\|}{\|x\|} \leqslant \kappa(A)\left(\frac{\|\Delta A\|}{\|A\|} + \frac{\|\Delta b\|}{\|b\|}\right).$$

In addition, $\kappa(A)$ is sharp in the sense that no smaller number will satisfy the inequality above for all A and b. Thus, $\kappa(A)$ measures how much the relative input error is amplified in the solution and $\log \kappa(A)$ measures the loss of precision. In Turing's words

> It is characteristic of ill-conditioned sets of equations that small percentage errors in the coefficients given may lead to large percentage errors in the solution.

When A is not invertible its condition number is not well defined. However, we can extend its definition by setting $\kappa(A) = \infty$ if A is singular. Matrices A with $\kappa(A)$ small are said to be *well-conditioned*, those with $\kappa(A)$ large are said to be *ill-conditioned*, and those with $\kappa(A) = \infty$ *ill-posed*.

Note that the set Σ of ill-posed problems has Lebesgue measure zero in the space $\mathbb{R}^{n^2}$. The distance of a matrix A to this set is closely related to $\kappa(A)$.

THEOREM 4.1 (Condition number theorem, ECKART and YOUNG [1936]). *For any $n \times n$ real matrix A one has*

$$\kappa(A) = \frac{\|A\|}{d_F(A, \Sigma)}.$$

Here d_F means distance in $\mathbb{R}^{n^2}$ with respect of the Frobenius norm $\|A\| = \sqrt{\sum a_{ij}^2}$.

The relationship between conditioning and distance to ill-posedness is a recurrent theme in numerical analysis (cf. DEMMEL [1987]). It will play a central role in our understanding of the condition of a linear program.

Reasonably enough, $\kappa(A)$ will appear in more elaborate round-off analysis in which errors may occur in all the operations. As an example, we mention such an analysis for Cholesky's method. If A is symmetric and positive definite we may solve the linear system $Ax = b$ by using Cholesky's factorization. If the computed solution is $(x + \Delta x)$ then one can prove that, for a round-off unit u sufficiently small,

$$\frac{\|\Delta x\|}{\|x\|} \leqslant 3n^3 u \kappa(A).$$

REMARK 4.2. Note that a bound as the one above for the relative forward error of an input $a \in \mathbb{R}^n$ in the form of an expression in its size n, its condition number $\kappa(a)$, and the round-off unit u may not be computable for a particular input a since we may not know $\kappa(a)$. Yet, such bounds allow us to compare algorithms with respect to stability. The fastest the expression tends to zero with u, the more stable the algorithm is.

Numerical linear algebra is probably one of the best developed areas in numerical analysis. The interested reader can find more about it in, for instance, the introductory books (DEMMEL [1997], TREFETHEN and BAU III [1997]) or in the more advanced (HIGHAM [1996]).

4.3. *Condition-based complexity*

Occasionally, an exact solution of a linear system is not necessary and an approximation, up to some predetermined ε, is sought instead. Typically, to find such approximate solution, the algorithm proceeds by iterating some basic step until the desired accuracy is reached. This kind of algorithms, called *iterative*, is ubiquitous in numerical analysis. Their cost, in contrast with the so-called *direct* methods whose cost depends only on the input size, may depend on ε and on the input itself.

The above applies for algorithms with both finite and infinite precision. An interesting point is that more often than not the cost of iterative algorithms appears to depend on the condition of the input.

For linear equation solving one may consider the use of iterative methods (for instance, when n is large and A is sparse). One such method is the conjugate gradient (cf. DEMMEL [1997], TREFETHEN and BAU III [1997]). It begins with a candidate solution $x_0 \in \mathbb{R}^n$ of the system $Ax = b$, A a real symmetric positive definite matrix. Then, it constructs a sequence of iterates $x_0, x_1, x_2, \ldots$ converging to the solution x^* of the system and satisfying

$$\|x_j - x^*\|_A \leqslant 2\left(\frac{\sqrt{\kappa(A)} - 1}{\sqrt{\kappa(A)} + 1}\right)^j \|x_0 - x^*\|_A.$$

Here the A-norm $\|v\|_A$ of a vector v is defined as $\|v\|_A = (v^{\mathrm{T}} A v)^{1/2}$. One concludes that for

$$j = \mathcal{O}\left(\sqrt{\kappa(A)} \log \frac{1}{\varepsilon}\right)$$

one has $\|x_j - x^*\|_A \leqslant \varepsilon \|x_0 - x^*\|_A$. Notice the dependence of the number of iterations on both $\kappa(A)$ and ε.

Probably the most interesting class of algorithms for which a condition based complexity analysis seems necessary is the family of so-called *path-following* algorithms. Interior-point methods belong to such class. So we will be interested in elaborating on condition based complexity for these methods. But before doing that we need to define a condition number for linear conic systems.

4.4. Linear conic systems

For notational simplicity, in the following we consider homogeneous linear conic systems. The primal and dual forms of those systems are thus

$$Ax = 0, \quad x \geqslant 0, \ x \neq 0 \tag{4.1}$$

and

$$A^{\mathrm{T}} y \leqslant 0, \quad y \neq 0, \tag{4.2}$$

respectively.

4.4.1. $C(A)$

RENEGAR [1994], RENEGAR [1995a], RENEGAR [1995b] introduced a condition number for the pair of linear conic systems above which used the idea, contained in the Condition number theorem of linear algebra, of conditioning as inverse to the distance to ill-posedness. Let $\mathcal{D}$ (resp. $\mathcal{P}$) denote the set of matrices A for which (4.2) (resp. (4.1)) is feasible and let Σ be the boundary between $\mathcal{D}$ and $\mathcal{P}$. Renegar defined

$$C(A) = \frac{\|A^{\mathrm{T}}\|}{d(A, \Sigma)},$$

where both numerator and denominator are for the operator norm with respect to the Euclidean norm in both $\mathbb{R}^m$ and $\mathbb{R}^n$.[2] It turns out that the condition number thus defined naturally appears in a variety of bounds related with the problem above – besides, of course, those in error analysis. For instance, it appears when studying the relative distance of solutions to the boundary of the solution set, the number of necessary iterations to solve the problem (for iterative algorithms such as interior-point or ellipsoid), or – as expected – in different error estimates in the round-off analysis of these algorithms, are all bounded by expressions in which $C(A)$ appears as a parameter. References for the above are: RENEGAR [1995b], FREUND and VERA [1999b], FREUND and VERA [1999a], VERA [1998], CUCKER and PEÑA [2001]. In Section 5.3 below we will describe a complexity and round-off analysis for an interior point algorithm done in terms of $C(A)$.

The following easy to prove proposition will be used in Section 5.3.

PROPOSITION 4.1. *Assume that A does not have zero columns. Let $\overline{A}$ be the matrix obtained by scaling each column a_k of A by a positive number so that $\|a_k\| = 1$. Then*

$$C(\overline{A}) \leqslant nC(A).$$

For a detailed discussion on the distance to ill-posedness and condition numbers see RENEGAR [1995b], PEÑA [2000].

---

[2] The consideration of the Euclidean norm in both $\mathbb{R}^n$ and $\mathbb{R}^m$ is inessential. Other norms may be considered as well and we will see an instance of this in Section 5.3.

4.4.2. $\mathcal{C}(A)$

The condition number $\mathcal{C}(A)$, a very close relative of $C(A)$, enjoys all the good properties of $C(A)$ and has some additional ones including some nice geometric interpretation. Let a_k denote the kth row of A^{T}, $k = 1, \dots, n$, and $y \in S^{m-1}$, the unit sphere in $\mathbb{R}^m$. Define

$$f_k(y) = \frac{\langle a_k, y\rangle}{\|a_k\|}$$

(here $\langle\ ,\ \rangle$ denotes the standard inner product in $\mathbb{R}^m$ and $\|\ \|$ its induced norm) and $D = \min_{y\in S^{m-1}} \max_{1\leqslant k\leqslant n} f_k(y)$. We define $\mathcal{C}(A)$ to be

$$\mathcal{C}(A) = \frac{1}{|D|}.$$

For any vector $y \in \mathbb{R}^m$, $y \neq 0$, let $\theta_k(A, y) \in [0, \pi]$ be the angle between y and a_k and

$$\theta(A, y) = \min_{k\leqslant n} \theta_k(A, y).$$

Denote by $\bar{y}$ any vector satisfying

$$\theta(A) = \theta(A, \bar{y}) = \max_{\substack{y\in\mathbb{R}^m\\ y\neq 0}} \theta(A, y).$$

Then

$$\begin{aligned}
\cos\theta(A) &= \cos\Bigl(\max_{\substack{y\in\mathbb{R}^m\\ y\neq 0}} \min_{k\leqslant n} \theta_k(A, y)\Bigr)\\
&= \cos\Bigl(\max_{\substack{y\in\mathbb{R}^m\\ y\neq 0}} \min_{k\leqslant n} \arccos f_k\Bigl(\frac{y}{\|y\|}\Bigr)\Bigr)\\
&= D
\end{aligned}$$

and we conclude that

$$\mathcal{C}(A) = \frac{1}{|\cos(\theta(A))|}.$$

The number $\mathcal{C}(A)$ captures several features related with the feasibility of the system $A^{\mathrm{T}}y \leqslant 0$. We next briefly state them. Proofs of these results can be found in CHEUNG and CUCKER [2001].

Let $\mathrm{Sol}(A) = \{y \mid A^{\mathrm{T}}y \leqslant 0,\ y \neq 0\}$ and $\mathcal{D} = \{A \in \mathbb{R}^{m\times n} \mid \mathrm{Sol}(A) \neq \emptyset\}$.

LEMMA 4.1. *Let $y \in \mathbb{R}^m$ and $\bar{y}$ as above. Then,*

(i) $\langle a_k, y\rangle \leqslant 0 \Leftrightarrow \cos\theta_k(A, y) \leqslant 0 \Leftrightarrow \theta_k(A, y) \geqslant \frac{\pi}{2}$,

(ii) $y \in \mathrm{Sol}(A) \Leftrightarrow \theta(A, y) \geqslant \frac{\pi}{2} \Leftrightarrow \cos\theta(A, y) \leqslant 0$, *and*

(iii) $A \in \mathcal{D} \Leftrightarrow \bar{y} \in \mathrm{Sol}(A)$.

A version of the Condition number theorem holds for $\mathcal{C}(A)$. Let

$$\varrho(A)=\sup\left\{\Delta \,\middle|\, \max_{k\leqslant n}\frac{\|a'_k-a_k\|}{\|a_k\|}<\Delta\Rightarrow(A\in\mathcal{D}\Leftrightarrow A'\in\mathcal{D})\right\},$$

where A' denotes the matrix with a'_k as its kth column.

THEOREM 4.2. *For all* $A\in\mathbb{R}^{m\times n}$, $\mathcal{C}(A)=1/\varrho(A)$.

We already mentioned that $\mathcal{C}$ is a close relative to Renegar's condition number C. Actually, one can prove that, for all matrix A, $\mathcal{C}(A)\leqslant\sqrt{n}C(A)$. Moreover, there is no converse of this in the sense that there is no function $f(m,n)$ such that $C(A)\leqslant f(m,n)\mathcal{C}(A)$ for all $m\times n$ matrices A. This shows that $C(A)$ can be arbitrarily larger than $\mathcal{C}(A)$. However, the following relation holds.

PROPOSITION 4.2.

$$C(A)\leqslant\frac{\|A^{\mathrm{T}}\|}{\min_k\|a_k\|}\mathcal{C}(A).$$

Therefore, restricted to the set of matrices A such that $\|a_k\|=1$ for $k=1,\dots,n$, one has

$$\mathcal{C}(A)\leqslant\sqrt{n}C(A)\leqslant n\mathcal{C}(A).$$

We now note that if A is arbitrary and $\overline{A}$ is the matrix whose kth column is

$$\bar{a}_k=\frac{a_k}{\|a_k\|}$$

then $\mathcal{C}(\overline{A})=\mathcal{C}(A)$ since $\mathcal{C}$ is invariant by column-scaling and, by Proposition 4.1, $C(\overline{A})\leqslant nC(A)$. Thus, $\mathcal{C}(A)$ is closely related to $C(\overline{A})$ for a normalization $\overline{A}$ of A which is easy to compute and does not increase too much $C(A)$.

The last feature of $\mathcal{C}(A)$ we mention in this section relates the probabilistic behaviour of $\mathcal{C}(A)$ (for random matrices A) with a classical problem in geometric probability.

Assume that the columns of A have all norm 1. Then it is easy to prove that

$$\theta(A)=\inf\{\theta\text{: the union of the circular caps with centers } a_k \text{ and angular radius } \theta \text{ covers } S^{m-1}\}.$$

Thus, if the columns a_k of A are randomly drawn from S^{m-1}, independently and with a uniform distribution, the random variable $\theta(A)$ (and a fortiori $\mathcal{C}(A)$) is related to the problem of covering the sphere with random circular caps. The latter is a classical problem in geometric probability (cf. HALL [1988], SOLOMON [1978]). The aspect most studied of this problem is to estimate, for given θ and n, the probability that n circular caps of angular radius θ cover S^{m-1}. A full solution of this problem is, as today, unknown. Partial results and some asymptotics can be found in GILBERT [1966], MILES [1969] and JANSON [1986].

The covering problem can be explicitly solved for the special value $\theta = \frac{\pi}{2}$ in which case (cf. Theorem 1.5 in HALL [1988]), the probability that n circular caps cover S^{m-1} is equal to

$$1 - \frac{1}{2^{n-1}} \sum_{k=0}^{m-1} \binom{n-1}{k}.$$

This has some immediate consequences in our context. In the following, by "$A^{\mathrm{T}} y \leqslant 0$ is feasible" we mean that $A^{\mathrm{T}} y \leqslant 0$ has nonzero solutions.

PROPOSITION 4.3. *Let A be a random matrix whose n columns are randomly drawn in S^{m-1} independently and uniformly distributed. Then,*

$$\mathbf{P}(A^{\mathrm{T}} y \leqslant 0 \textit{ is feasible}) = \frac{1}{2^{n-1}} \sum_{k=0}^{m-1} \binom{n-1}{k}.$$

Consequently, $\mathbf{P}(A^{\mathrm{T}} y \leqslant 0$ is feasible$) = 1$ if $n \leqslant m$, $\mathbf{P}(A^{\mathrm{T}} y \leqslant 0$ is feasible$) \to 0$ if m is fixed and $n \to \infty$ and $\mathbf{P}(A^{\mathrm{T}} y \leqslant 0$ is feasible$) = \frac{1}{2}$ when $m = 2n$.

PROOF. From (ii) and (iii) of Lemma 4.1 it follows that $A \in \mathcal{D}$ if and only if $\theta(A) \geqslant \frac{\pi}{2}$. And the latter is equivalent to say that the n circular caps with centers a_k and angular radius $\frac{\pi}{2}$ do not cover S^{m-1}. Thus, the first statement follows. The rest of the proposition is trivial. □

4.4.3. $\bar{\chi}_A$

Another used condition measure is $\bar{\chi}_A$, which is defined, for full-rank matrices A with $n \geqslant m$, as follows.

Let $\mathcal{D}$ be the set of all positive definite $n \times n$ diagonal matrices. Then, $\bar{\chi}_A$ is defined by

$$\bar{\chi}_A = \sup\left\{ \left\| A^{\mathrm{T}} \left(A D A^{\mathrm{T}} \right)^{-1} A D \right\| \colon D \in \mathcal{D} \right\},$$

where we are using the 2-norm and the induced operator norm. Since $(ADA^{\mathrm{T}})^{-1} ADc$ minimizes $\| D^{1/2} (A^{\mathrm{T}} y - c) \|$, an equivalent way to define these parameters is in terms of weighted least squares:

$$\bar{\chi}_A = \sup\left\{ \frac{\|A^{\mathrm{T}} y\|}{\|c\|} \colon y \text{ minimizes } \left\| D^{1/2} \left(A^{\mathrm{T}} y - c \right) \right\| \text{ for some } c \in \mathbb{R}^n,\ D \in \mathcal{D} \right\}. \tag{4.3}$$

Observe that this parameter is invariant when A is scaled by a constant. More generally, it is invariant if A is replaced by RA for any $m \times m$ nonsingular matrix R. This means that $\bar{\chi}_A$ actually depends on $\mathcal{N}(A)$, the nullspace of A, rather than on A itself.

LEMMA 4.2. *Let $\mathcal{B}$ denote the set of all bases (nonsingular $m \times m$ submatrices) of A. Then,*

$$\bar{\chi}_A \leqslant \max\left\{ \left\| A^{\mathrm{T}} B^{-\mathrm{T}} \right\| \colon B \in \mathcal{B} \right\}.$$

PROOF. We use the characterization (4.3) and analyze those vectors y that are candidates for the solution of the least-squares problems there. Fix c and consider the polyhedral subdivision of $\mathbb{R}^m$ induced by the hyperplanes $a_j^{\mathrm{T}} y = c_j$, $j = 1, \ldots, n$. A least-squares solution y must lie in one of these polyhedra. If it lies in a bounded polyhedron, it can be written as a convex combination of the vertices of the subdivision. If not, it can be expressed as the sum of a convex combination of vertices of the subdivision and a nonnegative combination of extreme rays of the subdivision. But removing the second summand, if present, moves y closer to at least one of the hyperplanes and no further from any of them, and so decreases the least-squares objective no matter what D is. Thus all least-squares solutions lie in the convex hull of the vertices of the subdivision. Each such vertex has the form $B^{-\mathrm{T}} c_B$ for some basis, where c_B is the corresponding subvector of c, and thus

$$\|A^{\mathrm{T}} y\| / \|c\| \leqslant \max\left\{ \left\| A^{\mathrm{T}} B^{-\mathrm{T}} c_B \right\| / \|c\| \right\} \leqslant \max\left\{ \left\| A^{\mathrm{T}} B^{-\mathrm{T}} \right\| \right\}. \qquad \square$$

The following lemma is from VAVASIS and YE [1995].

LEMMA 4.3. *Let B be a basis of A. Then, $\|A^{\mathrm{T}} B^{-\mathrm{T}}\| \leqslant \bar{\chi}_A$.*

PROOF. Let D_ε be the matrix with ones in positions corresponding to B and $0 < \varepsilon \ll 1$ (very small) in the remaining diagonal positions. In the limit as $\varepsilon \to 0^+$, we have that $A D_\varepsilon A^{\mathrm{T}}$ tends to $B B^{\mathrm{T}}$. Since B is invertible, $(A D_\varepsilon A^{\mathrm{T}})^{-1}$ tends to $(B B^{\mathrm{T}})^{-1}$. Also, $A D_\varepsilon$ tends to a matrix with a copy of B, and the remaining columns filled in by zeros. The resulting product in the definition in the limit is precisely $[A^{\mathrm{T}} B^{-\mathrm{T}}, 0]$. $\square$

Combining Lemmas 4.2 and 4.3 we deduce the following.

THEOREM 4.3. *Let $\mathcal{B}$ denote the set of all bases of A. Then*

$$\bar{\chi}_A = \max\left\{ \left\| B^{-1} A \right\| \colon B \in \mathcal{B} \right\}.$$

COROLLARY 4.1. *Let $n \geqslant m$. For all full-rank $n \times m$ matrices A, $\bar{\chi}_A$ is finite.*

REMARK 4.3. It appears that DIKIN [1967] first proved the finiteness of $\bar{\chi}_A$ using the Cauchy–Binet formula. Later independent proofs were given by BEN-TAL and TEBOULLE [1990], STEWART [1989], and TODD [1990]. FORSGREN [1995] describes the history and gives some extensions of this result.

4.4.4. $\sigma(A)$

Let $A \in \mathbb{R}^{m \times n}$. Define $\mathcal{P} = \{x \in \mathbb{R}^n \mid Ax = 0,\ x \geqslant 0,\ x \neq 0\}$, i.e. the solution set of the primal. Similarly, let $\mathcal{D} = \{s \in \mathbb{R}^n \mid s = A^{\mathrm{T}} y$ for some $y,\ s \geqslant 0,\ s \neq 0\}$. There is a unique partition $[B, N]$ of the columns of $\{1, \ldots, n\}$, s.t.

$$\mathcal{P} = \left\{ x \in \mathbb{R}^n \mid Ax = 0,\ x_B \geqslant 0,\ x_N = 0,\ x \neq 0 \right\}$$

and

$$\mathcal{D} = \{s \in \mathbb{R}^n \mid s = A^{\mathrm{T}} y \text{ for some } y,\ s_N \geqslant 0,\ s_B = 0,\ s \neq 0\}.$$

Let

$$\sigma_P = \min_{j \in B} \max_{x \in \mathcal{P}} \frac{x_j}{\|x\|_1},$$

$$\sigma_D = \min_{j \in N} \max_{s \in \mathcal{D}} \frac{s_j}{\|s\|_1}.$$

If $\mathcal{P}$ ($\mathcal{D}$) is empty, σ_P (resp. σ_D) is set to be 1. We define

$$\sigma(A) = \min\{\sigma_P, \sigma_D\}$$

4.4.5. $\mu(A)$

The last condition measure we mention is $\mu(A)$.

Let $\mathcal{H}_A = \{Ax \in \mathbb{R}^m \mid x \geqslant 0,\ \|x\|_1 = 1\}$, i.e. the convex hull of the column vectors of A. Let

$$\operatorname{sym}(A) = \max\{t \in \mathbb{R} \mid -tv \in \mathcal{H}_A \text{ for all } v \in \mathcal{H}_A\}.$$

We then define

$$\mu(A) = \frac{1}{\operatorname{sym}(A)}.$$

$\mu(A)$ is only defined for the case the system (FP) has an interior point. In this case, $\mu(A)$ and $\sigma(A)$ are closely related.

THEOREM 4.4. *It* (FP) *has an interior point,* $\mu(A) = 1/\sigma(A) - 1$.

4.5. Relations between condition measures for linear programming

Several relationships are known between the condition measures introduced above. We summarize them in Table 4.1. If the cell on row i and column j is No it means that the condition number in column j carries no upper bound information about condition number in row i. The symbol ? means that no result is known.

TABLE 4.1

	$C(A)$	$\mathcal{C}(A)$	$\bar{\chi}_A$	$1/\sigma(A)$	$\mu(A)$
$C(A)$		No	No	No	No
$\mathcal{C}(A)$	$\mathcal{C}(A) \leqslant \sqrt{n}C(A)$		No	No	No
$\bar{\chi}_A$	No	No		No	No
$1/\sigma(A)$	?	No	$\frac{1}{\sigma(A)} \leqslant \bar{\chi}_A + 1$		$\sigma(A) = \frac{1}{1+\mu(A)}$
$\mu(A)$	$\mu(A) \leqslant C(A)$	No	$\mu(A) \leqslant \bar{\chi}_A$	$\mu(A) = \frac{1}{\sigma(A)} - 1$	

5. Condition-based analysis of interior-point algorithms

5.1. *Interior point algorithms*

There are many interior point algorithms for LP. A popular one is the symmetric primal–dual path-following algorithm.

Consider a linear program in the standard form (LP) and (LD) described in Section 2.4.

Let $\mathcal{F} = \mathcal{F}_P \times \mathcal{F}_D$ and assume that $\mathring{\mathcal{F}} = \mathring{\mathcal{F}}_P \times \mathring{\mathcal{F}}_D \neq \emptyset$, i.e. both $\mathring{\mathcal{F}}_P \neq \emptyset$ and $\mathring{\mathcal{F}}_D \neq \emptyset$, and denote (x^*, y^*, s^*) an optimal solution.

DEFINITION 5.1. We define the *central path* in the primal–dual form to be

$$\mathcal{C} = \left\{ (x, y, s) \in \mathring{\mathcal{F}} \colon\ Xs = \frac{x^{\mathrm{T}} s}{n} e \right\}.$$

Points in the central path are said to be *perfectly centered.*

The central path has a well-defined geometric interpretation. Let the *primal logarithmic barrier function* be

$$(x, \mu) \mapsto -\mu \sum_{j=1}^{n} \log x_j .$$

For any $\mu > 0$ one can derive a point in the central path simply by minimizing the primal LP with the logarithmic barrier function above added, i.e. by solving

$$\begin{array}{lll} \text{(P)} & \text{minimize} & c^{\mathrm{T}} x - \mu \sum_{j=1}^{n} \log x_j \\ & \text{s.t.} & Ax = b, \quad x \geqslant 0. \end{array}$$

Let $x = x(\mu) \in \mathring{\mathcal{F}}_P$ be the (unique) minimizer of (P). Denote $f(x) = c^{\mathrm{T}} x - \mu \sum_{j=1}^{n} \log x_j$. Then $\nabla f(x) = c - \mu(1/x_1, 1/x_2, \ldots, 1/x_n)^{\mathrm{T}}$. The first order necessary optimality condition for (P) is $\nabla f(x) \in \mathcal{R}(A^{\mathrm{T}})$. That is, $c - \mu(1/x_1, 1/x_2, \ldots, 1/x_n)^{\mathrm{T}} = A^{\mathrm{T}} y$ for some $y \in \mathbb{R}^m$. Let $s = c - A^{\mathrm{T}} y = \mu(1/x_1, 1/x_2, \ldots, 1/x_n)^{\mathrm{T}}$. Then, $Xs = \mu e$ and, for some $y \in \mathbb{R}^m$ and some $s \in \mathbb{R}^n$, the triple (x, y, s) satisfies the optimality conditions

$$\begin{aligned} Xs &= \mu e, \\ Ax &= b, \\ -A^{\mathrm{T}} y - s &= -c. \end{aligned} \tag{5.1}$$

Now consider the *dual logarithmic barrier function* $(s, \mu) \mapsto \mu \sum_{j=1}^{n} \log s_j$ and the maximization problem

$$\begin{array}{lll} \text{(D)} & \text{maximize} & b^{\mathrm{T}} y + \mu \sum_{j=1}^{n} \log s_j \\ & \text{s.t.} & A^{\mathrm{T}} y + s = c, \quad s \geqslant 0. \end{array}$$

Let $(y, s) = (y(\mu), s(\mu)) \in \mathring{\mathcal{F}}_D$ be the (unique) minimizer of (D). Then, just as above, for some $x \in \mathbb{R}^n$, the triple (x, y, s) satisfies the optimality conditions (5.1) as well. Thus, both minimizers $x(\mu)$ and $(y(\mu), s(\mu))$ are on the central path with $x(\mu)^{\mathrm{T}} s(\mu) = n\mu$.

The key idea of interior-point methods rely on the fact that, as μ tends to zero, the triple $(x(\mu), y(\mu), s(\mu))$ tends to (x^*, y^*, s^*). Therefore, "following" the central path one is eventually lead to the optimal solution. This follow up is performed by constructing a sequence of points in $\mathcal{C}$ for decreasing values of μ. To do this, however, one faces the following problems:

(i) Since the equations defining $\mathcal{C}$ are nonlinear one can not compute points in $\mathcal{C}$. Therefore, the sequence above is constructed "near" $\mathcal{C}$. Actually, the distance between points in this sequence and the central path decreases with μ.
(ii) A stopping criterium is necessary allowing to jump from one point close to $\mathcal{C}$ (for a value of μ small enough) to an exact solution (x^*, y^*, s^*).
(iii) An initial point in $\mathcal{C}$ (or close enough to $\mathcal{C}$) is necessary.

In this section we will focus on the ways of dealing with point (i) above. Points (ii) and (iii) will be dealt with in Sections 5.2 and 5.3, respectively.

Actually, we next describe and analyze a predictor-corrector interior-point algorithm for linear programming which receives as input, in addition to the triple (A, b, c), an initial point as described in (iii) above. In some sense the algorithm follows the central path

$$\mathcal{C} = \left\{(x, y, s) \in \mathring{\mathcal{F}}\colon\ Xs = \mu e \text{ where } \mu = \frac{x^{\mathrm{T}} s}{n}\right\}$$

in primal–dual form. It generates a sequence of points (x, y, s) which keeps close to $(x(\mu), y(\mu), s(\mu))$ while μ is decreased at each iteration.

Once we have a pair $(x, y, s) \in \mathring{\mathcal{F}}$ with $\mu = x^{\mathrm{T}} s/n$, we can generate a new iterate (x^+, y^+, s^+). We now explain how this is done. First, fix $\gamma \geqslant 0$ and solve for d_x, d_y and d_s the system of linear equations

$$\begin{aligned} Sd_x + Xd_s &= \gamma\mu e - Xs, \\ Ad_x &= 0, \\ -A^{\mathrm{T}} d_y - d_s &= 0. \end{aligned} \tag{5.2}$$

Let $d := (d_x, d_y, d_s)$. To emphasize the dependence of d on the current pair (x, s) and the parameter γ, we write $d = d(x, s, \gamma)$. Note that $d_x^{\mathrm{T}} d_s = -d_x^{\mathrm{T}} A^{\mathrm{T}} d_y = 0$ here.

The system (5.2) is the Newton step starting from (x, y, s) which helps to find the point on the central path with duality gap $\gamma n\mu$. If $\gamma = 0$, it steps toward the optimal solution characterized by the system of equations (2.2); if $\gamma = 1$, it steps toward the central path point $(x(\mu), y(\mu), s(\mu))$ characterized by the system of equations (5.1); if $0 < \gamma < 1$, it steps toward a central path point with a smaller complementarity gap. In our algorithm we will only use the values 0 and 1 for γ.

DEFINITION 5.2. Let $\eta \in (0, 1)$. We define the *central neighbourhood* of radius η to be

$$\mathcal{N}(\eta) = \left\{ (x, y, s) \in \mathring{\mathcal{F}}\colon \ \|Xs - \mu e\| \leqslant \eta\mu \text{ where } \mu = \frac{x^{\mathrm{T}}s}{n} \right\}.$$

The algorithm generates a sequence of iterates in $\mathcal{N}(\eta_0)$, where $\eta_0 \in [0.2, 0.25]$. Actually, it also generates intermediate iterates in $\mathcal{N}(2\eta_0)$.

Given $(x, y, s) \in \mathcal{N}(\eta)$ and $\gamma \in [0, 1]$, the direction $d = (d_x, d_y, d_s)$ is generated from (5.2). Having obtained the search direction d, we let

$$\begin{aligned} x(\theta) &:= x + \theta d_x, \\ y(\theta) &:= y + \theta d_y, \\ s(\theta) &:= s + \theta d_s. \end{aligned} \tag{5.3}$$

We will frequently let the next iterate be $(x^+, s^+) = (x(\bar{\theta}), s(\bar{\theta}))$, where $\bar{\theta}$ is as large as possible so that $(x(\theta), s(\theta))$ remains in the neighborhood $\mathcal{N}(\eta)$ for all $\theta \in [0, \bar{\theta}]$.

Let $\mu(\theta) = x(\theta)^{\mathrm{T}}s(\theta)/n$ and $X(\theta) = \operatorname{diag}(x(\theta))$. In order to get bounds on $\bar{\theta}$, we first note that

$$\mu(\theta) = (1 - \theta)\mu + \theta\gamma\mu, \tag{5.4}$$

$$X(\theta)s(\theta) - \mu(\theta)e = (1 - \theta)(Xs - \mu e) + \theta^2 D_x d_s, \tag{5.5}$$

where $D_x = \operatorname{diag}(d_x)$. Thus $D_x d_s$ is the second-order term in Newton's method to compute a new point of $\mathcal{C}$. Hence we can usually choose a larger $\bar{\theta}$ (and get a larger decrease in the duality gap) if $D_x d_s$ is smaller. We next obtain several bounds on the size of $D_x d_s$. Before doing so, we remark that $\bar{\theta}$ can be computed as a specific root of a quartic polynomial on θ. Indeed, by definition, $\bar{\theta}$ is the largest positive θ satisfying

$$\left\| X(\theta)s(\theta) - \mu(\theta)e \right\| \leqslant 2\eta_0\mu(\theta) \quad \text{for all } \theta \leqslant \bar{\theta},$$

where $x(\theta)$ and $s(\theta)$ are defined in (5.3) with $\gamma = 0$. This inequality is equivalent to

$$\left\| X(\theta)s(\theta) - \mu(\theta)e \right\|^2 \leqslant 4\eta_0^2\mu(\theta)^2 \quad \text{for all } \theta \leqslant \bar{\theta}.$$

From Eqs. (5.4) and (5.5), this is equivalent to

$$\left\| (1 - \theta)(Xs - \mu e) + \theta^2 D_x d_s \right\|^2 \leqslant 4\eta_0^2(1 - \theta)^2\mu^2 \quad \text{for all } \theta \leqslant \bar{\theta}.$$

Define

$$f(\theta) = \left\| (1 - \theta)(Xs - \mu e) + \theta^2 D_x d_s \right\|^2 - 4\eta_0^2(1 - \theta)^2\mu^2.$$

This is a quartic polynomial on θ. Since $\|Xs - \mu e\| \leqslant \eta_0\mu$, $f(0) = \|(Xs - \mu e)\|^2 - 4\eta_0^2\mu^2 < 0$. In addition, it can be proved that $f(1) > 0$. Thus, $f(\theta)$ has at least one root in $(0, 1)$ and $\bar{\theta}$ is the smallest root of f in $(0, 1)$.

We now return to the problem of obtaining bounds on the size of $D_x d_s$. First, it is helpful to re-express $D_x d_s$. Let

$$\begin{aligned} p &:= X^{-0.5} S^{0.5} d_x, \\ q &:= X^{0.5} S^{-0.5} d_s, \\ r &:= (XS)^{-0.5}(\gamma \mu e - Xs). \end{aligned} \tag{5.6}$$

Note that $p+q=r$ and $p^{\mathrm{T}}q=0$ so that p and q represent an orthogonal decomposition of r.

LEMMA 5.1. *With the notations above,*

(i)

$$\|Pq\| \leqslant \frac{\sqrt{2}}{4}\|r\|^2;$$

(ii)

$$-\frac{\|r\|^2}{4} \leqslant p_j q_j \leqslant \frac{r_j^2}{4} \quad \textit{for each } j.$$

The bounds in Lemma 5.1 cannot be improved by much in the worst case. To see that consider the case where

$$\begin{aligned} r &= e = (1, 1, \ldots, 1)^{\mathrm{T}}, \\ p &= \big(1/2, 1/2, \ldots, 1/2, (1+\sqrt{n})/2\big)^{\mathrm{T}}, \quad \text{and} \\ q &= \big(1/2, 1/2, \ldots, 1/2, (1-\sqrt{n})/2\big)^{\mathrm{T}}. \end{aligned}$$

To use Lemma 5.1 we also need to bound r. The following result is useful:

LEMMA 5.2. *Let r be given by* (5.6). *Then*

(i) *If $\gamma = 0$, then $\|r\|^2 = n\mu$;*

(ii) *If $\eta \in (0, 1)$, $\gamma = 1$ and $(x, y, s) \in \mathcal{N}(\eta)$, then $\|r\|^2 \leqslant \eta^2\mu/(1-\eta)$.*

We now describe and analyze an algorithm that takes a single "corrector" step to the central path after each "predictor" step to decrease μ. Although it is possible to use more general values of η_0, we will work with nearly-centered pairs in $\mathcal{N}(\eta_0)$ with $\eta_0 \in [0.2, 0.25]$ (iterates after the corrector step), and intermediate pairs in $\mathcal{N}(2\eta_0)$ (iterates after a predictor step). As we noted above, this algorithm will not compute an optimal solution (x^*, y^*, s^*) but will stop when the duality gap is smaller than a positive number ε.

ALGORITHM 5.1.

input $A \in \mathbb{R}^{m\times n}$, $b \in \mathbb{R}^m$, $c \in \mathbb{R}^n$, $\varepsilon > 0$ and $(x^0, y^0, s^0) \in \mathcal{N}(\eta_0)$ with $\eta_0 \in [0.2, 1/4]$.

Set $k := 0$.

While $(x^k)^{\mathrm{T}} s^k > \varepsilon$ **do:**

1. `Predictor step:` set $(x, y, s) = (x^k, y^k, s^k)$ and compute $d = d(x, s, 0)$ from (5.2); compute the largest $\bar{\theta}$ so that

$$\big(x(\theta), y(\theta), s(\theta)\big) \in \mathcal{N}(2\eta_0) \quad \text{for } \theta \in [0, \bar{\theta}].$$

2. `Corrector step:` set $(x', y', s') = (x(\bar{\theta}), y(\bar{\theta}), s(\bar{\theta}))$ and compute $d' = d(x', s', 1)$ from (5.2); set $(x^{k+1}, y^{k+1}, s^{k+1}) = (x' + d'_x, y' + d'_y, s' + d'_s)$.
3. Let $k := k + 1$.

Our main result concerning the algorithm above is the following.

THEOREM 5.1. *Let* $\eta_0 \in [0.2, 0.25]$. *Then Algorithm* 5.1 *will terminate in at most* $\mathcal{O}(\sqrt{n}\log((x^0)^{\mathrm{T}}s^0/\varepsilon))$ *iterations yielding*

$$c^{\mathrm{T}}x^k - b^{\mathrm{T}}y^k \leqslant \varepsilon.$$

Towards the proof of Theorem 5.1 we first prove some lemmas.

The following lemma proves that after a corrector step, the iterate will return to the smaller neighborhood of the central path.

LEMMA 5.3. *For each k,* $(x^k, y^k, s^k) \in \mathcal{N}(\eta_0)$.

PROOF. The claim holds for $k = 0$ by hypothesis. For $k > 0$, let $(x', y', s') \in \mathcal{N}(2\eta_0)$ be the result of the predictor step at the kth iteration and let $d' = d(x', s', 1)$, as in the description of the algorithm. For $\theta \in [0, 1]$, let $x'(\theta)$ and $s'(\theta)$ be defined as in (5.3) and p', q' and r' as in (5.6) using x', s' and d'. Note that

$$(x', s') = \big(x'(0), s'(0)\big) \quad \text{and} \quad \big(x^{k+1}, s^{k+1}\big) = \big(x'(1), s'(1)\big).$$

Let $\mu'(\theta) := x'(\theta)^{\mathrm{T}}s'(\theta)/n$ for all $\theta \in [0, 1]$ with $\mu' := \mu'(0) = (x')^{\mathrm{T}}s'/n$ and $\mu^{k+1} := \mu'(1) = (x^{k+1})^{\mathrm{T}}s^{k+1}/n$.

From (5.4),

$$\mu'(\theta) = \mu' \quad \text{for all } \theta, \tag{5.7}$$

and, in particular, $\mu^{k+1} = \mu'$. From (5.5),

$$\begin{aligned} X'(\theta)s'(\theta) - \mu'(\theta)e &= (1-\theta)(X's' - \mu'e) + \theta^2 D'_x d'_s \\ &= (1-\theta)(X's' - \mu'e) + \theta^2 P'q', \end{aligned} \tag{5.8}$$

where $X'(\theta) = \operatorname{diag}(x'(\theta))$, etc. But by Lemmas 5.1(i) and 5.2(ii) and $(x', y', s') \in \mathcal{N}(2\eta)$ with $\eta = 1/4$,

$$\|P'q'\| \leqslant \frac{\sqrt{2}}{4}\|r'\|^2 \leqslant \frac{\sqrt{2}}{4}\frac{(2\eta)^2}{1-2\eta}\mu' < \frac{1}{4}\mu'.$$

It follows that

$$\big\|X'(\theta)s'(\theta) - \mu'e\big\| \leqslant (1-\theta)\frac{\mu'}{2} + \theta^2\frac{\mu'}{4} \leqslant \frac{1}{2}\mu'. \tag{5.9}$$

Thus $X'(\theta)s'(\theta) \geqslant \frac{\mu'}{2}e > 0$ for all $\theta \in [0, 1]$, and this implies that $x'(\theta) > 0$, $s'(\theta) > 0$ for all such θ by *continuity*. In particular, $x^{k+1} > 0$, $s^{k+1} > 0$, and (5.9) gives $(x^{k+1}, y^{k+1}, s^{k+1}) \in \mathcal{N}(1/4)$ as desired when we set $\theta = 1$. □

Now let $(x, y, s) = (x^k, y^k, s^k)$, $d = d(x, s, 0)$, $\mu = \mu^k = x^{\mathrm{T}}s/n$, and p, q and r be as in (5.6); these quantities all refer to the predictor step at iteration k. By (5.4),

$$\begin{aligned} \mu' &= (1 - \bar{\theta})\mu, \quad \text{or} \\ \mu^{k+1} &= (1 - \bar{\theta})\mu^k. \end{aligned} \tag{5.10}$$

Hence the improvement in the duality gap at the kth iteration depends on the size of $\bar{\theta}$.

LEMMA 5.4. *With the notation above, the step-size in the predictor step satisfies*

$$\bar{\theta} \geqslant \frac{2}{1 + \sqrt{1 + 4\|Pq/\mu\|/\eta}}.$$

PROOF. By (5.5) applied to the predictor step,

$$\begin{aligned} \big\|X(\theta)s(\theta) - \mu(\theta)e\big\| &= \big\|(1-\theta)(Xs - \mu e) + \theta^2 Pq\big\| \\ &\leqslant (1-\theta)\|Xs - \mu e\| + \theta^2\|Pq\| \\ &\leqslant (1-\theta)\eta\mu + \theta^2\|Pq\|, \end{aligned}$$

the last inequality by Lemma 5.3. We see that, for

$$0 \leqslant \theta \leqslant \frac{2}{1 + \sqrt{1 + 4\|Pq/\mu\|/\eta}},$$

$$\begin{aligned} \big\|X(\theta)s(\theta) - \mu(\theta)e\big\|/\mu &\leqslant (1-\theta)\eta + \theta^2\|Pq/\mu\| \\ &\leqslant 2\eta(1-\theta). \end{aligned}$$

This is because the quadratic term in θ

$$\|Pq/\mu\|\theta^2 + \eta\theta - \eta \leqslant 0$$

for θ between zero and the root

$$\frac{-\eta + \sqrt{\eta^2 + 4\|Pq/\mu\|\eta}}{2\|Pq/\mu\|} = \frac{2}{1 + \sqrt{1 + 4\|Pq/\mu\|/\eta}}.$$

Thus,

$$\big\|X(\theta)s(\theta) - \mu(\theta)e\big\| \leqslant 2\eta(1-\theta)\mu = 2\eta\mu(\theta)$$

or $(x(\theta), y(\theta), s(\theta)) \in \mathcal{N}(2\eta)$ for

$$0 \leqslant \theta \leqslant \frac{2}{1 + \sqrt{1 + 4\|Pq/\mu\|/\eta}}.$$

□

We can now prove our main result in this section.

PROOF OF THEOREM 5.1. Using Lemmas 5.1(i) and 5.2(i), we have

$$\|Pq\| \leqslant \frac{\sqrt{2}}{4}\|r\|^2 = \frac{\sqrt{2}}{4}n\mu,$$

so that

$$\bar{\theta} \geqslant \frac{2}{1+\sqrt{1+\sqrt{2}n/\eta}} = \frac{2}{1+\sqrt{1+4\sqrt{2}n}}$$

at each iteration. Then (5.10) and Lemma 5.4 imply that

$$\mu^{k+1} \leqslant \left(1 - \frac{2}{1+\sqrt{1+4\sqrt{2}n}}\right)\mu^k$$

for each k. This yields the desired result. □

5.2. An example of complexity analysis: the Vavasis–Ye method

The path-following algorithm described earlier takes small steps along the *central path* until they are "sufficiently" close to an optimum. Once sufficiently close, a "rounding" procedure such as KHACHIJAN's [1979] or least-squares computation such as YE's [1992b] is used to obtain an exact optimum.

Here, we present a different method, the Vavasis–Ye method, that interleaves small steps with longer *layered least-squares* (LLS) steps to follow the central path. Thus, the Vavasis–Ye method is always at least as efficient as existing path-following interior point methods, and the last LLS step moves directly to an exact optimum. The algorithm, which will be called "layered-step interior point" (LIP), terminates in a finite number of steps. Furthermore, the total number of iterations depends only on A: the running time is $\mathcal{O}(n^{3.5}c(A))$ iterations, where $c(A)$ is defined below. This is in contrast to other interior point methods, whose complexity depend on the vectors b and c as well as on the matrix A. This is important because there are many classes of problems in which A is "well-behaved" but b and c are arbitrary vectors.

An important feature of the LIP method is that it requires the knowledge of $\bar{\chi}_A$ or at least of some bound for it. In this section we thus assume we know $\bar{\chi}_A$. We will discuss this assumption in more detail in Remark 5.1.

In order to provide intuition on how the LIP algorithm works we consider the linear programming problem presented in Fig. 5.1. The problem solved in this figure is the dual-form problem:

$$\begin{aligned} \text{minimize} \quad & 2y_1 + 5y_2 \\ \text{subject to} \quad & y_1 \geqslant 0, \\ & y_1 \leqslant 1, \\ & y_2 \geqslant 0, \\ & y_2 \leqslant 1, \\ & y_1 + 2y_2 \geqslant \varepsilon, \end{aligned} \tag{5.11}$$

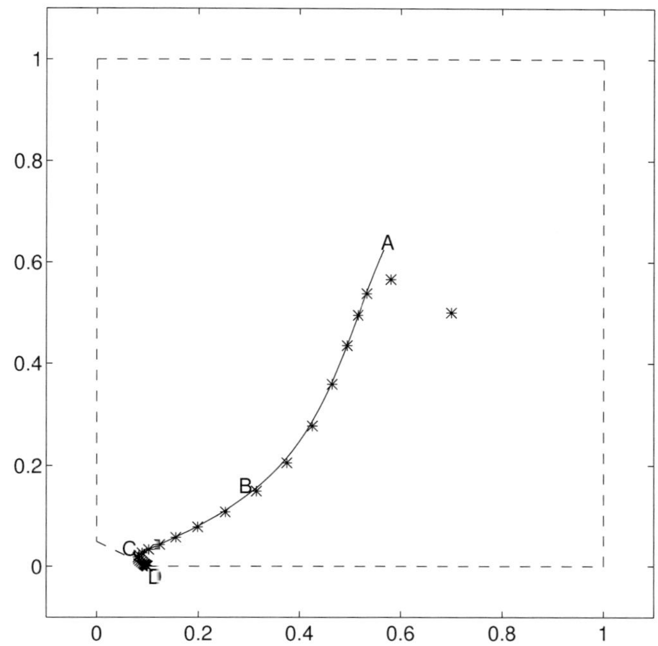

FIG. 5.1. Example of an interior point method.

where $\varepsilon = 0.1$. The dashed line indicates the boundaries of the feasible region defined by these constraints. The point A, known as the *analytic center*, is the (unique) maximizer of the barrier function

$$\log(y_1(1-y_1)) + \log(y_2(1-y_2)) + \log(y_1 + 2y_2 - \varepsilon)$$

in the feasible region. The solid line is the *central path*, which we defined in Section 5.1. It connects the point A to the point D, the optimum solution. The asterisks show the iterates of a conventional path-following interior point method. Such a method generates iterates approximately on the central path with spacing between the iterates decreasing geometrically. Once the iterates are sufficiently close to the optimum D, where "sufficiently close" means that the current iterate is closer to D than to any other vertex of the feasible region, then the exact optimum may be computed. Fig. 5.2 provides a close-up view of the region near $(0,0)$ from this problem.

Now, consider what happens as we let ε get smaller in this problem. This creates a "near degeneracy" near $(0,0)$, and the diagonal segment in Figs. 5.1 and 5.2 gets shorter and shorter. This means that the interior point method described in Section 5.1 must take more and more iterations in order to distinguish the optimal vertex $(\varepsilon, 0)$ from the nearby vertex $(0, \varepsilon/2)$. Actually, it can be proved that the method described in Section 5.1 applied to this problem would require $\mathcal{O}(|\log \varepsilon|)$ iterations. Note that ε is a component of the right-hand side vector c; this explains why the methods as those described in Section 5.1 have complexity depending on c. To better understand the statements above assume for a while that $\varepsilon \in \mathbb{Q}$ and consider the bit cost model

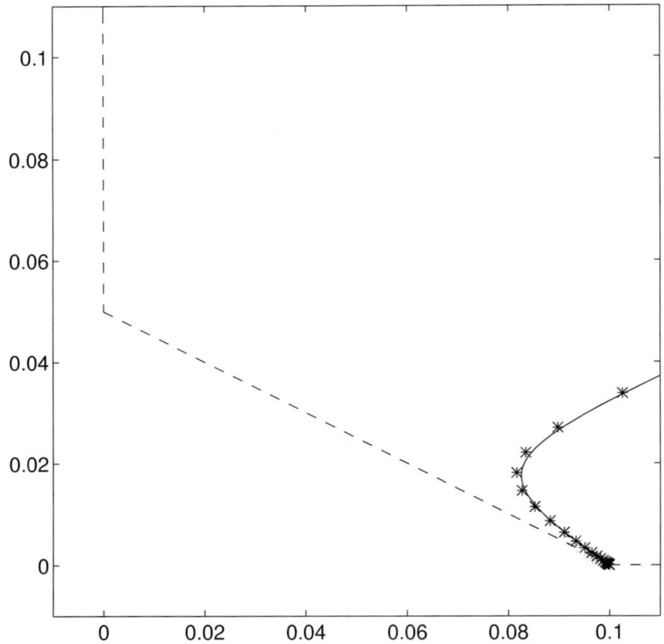

FIG. 5.2. A close-up of the lower-left corner of Fig. 5.1.

of computation. Then the running time of these methods is written as $\mathcal{O}(\sqrt{n}L)$ (or sometimes $\mathcal{O}(nL)$) iterations, where L is the total number of bits required to write A, b, c. Thus, for this example, L will depend logarithmically on ε.

In contrast with the above, the LIP method will jump from point B directly to point C without intervening iterations. This is accomplished by solving a weighted least-squares problem involving the three constraints $y_1 \geqslant 0$, $y_2 \geqslant 0$, $y_1 + 2y_2 \geqslant \varepsilon$, which are "near active" at optimum. In a more complex example, there would be several "layers" involved in the least-squares computation. When point C is reached, the LIP method takes small interior point steps, the number of which depends only on the constraint matrix

$$A^{\mathrm{T}} = \begin{pmatrix} 1 & 0 \\ -1 & 0 \\ 0 & 1 \\ 0 & -1 \\ 1 & 2 \end{pmatrix}.$$

After these small steps, the LIP method jumps again directly to the optimal point D.

An observation concerning this figure is that the central path appears to consist of a curved segment from A to B, an approximately straight segment from B to C, a curved segment near C, and another approximately straight segment from near C to D. Indeed, a consequence of the LIP method is a new characterization of the central path as being composed of at most n^2 curved segments alternating with approximately straight segments. For a problem with no near-degeneracies, there is only one curved segment followed by an approximately straight segment leading to the optimum.

As we discussed in Section 5.1, in an interior point method, points in the central-path are never computed because there is no finite algorithm to solve the nonlinear equations (5.1). Therefore, one defines *approximate centering*. Many definitions are possible, but we use the following proximity measure. Let $\mu > 0$ be fixed, and let (x, y, s) be an interior point feasible for both the primal and the dual. Then we define the proximity measure

$$\eta(x, y, s, \mu) = \|SXe/\mu - e\|$$

(e.g., KOJIMA, MIZUNO and YOSHISE [1989] and MONTEIRO and ADLER [1989]). Note that we then have

$$\mathcal{N}(\eta) = \left\{(x, y, s) \in \mathring{\mathcal{F}}\colon \eta\left(x, y, s, \frac{x^{\mathrm{T}}s}{n}\right) \leqslant \eta\right\}.$$

5.2.1. Layered least squares

The LIP method uses standard path-following primal–dual interior point steps, plus an occasional use of another type of step, the *layered least-squares* (LLS) step. In Section 5.1 we described the standard primal–dual interior point steps. Now we describe the layered step. We remark that in all that follows we still assume the availability of an initial point near the central path.

Let x and s be arbitrary n-vectors, Δ be an arbitrary $n \times n$ positive definite diagonal matrix, and $J_1, \ldots, J_p$ an arbitrary partition of the index set $\{1, \ldots, n\}$. In Section 5.2.2 we will specialize all these objects to the case of the primal–dual interior point method.

Let $A_1, \ldots, A_p$ be the partitioning of the columns of A according to $J_1, \ldots, J_p$. Similarly, let $\Delta_1, \ldots, \Delta_p$ be diagonal matrices that are the submatrices of Δ indexed by $J_1, \ldots, J_p$. Vectors s and x are partitioned in the same way.

The dual LLS step is defined as follows. Define $L_0^{\mathrm{D}} = \mathcal{R}(A^{\mathrm{T}})$. Then for $k = 1, \ldots, p$ define

$$L_k^{\mathrm{D}} := \left\{\text{minimizers } \delta s \text{ of } \left\|\Delta_k^{-1}(\delta s_k + s_k)\right\| \text{ subject to } \delta s \in L_{k-1}^{\mathrm{D}}\right\} \tag{5.12}$$

so that $L_0^{\mathrm{D}} \supset L_1^{\mathrm{D}} \supset \cdots \supset L_p^{\mathrm{D}}$. Finally, let δs^*, the *dual LLS step*, be the unique element in L_p^{D} (it will follow from the next paragraph that this vector is unique).

We claim that δs^* depends linearly on the input vector s. To see this, let us write out the Lagrange multiplier conditions defining the dual LLS step. They are:

$$\begin{aligned}
&A_1\Delta_1^{-2}\delta s_1^* + A_1\Delta_1^{-2}s_1 = \mathbf{0},\\
&A_2\Delta_2^{-2}\delta s_2^* + A_2\Delta_2^{-2}s_2 = A_1\lambda_{2,1},\\
&\qquad\vdots\\
&A_p\Delta_p^{-2}\delta s_p^* + A_p\Delta_p^{-2}s_p = A_1\lambda_{p,1} + \cdots + A_{p-1}\lambda_{p,p-1},\\
&\qquad\qquad \delta s^* = A^{\mathrm{T}}\delta y.
\end{aligned} \tag{5.13}$$

We see that if $(\delta s^*, s)$ and $(\delta\hat{s}^*, \hat{s})$ are two solutions for (5.13), then so is $(\alpha\delta s^* + \beta\delta\hat{s}^*, \alpha s + \beta\hat{s})$ for any choice of α, β (and for some choice of Lagrange multipliers). Thus, there is a linear relation between δs^* and s, i.e. the set of all possible $(\delta s^*, s)$ satisfying (5.13) is the solution space to some system of equations of the form $Q\delta s^* = Ps$.

In fact, we claim this relation is actually a linear function mapping s into δs^* (i.e. Q above is invertible and, changing P if necessary, it may be taken to be the identity matrix). If $s_1 = \mathbf{0}$, then we must have $\delta s_1^* = \mathbf{0}$, since $\Delta_1^{-1}\delta s_1^* \in \mathcal{N}(A_1\Delta_1^{-1})$ from the first equation of (5.13) and also $\Delta_1^{-1}\delta s_1^* \in \mathcal{R}((A_1\Delta_1^{-1})^{\mathrm{T}})$. If in addition $s_2 = \mathbf{0}$, then we have $(-\lambda_{2,1}; \Delta_2^{-1}\delta s_2^*) \in \mathcal{N}((A_1, A_2\Delta_2^{-1}))$ from the second equation of (5.13) and also $(\delta s_1^*; \Delta_2^{-1}\delta s_2^*) \in \mathcal{R}((A_1, A_2\Delta_2^{-1})^{\mathrm{T}})$. Thus, these two vectors are orthogonal, which implies $\delta s_2^* = \mathbf{0}$ since $\delta s_1^* = \mathbf{0}$. Proceeding in this manner by induction, we conclude that $s = \mathbf{0}$ maps to $\delta s^* = \mathbf{0}$. Thus, there is a matrix P (which depends on A, Δ, and $J_1, \ldots, J_p$) mapping s to δs^* and defining the LLS step as claimed above.

The LLS step for the primal is similar, except we work with the nullspace of A, Δ instead of Δ^{-1}, and the opposite order of the partitions. We define the primal layered least-squares step for the same coefficient matrix A, partition J, and weight matrix Δ as follows. For a given input vector x, we construct a sequence of nested subspaces $L_0^{\mathrm{P}} \subset L_1^{\mathrm{P}} \subset \cdots \subset L_p^{\mathrm{P}}$. We define $L_p^{\mathrm{P}} = \mathcal{N}(A)$. Then, for each $k = p, p-1, \ldots, 2, 1$, we define the subspace

$$L_{k-1}^{\mathrm{P}} := \{\text{minimizers } \delta x \text{ of } \|\Delta_k(\delta x_k + x_k)\| \text{ subject to } \delta x \in L_k^{\mathrm{P}}\}.$$

As above, there is a unique minimizer in L_0^{P}. We let δx^* be this minimizer.

The LLS step may be thought of as weighted least-squares with the weights in J_1 "infinitely higher" than the weights of J_2, and so on. We now formalize this idea by showing that the two LLS steps are in fact the limiting case of an ordinary weighted least-squares problem.

LEMMA 5.5. *Consider the dual LLS problem with input data given by an arbitrary partition $J_1, \ldots, J_p$, positive definite diagonal weights Δ, $m \times n$ coefficient matrix A of rank m, and an input data vector s. Let Ξ_J denote the $n \times n$ diagonal matrix whose entries are as follows. For $k = 1, \ldots, p$, for indices $i \in J_k$, the (i, i) entry of Ξ_J is 2^k. Let δs^* be the dual LLS solution, and let $\delta s(t)$ (where t is a natural number) be the solution to the ordinary least-squares problem*

$$\begin{aligned} &\textit{minimize} && \|\Delta^{-1}\Xi_J^{-t}(\delta s + s)\| \\ &\textit{subject to} && \delta s = A^{\mathrm{T}}\delta y \quad (\textit{or } \delta s \in \mathcal{R}(A^{\mathrm{T}})). \end{aligned}$$

Then

$$\lim_{t\to\infty} \delta s(t) = \delta s^*.$$

We have a similar lemma for the primal LLS step.

LEMMA 5.6. *Consider a primal LLS problem with input data given by an arbitrary partition $J_1, \ldots, J_p$, positive definite diagonal weights Δ, $m \times n$ coefficient matrix A*

of rank m, and an input data vector x. Let Ξ_J denote the $n \times n$ diagonal matrix defined in Lemma 5.5. Let δx^* be the primal LLS solution, and let $\delta x(t)$ (where t is a natural number) be the solution to

$$\begin{aligned} &\textit{minimize} \quad \left\| \Delta \Xi_J^t (\delta x + x) \right\| \\ &\textit{subject to} \quad A\delta x = \mathbf{0} \quad (\textit{or } \delta x \in \mathcal{N}(A)). \end{aligned}$$

Then

$$\lim_{t \to \infty} \delta x(t) = \delta x^*.$$

Because of Lemmas 5.5 and 5.6, we introduce infinite exponents to define a general LLS step pair for any given vector x. We write

$$\begin{aligned} \textit{PLLS}(x)\text{:} \quad &\text{minimize} \quad \left\| \Delta \Xi_J^{\infty} (\delta x + x) \right\| \\ &\text{subject to} \quad A\delta x = \mathbf{0} \quad (\text{or } \delta x \in \mathcal{N}(A)), \end{aligned} \tag{5.14}$$

and denote by δx^* the minimizer of this problem, which we call the *primal LLS step*.

Similarly, for any given vector s, define the *dual LLS step* to be the minimizer δs^* of the problem

$$\begin{aligned} \textit{DLLS}(s)\text{:} \quad &\text{minimize} \quad \left\| \Delta^{-1} \Xi_J^{-\infty} (\delta s + s) \right\|, \\ &\text{subject to} \quad \delta s = A^{\mathrm{T}} \delta y \quad (\text{or } \delta s \in \mathcal{R}(A^{\mathrm{T}})). \end{aligned} \tag{5.15}$$

Thus, we regard the LLS solutions as weighted least-squares problems with infinite spacing between weights. Note that the choice of δy^* is also uniquely determined (i.e. there is a unique solution to $A^{\mathrm{T}} \delta y^* = \delta s^*$) because A is assumed to have full rank.

In addition, we have the following properties of the LLS step based on our early discussions:

(1) Vector δx^* (δs^*) is uniquely determined and linear in x (resp. in s). In particular, $\mathbf{0}$ maps to $\mathbf{0}$.

(2) If $x_p = \cdots = x_k = \mathbf{0}$ ($s_1 = \cdots = s_k = \mathbf{0}$), then $\delta x_p^* = \cdots = \delta x_k^* = \mathbf{0}$ (resp. $\delta s_1^* = \cdots = \delta s_k^* = \mathbf{0}$). Note the reverse order between the primal and dual steps.

(3) If $x \in \mathcal{N}(A)$ ($s \in \mathcal{R}(A^{\mathrm{T}})$), then $\delta x^* = -x$ (resp. $\delta s^* = -s$). This is because $\delta x = -x$ make the objective value of (5.14) equal 0, which must be optimal.

Another important property of the points δx^* and δs^* relates them (and x, s) to the condition of the input matrix A. We state this in the next proposition.

PROPOSITION 5.1.

(i) $\|\delta s^*\| \leqslant \bar{\chi}_A \|s\|$,

(ii) $\|\delta s^* + s\| \leqslant (\bar{\chi}_A + 1)\|s\|$,

(iii) $\|\delta x^* + x\| \leqslant \bar{\chi}_A \|x\|$,

(iv) $\|\delta x^*\| \leqslant (\bar{\chi}_A + 1)\|x\|$.

PROOF. The point δs^* is the solution of

$$\begin{aligned} &\text{minimize} \quad \left\| \Delta^{-1} \Xi_J^{-\infty} (\delta s + s) \right\| \\ &\text{subject to} \quad \delta s = A^{\mathrm{T}} \delta y \quad (\text{or } \delta s \in \mathcal{R}(A^{\mathrm{T}})). \end{aligned}$$

Let δy^* be a vector in $\mathbb{R}^n$, s.t. $\delta s^* = A^{\mathrm{T}}\delta y^*$. Then δy^* minimizes $\|\Delta^{-1}\Xi^{-\infty}(A^{\mathrm{T}}\delta y + s)\|$. Since, by equality (4.3),

$$\bar{\chi}_A = \sup\left\{\frac{\|A^{\mathrm{T}}y\|}{\|c\|}\colon\ y \text{ minimizes } \left\|D\left(A^{\mathrm{T}}y - c\right)\right\| \text{ for some } c \in \mathbb{R}^n,\ D \in \mathcal{D}\right\}$$

and $\Delta^{-1}\Xi^{-\infty} \in \mathcal{D}$ it follows that $\|A^{\mathrm{T}}\delta y^*\|/\|s\| \leqslant \bar{\chi}_A$. That is, $\|\delta s^*\| = \|A^{\mathrm{T}}\delta y^*\| \leqslant \bar{\chi}_A\|s\|$. This shows (i). Part (ii) follows from (i).

To prove (iii), consider $\delta x(t)$ defined in Lemma 5.6. For any natural number t, let $D_t = \Delta\Xi_J^t$. Then, we have

$$\delta x(t) = \left(I - D_t^{-2}A^{\mathrm{T}}\left(AD_t^{-2}A^{\mathrm{T}}\right)^{-1}A\right)(-x),$$

or

$$\delta x(t) + x = D_t^{-2}A^{\mathrm{T}}\left(AD_t^{-2}A^{\mathrm{T}}\right)^{-1}Ax.$$

Thus,

$$\left\|\delta x(t) + x\right\| \leqslant \bar{\chi}_A\|x\|.$$

Since $\lim_{t\to\infty}\delta x(t) = \delta x^*$, we must have (iii). Again, (iv) follows (iii). □

We end this subsection by presenting a lemma which is an enhancement of Proposition 5.1 and an important tool used in proving Proposition 5.2.

LEMMA 5.7. (a) *Consider the minimizer δx_0^* of the general primal LLS problem PLLS(x_0) in* (5.14) *for any given x_0. Let E be the $n \times n$ diagonal matrix with 1's in positions corresponding to $J_k \cup \cdots \cup J_p$ and 0's elsewhere. Let $\|u\|_E$ denote the seminorm $(u^{\mathrm{T}}Eu)^{1/2}$. Then*

$$\left\|\delta x_0^* + x_0\right\|_E \leqslant \bar{\chi}_A \cdot \|x_0\|_E$$

and

$$\left\|\delta x_0^*\right\|_E \leqslant (\bar{\chi}_A + 1) \cdot \|x_0\|_E.$$

Furthermore, for an x such that $x_0 - x \in \mathcal{N}(A)$ and such that δx^* is the minimizer of PLLS(x),

$$\|\delta x^* + x\|_E \leqslant \bar{\chi}_A \cdot \|x_0\|_E.$$

(b) *Consider the minimizer δs_0^* of the general dual LLS problem DLLS(s_0) in* (5.15) *for any given s_0. Let D be the $n \times n$ diagonal matrix with 1's in positions corresponding to $J_1 \cup \cdots \cup J_k$ and 0's elsewhere. Let $\|u\|_D$ denote the seminorm $(u^{\mathrm{T}}Du)^{1/2}$. Then*

$$\left\|\delta s_0^* + s_0\right\|_D \leqslant (\bar{\chi}_A + 1) \cdot \|s_0\|_D$$

and

$$\left\|\delta s_0^*\right\|_D \leqslant \bar{\chi}_A \cdot \|s_0\|_D.$$

Furthermore, for an s such that $s_0 - s \in \mathcal{R}(A^{\mathrm{T}})$ and such that δs^* is the minimizer of DLLS(s),

$$\|\delta s^* + s\|_D \leqslant (\bar{\chi}_A + 1) \cdot \|s_0\|_D.$$

5.2.2. One step of the algorithm

We now describe one iteration of the LIP method. Assume at the beginning of the iteration that we have a current approximate centered point, i.e. a point (x, y, s, μ) with $\eta(x, y, s, \mu) \leqslant \eta_0$.

Recall that $\eta(\cdot,\cdot,\cdot,\cdot)$ was defined as proximity to the central path. We assume throughout the rest of this section that $\eta_0 = 0.2$. Partition the index set $\{1, \ldots, n\}$ into p *layers* $J_1, \ldots, J_p$ by using s and x as follows. Let

$$\delta_i = \sqrt{\mu s_i / x_i},$$

where $\mu = x^{\mathrm{T}} s / n$. Let

$$\Delta = \operatorname{diag}(\delta) = \mu^{1/2} S^{1/2} X^{-1/2}. \tag{5.16}$$

Note that if (x, y, s) is perfectly centered, then $SXe = \mu e$, i.e. $\delta = s = \mu X^{-1} e$. Now find a permutation π that sorts these quantities in increasing order:

$$\delta_{\pi(1)} \leqslant \delta_{\pi(2)} \leqslant \cdots \leqslant \delta_{\pi(n)}.$$

Let

$$g = 128(1 + \eta_0) n^2 (\bar{\chi}_A + 1) \tag{5.17}$$

be a "gap size" parameter. Find the leftmost ratio-gap of size greater than g in the sorted slacks, i.e. find the smallest i such that $\delta_{\pi(i+1)}/\delta_{\pi(i)} > g$. Then let $J_1 = \{\pi(1), \ldots, \pi(i)\}$. Now, put $\pi(i+1), \pi(i+2), \ldots$ in J_2, until another ratio-gap greater than g is encountered, and so on. Thus, the values of δ_i for constraints indexed by J_k for any k are within a factor of $g^{|J_k|} \leqslant g^n$ of each other, and are separated by more than a factor of g from constraints in J_{k+1}.

To formalize this, define

$$\theta_k = \max_{i \in J_k} \delta_i \tag{5.18}$$

and

$$\phi_k = \min_{i \in J_k} \delta_i. \tag{5.19}$$

The construction above ensures that for each k,

$$\theta_k < \phi_{k+1}/g$$

and that

$$\phi_k \leqslant \theta_k \leqslant g^n \phi_k.$$

By the assumption of approximate centrality, each diagonal entry of SX is between $\mu(1-\eta_0)$ and $\mu(1+\eta_0)$. Thus, we have the two inequalities

$$\mu^{1/2}\sqrt{1-\eta_0} \leqslant \|S^{1/2}X^{1/2}\| \leqslant \mu^{1/2}\sqrt{1+\eta_0} \tag{5.20}$$

and

$$\mu^{-1/2}(1+\eta_0)^{-1/2} \leqslant \|S^{-1/2}X^{-1/2}\| \leqslant \mu^{-1/2}(1-\eta_0)^{-1/2}, \tag{5.21}$$

which will be used frequently during the upcoming analysis.

Here we explain one main-loop iteration of our algorithm. The iteration begins with a current iterate (x, y, s) that is feasible and approximately centered with $\eta(x, y, s, \mu) \leqslant \eta_0$. Let δx^* be the minimizer of *PLLS*(x) and $(\delta y^*, \delta s^*)$ the minimizer of *DLLS*(s). In other words, δx^* minimizes $\Delta \Xi_J^{\infty}(\delta x + x)$ subject to $\delta x \in \mathcal{N}(A)$, and δs^* minimizes $\Delta^{-1}\Xi_J^{-\infty}(\delta s + s)$ subject to $\delta s \in \mathcal{R}(A^{\mathrm{T}})$. We also let

$$x^* = x + \delta x^*, \quad y^* = y + \delta y^*, \quad \text{and} \quad s^* = s + \delta s^*.$$

For $k = 1, \ldots, p$, define γ_k^{P} and γ_k^{D} as follows:

$$\gamma_k^{\mathrm{P}} = \|\Delta_k \delta x_k^*\|_{\infty}/\mu = \|\Delta_k(x_k^* - x_k)\|_{\infty}/\mu, \tag{5.22}$$

$$\gamma_k^{\mathrm{D}} = \|\Delta_k^{-1}\delta s_k^*\|_{\infty} = \|\Delta_k^{-1}(s_k^* - s_k)\|_{\infty}. \tag{5.23}$$

Notice the use of the infinity-norm in this definition.

There are three possibilities for one main-loop iteration of our algorithm, depending on the following cases.

Case I. There is a layer k such that

$$\gamma_k^{\mathrm{P}} \geqslant \frac{1}{4\sqrt{n}} \quad \text{and} \quad \gamma_k^{\mathrm{D}} \geqslant \frac{1}{4\sqrt{n}}. \tag{5.24}$$

In this case, for one main-loop iteration we take a number of ordinary path-following interior points steps (i.e. those described in Section 5.1).

Case II. There is a layer k such that

$$\gamma_k^{\mathrm{P}} < \frac{1}{4\sqrt{n}} \quad \text{and} \quad \gamma_k^{\mathrm{D}} < \frac{1}{4\sqrt{n}}. \tag{5.25}$$

This case will never happen due to the choice of g.

Case III. For every layer k, either

$$\gamma_k^{\mathrm{P}} \geqslant \frac{1}{4\sqrt{n}} \quad \text{and} \quad \gamma_k^{\mathrm{D}} < \frac{1}{4\sqrt{n}} \tag{5.26}$$

holds, or

$$\gamma_k^{\mathrm{P}} < \frac{1}{4\sqrt{n}} \quad \text{and} \quad \gamma_k^{\mathrm{D}} \geqslant \frac{1}{4\sqrt{n}} \tag{5.27}$$

holds. In this case, for each layer in the LLS step, we define $\varepsilon_k^{\mathrm{P}}$ to be the scaled primal residual, that is,

$$\varepsilon_k^{\mathrm{P}} = \left\| \Delta_k (\delta x_k^* + x_k) \right\|_\infty / \mu = \left\| \Delta_k x_k^* \right\|_\infty / \mu. \tag{5.28}$$

Also, we define the dual residual to be

$$\varepsilon_k^{\mathrm{D}} = \left\| \Delta_k^{-1} (\delta s_k^* + s_k) \right\|_\infty = \left\| \Delta_k^{-1} s_k^* \right\|_\infty. \tag{5.29}$$

Finally, we define α_k for each layer. If (5.26) holds, we take

$$\alpha_k = \min\bigl(1, 8\varepsilon_k^{\mathrm{P}} \sqrt{n}\bigr). \tag{5.30}$$

Else if (5.27) holds, we take

$$\alpha_k = \min\bigl(1, 8\varepsilon_k^{\mathrm{D}} \sqrt{n}\bigr). \tag{5.31}$$

Then we define

$$\bar{\alpha} = \max\{\alpha_1, \ldots, \alpha_p\}. \tag{5.32}$$

Now we take a step defined by the primal and dual LLS directions; we compute a new feasible iterate

$$x^+ = x + (1 - \bar{\alpha})\delta x^* = \bar{\alpha} x + (1 - \bar{\alpha}) x^*,$$
$$y^+ = y + (1 - \bar{\alpha})\delta y^* = \bar{\alpha} y + (1 - \bar{\alpha}) y^*,$$

and

$$s^+ = s + (1 - \bar{\alpha})\delta s^* = \bar{\alpha} s + (1 - \bar{\alpha}) s^* = A^{\mathrm{T}} y^+ - \mathbf{c}.$$

If $\bar{\alpha} = 0$, this is the termination of our algorithm: (x^*, y^*, s^*) is an optimal solution pair for the primal and dual problems as proved in Corollary 5.1 below. Else we set $\mu^+ = \mu\bar{\alpha}$. In Theorem 5.2 below we claim that $\eta(x^+, y^+, s^+, \mu^+) < 0.65$. One can prove that from this inequality it follows that two Newton steps with this fixed μ^+ restore the proximity to 0.2.

After these two Newton steps to restore approximate centering, we take a number of ordinary interior point steps. This concludes the description of Case III.

To summarize, a main-loop iteration of Algorithm LIP reduces μ using one of the three cases in this subsection until the termination criterion in Case III (i.e. the equality $\bar{\alpha} = 0$) holds.

EXAMPLE. We return to Example 5.11, with the hypothesis that ε is very close to zero. This example is illustrated by Fig. 5.1. Assume that μ is chosen so that $|\varepsilon| \ll \mu \ll 1$ and consider the point B (see Fig. 5.1) in the central path corresponding to this μ. Two distinct layers of constraints are observed: those with "small" slacks, namely, $y_1 \geqslant 0$, $y_2 \geqslant 0$ and $y_1 + 2y_2 \geqslant 0$, and the remaining two constraints $y_1 \leqslant 1$, $y_2 \leqslant 1$. In other words, $J_1 = \{1, 3, 5\}$ and $J_2 = \{2, 4\}$.

The central path point for the dual is $(y_1(\mu), y_2(\mu)) \approx (0.791\mu, 0.283\mu)$ with $s(\mu) = A^T y(\mu) - \mathbf{c} \approx (0.791\mu, 1, 0.283\mu, 1, 1.358\mu)$ and $x(\mu) = \mu S(\mu)^{-1}\varepsilon \approx (1.264, \mu, 3.527, \mu, 0.736)$.

Now, the dual LLS step can be derived as follows. We want to find δy^* so that $\|\Delta_1^{-1}(A_1^T(y + \delta y^*) - \mathbf{c}_1)\|$ is minimized (because $\delta s_1^* = A_1^T \delta y^*$ and $A_1^T y - \mathbf{c}_1 = s_1$ by definition). Since A_1^T has rank 2, this first layer completely determines δy^* and therefore δs^* as well. The solution to the dual LLS problem is $\delta y^* = -y^* + \mathcal{O}(\varepsilon)$ since $\|\mathbf{c}_1\| = \mathcal{O}(\varepsilon)$. Thus, $\delta s^* = (-s_1 + \mathcal{O}(\varepsilon), \mathcal{O}(\mu), -s_3 + \mathcal{O}(\varepsilon), \mathcal{O}(\mu), -s_5 + \mathcal{O}(\mu))$. As for the primal LLS step, we have $\delta x_2^* = (-\mu, -\mu)$, and then $\delta x_1^* = (\mathcal{O}(\mu), \mathcal{O}(\mu), \mathcal{O}(\mu))$.

Now, we have $\gamma_1^D = \mathcal{O}(1)$ and $\gamma_2^D = \mathcal{O}(\mu)$. Also, $\gamma_1^P = \mathcal{O}(\mu)$ and $\gamma_2^P = \mathcal{O}(1)$. Therefore, we are in Case III.

Next, we have $\varepsilon_1^D = \mathcal{O}(\varepsilon/\mu)$ and $\varepsilon_2^D = \mathcal{O}(\mu)$. Also, $\varepsilon_1^P = \mathcal{O}(1)$ and $\varepsilon_2^P = 0$. This means that $\alpha_1 = \mathcal{O}(\varepsilon/\mu)$ and $\alpha_2 = 0$, and hence $\bar{\alpha} = \mathcal{O}(\varepsilon/\mu)$. Therefore, we can decrease the central path parameter by a factor of ε/μ, i.e. down to $\mathcal{O}(\varepsilon)$. Thus, as explained at the beginning of Section 5.2, no matter how small ε is, we can always follow the central path with a straight-line segment until we are in a neighborhood of the near-degeneracy. Also, observe that if $\varepsilon = 0$ then $\bar{\alpha} = 0$. Therefore the dual LLS step will step exactly at (0, 0) of the dual (optimal), and the primal LLS step will land at a primal optimal solution as well (but note the primal is degenerate in this case).

We can show that (x^+, y^+, s^+) is approximately centered in Case III above.

THEOREM 5.2. *Let $\alpha \in [\bar{\alpha}, 1]$ be chosen arbitrarily, where $\bar{\alpha}$ is defined by* (5.32) *and Case* III *above holds. In the case that $\bar{\alpha} = 0$, assume further that $\alpha > 0$. Then (x, y, s), defined by $(x, y, s) = \alpha(x, y, s) + (1 - \alpha)(x^*, y^*, s^*)$, is a strictly feasible point. Furthermore,*

$$\eta(x, y, s, \mu\alpha) \leqslant \eta_0 + 3\sqrt{1 + \eta_0}/8 + 1/32.$$

REMARK. If $\eta_0 = 0.2$, then the right-hand side of the preceding inequality is at most 0.65.

Moreover, we have the following.

COROLLARY 5.1. *If $\bar{\alpha} = 0$ in Case* III *of Algorithm LIP, then (x^*, y^*, s^*) is a primal–dual optimal pair.*

From previous results and the crossover analysis in the next subsection, essentially each main loop of the algorithm needs to reduce the duality gap by a factor of $(n \cdot \bar{\chi}_A)^n$ at most. Since each step of the classical predictor-corrector algorithm described earlier reduces the gap by a constant factor, we have the following result.

PROPOSITION 5.2. *There are $\mathcal{O}(\sqrt{n} \cdot c_1(A))$ classical interior-point algorithm steps per main loop iteration, where $c_1(A) = \mathcal{O}(n(\log \bar{\chi}_A + \log n))$.*

5.2.3. Crossover events and LIP's complexity

Theorem 5.2 and Corollary 5.1 in the last subsection prove that Algorithm LIP is valid in the sense that it accurately tracks the central path, and terminates only when an optimum is reached. In this subsection, we describe a key idea used to show that the number of main-loop iterations required by Algorithm LIP is finite, and, in particular, is bounded by $\mathcal{O}(n^2)$.

DEFINITION 5.3. Given an LP problem in primal–dual form, and given a current approximately centered iterate (x, y, s, μ), we say that the 4-tuple (μ, μ', i, j) defines a *crossover event* for $\mu' > 0$, and for $i, j \in \{1, \dots, n\}$ if, for some k,

$$i \in J_1 \cup \cdots \cup J_k \quad \text{and} \quad j \in J_k \cup \cdots \cup J_p \tag{5.33}$$

and for all $\mu'' \in (0, \mu']$,

$$s_i(\mu'') \geqslant 5g^n s_j(\mu''). \tag{5.34}$$

Note that one must have $i \neq j$ for (5.34) to hold, since $g > 1$. Notice also that $\mu' < \mu$ since (5.34) cannot hold for $\mu'' = \mu$.

DEFINITION 5.4. We say that two crossover events (μ, μ', i, j) and $(\mu_1, \mu_1', i_1, j_1)$ are *disjoint* if $(\mu', \mu) \cap (\mu_1', \mu_1) = \emptyset$.

LEMMA 5.8. *Let $(\mu_1, \mu_1', i_1, j_1), \dots, (\mu_t, \mu_t', i_t, j_t)$ be a sequence of t disjoint crossover events for a particular instance of primal–dual LP. Then $t \leqslant n(n-1)/2$.*

The main theorem here is that every pass through the main loop of Algorithm LIP causes at least one crossover event to occur, until $\bar{\alpha} = 0$ (see the original paper for details of the proof). This sequence of crossover events is clearly a disjoint sequence since μ decreases monotonically throughout Algorithm LIP. Therefore, there can be at most $n(n-1)/2$ main-loop iterations of Algorithm LIP before termination.

THEOREM 5.3. *Consider one main loop iteration of Algorithm LIP. (If Case* III *holds for the iteration, assume further that $\bar{\alpha} > 0$, i.e. the algorithm does not terminate.) This iteration causes a crossover event to take place.*

A more careful analysis shows that there are a total of at most $n^2/4$ iterations of Algorithm LIP. Thus, using Proposition 5.2, the total number of small step iterations over all $\mathcal{O}(n^2)$ main loop iterations is $\mathcal{O}(n^{3.5}c(A))$ where

$$c(A) = 2c_0\big(\log(\bar{\chi}_A) + 2\log n + \text{const}\big). \tag{5.35}$$

Here c_0 is a universal constant bounded above by 10. Each small step requires solution of a system of linear equations, which takes $\mathcal{O}(m^2 n)$ arithmetic operations.

It should be noted that this complexity bound is independent of the size of ε that is given as input data to Algorithm LIP. This is an important point for an initialization procedure to generate an initial point to start the algorithm.

REMARK 5.1. The complexity bound for the LIP method is not the first for linear programming that depends only on A; TARDOS [1986] earlier proposed such a method. Tardos' method, however, "probably should be considered a purely theoretical contribution" (TARDOS [1986], p. 251) because it requires the solution of n complete LP's. In contrast, the LIP method is a fairly standard kind of interior point method accompanied by an acceleration step; accordingly, we believe that it is quite practical.

Another major difference is that we regard our method as primarily a real-number method. Our method is the first polynomial-time linear programming algorithm that also has a complexity bound depending only on A in the real-number model of computation for finding an optimum. In contrast, Tardos uses the assumption of integer data in a fairly central way: an important tool in TARDOS [1986] is the operation of rounding down to the nearest integer. It is not clear how to generalize the rounding operation to noninteger data.

The biggest barrier preventing Algorithm LIP from being fully general-purpose is the fact that we do not know how to compute or obtain good upper bounds on χ_A or $\bar{\chi}_A$. There are several places in our algorithm where explicit knowledge of $\bar{\chi}_A$ is used – the most crucial use of this knowledge is in the formula for g given by (5.17), which determines the spacing between the layers. Note that we do not need to know these parameters exactly – upper bounds will suffice.

The only known algorithm for computing these parameters is implicit in the results of STEWART [1989] and O'LEARY [1990] and requires exponential-time. We suspect that computing them, or even getting a good upper bound, may be a hard problem. KHACHIJAN [1994] has shown that it is NP-hard to compute or approximate χ_A, and his results may extend to $\bar{\chi}_A$.

A very straightforward "guessing" algorithm was suggested by J. Renegar. Simply guess $\bar{\chi}_A = 100$ and run the entire LIP algorithm. If it fails in any way, then guess $\bar{\chi}_A = 100^2$ and try again. Repeatedly square the guesses, until Algorithm LIP works. (Note that, since Algorithm LIP produces complementary primal and dual solutions, we have a certificate concerning whether it terminates accurately.) This multiplies the running time by a factor of $\log\log\bar{\chi}_A$.

Another possibility would be to obtain a bound on $\bar{\chi}_A$ for special classes of linear programs that occur in practice. For example, Vavasis and Ye carried out this process for min-cost flow problems where $\bar{\chi}_A \leqslant O(mn)$. Other good candidates for special-case analysis would be linear programs arising in scheduling, optimization problems involving finite-element subproblems, and LP relaxations of combinatorial optimization problems.

The idea of partitioning the slacks into layers based on their relative sizes has been proposed by KALISKI and YE [1993] and TONE [1993], who both propose a decomposition into two layers. The interest of these authors is in improving the running-time of computing one iteration of an interior point method, rather than in obtaining new bounds on the number of iterations. WRIGHT [1976] in her PhD thesis also proposed the partitioning idea in the more general setting of barrier methods for constrained optimization.

5.3. An example of round-off (and complexity) analysis

We next describe the main features of the analysis of a finite-precision interior-point algorithm. This analysis includes both complexity and round-off.

Recall, for a matrix $A \in \mathbb{R}^{m\times n}$ we considered in Section 4.4 the systems (4.1) and (4.2) given respectively by

$$Ax = 0, \quad x \geqslant 0, \ x \neq 0$$

and

$$A^{\mathrm{T}}y \leqslant 0, \quad y \neq 0.$$

In CUCKER and PEÑA [2001], an algorithm is described which, for well-posed pairs, decides which of (4.1) and (4.2) has a strict solution (i.e. one for which the inequalities above are strict in all components) and produce such a solution. A key feature of that algorithm is that it does not assume infinite precision for real number arithmetic. This kind of computations are performed with finite precision. The round-off unit, though, varies during the computation.

The assumption of finite precision sets some limitations on the kind of results one may obtain. If system (4.2) has strict solutions then one may obtain, after sufficiently refining the precision, a strict solution $y \in \mathbb{R}^m$ of (4.2). On the other hand, if the system having a strict solution is (4.1), then there is no hope to exactly compute one such solution x since the set of solutions is thin in $\mathbb{R}^n$ (i.e. has empty interior). In such case there is no way to ensure that the errors produced by the use of finite precision will not move any candidate for a solution out of this set. One can however compute good approximations, namely, forward-approximate solutions.

DEFINITION 5.5. Let $\gamma \in (0, 1)$. A point $x \in \mathbb{R}^n$ is a *γ-forward solution* of the system $Ax = 0$, $x \geqslant 0$, if $x \neq 0$, and there exists $\bar{x} \in \mathbb{R}^n$ such that

$$A\bar{x} = 0, \quad \bar{x} \geqslant 0$$

and, for $i = 1, \ldots, n$,

$$|x_i - \bar{x}_i| \leqslant \gamma x_i.$$

The point $\bar{x}$ is said to be an *associated solution* for x. A point is a *forward-approximate solution* of $Ax = 0$, $x \geqslant 0$ if it is a γ-forward solution of the system for some $\gamma \in (0, 1)$.

In case system (4.1) has a strict solution, the algorithm in CUCKER and PEÑA [2001] finds a forward approximate solution. Actually, if the desired accuracy γ of this approximation is given to the algorithm, the returned solution is a γ-forward solution.

The main result in CUCKER and PEÑA [2001] can be stated as follows (recall Remark 4.1 for the notion of total cost in the statement).

THEOREM 5.4. *There exists a finite precision algorithm which, with input a matrix $A \in \mathbb{R}^{m\times n}$ and a number $\gamma \in (0, 1)$, finds either a strict γ-forward solution $x \in \mathbb{R}^n$ of*

$Ax = 0$, $x \geqslant 0$, or a strict solution $y \in \mathbb{R}^m$ of the system $A^{\mathrm{T}} y \leqslant 0$. The round-off unit varies during the execution of the algorithm. The finest required precision is

$$u = \frac{1}{\mathbf{c}(m+n)^{12} C(A)^2},$$

where $\mathbf{c}$ is a universal constant. The number of main (interior-point) iterations of the algorithm is bounded by

$$\mathcal{O}\big((m+n)^{1/2}\big(\log(m+n) + \log\big(C(A)\big) + |\log\gamma|\big)\big).$$

The number of arithmetic operations performed by the algorithm is bounded by $\mathcal{O}((m + n)^{3.5}(\log(m+n) + \log C(A) + |\log\gamma|))$. The total cost of the algorithm is

$$\mathcal{O}\big((m+n)^{3.5}\big(\log(m+n) + \log C(A) + |\log\gamma|\big)^3\big).$$

The bounds above are for the case (4.1) *is strictly feasible. If, instead,* (4.2) *is strictly feasible, then similar bounds hold without the $|\log\gamma|$ terms.*

Note that if none of (4.1) and (4.2) has a strict solution then the problem is *ill-posed*. That is, either system can be made infeasible (i.e. without solutions) by making arbitrarily small perturbations on A. When this occurs, we do not expect the algorithm to yield any solution. Indeed, if A is ill-posed then the algorithm will not halt.

REMARK 5.2. The analysis in CUCKER and PEÑA [2001] was done for $C(A)$ defined with respect to the operator norm $\| \; \|_{1,\infty}$ given by

$$\|A\|_{1,\infty} := \sup\big\{\|Ax\|_1 \colon \|x\|_\infty \leqslant 1\big\}.$$

Thus, in what follows, we will use this norm as well. In addition, we will assume the matrix A has been normalized such that, for $k = 1, \ldots, n$

$$\|Ae_k\|_1 = 1.$$

This is equivalent to assume that, if a_k denotes the kth column of A, $\|a_k\|_1 = 1$ for $k = 1, \ldots, n$. This assumption is trivial from a computational viewpoint; it takes a few operations to reduce the matrix to have this form. The condition number of the new matrix may have changed though but not too much (cf. Proposition 4.1). Notice that, as a consequence of our assumption, $1 \leqslant \|A\|_{1,\infty} \leqslant n$.

We next explain the main ideas behind the proof of Theorem 5.4. For this we rely on the intuitions provided in Section 5.1.

We approach the feasibility problems $Ax = 0$, $x \geqslant 0$ and $A^{\mathrm{T}} y \leqslant 0$ by studying the related pair of optimization problems

$$\begin{aligned} \min \quad & \|\tilde{x}\|_1 \\ \text{s.t.} \quad & Ax + \tilde{x} = 0, \\ & x \geqslant 0, \\ & \|x\|_\infty \leqslant 1, \end{aligned} \tag{5.36}$$

and

$$\begin{array}{ll} \min & \|\tilde{y}\|_1 \\ \text{s.t.} & A^{\mathrm{T}} y + \tilde{y} \leqslant 0, \\ & \tilde{y} \leqslant 0, \\ & \|y\|_\infty \leqslant 1. \end{array} \tag{5.37}$$

Denoting by e_m the vector in $\mathbb{R}^m$ whose components are all 1 we can recast these problems as the following primal–dual pair of linear programs:

$$\begin{array}{ll} \min & e_m^{\mathrm{T}} x' + e_m^{\mathrm{T}} x'' \\ \text{s.t.} & \begin{bmatrix} A & I_m & -I_m & \\ I_n & & & I_n \end{bmatrix} \begin{bmatrix} x \\ x' \\ x'' \\ x''' \end{bmatrix} = \begin{bmatrix} 0 \\ e_n \end{bmatrix}, \\ & x, x', x'', x''' \geqslant 0, \end{array} \tag{5.38}$$

and

$$\begin{array}{ll} \max & e_n^{\mathrm{T}} y' \\ \text{s.t.} & \begin{bmatrix} A^{\mathrm{T}} & I_n \\ I_m & \\ -I_m & \\ & I_n \end{bmatrix} \begin{bmatrix} y \\ y' \end{bmatrix} + \begin{bmatrix} s \\ s' \\ s'' \\ s''' \end{bmatrix} = \begin{bmatrix} 0 \\ e_m \\ e_m \\ 0 \end{bmatrix}, \\ & s, s', s'', s''' \geqslant 0. \end{array} \tag{5.39}$$

We shall apply a primal–dual interior-point method to the pair (5.38), (5.39). A basic feature of interior-point methods is to generate iterates that are pushed away from the boundary of the feasible region. In addition, for the pair (5.38), (5.39), it is obvious that at any optimal solution the variables x', x'', y' are all zero. Hence it is intuitively clear that an interior-point algorithm applied to (5.38), (5.39) will yield a strict solution for either $Ax = 0$, $x \geqslant 0$ or $A^{\mathrm{T}} y \leqslant 0$ provided the pair of systems is well-posed (i.e. $\rho(A) > 0$). Propositions 5.6–5.9 below formalize this statement.

In order to simplify the exposition, we will use the following notation

$$\vec{x} = \begin{bmatrix} x \\ x' \\ x'' \\ x''' \end{bmatrix}, \quad \vec{s} = \begin{bmatrix} s \\ s' \\ s'' \\ s''' \end{bmatrix}, \quad \vec{y} = \begin{bmatrix} y \\ y' \end{bmatrix},$$

and

$$\mathcal{A} = \begin{bmatrix} A & I_m & -I_m & \\ I_n & & & I_n \end{bmatrix}, \quad \vec{b} = \begin{bmatrix} 0 \\ e_n \end{bmatrix}, \quad \vec{c} = \begin{bmatrix} 0 \\ e_m \\ e_m \\ 0 \end{bmatrix}.$$

Thus, we can write the primal–dual pair (5.38), (5.39) in the more compact and standard linear programming form

$$\begin{aligned} \min \quad & \vec{c}^{\mathrm{T}}\vec{x} \\ \text{s.t.} \quad & \mathcal{A}\vec{x} = \vec{b}, \\ & \vec{x} \geqslant 0, \end{aligned} \tag{5.40}$$

and

$$\begin{aligned} \max \quad & \vec{b}^{\mathrm{T}}\vec{y} \\ \text{s.t.} \quad & \mathcal{A}^{\mathrm{T}}\vec{y} + \vec{s} = \vec{c}, \\ & \vec{s} \geqslant 0. \end{aligned} \tag{5.41}$$

Recall that the *central path* $\mathcal{C}$ of the pair (5.40), (5.41) is the set of solutions of the nonlinear system of equations

$$\begin{aligned} \mathcal{A}\vec{x} &= \vec{b}, \\ \mathcal{A}^{\mathrm{T}}\vec{y} + \vec{s} &= \vec{c}, \\ \vec{X}\vec{S}e_{2(m+n)} &= \mu e_{2(m+n)}, \end{aligned} \tag{5.42}$$

with $\vec{x}, \vec{s} \geqslant 0$ for all values of the parameter $\mu > 0$.

Let w denote a generic point $(\vec{x}, \vec{y}, \vec{s})$ and for such a point define

$$\mu(w) := \frac{e_{2(m+n)}^{\mathrm{T}}\vec{X}\vec{S}e_{2(m+n)}}{2(m+n)} = \frac{1}{2(m+n)}\sum_{i=1}^{2(m+n)} \vec{x}_i\vec{s}_i.$$

Note that if $w \in \mathcal{C}$ for a certain value of μ then $\mu(w) = \mu$. We may sometimes write μ for $\mu(w)$ when w is clear from the context. We now briefly describe the algorithm. This is essentially a standard primal–dual short-step algorithm (cf. MONTEIRO and ADLER [1989] or WRIGHT [1997], Chapter 5) enhanced with two additional features. One of these features is the stopping criteria and the other one is the presence of finite precision and the adjustment of this precision as the algorithm progresses. To ensure the correctness of the algorithm, the precision will be set to

$$\phi\big(\mu(w)\big) := \min\big\{\mu(w)^2, 1\big\}\frac{1}{\mathbf{c}(m+n)^{12}}$$

at each iteration. Here $\mathbf{c}$ is a universal constant.

Let $\eta = 1/4$ and $\xi = 1/12$. Also, if M is a matrix, $\sigma_{\min}(M)$ denotes its smallest singular value. Finally, in the sequel, we skip the subindices denoting dimension in the vectors e_m.

input (A, γ)

(i) Set the machine precision to $u := 1/\mathbf{c}(m+n)^{12}$

$$K := \frac{2mn}{\eta},$$

$$w := \big(\tfrac{1}{2}e, Ke, \tfrac{1}{2}Ae + Ke, \tfrac{1}{2}e, 0, -2Ke, 2Ke, e, e, 2Ke\big)$$

(ii) Set the machine precision to $u := \phi(\mu(w))$.
(iii) If $A^{\mathrm{T}} y < -2u(\lceil \log_2 m \rceil + 1)e$ then HALT and return y as a feasible solution for $A^{\mathrm{T}} y < 0$.
(iv) If $\sigma_{\min}(X^{1/2} S^{-1/2} A^{\mathrm{T}}) > \frac{3(m+n)\mu(w)^{1/2}}{\gamma(1-2\eta)}$, then HALT and return x as a γ-forward solution for $Ax = 0$, $x > 0$.
(v) Set $\bar{\mu} := \big(1 - \frac{\xi}{\sqrt{2(m+n)}}\big)\mu(w)$.
(vi) Update w by solving a linearization of (5.42) for $\mu = \bar{\mu}$.
(vii) Go to (ii).

A key step in the analysis of the algorithm above is understanding the properties of the update of w performed at step (vi). The update w^+ is defined as $w^+ = w - \Delta w$ where $\Delta w = (\Delta\vec{x}, \Delta\vec{y}, \Delta\vec{s})$ solves the linear system

$$\begin{bmatrix} \mathcal{A} & & \\ & \mathcal{A}^{\mathrm{T}} & I \\ \vec{S} & & \vec{X} \end{bmatrix} \begin{bmatrix} \Delta\vec{x} \\ \Delta\vec{y} \\ \Delta\vec{s} \end{bmatrix} = \begin{bmatrix} 0 \\ 0 \\ \vec{X}\vec{S}e - \bar{\mu}e \end{bmatrix}. \tag{5.43}$$

Because of round-off errors, the computed Δw will not satisfy the first two constraints in (5.43). The following notation will help to quantify this loss of accuracy.

Given a point $w = (\vec{x}, \vec{y}, \vec{s})$, let us denote by $\underline{w} := (\underline{\vec{x}}, \vec{y}, \underline{\vec{s}})$ the vector

$$\big(x, x', Ax + x', e - x, y, y', -A^{\mathrm{T}} y - y', e - y, e + y, -y'\big).$$

If (5.43) were solved with exact arithmetic, then the new iterate would satisfy $w^+ = \underline{w^+}$. This will not be the case because of the finite precision on the computations. However, the difference between these vectors can be bounded componentwise, as the following proposition states.

PROPOSITION 5.3. *If step* (vi) *above is performed with round-off unit* $\phi(\mu(w))$ *then the new iterate* w^+ *satisfies*

$$\|w^+ - \underline{w^+}\|_\infty \leqslant \frac{\eta \min\{\mu(w^+), 1\}}{20(m+n)^2}. \tag{5.44}$$

The considerations above suggest defining the following two enlargements of the central path.

DEFINITION 5.6. Recall that, given $\eta \in (0, \frac{1}{4}]$, the *central neighborhood* $\mathcal{N}(\eta)$ is defined as the set of points $w = (\vec{x}, \vec{y}, \vec{s})$, with $\vec{x}, \vec{s} > 0$, such that the following constraints hold

$$\begin{aligned} \mathcal{A}\vec{x} &= \vec{b}, \\ \mathcal{A}^{\mathrm{T}}\vec{y} + \vec{s} &= \vec{c} \\ \|\vec{X}\vec{S}e - \mu(w)e\| &\leqslant \eta\mu(w). \end{aligned}$$

The *extended central neighborhood* $\overline{\mathcal{N}(\eta)}$ is thus defined by

$$\overline{\mathcal{N}(\eta)} := \left\{ w\colon\ \underline{w} \in \mathcal{N}(\eta) \text{ and } \|w - \underline{w}\|_\infty \leqslant \frac{\eta \min\{\mu(w), 1\}}{20(m+n)^2} \right\}.$$

REMARK 5.3. Ideally, one would like to generate a sequence of points on the central path $\mathcal{C}$ with values of μ decreasing to zero. Limitations on our ability to solve nonlinear systems have led interior-point methods to generate the sequence above in the central neighborhood $\mathcal{N}(\eta)$. The additional limitations arising from the use of finite precision lead us to generate this sequence in the extended central neighborhood $\overline{\mathcal{N}(\eta)}$.

We are now ready to present the stepping stones towards the proof of Theorem 5.4.

PROPOSITION 5.4. *Let $w \in \overline{\mathcal{N}(\eta)}$ and suppose $w^+ = w - \Delta w$ is obtained as described above (with finite precision). Then*

(i)

$$\left(1 - \frac{3\xi}{2\sqrt{2(m+n)}}\right)\mu(w) \leqslant \mu(w^+) \leqslant \left(1 - \frac{\xi}{2\sqrt{2(m+n)}}\right)\mu(w),$$

and

(ii) $w^+ \in \overline{\mathcal{N}(\eta)}$.

PROPOSITION 5.5. *For $K \geqslant 2mn/\eta$ the point $w_0 = (\vec{x}, \vec{y}, \vec{s})$, defined as follows, belongs to $\mathcal{N}(\eta)$:*

$$x = x''' = \tfrac{1}{2}e, \quad x' = Ke, \quad x'' = \tfrac{1}{2}Ae + Ke,$$
$$y = 0, \quad s' = s'' = e, \quad s = s''' = -y' = 2Ke.$$

In addition, if this point is computed with round-off unit $\mathbf{c}(m+n)^{-12}$, the resulting point belongs to $\overline{\mathcal{N}(\eta)}$.

PROPOSITION 5.6. *If in step* (iii) *the algorithm above yields* (*with round-off errors*) $A^{\mathrm{T}}y < -2u(\lceil \log_2 m \rceil + 1)e$ *then y is a strict solution of* (4.2), *i.e.* $A^{\mathrm{T}}y < 0$.

PROPOSITION 5.7. *Assume $w \in \overline{\mathcal{N}(\eta)}$. If in step* (iv) *the algorithm above yields* (*with round-off errors*)

$$\sigma_{\min}\left(X^{1/2}S^{-1/2}A^{\mathrm{T}}\right) > \frac{3(m+n)\mu(w)^{1/2}}{\gamma(1-2\eta)}$$

then x is a γ-forward solution of (4.1) *and the projection*

$$\bar{x} = x - XS^{-1}A^{\mathrm{T}}\left(AXS^{-1}A^{\mathrm{T}}\right)^{-1}Ax$$

is an associated solution for x.

PROPOSITION 5.8. *Assume* (4.2) *is strictly feasible and $w \in \overline{\mathcal{N}(\eta)}$. If $\mu(w) \leqslant \rho(A)/20(n+m)^2$ then $A^{\mathrm{T}}y < -4u(\lceil \log_2 m \rceil + 1)e$ holds exactly and the algorithm halts in step* (iii).

PROPOSITION 5.9. *Let $\gamma \in (0,1)$, assume $w \in \overline{\mathcal{N}(\eta)}$ and* (4.1) *is strictly feasible. If*

$$\mu(w) \leqslant \frac{(1-2\eta)^2 \rho(A)}{20(m+n)^{5/2}} \left(1+\frac{1}{\gamma}\right)^{-1}$$

then $\sigma_{\min}(X^{1/2}S^{-1/2}A^{\mathrm{T}}) > 4(m+n)\mu(w)^{1/2}/\gamma(1-2\eta)$ holds exactly and the algorithm halts in step (iv).

Note that Propositions 5.4 and 5.5 contain the geometric description of the algorithm. Basically, one begins with a point $w_0 \in \overline{\mathcal{N}(\eta)}$ and iteratively constructs a sequence of points w_k in $\overline{\mathcal{N}(\eta)}$. Since the associated sequence $\mu(w_k)$ tends to zero these points follow the central path while it approaches a solution and become increasingly closer to it.

PROOF OF THEOREM 5.4. The correctness of the algorithm is ensured by Propositions 5.6 and 5.7.

Concerning complexity, we first bound the number of iterations. If w_k denotes the value of w at the kth iteration of the algorithm we have that, by Proposition 5.4, $\mu(w_k) \leqslant \mu(w_0)\alpha^k$ with $\alpha = (1-\delta/2\sqrt{2(m+n)})$.

Assume (4.1) is strictly feasible. Then, by Proposition 5.9, when

$$\mu(w_k) \leqslant \varepsilon_P = \frac{(1-\eta)\rho(A)}{12(m+n)^2}\left(1+\frac{1}{\gamma}\right)^{-1}$$

the algorithm will halt. From here it follows that the number of iterations performed by the algorithm before it halts is no more than

$$\frac{\log \mu(w_0) + |\log \varepsilon_P|}{|\log \alpha|}.$$

But $|\log\alpha| \geqslant \delta/2\sqrt{2(m+n)}$ from which, using that $\mu(w_0) = \mathcal{O}(mn)$ (this follows from the assumption on A we did in Remark 5.2) it follows that the number of iterations done by the algorithm is bounded by

$$\mathcal{O}\big(\sqrt{m+n}\big(\log(m+n) + \big|\log\rho(A)\big| + |\log\gamma|\big)\big).$$

In case (4.2) is strictly feasible the same reasoning (using the bound given by Proposition 5.8 in the place of ε_P) applies to obtain a similar expression but without the term $|\log\gamma|$.

The arithmetic complexity of the algorithm is dominated by the cost of step (vi). This step amounts to compute the QR factorization of a $2(m+n) \times (m+n)$ matrix and the latter can be done with $4(m+n)^3$ arithmetic operations (cf. BJÖRCK [1996], Section 2.4.1 or DEMMEL [1997], Section 3.4). Multiplying this bound by the number of iterations it follows that the arithmetic complexity of the algorithm is bounded by

$$\mathcal{O}\big((m+n)^{7/2}\big(\log(m+n) + \big|\log\rho(A)\big| + |\log\gamma|\big)\big)$$

if (4.1) is strictly feasible and by a similar expression but without the term $|\log\gamma|$ if (4.2) is.

Since by Propositions 5.8 and 5.9 the value of $\mu(w)$ during the execution of the algorithm is not smaller than $\Omega(\rho(A)(m+n)^{-2}\gamma)$ it follows that $|\log\phi(\mu(w))|$ remains bounded above by

$$\mathcal{O}\big(\log n + \big|\log\rho(A)\big| + |\log\gamma|\big)$$

(without the term $|\log\gamma|$ if (4.2) is strictly feasible). The bound for the total cost now follows using that $\|A\|_{1,\infty} \geqslant 1$ and therefore $|\log\rho(A)| \leqslant \log(C(A))$. □

6. Probabilistic analysis of condition measures

6.1. Introduction

Condition numbers have two probably disturbing features. One is that, in general, we do not have an a priori knowledge of the condition of an input. Unlike the size of a, which is straightforwardly obtained from a, the condition number of a seems to require a computation which is not easier than solving the problem for which a is an instance. The other is that there are no bounds on their magnitude as a function of the input size. They may actually be infinite.

A reasonable way to cope with these features is to assume a probability measure on the space of inputs and to estimate the expected value (and if possible, other moments) of the condition number (or of its logarithm). Such an estimate may be used to get average complexity bounds or similar bounds in round-off analysis.

We say that a random matrix is *Gaussian* when its entries (real and imaginary parts of the entries for the complex case) are i.i.d. $N(0,1)$ random variables. Let log denote the logarithm to the base 2. A basic result concerning average condition in linear algebra is the following.

THEOREM 6.1. *Let A be a $n \times n$ Gaussian matrix. Then the expected value of $\log(\kappa(A))$ satisfies*

$$\mathbf{E}\big(\log\big(\kappa(A)\big)\big) = \log n + c + \mathrm{o}(1) \quad \textit{when } n \to \infty,$$

where $c \approx 1.537$ for real matrices and $c \approx 0.982$ for complex matrices.

In this section we extend Theorem 6.1 to various condition measures for linear programming.

THEOREM 6.2. *Let A be a Gaussian $m \times n$ matrix. Then*

$$\mathbf{E}\big(\log\mathscr{C}(A)\big) = \begin{cases} \mathcal{O}(\min\{n, m\log n\}) & \text{if } n > m,\\ \mathcal{O}(\log n) & \text{otherwise,}\end{cases}$$

$$\mathbf{E}\big(\log C(A)\big) = \begin{cases} \mathcal{O}(\min\{n, m\log n\}) & \text{if } n > m,\\ \mathcal{O}(\log m) & \text{otherwise}\end{cases}$$

and, if $n > m > 3$,

$$\mathbf{E}(\log\bar{\chi}_A), \mathbf{E}\big(\log\sigma(A)\big) = \mathcal{O}\big(\min\{n, m\log n\}\big).$$

6.2. *Linear systems and linear cones*

To prove Theorem 6.2 we relate $\mathcal{C}(A)$ and $\bar{\chi}_A$ with the condition number (in the linear algebra sense) of the $m \times m$ submatrices of A. For the rest of this section assume $n \geqslant m$.

Let B be an $m \times m$ real matrix. We define $\mu(B) = \|B^{-1}\| \|B\|_F$ where $\|B^{-1}\|$ denotes the operator norm of B and $\|B\|_F = (\sum_{i,j \leqslant m} b_{ij}^2)^{1/2}$ its Frobenius norm. Note the closeness between $\mu(B)$ and $\kappa(B)$ defined in Section 4.2.

THEOREM 6.3. *Let $\mathcal{B}$ be the set of all $m \times m$ submatrices of A. Then*
(i) $\mathcal{C}(A) \leqslant \max_{B \in \mathcal{B}} \mu(B)$.
(ii) $\bar{\chi}_A \leqslant (\|A\|/\min_{1 \leqslant k \leqslant n} \|a_k\|) \max_{B \in \mathcal{B}} \mu(B)$.

6.2.1. *Proof of Theorem* 6.3(i)

Let $I = \{1, 2, \ldots, n\}$, $\bar{I} = \{i \in I \mid \theta_i(A, \bar{y}) = \theta(A, \bar{y})\}$ and $\overline{B}$ be the submatrix of A corresponding to the index set $\bar{I}$. Denote by y^* any vector in $\mathbb{R}^m$ such that $\theta(\overline{B}, y^*) = \sup_{y \in \mathbb{R}^m} \theta(\overline{B}, y)$. In addition, for any $m \times m$ matrix B, let $v(B) = (\|b_1\|, \|b_2\|, \ldots, \|b_m\|)^{\mathrm{T}}$. Note that $\|v(B)\| = \|B\|_F$.

LEMMA 6.1. *Let B' be any matrix obtained by removing some columns from $\overline{B}$. Then*

$$B'\left(\mathcal{C}(A)\frac{\bar{y}}{\|\bar{y}\|}\right) = \begin{cases} v(B') & \text{if } \theta(A) \leqslant \frac{\pi}{2}, \\ -v(B') & \text{otherwise.} \end{cases}$$

PROOF. By definition of $\overline{B}$, $\theta_i(\overline{B}, \bar{y}) = \theta(A, \bar{y})$ for $i \in \bar{I}$. Therefore, for any $i \in \bar{I}$,

$$\begin{aligned} a_i\left(\mathcal{C}(A)\frac{\bar{y}}{\|\bar{y}\|}\right) &= \|a_i\| \left\|\left(\mathcal{C}(A)\frac{\bar{y}}{\|\bar{y}\|}\right)\right\| \cos\theta_i(A, \bar{y}) \\ &= \|a_i\| \mathcal{C}(A) \cos\theta(A) \\ &= \begin{cases} \|a_i\| & \text{if } \cos\theta(A) \geqslant 0, \\ -\|a_i\| & \text{if } \cos\theta(A) < 0. \end{cases} \end{aligned}$$

The statement follows from the definition of $v(B')$. □

LEMMA 6.2. *If B is an invertible square matrix then $\|B^{-1}v(B)\| \leqslant \mu(B)$.*

PROOF. $\|B^{-1}v(B)\| \leqslant \|B^{-1}\| \|v(B)\| = \|B^{-1}\| \|B\|_F = \mu(B)$. □

LEMMA 6.3. *If $|\bar{I}| \geqslant m$ then $\mathcal{C}(A) \leqslant \max_{B \in \mathcal{B}} \mu(B)$.*

PROOF. Let B' be any $m \times m$ submatrix of $\overline{B}$. If B' is not invertible then $\mu(B') = \infty$ and the result is trivially true. Otherwise, we apply Lemmas 6.1 and 6.2 to deduce that $\mathcal{C}(A) = \|B'^{-1}v(B')\| \leqslant \mu(B')$. Since $B' \in \mathcal{B}$ the proof is completed. □

LEMMA 6.4. $\mathcal{C}(A) \leqslant \mathcal{C}(\overline{B})$.

PROOF. First assume that $A \in \mathcal{F}$. Then, by (ii) and (iii) in Lemma 4.1, $\theta(A) > \pi/2$. By definition of $\overline{B}$, $\theta_i(\overline{B}, \bar{y}) = \theta(A)$ for $i \in \bar{I}$. Thus, $\theta(\overline{B}, \bar{y}) = \theta(A)$ and hence $\theta(\overline{B}, y^*) \geqslant \theta(\overline{B}, \bar{y}) > \pi/2$. Since $|\cos t|$ increases on $[\pi/2, \pi]$

$$\mathcal{C}(A) = \frac{1}{|\cos\theta(A)|} = \frac{1}{|\cos\theta(\overline{B}, \bar{y})|} \leqslant \frac{1}{|\cos\theta(\overline{B}, y^*)|} = \mathcal{C}(\overline{B}).$$

Now assume that $A \notin \mathcal{F}$. Then, by Lemma 4.1, $\theta(A) \leqslant \pi/2$. By definition of $\overline{B}$, $\theta_i(\overline{B}, \bar{y}) = \theta(A)$ for $i \in \bar{I}$. Since, for $i \in \bar{I}$, $\theta_i(\overline{B}, -\bar{y}) = \pi - \theta_i(\overline{B}, \bar{y})$ we deduce $\theta(\overline{B}, -\bar{y}) = \pi - \theta(A) \geqslant \pi/2$. In particular $|\cos\theta(\overline{B}, -\bar{y})| = |\cos\theta(A)|$. By definition of y^*, $\theta(\overline{B}, y^*) \geqslant \theta(\overline{B}, -\bar{y}) \geqslant \pi/2$. Again, since $|\cos t|$ increases on $[\pi/2, \pi]$, we have

$$\mathcal{C}(A) = \frac{1}{|\cos\theta(A, \bar{y})|} = \frac{1}{|\cos\theta(\overline{B}, -\bar{y})|} \leqslant \frac{1}{|\cos\theta(\overline{B}, y^*)|} = \mathcal{C}(\overline{B}). \qquad \square$$

LEMMA 6.5. *Suppose $|I| < m$. Let B' be any $m \times m$ matrix submatrix of A containing $\overline{B}$ as a submatrix. Then $\mathcal{C}(\overline{B}) \leqslant \mathcal{C}(B')$.*

PROOF. Let y' be any vector in $\mathbb{R}^m$ such that $\theta(B', y') = \sup_{y \in \mathbb{R}^m} \theta(B', y)$. Since B' has more columns that $\overline{B}$, $\theta(B', y') \leqslant \theta(\overline{B}, y^*)$. Since the number n of constraints is less than the number m of variables both systems $B'y < 0$ and $\overline{B}y < 0$ are feasible. By Lemma 4.1, we have $\theta(B', y') > \pi/2$ and $\theta(\overline{B}, y^*) > \pi/2$. Once again, since $f(t) = |\cos t|$ increases on $[\pi/2, \pi]$, we have

$$\mathcal{C}(\overline{B}) = \frac{1}{|\cos\theta(\overline{B}, y^*)|} \leqslant \frac{1}{|\cos\theta(B', y')|} = \mathcal{C}(B'). \qquad \square$$

LEMMA 6.6. *If $|I| < m$ then $\mathcal{C}(A) \leqslant \max_{B \in \mathcal{B}} \mathcal{C}(B)$.*

PROOF. Let B' be any $m \times m$ matrix submatrix of A containing $\overline{B}$ as a submatrix. By Lemmas 6.4 and 6.5, we have $\mathcal{C}(A) \leqslant \mathcal{C}(\overline{B}) \leqslant \mathcal{C}(B')$. Therefore, $\mathcal{C}(A) \leqslant \max_{B \in \mathcal{B}} \mathcal{C}(B)$. $\square$

LEMMA 6.7. *If B is a $m \times m$ matrix then $\mathcal{C}(B) \leqslant \mu(B)$.*

PROOF. Let $y = -B^{-1}v(B)$. Then $By = -v(B)$. So, $\cos\theta_i(B, y)\|y\| = -1$ for $i = 1, 2, \ldots, m$. So, $\cos\theta(B, y)\|y\| = -1$, i.e. $\cos\theta(B, y)\|B^{-1}v(B)\| = -1$. By definition, $\theta(B, \bar{y}) \geqslant \theta(B, y)$. So $\cos\theta(B, \bar{y})\|B^{-1}v(B)\| \leqslant -1$. And thus, $|\cos\theta(B, \bar{y})| \times \|B^{-1}v(B)\| \geqslant 1$. As a result,

$$\mathcal{C}(B) = \frac{1}{|\cos\theta(B, \bar{y})|} \leqslant \left\|B^{-1}v(B)\right\| \leqslant \mu(B). \qquad \square$$

The proof of Theorem 6.3(i) follows from Lemmas 6.3, 6.6, and 6.7.

6.2.2. Proof of Theorem 6.3(ii)

Let $\mathcal{B}$ be the set of bases of A. Then, by Theorem 4.3,

$$\begin{aligned}\bar{\chi}_A &= \max_{B\in\mathcal{B}}\|B^{-1}A\| \leqslant \max_{B\in\mathcal{B}}\|B^{-1}\|\|A\| = \|A\|\max_{B\in\mathcal{B}}\|B^{-1}\| \\ &= \|A\|\max_{B\in\mathcal{B}}\frac{\mu(B)}{\|B\|_F} \leqslant \|A\|\max_{B\in\mathcal{B}}\mu(B)\frac{1}{\min_{B\in\mathcal{B}}\|B\|_F}.\end{aligned}$$

But $\|B\|_F \geqslant \|a_i\|$ for all $i \leqslant n$ such that a_i is a column of B. Therefore

$$\min_{B\in\mathcal{B}}\|B\|_F \geqslant \min_{i\leqslant n}\|a_i\|$$

and the proof is completed.

6.3. Proof of Theorem 6.2

LEMMA 6.8. *Let $n \geqslant m$. For a $m\times n$ Gaussian matrix A,*

$$\mathbf{E}\Bigl(\log\max_{B\in\mathcal{B}}\mu(B)\Bigr) \leqslant 2\min\{n, m\log n\} + \tfrac{5}{2}\log m + 2.$$

PROOF. For all $t > 1$ (cf. BLUM, CUCKER, SHUB and SMALE [1998], Theorem 2 in Chapter 11),

$$\operatorname{Prob}\bigl(\mu(B) > t\bigr) \leqslant m^{5/2}(1/t).$$

Consequently

$$\operatorname{Prob}\bigl(\sqrt{\mu(B)} > t\bigr) = \operatorname{Prob}\bigl(\mu(B) > t^2\bigr) \leqslant m^{5/2}(1/t)^2$$

and

$$\begin{aligned}\mathbf{E}\bigl(\sqrt{\mu(B)}\bigr) &= \int_0^\infty \operatorname{Prob}\bigl(\sqrt{\mu(B)} > t\bigr)\,\mathrm{d}t \leqslant \int_0^{m^{5/4}} 1\,\mathrm{d}t + \int_{m^{5/4}}^\infty m^{5/2}(1/t)^2\,\mathrm{d}t \\ &= m^{5/4} + m^{5/2}\int_{m^{5/4}}^\infty (1/t)^2\,\mathrm{d}t = 2m^{5/4}.\end{aligned}$$

Finally,

$$\begin{aligned}\mathbf{E}\Bigl(\log\max_{B\in\mathcal{B}}\mu(B)\Bigr) &= \mathbf{E}\Bigl(2\log\sqrt{\max_{B\in\mathcal{B}}\mu(B)}\Bigr) \leqslant 2\log\mathbf{E}\Bigl(\sqrt{\max_{B\in\mathcal{B}}\mu(B)}\Bigr) \\ &\leqslant 2\log\mathbf{E}\Bigl(\max_{B\in\mathcal{B}}\sqrt{\mu(B)}\Bigr) \leqslant 2\log\binom{n}{m}\mathbf{E}\bigl(\sqrt{\mu(B)}\bigr) \\ &\leqslant 2\log\bigl(\min\{2^n, n^m\}2m^{5/4}\bigr) \\ &= 2\min\{n, m\log n\} + \tfrac{5}{2}\log m + 2.\end{aligned}$$

□

6.3.1. Proof of Theorem 6.2 *for* $\mathcal{C}(A)$
Case I: $n > m$. Apply Theorem 6.3(i) and Lemma 6.8.

Case II: $n \leqslant m$. If $n = m$ then one proves as above that $\mathbf{E}(\sqrt{\mathcal{C}(A)}) \leqslant 2n^{5/4}$. Therefore, $\mathbf{E}(\log \mathcal{C}(A)) \leqslant 2 \log \mathbf{E}(\sqrt{\mathcal{C}(A)}) \leqslant \frac{5}{2} \log n + 2$.

Assume now that $n < m$. Then the vectors $a_i/\|a_i\|$ are n independent vectors uniformly distributed on the unit sphere in $\mathbb{R}^m$. Almost surely, they span a linear subspace H of $\mathbb{R}^m$ of dimension n. Conditioned to belong to a given, non random, subspace H of dimension n one can prove that the vectors $a_i/\|a_i\|$ are independent and uniformly distributed on the unit sphere in H. Since $\bar{y}$ must lie in H as well we have that $\mathbf{E}(\log \mathcal{C}(A) \mid a_i \in H \text{ for all } i = 1, \ldots, n)$ is the same as $\mathcal{E}_n = \mathbf{E}(\log \mathcal{C}(B))$ for Gaussian $n \times n$ matrices B. Then,

$$\mathbf{E}\big(\log \mathcal{C}(A)\big) = \mathbf{E}\big(\mathbf{E}\big(\log \mathcal{C}(A) \mid a_i \in H \text{ for all } i = 1, \ldots, n\big)\big) = \mathbf{E}(\mathcal{E}_n) = \mathcal{E}_n.$$

6.3.2. Proof of Theorem 6.2 *for* $C(A)$
We apply Proposition 4.2 to deduce

$$\mathbf{E}\big(\log C(A)\big) \leqslant \mathbf{E}\big(\log \mathcal{C}(A)\big) + \mathbf{E}\big(\log(\|A\|)\big) + \mathbf{E}\Big(\log \max_{i \in I}\big(\|a_i\|^{-1}\big)\Big).$$

Since $\mathbf{E}(a_{ij}^2) = 1$ and the entries a_{ij} are independently drawn we have $\mathbf{E}(\|A\|_F^2) = nm$. Using that $\|A\| \leqslant \|A\|_F$ we get

$$\begin{aligned}\mathbf{E}\big(\log(\|A\|)\big) &\leqslant \mathbf{E}\big(\log(\|A\|_F)\big) = \tfrac{1}{2}\mathbf{E}\big(\log(\|A\|_F^2)\big) \leqslant \tfrac{1}{2} \log \mathbf{E}\big((\|A\|_F^2)\big) \\ &= \frac{\log n + \log m}{2}.\end{aligned}$$

Also, for $i = 1, \ldots, n$, since $|a_{i1}| \leqslant \|a_i\|_2$, we have $\|a_i\|_2^{-1/2} \leqslant |a_{i1}|^{-1/2}$. But a_{i1} is Gaussian and it is known (see Lemma 5 in TODD, TUNCCEL and YE [2001]) that $\mathbf{E}(|a_{i1}|^{-1/2}) \leqslant 1 + (2/\sqrt{2\pi}) \leqslant 2$. Consequently

$$\mathbf{E}\Big(\max_{i \in I}\big(\|a_i\|^{-1/2}\big)\Big) \leqslant \sum_{i \in I} \mathbf{E}\big(\big(\|a_i\|^{-1/2}\big)\big) \leqslant \sum_{i \in I} \mathbf{E}\big(\big(|a_{i1}|^{-1/2}\big)\big) \leqslant 2n$$

and

$$\mathbf{E}\Big(\log \max_{i \in I}\big(\|a_i\|^{-1}\big)\Big) = 2\mathbf{E}\Big(\log \max_{i \in I}\big(\|a_i\|^{-1/2}\big)\Big) \leqslant 2 \log n + 2.$$

We conclude that

$$\begin{aligned}\mathbf{E}\big(\log C(A)\big) &\leqslant \mathbf{E}\big(\log \mathcal{C}(A)\big) + \frac{\log m + \log n}{2} + 2 \log n + 2 \\ &= \mathbf{E}\big(\log \mathcal{C}(A)\big) + \tfrac{5}{2} \log n + \tfrac{1}{2} \log m + 2.\end{aligned}$$

6.3.3. Proof of Theorem 6.2 *for* $\bar{\chi}_A$
Use Theorem 6.3(ii), Lemma 6.8 and the argument in Section 6.3.2.

6.3.4. Proof of Theorem 6.2 *for* $\sigma(A)$
Use the inequality $1/\sigma(A) \leqslant \bar{\chi}_A + 1$.

7. Semidefinite programming algorithms and analyses

Recall that $\mathcal{M}^n$ denotes the set of symmetric matrices in $\mathbb{R}^{n\times n}$. Let $\mathcal{M}^n_+$ denote the set of positive semi-definite matrices and $\mathring{\mathcal{M}}^n_+$ the set of positive definite matrices in $\mathcal{M}^n$. The goal of this section is to describe an interior-point algorithm to solve the semi-definite programming problems (SDP) and (SDD) presented in Section 2.5.

(SDP) and (SDD) are analogues to linear programming (LP) and (LD). Actually (LP) and (LD) can be expressed as semi-definite programs by defining

$$C = \mathrm{diag}(c), \qquad A_i = \mathrm{diag}(a_i), \qquad b = b,$$

where a_i is the ith row of matrix A. Many of the theorems and algorithms used in LP have analogues in SDP. However, while interior-point algorithms for LP are generally considered competitive with the simplex method in practice and outperform it as problems become large, interior-point methods for SDP outperform other methods on even small problems.

Denote the primal feasible set by $\mathcal{F}_p$ and the dual by $\mathcal{F}_d$. We assume that both $\mathring{\mathcal{F}}_p$ and $\mathring{\mathcal{F}}_d$ are nonempty. Thus, we recall from Section 2.5, the optimal solution sets for both (SDP) and (SDD) are bounded and the central path exists. Let $z^* \in \mathbb{R}$ denote the optimal value and $\mathcal{F} = \mathcal{F}_p \times \mathcal{F}_d$. In this section, we are interested in finding an *ε-approximate solution* for the SDP problem, i.e. an element (X, y, S) satisfying

$$C \bullet X - b^{\mathrm{T}} y = S \bullet X \leqslant \varepsilon.$$

For simplicity, we assume the availability of a point (X^0, y^0, S^0) in the central path satisfying

$$(X^0)^{0.5} S^0 (X^0)^{0.5} = \mu^0 I \quad \text{and} \quad \mu^0 = X^0 \bullet S^0 / n.$$

We will use it as our initial point throughout this section.

Let $X \in \mathring{\mathcal{F}}_p$, $(y, S) \in \mathring{\mathcal{F}}_d$, and $z \leqslant z^*$. Then consider the *primal potential function*

$$\mathcal{P}(X, z) = (n+\rho)\log(C \bullet X - z) - \log\det X,$$

and the *primal–dual potential function*

$$\psi(X, S) = (n+\rho)\log(S \bullet X) - \log\det XS,$$

where $\rho = \sqrt{n}$. Note that if $z = b^{\mathrm{T}} y$ then $S \bullet X = C \bullet X - z$, and we have

$$\psi(x, s) = \mathcal{P}(x, z) - \log\det S.$$

Define the "∞-norm" (which is the traditional l_2 operator norm for matrices) on $\mathcal{M}^n$ by

$$\|X\|_\infty := \max_{j\in\{1,\dots,n\}} \{|\lambda_j(X)|\},$$

where $\lambda_j(X)$ is the jth eigenvalue of X. Also, define the "Euclidean" norm (which is the traditional Frobenius norm) by

$$\|X\| := \|X\|_F = \sqrt{X \bullet X} = \sqrt{\sum_{j=1}^{n} (\lambda_j(X))^2}.$$

We rename these norms because they are perfect analogues to the norms of vectors used in LP.

Furthermore, note that, for $X \in \mathcal{M}^n$,

$$\operatorname{tr}(X) = \sum_{j=1}^{n} \lambda_j(X) \quad \text{and} \quad \det(I + X) = \prod_{j=1}^{n} \bigl(1 + \lambda_j(X)\bigr).$$

We have the following lemma.

LEMMA 7.1. *Let $X \in \mathcal{M}^n$ and $\|X\|_\infty < 1$. Then,*

$$\operatorname{tr}(X) \geqslant \log\det(I + X) \geqslant \operatorname{tr}(X) - \frac{\|X\|^2}{2(1 - \|X\|_\infty)}.$$

7.1. Potential reduction algorithm

Consider a point $(X^k, y^k, S^k) \in \mathring{\mathcal{F}}$. Fix $z^k = b^{\mathrm{T}} y^k$. Then the gradient matrix of the primal potential function at X^k is

$$\nabla \mathcal{P}(X^k, z^k) = \frac{n + \rho}{S^k \bullet X^k} C - (X^k)^{-1}.$$

A basic property of $\mathcal{P}$ is captured in the following proposition.

PROPOSITION 7.1. *Let $X^k \in \mathring{\mathcal{M}}^n_+$ and $X \in \mathcal{M}^n$ such that $\|(X^k)^{-0.5}(X - X^k) \cdot (X^k)^{-0.5}\|_\infty < 1$. Then, $X \in \mathring{\mathcal{M}}^n_+$ and*

$$\begin{aligned} &\mathcal{P}(X, z^k) - \mathcal{P}(X^k, z^k) \\ &\quad\leqslant \nabla \mathcal{P}(X^k, z^k) \bullet (X - X^k) + \frac{\|(X^k)^{-0.5}(X - X^k)(X^k)^{-0.5}\|^2}{2(1 - \|(X^k)^{-0.5}(X - X^k)(X^k)^{-0.5}\|_\infty)}. \end{aligned}$$

Let

$$\mathcal{A} = \begin{pmatrix} A_1 \\ A_2 \\ \cdots \\ A_m \end{pmatrix}.$$

Define

$$\mathcal{A}X = \begin{pmatrix} A_1 \bullet X \\ A_2 \bullet X \\ \cdots \\ A_m \bullet X \end{pmatrix} = b,$$

and

$$\mathcal{A}^{\mathrm{T}} y = \sum_{i=1}^{m} y_i A_i.$$

Now consider the following "ball-constrained" problem:

$$\begin{aligned} \text{minimize} \quad & \nabla\mathcal{P}(X^k, z^k) \bullet (X - X^k) \\ \text{s.t.} \quad & \mathcal{A}(X - X^k) = 0, \\ & \left\|(X^k)^{-0.5}(X - X^k)(X^k)^{-0.5}\right\| \leqslant \alpha < 1. \end{aligned}$$

Note that for any symmetric matrices $Q, T \in \mathcal{M}^n$ and $X \in \mathring{\mathcal{M}}^n_+$,

$$Q \bullet X^{0.5} T X^{0.5} = X^{0.5} Q X^{0.5} \bullet T \quad \text{and} \quad \|XQ\|_. = \|QX\|_. = \left\|X^{0.5} Q X^{0.5}\right\|_. .$$

Then we may rewrite the above problem as

$$\begin{aligned} \text{minimize} \quad & (X^k)^{0.5} \nabla\mathcal{P}(X^k, z^k)(X^k)^{0.5} \bullet (X' - I) \\ \text{s.t.} \quad & \mathcal{A}'(X' - I) = 0, \\ & \|X' - I\| \leqslant \alpha, \end{aligned}$$

where $X' = (X^k)^{-0.5} X (X^k)^{-0.5}$ and

$$\mathcal{A}' = \begin{pmatrix} A'_1 \\ A'_2 \\ \cdots \\ A'_m \end{pmatrix} := \begin{pmatrix} (X^k)^{0.5} A_1 (X^k)^{0.5} \\ (X^k)^{0.5} A_2 (X^k)^{0.5} \\ \cdots \\ (X^k)^{0.5} A_m (X^k)^{0.5} \end{pmatrix}.$$

Let X' be its minimizer and let $X^{k+1} = (X^k)^{0.5} X' (X^k)^{0.5}$. Then, we have a closed formula for X':

$$X' - I = -\alpha \frac{P^k}{\|P^k\|},$$

where

$$P^k = \pi_{\mathcal{A}'}(X^k)^{0.5} \nabla\mathcal{P}(X^k, z^k)(X^k)^{0.5} = (X^k)^{0.5} \nabla\mathcal{P}(X^k, z^k)(X^k)^{0.5} - \mathcal{A}'^{\mathrm{T}} y^k$$

or

$$P^k = \frac{n + \rho}{S^k \bullet X^k} (X^k)^{0.5} (C - \mathcal{A}^{\mathrm{T}} y^k)(X^k)^{0.5} - I,$$

and

$$y^k = \frac{S^k \bullet X^k}{n + \rho} (\mathcal{A}'\mathcal{A}'^{\mathrm{T}})^{-1} \mathcal{A}'(X^k)^{0.5} \nabla\mathcal{P}(X^k, z^k)(X^k)^{0.5}.$$

Here, $\pi_{\mathcal{A}'}$ is the projection operator onto the null space of $\mathcal{A}'$, and

$$\mathcal{A}'\mathcal{A}'^{\mathrm{T}} := \begin{pmatrix} A'_1 \bullet A'_1 & A'_1 \bullet A'_2 & \cdots & A'_1 \bullet A'_m \\ A'_2 \bullet A'_1 & A'_2 \bullet A'_2 & \cdots & A'_2 \bullet A'_m \\ \cdots & \cdots & \cdots & \cdots \\ A'_m \bullet A'_1 & A'_m \bullet A'_2 & \cdots & A'_m \bullet A'_m \end{pmatrix} \in \mathcal{M}^m.$$

Define X^{k+1} by

$$X^{k+1} - X^k = -\alpha \frac{(X^k)^{0.5} P^k (X^k)^{0.5}}{\|P^k\|}. \tag{7.1}$$

Then,

$$\begin{aligned}\nabla\mathcal{P}(X^k,z^k)\bullet(X^{k+1}-X^k) &= -\alpha\frac{\nabla\mathcal{P}(X^k,z^k)\bullet(X^k)^{0.5}P^k(X^k)^{0.5}}{\|P^k\|}\\ &= -\alpha\frac{(X^k)^{0.5}\nabla\mathcal{P}(X^k,z^k)(X^k)^{0.5}\bullet P^k}{\|P^k\|}\\ &= -\alpha\frac{\|P^k\|^2}{\|P^k\|} = -\alpha\|P^k\|.\end{aligned}$$

In view of Proposition 7.1 and the equality above we have

$$\mathcal{P}(X^{k+1},z^k)-\mathcal{P}(X^k,z^k)\leqslant -\alpha\|P^k\|+\frac{\alpha^2}{2(1-\alpha)}.$$

Thus, as long as $\|P^k\|\geqslant\beta>0$, we may choose an appropriate α such that

$$\mathcal{P}(X^{k+1},z^k)-\mathcal{P}(X^k,z^k)\leqslant -\delta$$

for some positive constant δ.

Now, we focus on the expression of P^k, which can be rewritten as

$$P(z^k):=P^k=\frac{n+\rho}{S^k\bullet X^k}(X^k)^{0.5}S(z^k)(X^k)^{0.5}-I$$

with

$$S(z^k)=C-\mathcal{A}^{\mathrm{T}}y(z^k) \tag{7.2}$$

as well as on the expressions

$$y(z^k):=y_2-\frac{S^k\bullet X^k}{n+\rho}y_1=y_2-\frac{C\bullet X^k-z^k}{n+\rho}y_1, \tag{7.3}$$

where y_1 and y_2 are given by

$$\begin{aligned}y_1&=(\mathcal{A}'\mathcal{A}'^{\mathrm{T}})^{-1}\mathcal{A}'I=(\mathcal{A}'\mathcal{A}'^{\mathrm{T}})^{-1}b,\\ y_2&=(\mathcal{A}'\mathcal{A}'^{\mathrm{T}})^{-1}\mathcal{A}'(X^k)^{0.5}C(X^k)^{0.5}.\end{aligned} \tag{7.4}$$

Regarding $\|P^k\|=\|P(z^k)\|$, we have the following lemma.

LEMMA 7.2. *Let*

$$\mu^k=\frac{S^k\bullet X^k}{n}=\frac{C\bullet X^k-z^k}{n}\quad\textit{and}\quad\mu=\frac{S(z^k)\bullet X^k}{n}.$$

If

$$\|P(z^k)\|<\min\left(\beta\sqrt{\frac{n}{n+\beta^2}},1-\beta\right), \tag{7.5}$$

then the following three inequalities hold:

$$\begin{aligned} &S(z^k) \succ 0, \\ &\|(X^k)^{0.5} S(z^k)(X^k)^{0.5} - \mu e\| < \beta\mu \quad \textit{and} \\ &\mu < (1 - 0.5\beta/\sqrt{n})\mu^k. \end{aligned} \tag{7.6}$$

PROOF. The proof is by contradiction. For example, if the first inequality of (7.6) is not true, then $(X^k)^{0.5} S(z^k)(X^k)^{0.5}$ has at least one eigenvalue less than or equal to zero, and

$$\|P(z^k)\| \geqslant 1.$$

The proofs of the second and third inequalities can be done similarly. □

Based on this lemma, we have the following potential reduction theorem.

THEOREM 7.1. *Given* $X^k \in \mathring{\mathcal{F}}_p$ *and* $(y^k, S^k) \in \mathring{\mathcal{F}}_d$, *let* $\rho = \sqrt{n}$, $z^k = b^{\mathrm{T}} y^k$, X^{k+1} *be given by* (7.1), *and* $y^{k+1} = y(z^k)$ *in* (7.3) *and* $S^{k+1} = S(z^k)$ *in* (7.2). *Then, either*

$$\psi(X^{k+1}, S^k) \leqslant \psi(X^k, S^k) - \delta$$

or

$$\psi(X^k, S^{k+1}) \leqslant \psi(X^k, S^k) - \delta,$$

where $\delta > 1/20$.

PROOF. If (7.5) does not hold, i.e.

$$\|P(z^k)\| \geqslant \min\left(\beta\sqrt{\frac{n}{n+\beta^2}}, 1-\beta\right),$$

then, since $\psi(X^{k+1}, S^k) - \psi(X^k, S^k) = \mathcal{P}(X^{k+1}, z^k) - \mathcal{P}(X^k, z^k)$,

$$\psi(X^{k+1}, S^k) - \psi(X^k, S^k) \leqslant -\alpha \min\left(\beta\sqrt{\frac{n}{n+\beta^2}}, 1-\beta\right) + \frac{\alpha^2}{2(1-\alpha)}.$$

Otherwise, from Lemma 7.2 the inequalities of (7.6) hold:

(i) The first of (7.6) indicates that y^{k+1} and S^{k+1} are in $\mathring{\mathcal{F}}_d$.

(ii) Using the second of (7.6) and applying Lemma 7.1 to the matrix $(X^k)^{0.5} S^{k+1} \times (X^k)^{0.5}/\mu$, we have

$$\begin{aligned} &n \log S^{k+1} \bullet X^k - \log\det S^{k+1} X^k \\ &\quad = n \log S^{k+1} \bullet X^k/\mu - \log\det(X^k)^{0.5} S^{k+1} (X^k)^{0.5}/\mu \\ &\quad = n \log n - \log\det(X^k)^{0.5} S^{k+1} (X^k)^{0.5}/\mu \\ &\quad \leqslant n \log n + \frac{\|(X^k)^{0.5} S^{k+1} (X^k)^{0.5}/\mu - I\|^2}{2(1 - \|(X^k)^{0.5} S^{k+1} (X^k)^{0.5}/\mu - I\|_\infty)} \\ &\quad \leqslant n \log n + \frac{\beta^2}{2(1-\beta)} \end{aligned}$$

$$\leqslant n \log S^k \bullet X^k - \log\det S^k X^k + \frac{\beta^2}{2(1-\beta)}.$$

(iii) According to the third of (7.6), we have

$$\sqrt{n}\left(\log S^{k+1} \bullet X^k - \log S^k \bullet X^k\right) = \sqrt{n} \log \frac{\mu}{\mu^k} \leqslant -\frac{\beta}{2}.$$

Adding the two inequalities in (ii) and (iii), we have

$$\psi\left(X^k, S^{k+1}\right) \leqslant \psi\left(X^k, S^k\right) - \frac{\beta}{2} + \frac{\beta^2}{2(1-\beta)}.$$

Thus, by choosing $\beta = 0.43$ and $\alpha = 0.3$ we have the desired result. □

Theorem 7.1 establishes an important fact: the *primal–dual* potential function ψ can be reduced by a constant no matter where X^k and y^k are. In practice, one can perform the line search to minimize the primal–dual potential function. This results in the following potential reduction algorithm.

ALGORITHM 7.1.
Input $X^0 \in \mathring{\mathcal{F}}_p$ and $(y^0, S^0) \in \mathring{\mathcal{F}}_d$.
Set $z^0 := b^{\mathrm{T}} y^0$. Set $k := 0$.
While $S^k \bullet X^k \geqslant \varepsilon$ **do**

1. Compute y_1 and y_2 from (7.4).
2. Set $y^{k+1} := y(\bar{z})$, $S^{k+1} := S(\bar{z})$, $z^{k+1} := b^{\mathrm{T}} y^{k+1}$ with
$$\bar{z} := \arg\min_{z \geqslant z^k} \psi\left(X^k, S(z)\right).$$
If $\psi(X^k, S^{k+1}) > \psi(X^k, S^k)$ then $y^{k+1} := y^k$, $S^{k+1} := S^k$, $z^{k+1} := z^k$.
3. Let $X^{k+1} := X^k - \bar{\alpha}(X^k)^{0.5} P(z^{k+1})(X^k)^{0.5}$ with
$$\bar{\alpha} := \arg\min_{\alpha \geqslant 0} \psi\left(X^k - \alpha\left(X^k\right)^{0.5} P\left(z^{k+1}\right)\left(X^k\right)^{0.5}, S^{k+1}\right).$$
4. Let $k := k + 1$.

Note that Theorem 7.1 ensures that, at each iteration of the algorithm, $\bar{\alpha} \geqslant 0.3$. A similar remark hold for $\bar{z}$ and z^k. The performance of the algorithm is described in the following corollary.

COROLLARY 7.1. *Let $\rho = \sqrt{n}$. Then, Algorithm 7.1 terminates in at most* $\mathrm{O}(\sqrt{n} \log(C \bullet X^0 - b^{\mathrm{T}} y^0)/\varepsilon)$ *iterations with*

$$C \bullet X^k - b^{\mathrm{T}} y^k \leqslant \varepsilon.$$

PROOF. In $\mathrm{O}(\sqrt{n} \log(S^0 \bullet X^0/\varepsilon))$ iterations

$$\begin{aligned} -\sqrt{n} \log\left(S^0 \bullet X^0/\varepsilon\right) &= \psi\left(X^k, S^k\right) - \psi\left(X^0, S^0\right) \\ &\geqslant \sqrt{n} \log S^k \bullet X^k + n \log n - \psi\left(X^0, S^0\right) \\ &= \sqrt{n} \log\left(S^k \bullet X^k / S^0 \bullet X^0\right). \end{aligned}$$

Thus,

$$\sqrt{n}\log\left(C \bullet X^k - b^{\mathrm{T}} y^k\right) = \sqrt{n}\log S^k \bullet X^k \leqslant \sqrt{n}\log\varepsilon,$$

i.e.

$$C \bullet X^k - b^{\mathrm{T}} y^k = S^k \bullet X^k \leqslant \varepsilon. \qquad \square$$

7.2. Primal–dual algorithm

Once we have a point $(X, y, S) \in \mathring{\mathcal{F}}$ with $\mu = S \bullet X/n$, we can apply the primal–dual Newton method to generate a new iterate X^+ and (y^+, S^+) as follows. Solve for d_X, d_y and d_S the system of linear equations:

$$\begin{aligned} D^{-1} d_X D^{-1} + d_S &= R := \gamma\mu X^{-1} - S, \\ \mathcal{A} d_X &= 0, \\ -\mathcal{A}^{\mathrm{T}} d_y - d_S &= 0, \end{aligned} \tag{7.7}$$

where

$$D = X^{0.5}\left(X^{0.5} S X^{0.5}\right)^{-0.5} X^{0.5}.$$

Note that $d_S \bullet d_X = 0$.

This system can be written as

$$\begin{aligned} d_{X'} + d_{S'} &= R', \\ \mathcal{A}' d_{X'} &= 0, \\ -\mathcal{A}'^{\mathrm{T}} d_y - d_{S'} &= 0, \end{aligned} \tag{7.8}$$

where

$$d_{X'} = D^{-0.5} d_X D^{-0.5}, \quad d_{S'} = D^{0.5} d_S D^{0.5}, \quad R' = D^{0.5}\left(\gamma\mu X^{-1} - S\right) D^{0.5},$$

and

$$\mathcal{A}' = \begin{pmatrix} A'_1 \\ A'_2 \\ \dots \\ A'_m \end{pmatrix} := \begin{pmatrix} D^{0.5} A_1 D^{0.5} \\ D^{0.5} A_2 D^{0.5} \\ \dots \\ D^{0.5} A_m D^{0.5} \end{pmatrix}.$$

Again, we have $d_{S'} \bullet d_{X'} = 0$, and

$$d_y = \left(\mathcal{A}'\mathcal{A}'^{\mathrm{T}}\right)^{-1} \mathcal{A}' R', \quad d_{S'} = -\mathcal{A}'^{\mathrm{T}} d_y, \quad \text{and} \quad d_{X'} = R' - d_{S'}.$$

Then, assign

$$d_S = \mathcal{A}^{\mathrm{T}} d_y \quad \text{and} \quad d_X = D(R - d_S)D.$$

Let

$$V^{1/2} = D^{-0.5} X D^{-0.5} = D^{0.5} S D^{0.5} \in \mathring{\mathcal{M}}^n_+.$$

Then, we can verify that $S \bullet X = I \bullet V$. We now state a lemma bounding the variation of ψ when we replace a point by its Newton iterate.

LEMMA 7.3. *Let the direction* (d_X, d_y, d_S) *be the solution of system* (7.7) *with* $\gamma = n/(n+\rho)$, *and let*

$$\theta = \frac{\alpha}{\|V^{-1/2}\|_\infty \|\frac{I\bullet V}{n+\rho} V^{-1/2} - V^{1/2}\|}, \tag{7.9}$$

where α *is a positive constant less than* 1. *Let*

$$X^+ = X + \theta d_X, \quad y^+ = y + \theta d_y, \quad \text{and} \quad S^+ = S + \theta d_S.$$

Then, we have $(X^+, y^+, S^+) \in \mathring{\mathcal{F}}$ *and*

$$\psi(X^+, S^+) - \psi(X, S) \leqslant -\alpha \frac{\|V^{-1/2} - \frac{n+\rho}{I\bullet V} V^{1/2}\|}{\|V^{-1/2}\|_\infty} + \frac{\alpha^2}{2(1-\alpha)}.$$

On the other hand, it is possible to prove the following lemma.

LEMMA 7.4. *Let* $V \in \mathring{\mathcal{M}}^n_+$ *and* $\rho \geqslant \sqrt{n}$. *Then,*

$$\frac{\|V^{-1/2} - \frac{n+\rho}{I\bullet V} V^{1/2}\|}{\|V^{-1/2}\|_\infty} \geqslant \sqrt{3/4}.$$

From Lemmas 7.3 and 7.4 we have

$$\psi(X^+, S^+) - \psi(X, S) \leqslant -\alpha\sqrt{3/4} - \frac{\alpha^2}{2(1-\alpha)} = -\delta$$

for a constant δ. This leads to Algorithm 7.2.

ALGORITHM 7.2.
Input $(X^0, y^0, S^0) \in \mathring{\mathcal{F}}$.
Set $\rho = \sqrt{n}$ and $k := 0$.
While $S^k \bullet X^k \geqslant \varepsilon$ **do**

1. Set $(X, S) = (X^k, S^k)$ and $\gamma = n/(n+\rho)$ and compute (d_X, d_y, d_S) from (7.7).
2. Let $X^{k+1} = X^k + \bar{\alpha} d_X$, $y^{k+1} = y^k - \bar{\alpha} d_y$, and $S^{k+1} = S^k + \bar{\alpha} d_S$, where

$$\bar{\alpha} = \arg\min_{\alpha \geqslant 0} \psi(X^k + \alpha d_X, S^k + \alpha d_S).$$

3. Let $k := k + 1$.

THEOREM 7.2. *Let* $\rho = \sqrt{n}$. *Then, Algorithm* 7.2 *terminates in at most* $\mathrm{O}(\sqrt{n}\log(S^0 \bullet X^0/\varepsilon))$ *iterations with*

$$C \bullet X^k - b^{\mathrm{T}} y^k \leqslant \varepsilon.$$

8. Notes

The term "complexity" was introduced by HARTMANIS and STEARNS [1965]. Also see GAREY and JOHNSON [1979] and PAPADIMITRIOU and STEIGLITZ [1982]. The NP theory was due to COOK [1971], KARP [1972] and LEVIN [1973]. The importance of P was observed by EDMONDS [1967].

Linear programming and the simplex method were introduced by DANTZIG [1951]. Other inequality problems and convexity theories can be seen in GRITZMANN and KLEE [1993], GRÖTSCHEL, LOVÁSZ and SCHRIJVER [1988], GRÜNBAUM [1967], ROCKAFELLAR [1970], and SCHRIJVER [1986]. Various complementarity problems can be found in COTTLE, PANG and STONE [1992]. The positive semi-definite programming, an optimization problem in nonpolyhedral cones, and its applications can be seen in NESTEROV and NEMIROVSKY [1993], ALIZADEH [1991], and BOYD, EL GHAOUI, FERON and BALAKRISHNAN [1994]. Recently, GOEMANS and WILLIAMSON [1995] obtained several breakthrough results on approximation algorithms using positive semi-definite programming. The KKT condition for nonlinear programming was given by KUHN and TUCKER [1961].

It was shown by KLEE and MINTY [1972] that the simplex method is not a polynomial-time algorithm. The ellipsoid method, the first polynomial-time algorithm for linear programming with rational data, was proven by KHACHIJAN [1979]; also see BLAND, GOLDFARB and TODD [1981]. The method was devised independently by SHOR [1977] and by NEMIROVSKY and YUDIN [1983]. The interior-point method, another polynomial-time algorithm for linear programming, was developed by KARMARKAR [1984]. It is related to the classical barrier-function method studied by FRISCH [1955] and FIACCO and MCCORMICK [1968]; see GILL, MURRAY, SAUNDERS, TOMLIN and WRIGHT [1986], and ANSTREICHER [1996]. For a brief LP history, see the excellent article by WRIGHT [1985].

NEMIROVSKY and YUDIN [1983] used the real computation model which was eventually fully developed by BLUM, SHUB and SMALE [1989]. The asymptotic convergence rate and ratio can be seen in LUENBERGER [1984], ORTEGA and RHEINBOLDT [1970], and TRAUB [1972]. Other complexity issues in numerical optimization were discussed in VAVASIS [1991].

Many basic numerical procedures listed in this article can be found in GOLUB and VAN LOAN [1989]. The ball-constrained quadratic problem and its solution methods can be seen in MORÉ [1977], SORENSON [1982], and DENNIS and SCHNABEL [1983]. The complexity result of the ball-constrained quadratic problem was proved by VAVASIS [1991] and YE [1992a], YE [1994].

The approach we presented for solving semidefinite programs is not "condition-based". For something of this kind see NESTEROV, TOOD and YE [1999].

References

ALIZADEH, F. (1991). Combinatorial optimization with interior point methods and semi-definite matrices. PhD thesis, University of Minnesota, Minneapolis, MN, USA.

ANSTREICHER, K.M. (1996). On long step path following and SUMT for linear and quadratic programming. *SIAM J. Optim.* **6**, 33–46.

BALCÁZAR, J.L., DÍAZ, J., GABARRÓ, J. (1988). *Structural Complexity I*, EATCS Monographs on Theoretical Computer Science **11** (Springer-Verlag, Berlin).

BEN-TAL, A., TEBOULLE, M. (1990). A geometric property of the least squares solutions of linear equations. *Linear Algebra Appl.* **139**, 165–170.

BJÖRCK, A. (1996). *Numerical Methods for Least Squares Problems* (SIAM, Philadelphia, PA).

BLAND, R.G., GOLDFARB, D., TODD, M.J. (1981). The ellipsoidal method: a survey. *Oper. Res.* **29**, 1039–1091.

BLUM, L., CUCKER, F., SHUB, M., SMALE, S. (1998). *Complexity and Real Computation* (Springer-Verlag, Berlin).

BLUM, L., SHUB, M., SMALE, S. (1989). On a theory of computation and complexity over the real numbers: NP-completeness, recursive functions and universal machines. *Bull. Amer. Math. Soc.* **21**, 1–46.

BORGWARDT, K.H. (1982). The average number of pivot steps required by the simplex method is polynomial. *Z. Oper. Res.* **26**, 157–177.

BOYD, S., EL GHAOUI, L., FERON, E., BALAKRISHNAN, V. (1994). *Linear Matrix Inequalities in System and Control Theory*, SIAM Stud. Appl. Math. **15** (Society of Industrial and Applied Mathematics (SIAM), Philadelphia, PA).

BÜRGISSER, P., CLAUSEN, M., SHOKROLLAHI, A. (1996). *Algebraic Complexity Theory* (Springer-Verlag, Berlin).

CHEUNG, D., CUCKER, F. (2001). A new condition number for linear programming. *Math. Program.* **91**, 163–174.

COOK, S. (1971). The complexity of theorem proving procedures. In: *3rd Annual ACM Symp. on the Theory of Computing*, pp. 151–158.

COTTLE, R., PANG, J.S., STONE, R.E. (1992). *The Linear Complementarity Problem, Chapter 5.9: Interior-point Methods* (Academic Press, New York), pp. 461–475.

CUCKER, F., PEÑA, J. (2001). A primal–dual algorithm for solving polyhedral conic systems with a finite-precision machine. *SIAM J. Optim.* **12**, 522–554.

DANTZIG, G.B. (1951). Maximization of a linear function of variables subject to linear inequalities. In: Koopmans, Tj.C. (ed.), *Activity Analysis of Production and Allocation* (John Wiley & Sons, New York), pp. 339–347.

DEMMEL, J. (1987). On condition numbers and the distance to the nearest ill-posed problem. *Numer. Math.* **51**, 251–289.

DEMMEL, J.W. (1997). *Applied Numerical Linear Algebra* (SIAM, Philadelphia, PA).

DENNIS, J.E., SCHNABEL, R.E. (1983). *Numerical Methods for Unconstrained Optimization and Nonlinear Equations* (Prentice-Hall, Englewood Cliffs, NJ).

DIKIN, I.I. (1967). Iterative solution of problems of linear and quadratic programming. *Dokl. Akad. Nauk SSSR* **174**, 747–748; Transl.: *Soviet Math. Dokl.* **8**, 674–675 (1967).

ECKART, C., YOUNG, G. (1936). The approximation of one matrix by another of lower rank. *Psychometrika* **1**, 211–218.

EDMONDS, J. (1967). Systems of distinct representatives and linear algebra. *J. Res. National Bureau of Standards, Ser. B. Math. Math. Phys.* **71B**, 241–245.

FIACCO, A.V., MCCORMICK, G.P. (1968). *Nonlinear Programming: Sequential Unconstrained Minimization Techniques* (J. Wiley & Sons, Chichester).

FORSGREN, A. (1995). On linear least-squares problems with diagonally dominant weight matrices. Technical Report TRITA-MAT-1995-OS2, Department of Mathematics, Royal Institute of Technology, Stockholm, Sweden.

FREUND, R.M., VERA, J.R. (1999a). Condition-based complexity of convex optimization in conic linear form via the ellipsoid algorithm. *SIAM J. Optim.* **10**, 155–176.

FREUND, R.M., VERA, J.R. (1999b). Some characterizations and properties of the "distance to ill-posedness" and the condition measure of a conic linear system. *Math. Program.* **86**, 225–260.

FRISCH, K.R. (1955). The logarithmic potential method for convex programming, Unpublished manuscript, Institute of Economics, University of Oslo.

GAREY, M., JOHNSON, D.S. (1979). *Computers and Intractability: A Guide to the Theory of NP-Completeness* (Freeman, New York).

GILBERT, E.N. (1966). The probability of covering a sphere with N circular caps. *Biometrika* **52**, 323–330.

GILL, E.N., MURRAY, W., SAUNDERS, M.A., TOMLIN, J.A., WRIGHT, M.H. (1986). On projected Newton barrier methods for linear programming and an equivalence to Karmarkar's projective method. *Math. Program.* **36**, 183–209.

GOEMANS, M.X., WILLIAMSON, D.P. (1995). Improved approximation algorithms for maximum cut and satisfiability problems using semidefinite programming. *J. ACM* **42**, 1115–1145.

GOLDSTINE, H.H. (1977). *A History of Numerical Analysis from the 16th through the 19th Century* (Springer-Verlag, Berlin).

GOLUB, G., VAN LOAN, C. (1989). *Matrix Computations* (John Hopkins Univ. Press).

GRITZMANN, P., KLEE, V. (1993). Mathematical programming and convex geometry. In: Gruber, P., Wills, J. (eds.), *Handbook of Convex Geometry* (North-Holland, Amsterdam), pp. 627–674.

GRÖTSCHEL, M., LOVÁSZ, L., SCHRIJVER, A. (1988). *Geometric Algorithms and Combinatorial Optimization* (Springer-Verlag, Berlin).

GRÜNBAUM, B. (1967). *Convex Polytopes* (John Wiley & Sons, New York).

HAIMOVICH, M. (1983). The simplex method is very good! – on the expected number of pivot steps and related properties of random linear programs, Preprint.

HALL, P. (1988). *Introduction to the Theory of Coverage Processes* (John Wiley & Sons, New York).

HARTMANIS, J., STEARNS, R.E. (1965). On the computational complexity of algorithms. *Trans. Amer. Math. Soc.* **117**, 285–306.

HIGHAM, N. (1996). *Accuracy and Stability of Numerical Algorithms* (SIAM, Philadelphia, PA).

JANSON, S. (1986). Random coverings in several dimensions. *Acta Math.* **156**, 83–118.

KALISKI, J.A., YE, Y. (1993). A short-cut potential reduction algorithm for linear programming. *Management Sci.* **39**, 757–773.

KARMARKAR, N. (1984). A new polynomial time algorithm for linear programming. *Combinatorica* **4**, 373–395.

KARP, R.M. (1972). Reducibility among combinatorial problems. In: Miller, R., Thatcher, J. (eds.), *Complexity of Computer Computations* (Plenum Press, New York), pp. 85–103.

KHACHIJAN, L.G. (1979). A polynomial algorithm in linear programming. *Dokl. Akad. Nauk SSSR* **244**, 1093–1096 (in Russian); English transl.: *Soviet Math. Dokl.* **20**, 191–194 (1979).

KHACHIJAN, L. (1994). On the complexity of approximating extremal determinants in matrices, Preprint.

KLEE, V., MINTY, G.J. (1972). How good is the simplex method. In: Shisha, O. (ed.), *Inequalities III* (Academic Press, New York), pp. 159–175.

KOJIMA, M., MIZUNO, S., YOSHISE, A. (1989). A polynomial-time algorithm for a class of linear complementarity problems. *Math. Program.* **44**, 1–26.

KUHN, H.W., TUCKER, A.W. (1961). Nonlinear programming. In: Neyman, J. (ed.), *Proceedings of the 2nd Berkeley Symposium on Mathematical Statistics and Probability* (University of California Press), pp. 481–492.

LEVIN, L. (1973). Universal sequential search problems. *Probl. Pered. Inform.* **9** (3), 265–266 (in Russian); English transl.: *Probl. Inform. Trans.* **9** (3) (1973); corrected translation in TRAKHTENBROT [1984].

LUENBERGER, D.G. (1984). *Linear and Nonlinear Programming*, 2nd edn. (Addison-Wesley, Reading, MA).

MILES, R.E. (1969). The asymptotic values of certain coverage probabilities. *Biometrika* **56**, 661–680.

MONTEIRO, R., ADLER, I. (1989). Interior path following primal–dual algorithms. Part I: Linear programming. *Math. Program.* **44**, 27–41.

MORÉ, J.J. (1977). The Levenberg–Marquardt algorithm: implementation and theory. In: Watson, G.A. (ed.), *Numerical Analysis* (Springer-Verlag, Berlin).

NEMIROVSKY, A.S., YUDIN, D.B. (1983). *Problem Complexity and Method Efficiency in Optimization* (John Wiley and Sons, Chichester); Transl. by E.R. Dawson from *Slozhnost' Zadach i Effektivnost' Metodov Optimizatsii*, 1979, Glavnaya redaktsiya fiziko-matematicheskoi literatury izdatelstva "Nauka".

NESTEROV, Y.E., NEMIROVSKY, A.S. (1993). *Interior Point Polynomial Methods in Convex Programming: Theory and Algorithms* (SIAM, Philadelphia, PA).

NESTEROV, YU., TOOD, M.J., YE, Y. (1999). Infeasible-start primal–dual methods and infeasibility detectors for nonlinear programming problems. *Math. Program.* **84**, 227–267.

O'LEARY, D.P. (1990). On bounds for scaled projections and pseudoinverses. *Linear Algebra Appl.* **132**, 115–117.

ORTEGA, J.M., RHEINBOLDT, W.C. (1970). *Iterative Solution of Nonlinear Equations in Several Variables* (Academic Press, New York).

PAPADIMITRIOU, C.H. (1994). *Computational Complexity* (Addison-Wesley, Reading, MA).

PAPADIMITRIOU, C.H., STEIGLITZ, K. (1982). *Combinatorial Optimization: Algorithms and Complexity* (Prentice-Hall, Englewood Cliffs, NJ).

PEÑA, J. (2000). Understanding the geometry of infeasible perturbations of a conic linear system. *SIAM J. Optim.* **10**, 534–550.

RENEGAR, J. (1994). Some perturbation theory for linear programming. *Math. Program.* **65**, 73–91.

RENEGAR, J. (1995a). Incorporating condition measures into the complexity theory of linear programming. *SIAM J. Optim.* **5**, 506–524.

RENEGAR, J. (1995b). Linear programming, complexity theory and elementary functional analysis. *Math. Program.* **70**, 279–351.

ROCKAFELLAR, R.T. (1970). *Convex Analysis* (Princeton University Press).

SCHRIJVER, A. (1986). *Theory of Linear and Integer Programming* (John Wiley & Sons, New York).

SHOR, N.Z. (1977). Cut-off method with space extension in convex programming problems. *Kibernetika* **6**, 94–95.

SMALE, S. (1983). On the average number of steps of the simplex method of linear programming. *Math. Program.* **27**, 241–262.

SOLOMON, H. (1978). *Geometric Probability*, Regional Conference Series in Appl. Math. **28** (SIAM, Philadelphia, PA).

SORENSON, D.C. (1982). Newton's method with a model trust region modification. *SIAM J. Numer. Anal.* **19**, 409–426.

STEWART, G.W. (1989). On scaled projections and pseudoinverses. *Linear Algebra Appl.* **112**, 189–193.

TARDOS, É. (1986). A strongly polynomial algorithm to solve combinatorial linear programs. *Oper. Res.* **34**, 250–256.

TODD, M.J. (1990). A Dantzig–Wolfe-like variant of Karmarkar's interior-point linear programming algorithm. *Oper. Res.* **38**, 1006–1018.

TODD, M.J., TUNCCEL, L., YE, Y. (2001). Characterizations, bounds and probabilistic analysis of two complexity measures for linear programming problems. *Math. Program.* **90**, 59–69.

TONE, K. (1993). An active-set strategy in an interior point method for linear programming. *Math. Program.* **59**, 345–360.

TRAKHTENBROT, B.A. (1984). A survey of Russian approaches to perebor (brute-force search) algorithms. *Ann. Hist. Comput.* **6**, 384–400.

TRAUB, J.F. (1972). Computational complexity of iterative processes. *SIAM J. Comput.* **1**, 167–179.

TREFETHEN, L.N., BAU III, D. (1997). *Numerical Linear Algebra* (SIAM, Philadelphia, PA).

TURING, A.M. (1948). Rounding-off errors in matrix processes. *Quart. J. Mech. Appl. Math.* **1**, 287–308.

VAVASIS, S.A. (1991). *Nonlinear Optimization: Complexity Issues* (Oxford University Press, New York).

VAVASIS, S.A., YE, Y. (1995). Condition numbers for polyhedra with real number data. *Oper. Res. Lett.* **17**, 209–214.

VERA, J.R. (1998). On the complexity of linear programming under finite precision arithmetic. *Math. Program.* **80**, 91–123.

VON NEUMANN, J., GOLDSTINE, H.H. (1947). Numerical inverting matrices of high order. *Bull. Amer. Math. Soc.* **53**, 1021–1099.

WRIGHT, M.H. (1976). Numerical methods for nonlinearly constrained optimization. PhD thesis, Stanford University.

WRIGHT, M.H. (1985). A brief history of linear programming. *SIAM News* **18**, 4.

WRIGHT, S. (1997). *Primal–Dual Interior-Point Methods* (SIAM, Philadelphia, PA).

YE, Y. (1992a). On an affine scaling algorithm for nonconvex quadratic programming. *Math. Program.* **52**, 285–300.

YE, Y. (1992b). On the finite convergence of interior-point algorithms for linear programming. *Math. Program.* **57**, 325–336.

YE, Y. (1994). Combining binary search and Newton's method to compute real roots for a class of real functions. *J. Complexity* **10**, 271–280.

Numerical Solution of Polynomial Systems by Homotopy Continuation Methods

T.Y. Li[1]

Department of Mathematics, Michigan State University, East Lansing, MI 48824-1027, USA

1. Introduction

Let $P(x) = 0$ be a system of n polynomial equations in n unknowns. Denoting $P = (p_1, \ldots, p_n)$, we want to find all isolated solutions of

$$\begin{aligned} p_1(x_1, \ldots, x_n) &= 0, \\ &\vdots \\ p_n(x_1, \ldots, x_n) &= 0 \end{aligned} \tag{1.1}$$

for $x = (x_1, \ldots, x_n)$. This problem is very common in many fields of science and engineering, such as formula construction, geometric intersection problems, inverse kinematics, power flow problems with PQ-specified bases, computation of equilibrium states, etc. Many of those applications has been well documented in TRAVERSO [1997]. Elimination theory based methods, most notably the Buchberger algorithm (BUCHBERGER [1985]) for constructing Gröbner bases, are the classical approach to solving (1.1), but their reliance on symbolic manipulation makes those methods seem somewhat limited to relatively small problems.

[1]Research was supported in part by the NSF under Grant DMS-0104009.

Foundations of Computational Mathematics
Special Volume (F. Cucker, Guest Editor) of
HANDBOOK OF NUMERICAL ANALYSIS, VOL. XI
P.G. Ciarlet (Editor)

Solving polynomial systems is an area where numerical computations arise almost naturally. Given the complexity of the problem, we must use standard machine arithmetic to obtain efficient programs. Moreover, by Galois theory explicit formulas for the solutions are unlikely to exist. We are concerned with the robustness of our methods and want to be sure that *all* isolated solutions are obtained, i.e. we want exhaustive methods. These criteria are met by homotopy continuation methods. GARCIA and ZANGWILL [1979] and DREXLER [1977] independently presented theorems suggesting that homotopy continuation could be used to find numerically the full set of isolated solutions of (1.1). During the last two decades, this method has been developed and proved to be a reliable and efficient numerical algorithm for approximating all isolated zeros of polynomial systems. Note that continuation methods are the method of choice to deal with numerical solutions of nonlinear systems of equations to achieve global convergence as illustrated by the extensive bibliography listed in ALLGOWER and GEORG [1990].

In the early stage, the homotopy continuation method for solving (1.1) is to define a trivial system $Q(x) = (q_1(x), \dots, q_n(x)) = 0$ and then follow the curves in the real variable t which make up the solution set of

$$0 = H(x,t) = (1-t)Q(x) + tP(x). \tag{1.2}$$

More precisely, if $Q(x) = 0$ is chosen correctly, the following three properties hold:

PROPERTY 0 (*Triviality*). The solutions of $Q(x) = 0$ are known.

PROPERTY 1 (*Smoothness*). The solution set of $H(x,t) = 0$ for $0 \leqslant t < 1$ consists of a finite number of smooth paths, each parametrized by t in $[0, 1)$.

PROPERTY 2 (*Accessibility*). Every isolated solution of $H(x,1) = P(x) = 0$ can be reached by some path originating at $t = 0$. It follows that this path starts at a solution of $H(x,0) = Q(x) = 0$.

When the three properties hold, the solution paths can be followed from the initial points (known because of property 0) at $t = 0$ to all solutions of the original problem $P(x) = 0$ at $t = 1$ using standard numerical techniques, see ALLGOWER and GEORG [1990], ALLGOWER and GEORG [1993] and ALLGOWER and GEORG [1997].

Several authors have suggested choices of Q that satisfy the three properties, see CHOW, MALLET-PARET and YORKE [1979], LI [1983], MORGAN [1986], WRIGHT [1985], LI and SAUER [1987], and ZULENER [1988] for a partial list. A typical suggestion is

$$\begin{aligned} q_1(x) &= a_1 x_1^{d_1} - b_1, \\ &\;\;\vdots \\ q_n(x) &= a_n x_n^{d_n} - b_n, \end{aligned} \tag{1.3}$$

where $d_1, \dots, d_n$ are the degrees of $p_1(x), \dots, p_n(x)$, respectively, and a_j, b_j are random complex numbers (and therefore nonzero, with probability one). So in one sense, the original problem we posed is solved. All solutions of $P(x) = 0$ are found at the end of the $d_1 \cdots d_n$ paths that make up the solution set of $H(x,t) = 0$, $0 \leqslant t < 1$.

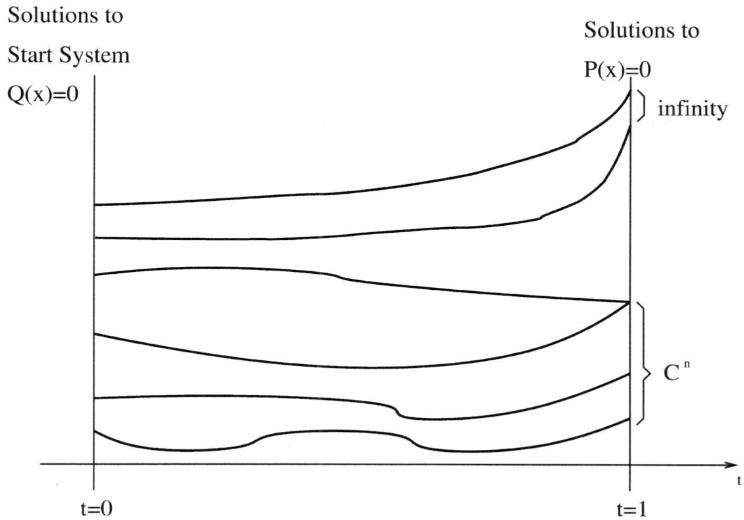

FIG. 1.1.

The book of MORGAN [1987] detailed many aspects of the above approach. A major part of this article will focus on the development afterwards that makes this method more convenient to apply.

The reason the problem is not satisfactorily solved by the above considerations is the existence of *extraneous paths*. Although the above method produces $d = d_1 \cdots d_n$ paths, the system $P(x) = 0$ may have fewer than d solutions. We call such a system *deficient*. In this case, some of the paths produced by the above method will be extraneous paths.

More precisely, even though Properties 0–2 imply that each solution of $P(x) = 0$ will lie at the end of a solution path, it is also consistent with these properties that some of the paths may diverge to infinity as the parameter t approaches 1 (the smoothness property rules this out for $t \to t_0 < 1$). In other words, it is quite possible for $Q(x) = 0$ to have more solutions than $P(x) = 0$. In this case, some of the paths leading from roots of $Q(x) = 0$ are extraneous, and diverge to infinity when $t \to 1$ (see Fig. 1.1).

Empirically, we find that most systems arising in applications are deficient. A great majority of the systems have fewer than, and in some cases only a small fraction of, the "expected number" of solutions. For a typical example of this sort, let's look at the following Cassou–Nogués system (TRAVERSO [1997])

$$
\begin{aligned}
p_1 &= 15b^4cd^2 + 6b^4c^3 + 21b^4c^2d - 144b^2c - 8b^2c^2e \\
&\quad - 28b^2cde - 648b^2d + 36b^2d^2e + 9b^4d^3 - 120, \\
p_2 &= 30b^4c^3d - 32cde^2 - 720b^2cd - 24b^2c^3e - 432b^2c^2 + 576ce - 576de \\
&\quad + 16b^2cd^2e + 16d^2e^2 + 16c^2e^2 + 9b^4c^4 + 39b^4c^2d^2 + 18b^4cd^3 \\
&\quad - 432b^2d^2 + 24b^2d^3e - 16b^2c^2de - 240c + 5184, \\
p_3 &= 216b^2cd - 162b^2d^2 - 81b^2c^2 + 1008ce - 1008de + 15b^2c^2de \\
&\quad - 15b^2c^3e - 80cde^2 + 40d^2e^2 + 40c^2e^2 + 5184, \\
p_4 &= 4b^2cd - 3b^2d^2 - 4b^2c^2 + 22ce - 22de + 261.
\end{aligned}
\tag{1.4}
$$

Since $d_1 = 7, d_2 = 8, d_3 = 6$ and $d_4 = 4$ for this system, the system $Q(x)$ in (1.3) will produce $d_1 \times d_2 \times d_3 \times d_4 = 7 \times 8 \times 6 \times 4 = 1344$ paths for the homotopy in (1.2). However, the system (1.4) has only 16 isolated zeros. Consequently, a major fraction of the paths are extraneous. Sending out 1344 paths in search of 16 solutions appears highly wasteful.

The choice of $Q(x)$ in (1.3) to solve the system $P(x) = 0$ requires an amount of computational effort proportional to $d_1 \times \cdots \times d_n$ and roughly, proportional to the size of the system. The main goal of this article is to derive methods for solving deficient systems for which the computational effort is instead proportional to the actual number of solutions.

This article is written for the readership of the numerical analysis community. Technical terms in algebraic geometry will therefore be confined to a minimum and highly technical proofs of major theorems will be omitted.

2. Linear homotopies

For deficient systems, there are some partial results that use algebraic geometry to reduce the number of extraneous paths with various degrees of success.

2.1. *Random product homotopy*

For a specific example that is quite simple, consider the system

$$
\begin{aligned}
p_1(x) &= x_1(a_{11}x_1 + \cdots + a_{1n}x_n) + b_{11}x_1 + \cdots + b_{1n}x_n + c_1 = 0,\\
&\ \vdots \\
p_n(x) &= x_1(a_{n1}x_1 + \cdots + a_{nn}x_n) + b_{n1}x_1 + \cdots + b_{nn}x_n + c_n = 0.
\end{aligned}
\tag{2.1}
$$

This system has total degree $d = d_1 \cdots d_n = 2^n$. Thus the "expected number of solutions" is 2^n, and the classical homotopy continuation method using the start system $Q(x) = 0$ in (1.3) sends out 2^n paths from 2^n trivial starting points. However, the system $P(x) = 0$ has only $n + 1$ isolated solutions (even fewer for special choices of coefficients). This is a deficient system, at least $2^n - n - 1$ paths will be extraneous. It is never known from the start which of the paths will end up to be extraneous, so they must all be followed to the end, representing wasted computation.

The random product homotopy was developed in LI, SAUER and YORKE [1987a], LI, SAUER and YORKE [1987b] to alleviate this problem. According to that technique, a more efficient choice for the trivial system $Q(x) = 0$ is

$$
\begin{aligned}
q_1(x) &= (x_1 + e_{11})(x_1 + x_2 + \cdots + x_n + e_{12}),\\
q_2(x) &= (x_1 + e_{21})(x_2 + e_{22}),\\
&\ \vdots \\
q_n(x) &= (x_1 + e_{n1})(x_n + e_{n2}).
\end{aligned}
\tag{2.2}
$$

Set

$$H(x,t) = (1-t)cQ(x) + tP(x), \quad c \in \mathbb{C}.$$

It is clear by inspection that for a generic choice of the complex number e_{ij}, $Q(x)=0$ has exactly $n+1$ roots. Thus there are only $n+1$ paths starting from $n+1$ starting points for this choice of homotopy. It is proved in LI, SAUER and YORKE [1987b] that Properties 0–2 hold for this choice of $H(x,t)$, for almost all complex numbers e_{ij} and c. Thus all solutions of $P(x)=0$ are found at the end of the $n+1$ paths. The result of LI, SAUER and YORKE [1987b] is then a mathematical result (that there can be at most $n+1$ solutions to (2.1)) and the basis of a numerical procedure for approximating the solutions.

The reason this works is quite simple. The solution paths of (1.2) which do not proceed to a solution of $P(x)=0$ in $\mathbb{C}^n$ diverge to infinity. If the system (1.2) is viewed in projective space

$$\mathbb{P}^n = \{(x_0,\dots,x_n) \in \mathbb{C}^{n+1}\backslash(0,\dots,0)\}/\sim,$$

where the equivalent relation "$\sim$" is given by $x \sim y$ if $x = cy$ for some nonzero $c \in \mathbb{C}$, the diverging paths simply proceed to a "point at infinity" in $\mathbb{P}^n$.

For a polynomial $f(x_1,\dots,x_n)$ of degree d, denote the associated homogeneous polynomial by

$$\tilde{f}(x_0,x_1,\dots,x_n) = x_0^d f\left(\frac{x_1}{x_0},\dots,\frac{x_n}{x_0}\right).$$

The solutions of $f(x)=0$ at infinity are those zeros of $\tilde{f}$ in $\mathbb{P}^n$ with $x_0=0$ and the remaining zeros of $\tilde{f}$ with $x_0 \neq 0$ are the solutions of $f(x)=0$ in $\mathbb{C}^n$ when x_0 is set to be 1.

Viewed in projective space $\mathbb{P}^n$ the system $P(x)=0$ in (2.1) has some roots at infinity. The roots at infinity make up a nonsingular variety, specifically the linear space $\mathbb{P}^{n-2}$ defined by $x_0=x_1=0$. A Chern class formula from intersection theory (e.g., FULTON [1984], 9.1.1 and 9.1.2) shows that the contribution of a linear variety of solutions of dimension e to the "total degree" $(d_1 \times \cdots \times d_n)$, or the total expected number of solutions, of the system is at least s, where s is the coefficient of t^e in the Maclaurin series expansion of

$$(1+t)^{e-n}\prod_{j=1}^{n}(1+d_j t).$$

In our case, $d_1=\cdots=d_n=2$, and $e=n-2$, hence,

$$\frac{(1+2t)^n}{(1+t)^2} = \frac{(1+t+t)^n}{(1+t)^2} = \frac{\sum_{j=0}^{n}(1+t)^{n-j}t^j\binom{n}{j}}{(1+t)^2} = \sum_{j=0}^{n}(1+t)^{n-j-2}t^j\binom{n}{j}$$

and $s=\sum_{j=0}^{n-2}\binom{n}{j}$, meaning there are at least $\sum_{j=0}^{n-2}\binom{n}{j}$ solutions of $P(x)=0$ at infinity. Thus there are at most

$$2^n - s = (1+1)^n - \sum_{j=0}^{n-2}\binom{n}{j} = n+1$$

solutions of $P(x) = 0$ in $\mathbb{C}^n$. The system $Q(x) = 0$ is chosen to have the same nonsingular variety at infinity, and this variety stays at infinity as the homotopy progresses from $t = 0$ to $t = 1$. As a result, the infinity solutions stay infinite, the finite solution paths stay finite, and no extraneous paths exist.

This turns out to be a fairly typical situation. Even though the system $P(x) = 0$ to be solved has isolated solutions, when viewed in projective space there may be large number of roots at infinity and quite often high-dimensional manifolds of roots at infinity. Extraneous paths are those that are drawn to the manifolds lying at infinity. If $Q(x) = 0$ can be chosen correctly, extraneous paths can be eliminated.

As another example, consider the algebraic eigenvalue problem,

$$Ax = \lambda x,$$

where

$$A = \begin{bmatrix} a_{11} & \cdots & a_{1n} \\ \vdots & & \vdots \\ a_{n1} & \cdots & a_{nn} \end{bmatrix}$$

is an $n \times n$ matrix. This problem is actually an n polynomial equations in the $n + 1$ variables $\lambda, x_1, \ldots, x_n$:

$$\begin{aligned} \lambda x_1 - (a_{11}x_1 + \cdots + a_{1n}x_n) &= 0, \\ &\vdots \\ \lambda x_n - (a_{n1}x_1 + \cdots + a_{nn}x_n) &= 0. \end{aligned}$$

Augmenting the system with a linear equation

$$c_1x_1 + \cdots + c_nx_n + c_{n+1} = 0,$$

where $c_1, \ldots, c_{n+1}$ are chosen at random, we have a polynomial system of $n + 1$ equations in $n + 1$ variables. This system has total degree 2^n. However, it can have at most n isolated solutions. So, the system is deficient. But the system $Q(x)$ in random product form:

$$\begin{aligned} q_1 &= (\lambda + e_{11})(x_1 + e_{12}), \\ q_2 &= (\lambda + e_{21})(x_2 + e_{22}), \\ &\vdots \\ q_n &= (\lambda + e_{n1})(x_n + e_{n2}), \\ q_{n+1} &= c_1x_1 + \cdots + c_nx_n + c_{n+1} \end{aligned}$$

has n isolated zeros for randomly chosen e_{ij}'s. This $Q(x)$ will produce n curves for the homotopy in (1.3) that proceed to all solutions of the eigenvalue problem. Implicit in this is the fact that the algebraic eigenvalue problem has at most n solutions. Moreover, the generic eigenvalue problem has exactly n solutions.

To be more precise, we state the main random product homotopy result, Theorem 2.2 of LI, SAUER and YORKE [1987b]. Let $V_\infty(Q)$ and $V_\infty(P)$ denote the variety of roots at infinity of $Q(x) = 0$ and $P(x) = 0$, respectively.

THEOREM 2.1. *If $V_\infty(Q)$ is nonsingular and contained in $V_\infty(P)$, then Properties* 1 *and* 2 *hold.*

Of course, Properties 1 and 2 are not enough. Without starting points, the path-following method cannot get started. Thus $Q(x)=0$ should also be chosen to be of random product forms, as in (2.2), which are trivial to solve because of their form.

This result was superseded by the result in LI and SAUER [1989]. While the complex numbers e_{ij} are chosen at random in LI, SAUER and YORKE [1987b] to ensure Properties 1 and 2, it was proved in LI and SAUER [1989] that e_{ij} can be any fixed numbers Properties 1 and 2 still hold as long as the complex number c is chosen at random. In fact, the result in LI and SAUER [1989] implies that the start system $Q(x)=0$ in Theorem 2.2 need not be in product form. It can be any chosen polynomial system as long as its zeros in $\mathbb{C}^n$ are known or easy to obtain and its variety of roots at infinity $V_\infty(Q)$ is nonsingular and contained in $V_\infty(P)$.

Theorem 2.1 in LI and WANG [1991] goes one step further. Even when the set $V_\infty(Q)$ of roots at infinity of $Q(x)=0$ has singularities, if the set is contained in $V_\infty(P)$ counting multiplicities, containment in the sense of *scheme* theory of algebraic geometry, then Properties 1 and 2 still hold. More precisely, let $I=\langle\tilde{q}_1,\dots,\tilde{q}_n\rangle$ and $J=\langle\tilde{p}_1,\dots,\tilde{p}_n\rangle$ be the homogeneous ideals spanned by homogenizations of q_i's and p_i's, respectively. For a point p at infinity, if the *local rings* I_p and J_p satisfy

$$I_p\subset J_p$$

then Properties 1 and 2 hold. However, this hypothesis can be much more difficult to verify than whether the set is nonsingular. This limits the usefulness of this approach for practical examples.

2.2. *m-homogeneous structure*

In MORGAN and SOMMESE [1987b], another interesting approach to reduce the number of extraneous paths is developed, using the concept of m-homogeneous structure.

The complex n-space $\mathbb{C}^n$ can be naturally embedded in the projective space $\mathbb{P}^n$. Similarly, the space $\mathbb{C}^{k_1}\times\cdots\times\mathbb{C}^{k_m}$ can be naturally embedded in $\mathbb{P}^{k_1}\times\cdots\times\mathbb{P}^{k_m}$. A point $(y_1,\dots,y_m)$ in $\mathbb{C}^{k_1}\times\cdots\times\mathbb{C}^{k_m}$ with $y_j=(y_1^{(j)},\dots,y_{k_j}^{(j)})$, $j=1,\dots,m$, corresponds to a point $(z_1,\dots,z_m)$ in $\mathbb{P}^{k_1}\times\cdots\times\mathbb{P}^{k_m}$ with $z_j=(z_0^{(j)},\dots,z_{k_j}^{(j)})$ and $z_0^{(j)}=1$, $j=1,\dots,m$. The set of such points in $\mathbb{P}^{k_1}\times\cdots\times\mathbb{P}^{k_m}$ is usually called the *affine space* in this setting. The points in $\mathbb{P}^{k_1}\times\cdots\times\mathbb{P}^{k_m}$ with at least one $z_0^{(j)}=0$ are called the *points at infinity*.

Let f be a polynomial in the n variables $x_1,\dots,x_n$. If we partition the variables into m groups $y_1=(x_1^{(1)},\dots,x_{k_1}^{(1)})$, $y_2=(x_1^{(2)},\dots,x_{k_2}^{(2)}),\dots,y_m=(x_1^{(m)},\dots,x_{k_m}^{(m)})$ with $k_1+\cdots+k_m=n$ and let d_i be the degree of f with respect to y_i (more precisely, to the variables in y_i), then we can define its m-homogenization as

$$\tilde{f}(z_1,\dots,z_m)=\left(z_0^{(1)}\right)^{d_1}\times\cdots\times\left(z_0^{(m)}\right)^{d_m}f\left(y_1/z_0^{(1)},\dots,y_m/z_0^{(m)}\right).$$

This polynomial is homogeneous with respect to each $z_j = (z_0^{(j)}, \dots, z_{k_j}^{(j)})$, $j = 1, \dots, m$. Here $z_i^{(j)} = x_i^{(j)}$, for $i \neq 0$. Such a polynomial is said to be *m-homogeneous*, and $(d_1, \dots, d_m)$ is its m-homogeneous degree. To illustrate this definition, let us consider the polynomial $p_j(x)$ in (2.1), for $j = 1, \dots, n$,

$$\begin{aligned} p_j(x) &= x_1(a_{j1}x_1 + \cdots + a_{jn}x_n) + b_{j1}x_1 + \cdots + b_{jn}x_n + c_j \\ &= a_{j1}x_1^2 + x_1(a_{j2}x_2 + \cdots + a_{jn}x_n + b_{j1}) + b_{j2}x_2 + \cdots + b_{jn}x_n + c_j. \end{aligned}$$

If we set $y_1 = (x_1)$, $y_2 = (x_2, \dots, x_n)$ and $z_1 = (x_0^{(1)}, x_1)$, $z_2 = (x_0^{(2)}, x_2, \dots, x_n)$, then the degree of $p_j(x)$ is two with respect to y_1 and is one with respect to y_2. Hence, its 2-homogenization with respect to z_1 and z_2 is

$$\begin{aligned} \tilde{p}_j(z_1, z_2) &= a_{j1}x_1^2x_0^{(2)} + x_1x_0^{(1)}\left(a_{j2}x_2 + \cdots + a_{jn}x_n + b_{j1}x_0^{(2)}\right) \\ &\quad + \left(x_0^{(1)}\right)^2\left(b_{j2}x_2 + \cdots + b_{jn}x_n + c_jx_0^{(2)}\right). \end{aligned}$$

When the system (2.1) is viewed in $\mathbb{P}^n$ with the homogenization

$$\begin{aligned} \tilde{p}_1(x_0, x_1, \dots, x_n) &= x_1(a_{11}x_1 + \cdots + a_{1n}x_n) \\ &\quad + (b_{11}x_1 + \cdots + b_{1n}x_n)x_0 + c_1x_0^2 = 0, \\ &\ \ \vdots \\ \tilde{p}_n(x_0, x_1, \dots, x_n) &= x_1(a_{n1}x_1 + \cdots + a_{nn}x_n) \\ &\quad + (b_{n1}x_1 + \cdots + b_{nn}x_n)x_0 + c_nx_0^2 = 0, \end{aligned}$$

its total degree, or the Bézout number, is $d = d_1 \times \cdots \times d_n = 2^n$. However, when the system is viewed in $\mathbb{P}^1 \times \mathbb{P}^{n-1} = \{(z_1, z_2) = ((x_0^{(1)}, x_1), (x_0^{(2)}, x_2, \dots, x_n))$ where $z_1 = (x_0^{(1)}, x_1) \in \mathbb{P}^1$ and $z_2 = (x_0^{(2)}, x_2, \dots, x_n) \in \mathbb{P}^{n-1}\}$ with 2-homogenization

$$\begin{aligned} \tilde{p}_1(z_1, z_2) &= a_{11}x_1^2x_0^{(2)} + x_1x_0^{(1)}\left(a_{12}x_2 + \cdots + a_{1n}x_n + b_{11}x_0^{(2)}\right) \\ &\quad + \left(x_0^{(1)}\right)^2\left(b_{12}x_2 + \cdots + b_{1n}x_n + c_1x_0^{(2)}\right), \\ &\ \ \vdots \\ \tilde{p}_n(z_1, z_2) &= a_{n1}x_1^2x_0^{(2)} + x_1x_0^{(1)}\left(a_{n2}x_2 + \cdots + a_{nn}x_n + b_{n1}x_0^{(2)}\right) \\ &\quad + \left(x_0^{(1)}\right)^2\left(b_{n2}x_2 + \cdots + b_{nn}x_n + c_nx_0^{(2)}\right), \end{aligned} \tag{2.3}$$

the Bézout number will be different. It is defined to be the coefficient of $\alpha_1^1\alpha_2^{n-1}$ in the product $(2\alpha_1 + \alpha_2)^n$, which is equal to $2n$.

In general, for an m-homogeneous system

$$\begin{aligned} \tilde{p}_1(z_1, \dots, z_m) &= 0, \\ &\ \ \vdots \\ \tilde{p}_n(z_1, \dots, z_m) &= 0, \end{aligned} \tag{2.4}$$

in $\mathbb{P}^{k_1} \times \cdots \times \mathbb{P}^{k_m}$ with $\tilde{p}_j$ having m-homogeneous degree $(d_1^{(j)}, \ldots, d_m^{(j)})$, $j = 1, \ldots, n$, with respect to $(z_1, \ldots, z_m)$, then the m-homogeneous Bézout number d of the system with respect to $(z_1, \ldots, z_m)$ is the coefficient of $\alpha_1^{k_1} \times \cdots \times \alpha_m^{k_m}$ in the product

$$\left(d_1^{(1)}\alpha_1 + \cdots + d_m^{(1)}\alpha_m\right)\left(d_1^{(2)}\alpha_1 + \cdots + d_m^{(2)}\alpha_m\right)\cdots\left(d_1^{(n)}\alpha_1 + \cdots + d_m^{(n)}\alpha_m\right), \tag{2.5}$$

see SHAFAREVICH [1977]. The classical Bézout theorem says the system (2.4) has no more than d isolated solutions, counting multiplicities, in $\mathbb{P}^{k_1} \times \cdots \times \mathbb{P}^{k_m}$. Applying this to our example in (2.3), the upper bound on the number of isolated solutions, in affine space and at infinity, is $2n$. When solving the original system in (2.1), we may choose the start system $Q(x)$ in the homotopy

$$H(x,t) = (1-t)cQ(x) + tP(x) = 0$$

in random product form to respect the 2-homogeneous structure of $P(x)$. For instance, we choose $Q(x) = (q_1(x), \ldots, q_n(x))$ to be

$$\begin{aligned} q_1(x) &= (x_1 + e_{11})(x_1 + e_{12})(x_2 + \cdots + x_n + e_{13}), \\ q_2(x) &= (x_1 + e_{21})(x_1 + e_{22})(x_2 + e_{23}), \\ &\vdots \\ q_n(x) &= (x_1 + e_{n1})(x_1 + e_{n2})(x_n + e_{n3}), \end{aligned} \tag{2.6}$$

which has the same 2-homogeneous structure as $P(x)$ with respect to the partition $y_1 = (x_1)$ and $y_2 = (x_2, \ldots, x_n)$. Namely, each $q_j(x)$ has degree two with respect to y_1 and degree one with respect to y_2. It is easy to see that for randomly chosen complex numbers e_{ij}, $Q(x) = 0$ has $2n$ solutions in $\mathbb{C}^n$ $(= \mathbb{C}^1 \times \mathbb{C}^{n-1})$ (thus, no solutions at infinity when viewed in $\mathbb{P}^1 \times \mathbb{P}^{n-1}$). Hence there are $2n$ paths emanating from $2n$ solutions of $Q(x) = 0$ for this choice of the homotopy. It was shown in MORGAN and SOMMESE [1987b] that Properties 1 and 2 hold for all complex number c except those lying on a finite number of rays starting at the origin. Thus, all solutions of $P(x) = 0$ are found at the end of $n+1$ paths. The number of extraneous paths, $2n - (n+1) = n-1$, is far less than the number of extraneous paths, $2^n - n - 1$, by using the classical homotopy with $Q(x) = 0$ in (1.3).

More precisely, we state the main theorem in MORGAN and SOMMESE [1987b].

THEOREM 2.2. *Let $Q(x)$ be a system of polynomials chosen to have the same m-homogeneous form as $P(x)$ with respect to certain partition of the variables $(x_1, \ldots, x_n)$. Assume $Q(x) = 0$ has exactly the Bézout number of nonsingular solutions with respect to this partition, and let*

$$H(x,t) = (1-t)cQ(x) + tP(x) = 0,$$

where $t \in [0,1]$ and $c \in \mathbb{C}^*$. If $c = re^{i\theta}$ for some positive $r \in \mathbb{R}$, then for all but finitely many θ, Properties 1 *and* 2 *hold.*

Notice that when the number of isolated zeros of $Q(x)$, having the same m-homogeneous structure of $P(x)$ with respect to a given partition of variables $(x_1, \ldots, x_n)$, reaches the corresponding Bézout number, then no other solutions of $Q(x)=0$ exist at infinity.

In general, if $x=(x_1,\dots,x_n)$ is partitioned into $x=(y_1,\dots,y_m)$ where

$$y_1=\left(x_1^{(1)},\dots,x_{k_1}^{(1)}\right),\quad y_2=\left(x_1^{(2)},\dots,x_{k_2}^{(2)}\right),\quad \dots,\quad y_m=\left(x_1^{(m)},\dots,x_{k_m}^{(m)}\right)$$

with $k_1+\dots+k_m=n$, and for polynomial system $P(x)=(p_1(x),\dots,p_n(x))$, $p_j(x)$ has degree $(d_1^{(j)},\dots,d_m^{(j)})$ with respect to $(y_1,\dots,y_m)$ for $j=1,\dots,n$, we may choose the start system $Q(x)=(q_1(x),\dots,q_n(x))$ where

$$q_j(x)=\prod_{i=1}^{m}\prod_{l=1}^{d_i^{(j)}}\left(c_{l1}^{(i)}x_1^{(i)}+\dots+c_{lk_i}^{(i)}x_{k_i}^{(i)}+c_{l0}^{(i)}\right),\quad j=1,\dots,n. \tag{2.7}$$

Clearly, $q_j(x)$ has degree $(d_1^{(j)},\dots,d_m^{(j)})$ with respect to $(y_1,\dots,y_m)$, the same degree structure of $p_j(x)$. Furthermore, it is not hard to see that for generic coefficients $Q(x)$ has exactly m-homogeneous Bézout number, with respect to this particular partition $x=(y_1,\dots,y_m)$, of nonsingular isolated zeros in $\mathbb{C}^n$. They are easy to obtain. In fact, the system $Q(x)$ in (2.6) is constructed according to this principle. In WAMPLER [1994], the product in (2.7) is modified to be more efficient to evaluate.

As mentioned earlier, solving system in (2.3) with start system $Q(x)$ in (2.6), there are still $n-1$ extraneous paths for the homotopy. This is because, even when viewed in $\mathbb{P}^1\times\mathbb{P}^{n-1}$, $P(x)$ has zeros at infinity. One can check in (2.3) that

$$S=\left\{\left(\left(x_0^{(1)},x_1\right),\left(x_0^{(2)},x_2,\dots,x_n\right)\right)\in\mathbb{P}^1\times\mathbb{P}^{n-1}\mid x_0^{(1)}=0,x_0^{(2)}=0\right\}$$

is a set of zeros of $P(x)$ at infinity. So, to lower the number of those extraneous paths further, we may choose the start system $Q(x)$ to have the same nonsingular variety of zeros at infinity S as $P(x)$ does, in addition to sharing the same 2-homogeneous structure of $P(x)$. For instance, the system $Q(x)=(q_1(x),\dots,q_n(x))$ where

$$\begin{aligned}
q_1(x)&=(x_1+e_{11})(x_1+x_2+\dots+x_n+e_{12}),\\
q_2(x)&=(x_1+e_{21})(x_1+x_2+e_{22}),\\
&\;\;\vdots\\
q_n(x)&=(x_1+e_{n1})(x_1+x_n+e_{n2})
\end{aligned}$$

shares the same 2-homogeneous structure of $P(x)$ with respect to the partition $y_1=(x_1)$ and $y_2=(x_2,\dots,x_n)$. Furthermore, when viewed in $(z_1,z_2)\in\mathbb{P}^1\times\mathbb{P}^{n-1}$ with $z_1=(x_0^{(1)},x_1)$ and $z_2=(x_0^{(2)},x_2,\dots,x_n)$, this system has the same set of zeros at infinity S as $P(x)$ does. The system $Q(x)=0$ also has $n+1$ solutions in $\mathbb{C}^n$ for generic e_{ji}'s, and there are no extraneous paths. The results in LI and WANG [1991] and MORGAN and SOMMESE [1987a] show that if $Q(x)$ in

$$H(x,t)=(1-t)cQ(x)+tP(x)=0$$

is chosen to have the same m-homogeneous structure as $P(x)$ and the set of zeros at infinity $V_\infty(Q)$ of $Q(x)$ is nonsingular and contained in the set of zeros at infinity

$V_\infty(P)$ of $P(x)$, then for $c = re^{i\theta}$ for positive $r \in \mathbb{R}$ and for all but finitely many θ Properties 1 and 2 hold.

Most often the zeros at infinity of an m-homogeneous polynomial system $\widetilde{P}(z_1, \dots, z_m)$ in $\mathbb{P}^{k_1} \times \cdots \times \mathbb{P}^{k_m}$ is hard to identify. Nevertheless, the choice of $Q(x) = 0$ in Theorem 2.2, assuming no zeros at infinity regardless of the structure of the zeros at infinity of $P(x)$, can still reduce the number of extraneous paths dramatically by simply sharing the same m-homogeneous structure of $P(x)$.

Let's consider the system

$$p_1(x) = x_1(a_{11}x_1 + \cdots + a_{1n}x_n) + b_{11}x_1 + \cdots + b_{1n}x_n + c_1 = 0,$$
$$\vdots$$
$$p_n(x) = x_1(a_{n1}x_1 + \cdots + a_{nn}x_n) + b_{n1}x_1 + \cdots + b_{nn}x_n + c_n = 0,$$

in (2.1) once again. This time we partition the variables $x_1, \dots, x_n$ into $y_1 = (x_1, x_2)$ and $y_2 = (x_3, \dots, x_n)$. For this partition, the 2-homogeneous degree structure of $p_j(x)$ stays the same, namely, the degree of $p_j(x)$ is two with respect to y_1 and is one with respect to y_2. However, the Bézout number with respect to this partition becomes the coefficient of $\alpha_1^2\alpha_2^{n-2}$ in the product $(2\alpha_1 + \alpha_2)^n$ according to (2.5). This number is

$$\binom{n}{2} \times 2^2 = 2n(n-1),$$

which is greater than the original Bézout number $2n$ with respect to the partition $y_1 = (x_1)$ and $y_2 = (x_2, \dots, x_n)$ when $n > 2$. If the start system $Q(x)$ is chosen to have the same m-homogeneous structure with respect to this partition, then, assuming $Q(x)$ has no zeros at infinity, we need to follow $2n(n-1)$ paths to find all $n+1$ isolated zeros of $P(x)$. This represents a much bigger amount of extraneous paths.

The m-homogeneous Bézout number is apparently highly sensitive to the chosen partition: different ways of partitioning the variables produce different Bézout numbers. By using Theorem 2.2, we usually follow the Bézout number (with respect to the chosen partition of variables) of paths for finding all the isolated zeros of $P(x)$. In order to minimize the number of paths need to be followed and hence avoid more extraneous paths, it's critically important to find a partition which provides the lowest Bézout number possible. In WAMPLER [1992], an algorithm for this purpose was given. By using this algorithm one can determine, for example, the partition $\mathcal{P} = \{(b), (c, d, e)\}$ which gives the lowest possible Bézout number 368 for the Cassou–Nogués system in (1.4). Consequently, a random product start system $Q(x)$, as in (2.7) for instance, can be constructed to respect the degree structure of the system with respect to this partition. The start system $Q(x)$ will have 368 isolated zeros in $\mathbb{C}^n$. Therefore only 368 homotopy paths need to be followed to find all 16 isolated zeros of the system, in contrast to following 1344 paths if we choose the start system $Q(x)$ as in (1.3).

We shall elaborate below the algorithm given in WAMPLER [1992] for the search of the partition of variables which provides the lowest corresponding m-homogeneous Bézout number of a polynomial system.

First of all, we need a systematic listing of all the possible partitionings of the variables $\{x_1, \dots, x_n\}$. This can be obtained by considering the reformulated problem:

how many different ways are there to partition n distinct items into m identical boxes for $m = 1, \ldots, n$? Denote those numbers by $g(n, m)$, $m = 1, \ldots, n$. Clearly, we have $g(n, n) = 1$, $g(n, 1) = 1$. Moreover, the recursive relation

$$g(n, m) = m \times g(n-1, m) + g(n-1, m-1)$$

holds, because for each of the $g(n-1, m)$ partitionings of $n-1$ items, we may add the nth item to any one of m boxes, plus for each of the $g(n-1, m-1)$ partitionings of $n-1$ items into $m-1$ boxes the nth item can only be in the mth box by itself. The numbers $g(n, m)$ are known as the *Stirling numbers of the second kind*, see, e.g., SELBY [1971]. Fig. 2.1 illustrates the process for $n = 1, \ldots, 4$. The partitionings can be listed by traversing the tree structure implied by Fig. 2.1 (WAMPLER [1992]).

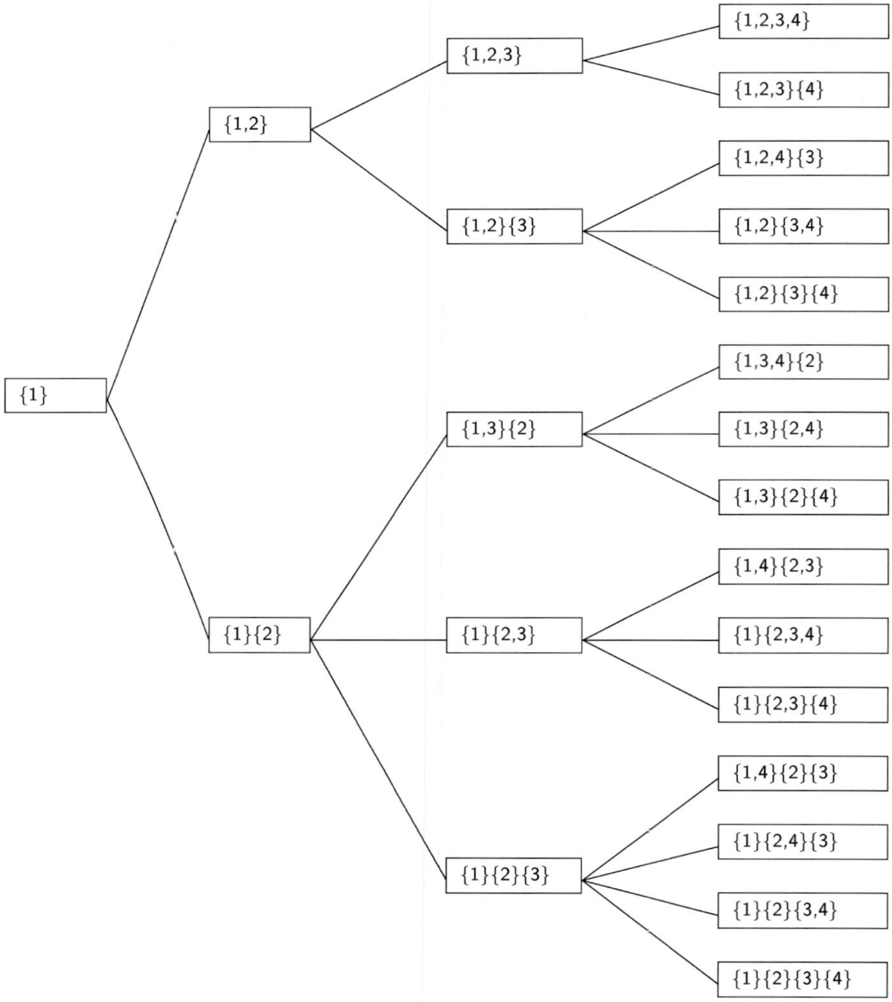

FIG. 2.1. Tree generating all groupings of 4 items.

For a given partition $y_1 = (x_1^{(1)}, \ldots, x_{k_1}^{(1)}), \ldots, y_m = (x_1^{(m)}, \ldots, x_{k_m}^{(m)})$ of the variables $\{x_1, \ldots, x_n\}$ of a polynomial system $P(x) = (p_1(x), \ldots, p_n(x))$ with $k_1 + \cdots + k_m = n$ and

$$d_{ij} = \text{degree of } p_i \text{ with respect to } y_j,$$

a straightforward computation of the Bézout number by expanding the product in (2.5) and finding the appropriate coefficient will not lead to an efficient algorithm except in the simplest cases. A simpler approach is given below.

It's easy to see that the Bézout number given in (2.5) equals the sum of all products of the form

$$d_{1\ell_1} \times d_{2\ell_2} \times \cdots \times d_{n\ell_n},$$

where among $\ell_1, \ldots, \ell_n$ each integer $j = 1, \ldots, m$ appears exactly k_j times. Equivalently, it is the sum of degree products over all possible ways to choose each row once while choosing k_j entries from each column j in the *degree matrix*

$$D = \begin{bmatrix} d_{11} & \cdots & d_{1m} \\ \vdots & \ddots & \vdots \\ d_{n1} & \cdots & d_{nm} \end{bmatrix}. \tag{2.8}$$

Thus, to calculate the Bézout number we may enumerate the permissible combinations in the degree matrix, form the corresponding degree products, and add them up. Since many of the degree products contain common factors, a method resembling the evaluation of a determinant via expansion by minors can be used to reduce the number of multiples, either down the column or across the rows of the degree matrix. The row expansion is generally more efficient, and we shall present only this alternative.

For *partition vector* $K = [k_1, \ldots, k_m]$, we form the degree products in degree matrix D in (2.8) as follows. First, in row 1 of D, suppose element d_{1j} is chosen. Then to complete the degree product we must choose one element from each of the remaining rows while only $k_j - 1$ elements from the jth column are included. So, a *minor* corresponding to d_{1j} is derived by deleting row 1 of D and decreasing k_j by 1. This *minor* has the corresponding Bézout number in its own right, with respect to the partition vector $K' = [k_1, \ldots, k_{j-1}, k_j - 1, k_{j+1}, \ldots, k_m]$. The *row expansion algorithm* for the Bézout number of degree matrix D with respect to the partition vector $K = [k_1, \ldots, k_m]$ is to compute the sum along the first row of each d_{1j} (where $k_j > 0$) times the Bézout number of the corresponding minor. The Bézout number of each minor is then computed recursively by the same row expansion procedure.

More precisely, let $b(D, \overline{K}, i)$ be the Bézout number of the degree matrix

$$D_i = \begin{bmatrix} d_{i1} & \cdots & d_{im} \\ \vdots & \ddots & \vdots \\ d_{n1} & \cdots & d_{nm} \end{bmatrix}$$

consisted of the last $n - i + 1$ rows of D in (2.8), with respect to the partition vector $\overline{K} = [\bar{k}_1, \ldots, \bar{k}_m]$. Here, of course, $\bar{k}_1 + \cdots + \bar{k}_m = n - i + 1$. Let $M(\overline{K}, j)$ be the partition vector derived by reducing $\bar{k}_j$ in $\overline{K}$ by 1, namely,

$$M(\overline{K}, j) = [\bar{k}_1, \ldots, \bar{k}_{j-1}, \bar{k}_j - 1, \bar{k}_{j+1}, \ldots, \bar{k}_m].$$

With the convention $b(D, \overline{K}, n+1) := 1$ the row expansion algorithm may be written as

$$b(D, \overline{K}, i) = \sum_{\substack{j=1 \\ \bar{k}_j \neq 0}}^{m} d_{ij} b\big(D, M(\overline{K}, j), i+1\big)$$

and the Bézout number of the original degree matrix D with respect to the partition vector $K = [k_1, \ldots, k_m]$ is simply $B = b(D, K, 1)$.

Note that if the degree matrix D is sparse, we may skip over computations where $d_{ij} = 0$ and avoid expanding the recursion below that branch.

EXAMPLE 2.1 (WAMPLER [1992]). For polynomial system $P(x) = (p_1(x), \ldots, p_4(x))$, $x = (x_1, x_2, x_3, x_4)$, let $y_1 = (x_1, x_2)$ and $y_2 = (x_3, x_4)$. So, the partition vector is $K = [2, 2]$. Let

$$D = \begin{bmatrix} d_{11} & d_{12} \\ d_{21} & d_{22} \\ d_{31} & d_{32} \\ d_{41} & d_{42} \end{bmatrix}$$

be the degree matrix. Then, by the row expansion algorithm, the Bézout number B of D with respect to K is,

$$\begin{aligned}
B &= d_{11} b\big(D, [1, 2], 2\big) + d_{12} b\big(D, [2, 1], 2\big) \\
&= d_{11}\big[d_{21} \cdot b\big(D, [0, 2], 3\big) + d_{22} \cdot b\big(D, [1, 1], 3\big)\big] \\
&\quad + d_{12}\big[d_{21} \cdot b\big(D, [1, 1], 3\big) + d_{22} \cdot b\big(D, [2, 0], 3\big)\big] \\
&= d_{11}\big[d_{21} d_{32} \cdot b\big(D, [0, 1], 4\big) + d_{22}\big(d_{31} \cdot b\big(D, [0, 1], 4\big) + d_{32} \cdot b\big(D, [1, 0], 4\big)\big)\big] \\
&\quad + d_{12}\big[d_{21}\big(d_{31} \cdot b\big(D, [0, 1], 4\big) + d_{32} \cdot b\big(D, [1, 0], 4\big)\big) \\
&\qquad + d_{22} \cdot \big(d_{31} \cdot b\big(D, [1, 0], 4\big)\big)\big] \\
&= d_{11}\big(d_{21} d_{32} d_{42} + d_{22}(d_{31} d_{42} + d_{32} d_{41})\big) \\
&\quad + d_{12}\big(d_{21}(d_{31} d_{42} + d_{32} d_{41}) + d_{22} d_{31} d_{41}\big).
\end{aligned}$$

EXAMPLE 2.2 (WAMPLER [1992]). Consider the system

$$\begin{aligned}
&x_1^2 + x_2 + 1 = 0, \\
&x_1 x_3 + x_2 + 2 = 0, \\
&x_2 x_3 + x_3 + 3 = 0.
\end{aligned}$$

There are five ways to partition the variables $\{x_1, x_2, x_3\}$. We list the degree matrices and Bézout numbers calculated by the row expansion algorithm for all five partitionings as follows:

(1) $\{x_1, x_2, x_3\}$

$$K = [3], \qquad D = \begin{bmatrix} 2 \\ 2 \\ 2 \end{bmatrix}.$$

Bézout number $= 8$.

(2) $\{x_1, x_2\}\{x_3\}$

$$K = [2, 1], \qquad D = \begin{bmatrix} 2 & 0 \\ 1 & 1 \\ 1 & 1 \end{bmatrix}.$$

Bézout number $= 4$.

(3) $\{x_1, x_3\}\{x_2\}$

$$K = [2, 1], \qquad D = \begin{bmatrix} 2 & 1 \\ 2 & 1 \\ 1 & 1 \end{bmatrix}.$$

Bézout number $= 8$.

(4) $\{x_1\}\{x_2, x_3\}$

$$K = [1, 2], \qquad D = \begin{bmatrix} 2 & 1 \\ 1 & 1 \\ 0 & 2 \end{bmatrix}.$$

Bézout number $= 6$.

(5) $\{x_1\}, \{x_2\}, \{x_3\}$

$$K = [1, 1, 1], \qquad D = \begin{bmatrix} 2 & 1 & 0 \\ 1 & 1 & 1 \\ 0 & 1 & 1 \end{bmatrix}.$$

Bézout number $= 5$.

Thus, the grouping $(\{x_1, x_2\}, \{x_3\})$ has the lowest Bézout number $(= 4)$ and will lead to the homotopy continuation with least extraneous paths.

When exhaustively searching for the partitioning of the variables which yields the minimal Bézout number of the system, there are several ways to speed up the process. For instance, as we sequentially test partitionings in search for minimal Bézout numbers, we can use the smallest one found so far to cut short unfavorable partitionings. Since the degrees are all nonnegative, the Bézout number is a sum of nonnegative degree products. If at any time the running subtotal exceeds the current minimal Bézout number, the calculation can be aborted and testing of the next partitioning can proceed. This can save a substantial amount of computation during an exhaustive search. See WAMPLER [1992] for more details.

While the number of partitionings to be tested grows rapidly with the number of variables, the exhaustive search can be easily parallelized by subdividing the tree of partitionings and distributing these branches to multiple processors for examination. Thus, continuing advances in both raw computer speed and in parallel machines will make progressively larger problems feasible.

In VERSCHELDE and HAEGEMANS [1993], a *generalized Bézout number* GB is developed in the GBQ-*algorithm*, in which the partition of variables is permitted to vary among the $p_j(x)$'s. Even lower Bézout number may be achieved when a proper partition structure of the variables for each individual polynomial $p_j(x)$, $j = 1, \ldots, n$, is chosen. This strategy can take a great advantage on certain sparse systems where an appropriate partition of variables is evident. For instance, for the polynomial system in LIU, LI and BAI [Preprint],

$$
\begin{aligned}
p_1(x) &= x_1x_2 + x_1^2x_2^2,\\
p_2(x) &= x_2x_3^2x_{15},\\
p_3(x) &= x_3x_4^2x_{10},\\
p_4(x) &= x_4^2x_5^2,\\
p_5(x) &= x_5^2x_6x_{12}^2,\\
p_6(x) &= x_6^2x_7^2x_{12} + x_6x_7x_{12} + x_6x_7^2x_{12}^2,\\
p_7(x) &= x_7x_8^2 = x_7^2x_8,\\
p_8(x) &= x_8x_9,\\
p_9(x) &= x_9^2x_{10} + x_9^2x_{10}^2,\\
p_{10}(x) &= x_{10}^2x_{11},\\
p_{11}(x) &= x_{11}x_{12},\\
p_{12}(x) &= x_{12}x_{13}x_{21}^2 + x_{12}x_{13}^2x_{21},\\
p_{13}(x) &= x_{13}x_{14},\\
p_{14}(x) &= x_{14}^2x_{15}x_{19}^2,\\
p_{15}(x) &= x_{15}x_{16}^2 + x_{15}^2x_{16},\\
p_{16}(x) &= x_{16}x_{17},\\
p_{17}(x) &= x_8x_{17}^2x_{18}^2 + x_8x_{17}x_{18}^2,\\
p_{18}(x) &= x_{18}^2x_{19} + x_{18}^2x_{19}^2,\\
p_{19}(x) &= x_{19}^2x_{20} + x_{19}x_{20},\\
p_{20}(x) &= x_{20}^2x_{21}^2 + x_{20}x_{21}^2,\\
p_{21}(x) &= x_6x_{21}^2,
\end{aligned}
\tag{2.9}
$$

an obvious partition of variables in each polynomial consists of the set of every variable in the polynomial, such as $(\{x_1\},\{x_2\})$ for $p_1(x)$, $(\{x_2\},\{x_3\},\{x_{15}\})$ for $p_2(x), \ldots,$ etc. Accordingly, the generalized Bézout number is 63,488, while the total degree is 79,626,240,000. An efficient algorithm to evaluate the generalized Bézout number for a given partition is proposed in LIU, LI and BAI [Preprint].

2.3. *Cheater's homotopy*

To organize our discussion in this section, we will at times use a notation that makes the coefficients and variables in the polynomial system $P(x) = 0$ explicit. Thus when the dependence on coefficients is important, we will consider the system $P(c,x)=0$ of n polynomial equations in n unknowns, where $c=(c_1,\ldots,c_M)$ are coefficients and $x=(x_1,\ldots,x_n)$ are unknowns.

A method called the *cheater's homotopy* has been developed in LI, SAUER and YORKE [1988], LI, SAUER and YORKE [1989] (a similar procedure can be found in MORGAN and SOMMESE [1989]) to deal with the problem when the system $P(c,x)=0$ is asked to be solved for several different values of the coefficients c.

The idea of the method is to theoretically establish Properties 1 and 2 by deforming a sufficiently generic system (in a precise sense to be given later) and then to "cheat" on Property 0 by using a preprocessing step. The amount of computation of preprocessing step may be large, but is amortized among the several solving characteristics of the problem.

We begin with an example. Let $P(x)$ be the system

$$\begin{aligned} p_1(x) &= x_1^3x_2^2+c_1x_1^3x_2+x_2^2+c_2x_1+c_3=0,\\ p_2(x) &= c_4x_1^4x_2^2-x_1^2x_2+x_2+c_5=0. \end{aligned} \tag{2.10}$$

This is a system of two polynomial equations in two unknowns x_1 and x_2. We want to solve this system of equations several times for various specific choices of $c=(c_1,\ldots,c_5)$.

It turns out that for any choice of coefficients c, system (2.10) has no more than 10 isolated solutions. More precisely, there is an open dense subset S of $\mathbb{C}^5$ such that for c belonging to S, there are 10 solutions of (2.10). Moreover, 10 is an upper bound for the number of isolated solutions for all c in $\mathbb{C}^5$. The total degree of the system is $6\times 5=30$, meaning that if we had taken a generic system of two polynomials in two variables of degree 5 and 6, there would be 30 solutions. Thus (2.10), with any choice of c, is a deficient system.

The classical homotopy using the start system $Q(x)=0$ in (1.3) produces $d=30$ paths, beginning at 30 trivial starting points. Thus there are (at least) 20 extraneous paths.

The cheater's homotopy continuation approach begins by solving (2.10) with *randomly-chosen* complex coefficients $\bar{c}=(\bar{c}_1,\ldots,\bar{c}_5)$; let X^* be the set of 10 solutions. No work is saved there, since 30 paths need to be followed, and 20 paths are wasted. However, the 10 elements of the set X^* are the seeds for the remainder of the process. In the future, for each choice of coefficients $c=(c_1,\ldots,c_5)$ for which the system (2.10) needs to be solved, we use the homotopy continuation method to follow a straight-line homotopy from the system with coefficient c^* to the system with coefficient c. We follow the 10 paths beginning at the 10 elements of X^*. Thus Property 0, that of having trivial-available starting points, is satisfied. The fact that Properties 1 and 2 are also satisfied is the content of Theorem 2.3 below. Thus for each fixed c, all 10 (or fewer) isolated solutions of (2.10) lie at the end of 10 smooth homotopy paths beginning at the seeds in X^*. After the foundational step of finding the seeds, the complexity of all further solving of (2.10) is proportional to the number of solutions 10, rather than the total degree 30.

Furthermore, this method requires no a priori analysis of the system. The first preprocessing step of finding the seeds establishes a sharp theoretical upper bound on the number of isolated solutions as a by-product of the computation; further solving of the system uses the optimal number of paths to be followed.

We earlier characterized a successful homotopy continuation method as having three properties: triviality, smoothness, and accessibility. Given an arbitrary system of polynomial equations, such as (2.10), it is not too hard (through generic perturbations) to find a family of systems with the last two properties. The problem is that one member of the family must be trivial to solve, or the path-following cannot get started. The idea of the cheater's homotopy is simply to "cheat" on this part of the problem, and run a preprocessing step (the computation of the seeds X^*) which gives us the triviality property in a roundabout way. Thus the name, the "cheater's homotopy".

A statement of the theoretical result we need follows. Let

$$\begin{aligned} &p_1(c_1,\ldots,c_M,x_1,\ldots,x_n)=0,\\ &\qquad\vdots\\ &p_n(c_1,\ldots,c_M,x_1,\ldots,x_n)=0, \end{aligned} \tag{2.11}$$

be a system of polynomial equations in the variables $c_1,\ldots,c_M,x_1,\ldots,x_n$. Write $P(c,x)=(p_1(c,x),\ldots,p_n(c,x))$. For each choice of $c=(c_1,\ldots,c_M)$ in $\mathbb{C}^M$, this is a system of polynomial equations in the variables $x_1,\ldots,x_n$. Let d be the total degree of the system for a generic choice of c.

THEOREM 2.3. *Let c belong to $\mathbb{C}^M$. There exists an open dense full-measure subset U of $\mathbb{C}^{n+M}$ such that for $(b_1^*,\ldots,b_n^*,c_1^*,\ldots,c_M^*)\in U$, the following holds:*

(a) *The set X^* of solutions $x=(x_1,\ldots,x_n)$ of*

$$\begin{aligned} &q_1(x_1,\ldots,x_n)=p_1\bigl(c_1^*,\ldots,c_M^*,x_1,\ldots,x_n\bigr)+b_1^*=0,\\ &\qquad\vdots\\ &q_n(x_1,\ldots,x_n)=p_n\bigl(c_1^*,\ldots,c_M^*,x_1,\ldots,x_n\bigr)+b_n^*=0 \end{aligned} \tag{2.12}$$

consists of d_0 isolated points, for some $d_0\leqslant d$.

(b) *The smoothness and accessibility properties hold for the homotopy*

$$\begin{aligned} H(x,t)=P\bigl(&n(1-t)c_1^*+tc_1,\ldots,(1-t)c_M^*+tc_M,x_1,\ldots,x_n\bigr)\\ &+(1-t)b^*, \end{aligned} \tag{2.13}$$

where $b^*=(b_1^*,\ldots,b_n^*)$. It follows that every solution of $P(c,x)=0$ is reached by a path beginning at a point of X^*.

A proof of this theorem can be found in LI, SAUER and YORKE [1989]. The theorem is used as part of the following procedure. Let $P(c,x)=0$ as in (2.11) denote the system to be solved for various values of the coefficients c.

CHEATER'S HOMOTOPY PROCEDURE.

(1) Choose complex number $(b_1^*, \ldots, b_n^*, c_1^*, \ldots, c_M^*)$ at random, and use the classical homotopy continuation method to solve $Q(x) = 0$ in (2.12). Let d_0 denote the number of solutions found (this number is bounded above by the total degree d). Let X^* denote the set of d_0 solutions.

(2) For each new choice of coefficients $c = (c_1, \ldots, c_M)$, follow the d_0 paths defined by $H(x, t) = 0$ in (2.13), beginning at the points in X^*, to find all solutions of $P(c, x) = 0$.

In step (1) above, for random complex numbers $(c_1^*, \ldots, c_M^*)$, using classical homotopy continuation methods to solve $Q(x) = 0$ in (2.12) may itself be computationally expensive. It is desirable that those numbers do not have to be random. For illustration, we regard the linear system

$$\begin{aligned} c_{11}x_1 + \cdots + c_{1n}x_n &= b_1, \\ &\vdots \\ c_{n1}x_1 + \cdots + c_{nn}x_n &= b_n, \end{aligned} \tag{2.14}$$

as a system of polynomial equations with degree one of each. For randomly chosen c_{ij}'s, (2.14) has a unique solution which is not available right away. However, if we choose $c_{ij} = \delta_{ij}$ (the Kronecker delta: $= 1$ if $i = j$, $= 0$ if $i \neq j$), the solution is obvious.

For this purpose, an alternative is suggested in LI and WANG [1992]. When a system $P(c, x) = 0$ with a particular parameter c^0 is solved, this c^0 may be assigned specifically instead of being chosen randomly, then for any parameter $c \in \mathbb{C}^M$ consider the nonlinear homotopy

$$H(a, x, t) = P\big(\big(1 - [t - t(1-t)a]\big)c^0 + \big(t - t(1-t)a\big)c, x\big) = 0. \tag{2.15}$$

It was shown in LI and WANG [1992] that for randomly chosen complex number a the solution paths of $H(a, x, t) = 0$ in (2.15), emanating from the solutions of $P(c^0, x) = 0$ will reach the isolated solutions of $P(c, x) = 0$ under the natural assumption that for generic c, $P(c, x)$ has the same number of isolated zeros in $\mathbb{C}^n$.

The most important advantage of the homotopy in (2.15) is that the parameter c^0 of the start system $P(c^0, x) = 0$ need not be chosen at random. We only require the system $P(c^0, x) = 0$ to have the same number of solutions as $P(c, x) = 0$ for generic c. Therefore, in some situations when the solutions of $P(c, x) = 0$ are easily available for a particular parameter c^0, the system $P(c^0, x) = 0$ may be used as the start system in (2.15) and the extra effort of solving $P(c, x) = 0$ for a randomly chosen c would not be necessary.

To finish, we give a non-trivial example of the use of the procedure discussed above. Consider the indirect position problem for revolute-joint kinematic manipulators. Each joint represents a one-dimensional choice of parameters, namely the angular position of the joint. If all angular positions are known, then of course the position and orientation of the end of the manipulator (the hand) are determined. The indirect position problem is the inverse problem: given the desired position and orientation of the hand, find a set of angular parameters for the (controllable) joints which will place the hand in the desired state.

The indirect position problem for six joints is reduced to a system of eight nonlinear equations in eight unknowns in TSAI and MORGAN [1985]. The coefficients of the equations depend on the desired position and orientation, and a solution of the system (an eight-vector) represents the sines and cosines of the angular parameters. Whenever the manipulator's position is changed, the system needs to be resolved with new coefficients. The equations are too long to repeat here, see the appendix of TSAI and MORGAN [1985]; suffice to say that it is a system of eight degree-two polynomial equations in eight unknowns which is quite deficient. The total degree of the system is $2^8 = 256$, but there are at most 32 isolated solutions.

The nonlinear homotopy (2.15) requires only 32 paths to solve the system with different set of parameters, see LI and WANG [1990], LI and WANG [1992]. The system contains 26 coefficients, and a specific set of coefficients is chosen for which the system has 32 solutions. For subsequent solving of the system, for any choice of the coefficients $c_1, \ldots, c_{26}$, all solutions can be found at the end of exactly 32 paths by using nonlinear homotopy in (2.15) with randomly chosen complex number a.

3. Combinatorial root count

3.1. Bernshteín's theorem

In the middle of 90's, a major computational breakthrough has emerged in solving polynomial systems by the homotopy continuation method. The new method takes a great advantage of the Bernshteín's theorem which provides a much tighter bound in general for the number of isolated zeros of a polynomial system in the algebraic tori $(\mathbb{C}^*)^n$ where $\mathbb{C}^* = \mathbb{C}\backslash\{0\}$. Based on this root count, the *polyhedral homotopy* was introduced in HUBER and STURMFELS [1995] to find all isolated zeros of polynomial systems.

We take the following example (HUBER and STURMFELS [1995]) as our point of departure. With $x = (x_1, x_2)$, consider the system $P(x) = (p_1(x), p_2(x))$ where

$$\begin{aligned} p_1(x_1, x_2) &= c_{11}x_1x_2 + c_{12}x_1 + c_{13}x_2 + c_{14} = 0, \\ p_2(x_1, x_2) &= c_{21}x_1x_2^2 + c_{22}x_1^2x_2 + c_{23} = 0. \end{aligned} \tag{3.1}$$

Here, $c_{ij} \in \mathbb{C}^*$. The monomials $\{x_1x_2, x_1, x_2, 1\}$ in p_1 can be written with their explicit exponents as $x_1x_2 = x_1^1x_2^1$, $x_1 = x_1^1x_2^0$, $x_2 = x_1^0x_2^1$ and $1 = x_1^0x_2^0$. The set of their exponents

$$S_1 = \{a = (0,0),\ b = (1,0),\ c = (1,1),\ d = (0,1)\}$$

is called the *support* of p_1, and its convex hull $Q_1 = \text{conv}(S_1)$ is called the *Newton polytope* of p_1. Similarly, p_2 has support $S_2 = \{e = (0,0),\ f = (2,1),\ g = (1,2)\}$ and Newton polytope $Q_2 = \text{conv}(S_2)$. With additional notations $x^q = x_1^{q_1}x_2^{q_2}$ where $q = (q_1, q_2)$, the system in (3.1) becomes

$$p_1(x) = \sum_{q \in S_1} c_{1,q}x^q, \qquad p_2(x) = \sum_{q \in S_2} c_{2,q}x^q.$$

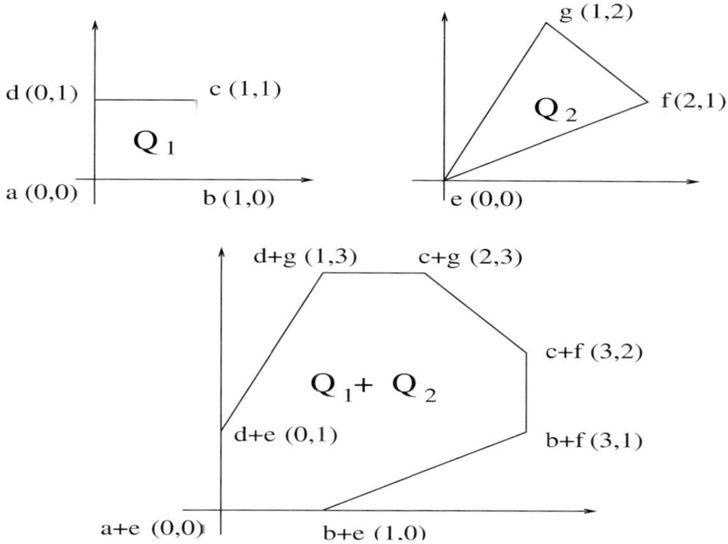

FIG. 3.1.

For polytopes $R_1, \ldots, R_k$ in $\mathbb{R}^n$, their *Minkowski sum* $R_1 + \cdots + R_k$ is defined by

$$R_1 + \cdots + R_n = \{r_1 + \cdots + r_k \mid r_j \in R_j,\ j = 1, \ldots, k\}.$$

(Polytopes Q_1, Q_2 and $Q_1 + Q_2$ for the system in (3.1) are shown in Fig. 3.1.) Let's consider the area of the convex polytope $\lambda_1 Q_1 + \lambda_2 Q_2$ with non-negative variables λ_1 and λ_2. First of all, the area of a triangle on the plane with vertices u, v and w is known to be

$$\frac{1}{2}\left|\det\begin{pmatrix} u - v \\ w - v \end{pmatrix}\right|. \tag{3.2}$$

To compute the area $f(\lambda_1, \lambda_2)$ of $\lambda_1 Q_1 + \lambda_2 Q_2$, we may partition the polytope $\lambda_1 Q_1 + \lambda_2 Q_2$ into a collection of mutually disjoint triangles, $A_1, A_2, \ldots, A_k$. If we choose those triangles in which none of their vertices is an interior point of the polytope $\lambda_1 Q_1 + \lambda_2 Q_2$, then all their vertices take the form $\lambda_1 r_1 + \lambda_2 r_2$ for certain $r_1 \in Q_1$ and $r_2 \in Q_2$. From (3.2), $f(\lambda_1, \lambda_2)$ is a second degree homogeneous polynomial in λ_1 and λ_2. Writing

$$f(\lambda_1, \lambda_2) = a_1\lambda_1^2 + a_2\lambda_2^2 + a_{12}\lambda_1\lambda_2,$$

the coefficient a_{12} of $\lambda_1\lambda_2$ in f is called the *mixed volume* of the polytopes Q_1 and Q_2, and is denoted by $\mathcal{M}(Q_1, Q_2)$.

Clearly,

$$\begin{aligned} a_{12} &= f(1,1) - f(1,0) - f(0,1) \\ &= \text{area of } (Q_1 + Q_2) - \text{ area of } (Q_1) - \text{ area of } (Q_2). \end{aligned}$$

The areas of $Q_1 + Q_2$, Q_1 and Q_2, as displayed in Fig. 3.1, are 6.5, 1 and 1.5, respectively, therefore the mixed volume $\mathcal{M}(Q_1, Q_2)$ of the polytopes Q_1 and Q_2 is

$$\mathcal{M}(Q_1, Q_2) = a_{12} = 6.5 - 1 - 1.5 = 4.$$

On the other hand, system (3.1) has two zeros $(x_0, x_1, x_2) = (0, 0, 1)$ and $(0, 1, 0)$ at infinity when viewed in $\mathbb{P}^2$; hence, it can have at most 4 isolated zeros in $(\mathbb{C}^*)^2$. This is the content of the Bernshteín theory: The number of isolated zeros of (3.1) in $(\mathbb{C}^*)^2$, counting multiplicities, is bounded above by the mixed volume of its Newton polytopes. Furthermore, when c_{ij}'s in (3.1) are chosen generically, these two numbers are exactly the same.

To state the Bernshteín theorem in general form, let the given polynomial system be $P(x) = (p_1(x), \dots, p_n(x)) \in \mathbb{C}[x]$ where $x = (x_1, \dots, x_n) \in \mathbb{C}^n$. With $x^a = x_1^{a_1} \cdots x_n^{a_n}$ and $a = (a_1, \dots, a_n)$, write

$$\begin{aligned} p_1(x) &= \sum_{a \in S_1} c^*_{1,a} x^a, \\ &\vdots \\ p_n(x) &= \sum_{a \in S_n} c^*_{n,a} x^a, \end{aligned} \tag{3.3}$$

where $S_1, \dots, S_n$ are fixed subsets of $\mathbb{N}^n$ with cardinals $k_j = \#S_j$, and $c^*_{j,a} \in \mathbb{C}^*$ for $a \in S_j$, $j = 1, \dots, n$. As before, S_j is the *support* of $p_j(x)$, its convex hull $Q_j = \operatorname{conv}(S_j)$ in $\mathbb{R}^n$ is the *Newton polytope* of p_j, and $S = (S_1, \dots, S_n)$ is the *support* of $P(x)$. For nonnegative variables $\lambda_1, \dots, \lambda_n$, let $\lambda_1 Q_1 + \cdots + \lambda_n Q_n$ denote the *Minkowski sum* of $\lambda_1 Q_1, \dots, \lambda_n Q_n$, that is

$$\lambda_1 Q_1 + \cdots + \lambda_n Q_n = \{\lambda_1 r_1 + \cdots + \lambda_n r_n \mid r_j \in Q_j,\ j = 1, 2, \dots, n\}.$$

Following similar reasonings for calculating the area of $\lambda_1 Q_1 + \lambda_2 Q_2$ of the system in (3.1), it can be shown that the n-dimensional volume, denoted by Vol_n, of the polytope $\lambda_1 Q_1 + \cdots + \lambda_n Q_n$ is a homogeneous polynomial of degree n in $\lambda_1, \dots, \lambda_n$. The coefficient of the monomial $\lambda_1 \cdots \lambda_n$ in this homogeneous polynomial is called the *mixed volume* of the polytopes $Q_1, \dots, Q_n$, denoted by $\mathcal{M}(Q_1, \dots, Q_n)$, or the mixed volume of the supports $S_1, \dots, S_n$ denoted by $\mathcal{M}(S_1, \dots, S_n)$. Sometimes, when no ambiguities exist, it is also called the mixed volume of $P(x)$.

We now embed the system $P(x)$ in (3.3) in the systems $P(c, x) = (p_1(c, x), \dots, p_n(c, x))$ where

$$\begin{aligned} p_1(c, x) &= \sum_{a \in S_1} c_{1,a} x^a, \\ &\vdots \\ p_n(c, x) &= \sum_{a \in S_n} c_{n,a} x^a, \end{aligned} \tag{3.4}$$

and the coefficients $c = (c_{j,a})$ with $a \in S_j$ for $j = 1, \ldots, n$ are taken to be a set of $m := k_1 + \cdots + k_n$ variables. That is, we regard $P(x) = P(c^*, x)$ for a set of specified values of coefficients $c^* = (c^*_{j,a})$ in (3.4).

In what follows, the total number of isolated zeros, counting multiplicities, of a polynomial system will be referred to as the *root count* of the system.

LEMMA 3.1 (HUBER [1995]). *For polynomial systems $P(c, x)$ in (3.4), there exists a polynomial system $G(c) = (g_1(c), \ldots, g_l(c))$ in the variables $c = (c_{j,a})$ for $a \in S_j$ and $j = 1, \ldots, n$ such that for those coefficients $c^* = (c^*_{j,a})$ for which $G(c^*) \neq 0$, the root count in $(\mathbb{C}^*)^n$ of the corresponding polynomial systems in (3.4) is a fixed number. And the root count in $(\mathbb{C}^*)^n$ of any other polynomial systems in (3.4) is bounded above by this number.*

A simple example that illustrates the assertion of the above lemma is the following 2×2 linear systems:

$$\begin{aligned} c_{11}x_1 + c_{12}x_2 &= b_1, \\ c_{21}x_1 + c_{22}x_2 &= b_2. \end{aligned} \tag{3.5}$$

Here, $c = (c_{11}, c_{12}, c_{21}, c_{22}, -b_1, -b_2)$. Let

$$G(c) = \det\begin{pmatrix} c_{11} & c_{12} \\ c_{21} & c_{22} \end{pmatrix} \times \det\begin{pmatrix} c_{11} & b_1 \\ c_{21} & b_2 \end{pmatrix} \times \det\begin{pmatrix} b_1 & c_{12} \\ b_2 & c_{22} \end{pmatrix}.$$

When the coefficient $c^* = (c^*_{11}, c^*_{12}, c^*_{21}, c^*_{22}, -b^*_1, -b^*_2)$ satisfies $G(c^*) = 0$, its corresponding system in (3.5) has no isolated solution in $(\mathbb{C}^*)^2$; otherwise, the system has a unique solution in $(\mathbb{C}^*)^2$.

Since the zeros of the polynomial $G(c)$ in Lemma 3.1 form an algebraic set with dimension smaller than m, its complement where $G(c) \neq 0$ is open and dense with full measure in $\mathbb{C}^m$. Therefore, polynomial systems $P(c, x)$ in (3.4) with $G(c) \neq 0$ are said to be *in general position*. For $P(x) = (p_1(x), \ldots, p_n(x))$ in general position with support $(S_1, \ldots, S_n)$, let $L(S_1, \ldots, S_n)$ be the fixed number of its isolated zeros in $(\mathbb{C}^*)^n$. This number satisfies the following properties:

(1) *(Symmetry)* $L(S_1, \ldots, S_n)$ remains invariant when S_i and S_j for $i \neq j$ exchange their positions.

(2) *(Shift invariant)* $L(S_1, \ldots, a + S_j, \ldots, S_n) = L(S_1, \ldots, S_n)$ for $a \in \mathbb{N}^n$.

Replacing $p_j(x)$ in $P(x) = (p_1(x), \ldots, p_n(x))$ by $x^a p_j(c, x)$ results in a new system with support $(S_1, \ldots, a + S_j, \ldots, S_n)$. Obviously, the number of its isolated zeros in $(\mathbb{C}^*)^n$ stays the same.

(3) *(Multilinear)* $L(S_1, \ldots, S_j + \overline{S}_j, \ldots, S_n) = L(S_1, \ldots, S_j, \ldots, S_n) + L(S_1, \ldots, \overline{S}_j, \ldots, S_n)$ for $\overline{S}_j \subset \mathbb{N}^n$.

Let $\overline{P}(x) = (p_1(x), \ldots, \bar{p}_j(x), \ldots, p_n(x))$ be a system in general position with support $(S_1, \ldots, \overline{S}_j, \ldots, S_n)$. Then replacing $p_j(x)$ in $P(x)$ by $p_j(x) \cdot \bar{p}_j(x)$ yields a system with support $(S_1, \ldots, S_j + \overline{S}_j, \ldots, S_n)$. It is clear that the number of isolated zeros of the resulting system in $(\mathbb{C}^*)^n$ is the sum of the number of those isolated zeros of $P(x)$ and $\overline{P}(x)$ in $(\mathbb{C}^*)^n$.

(4) (*Automorphism invariant*) $L(S_1,\dots,S_n) = L(US_1,\dots,US_n)$ where U is an integer entry $n\times n$ matrix with $\det U = \pm 1$ and $US_j = \{Ua \mid a \in S_j\}$ for $j = 1,\dots,n$.

Note that in writing $x^a = x_1^{a_1}\cdots x_n^{a_n}$, we regard $a = (a_1,\dots,a_n)$ as a column vector in convention. Let U_j be the jth column of $U = (u_{ij})$ and $x = y^U := (y^{U_1},\dots,y^{U_n})$, i.e.

$$x_j = y^{U_j} = y_1^{u_{1j}}\cdots y_n^{u_{nj}}, \quad j = 1,\dots,n.$$

This coordinate transformation yields

$$\begin{aligned} x^a &= x_1^{a_1}\cdots x_n^{a_n} \\ &= \left(y^{U_1}\right)^{a_1}\cdots\left(y^{U_n}\right)^{a_n} \\ &= y_1^{u_{11}a_1+\cdots+u_{1n}a_n}\cdots y_n^{u_{n1}a_1+\cdots+u_{nn}a_n} \\ &= y^{Ua}, \end{aligned} \tag{3.6}$$

and transforms the system $P(x)$ with support $(S_1,\dots,S_n)$ to $Q(y) = P(y^U)$ with support $(US_1,\dots,US_n)$. For a given isolated zeros y_0 of $Q(y)$ in $(\mathbb{C}^*)^n$, $x_0 = y_0^U$ is clearly an isolated zero of $P(x)$ in $(\mathbb{C}^*)^n$. On the other hand, since $\det U = \pm 1$, $V := U^{-1}$ is also an integer-entry matrix, and

$$x^V = \left(y^U\right)^V = y^{(UV)} = y.$$

Therefore, for an isolated zero x_0 of $P(x)$ in $(\mathbb{C}^*)^n$, $y_0 = x_0^V$ is an isolated zero of $Q(y)$ in $(\mathbb{C}^*)^n$. This one-to-one correspondence between isolated zeros of $Q(y)$ and $P(x)$ in $(\mathbb{C}^*)^n$ yields $L(S_1,\dots,S_n) = L(US_1,\dots,US_n)$.

Functions that take n finite subsets $S_1,\dots,S_n$ of $\mathbb{N}^n$ and return with a real number satisfying all the above four properties are rarely available. The mixed volume $\mathcal{M}(S_1,\dots,S_n)$, emerged in the early 20th century, happens to be one of them:

(1) (*Symmetry*) This property is obvious for $\mathcal{M}(S_1,\dots,S_n)$ by its definition.

(2) (*Shift invariant*) For $a \in \mathbb{N}^n$ and $Q_k = \operatorname{conv}(S_k)$ for $k = 1,\dots,n$,

$$\begin{aligned} &\mathrm{Vol}_n\left(\lambda_1 Q_1 + \cdots + \lambda_j(a + Q_j) + \cdots + \lambda_n Q_n\right) \\ &\quad = \mathrm{Vol}_n(\lambda_j a + \lambda_1 Q_1 + \cdots + \lambda_j Q_j + \cdots + \lambda_n Q_n) \\ &\quad = \mathrm{Vol}_n(\lambda_1 Q_1 + \cdots + \lambda_n Q_n). \end{aligned}$$

Hence, $\mathcal{M}(S_1,\dots,a+S_j,\dots S_n) = \mathcal{M}(S_1,\dots,S_n)$.

(3) (*Multilinear*) We shall prove this property for $\overline{S}_1 \subset \mathbb{N}^n$, i.e.

$$\mathcal{M}(S_1 + \overline{S}_1, S_2,\dots,S_n) = \mathcal{M}(S_1,\dots,S_n) + \mathcal{M}(\overline{S}_1,\dots S_n).$$

For positive $\alpha,\beta,\lambda_1,\dots,\lambda_n$ and $\overline{Q}_1 = \operatorname{conv}(\overline{S}_1)$,

$$\begin{aligned} &\mathrm{Vol}_n\left(\lambda_1(\alpha Q_1 + \beta\overline{Q}_1) + \lambda_2 Q_2 + \cdots + \lambda_n Q_n\right) \\ &\quad = \sum_{j_1+\cdots+j_n=n} a(\alpha,\beta,j_1,\dots,j_n)\lambda_1^{j_1}\cdots\lambda_n^{j_n} \end{aligned} \tag{3.7}$$

and

$$\begin{aligned}
&\mathrm{Vol}_n(\lambda_1\alpha Q_1 + \lambda_1\beta\overline{Q}_1 + \cdots + \lambda_n Q_n) \\
&\quad = \sum_{j_1+j_1'+\cdots+j_n=n} b(j_1, j_1', \ldots, j_n)(\lambda_1\alpha)^{j_1}(\lambda_1\beta)^{j_1'}\cdots\lambda_n^{j_n}.
\end{aligned} \tag{3.8}$$

Comparing the coefficients of $\lambda_1\cdots\lambda_n$ in (3.7) and (3.8) gives

$$a(\alpha, \beta, 1, \ldots, 1) = \alpha b(1, 0, 1, \ldots, 1) + \beta b(0, 1, \ldots, 1). \tag{3.9}$$

Letting (1) $\alpha = \beta = 1$, (2) $\alpha = 1, \beta = 0$, and (3) $\alpha = 0, \beta = 1$ in (3.9) respectively, yields

$$\begin{aligned}
\mathcal{M}(S_1 + \overline{S}_1, \ldots, S_n) &= a(1, \ldots, 1) = b(1, 0, 1, \ldots, 1) + b(0, 1, \ldots, 1) \\
&= a(1, 0, 1, \ldots, 1) + a(0, 1, \ldots, 1) \\
&= \mathcal{M}(S_1, \ldots, S_n) + \mathcal{M}(\overline{S}_1, \ldots, S_n).
\end{aligned}$$

(4) (*Automorphism invariant*) For linear transformation U,

$$\mathrm{Vol}_n\big(U(\lambda_1 Q_1 + \cdots + \lambda_n Q_n)\big) = |\det U|\,\mathrm{Vol}_n(\lambda_1 Q_1 + \cdots + \lambda_n Q_n).$$

Therefore, when $\det U = \pm 1$,

$$\begin{aligned}
\mathrm{Vol}_n\big(\lambda_1(U Q_1) + \cdots + \lambda_n(U Q_n)\big) &= \mathrm{Vol}_n\big(U(\lambda_1 Q_1 + \cdots + \lambda_n Q_n)\big) \\
&= \mathrm{Vol}_n(\lambda_1 Q_1 + \cdots + \lambda_n Q_n),
\end{aligned}$$

and consequently,

$$\mathcal{M}(U S_1, \ldots, U S_n) = \mathcal{M}(S_1, \ldots, S_n).$$

The above connection between $L(S_1, \ldots, S_n)$ and $\mathcal{M}(S_1, \ldots, S_n)$ suggested the establishment of the following Bernshteín theorem:

THEOREM 3.1 (BERNSHTEÍN [1975], Theorem A). *The number of isolated zeros, counting multiplicities, in $(\mathbb{C}^*)^n$ of a polynomial system $P(x) = (p_1(x), \ldots, p_n(x))$ with support $S = (S_1, \ldots, S_n)$ is bound above by the mixed volume $\mathcal{M}(S_1, \ldots, S_n)$. When $P(x)$ is in general position, it has exactly $\mathcal{M}(S_1, \ldots, S_n)$ isolated zeros in $(\mathbb{C}^*)^n$.*

In CANNY and ROJAS [1991], the root count in the above theorem was nicknamed the *BKK bound* after the works of BERNSHTEÍN [1975], KUSHNIRENKO [1976] and KHOVANSKIÍ [1978]. In general, it provides a much tighter bound compared to variant Bézout bounds such as those given in the last section. An apparent limitation of this theorem, important in practice, is that it only counts the number of isolated zeros of a polynomial system in $(\mathbb{C}^*)^n$ rather than all isolated zeros in affine space $\mathbb{C}^n$. This problem was first attempted in ROJAS [1994], a bound for the root count in $\mathbb{C}^n$ was obtained via the notion of the *shadowed* sets. Later, a significantly much tighter bound was given in the following

THEOREM 3.2 (LI and WANG [1997]). *The root count in $\mathbb{C}^n$ of a polynomial system $P(x) = (p_1(x), \ldots, p_n(x))$ with supports $S = (S_1, \ldots, S_n)$ is bounded above by the mixed volume $\mathcal{M}(S_1 \cup \{0\}, \ldots, S_n \cup \{0\})$.*

In other words, the root count of a polynomial system $P(x) = (p_1(x), \ldots, p_n(x))$ in $\mathbb{C}^n$ is bounded above by the root count in $(\mathbb{C}^*)^n$ of the polynomial system $\overline{P}(x)$ in general position obtained by augmenting constant terms to those p_j's in $P(x)$ in which the constant terms are absent. As a corollary, when $0 \in S_j$ for all $j = 1, \ldots, n$, namely, all $p_j(x)$'s in $P(x)$ have constant terms, then the mixed volume $\mathcal{M}(S_1, \ldots, S_n)$ of $P(x)$ is a bound for its root count in $\mathbb{C}^n$, more than just a root count in $(\mathbb{C}^*)^n$. This theorem was further extended in several different ways, see HUBER and STURMFELS [1997] and ROJAS and WANG [1996].

3.2. *Mixed volume and mixed subdivisions*

Let us consider the system in (3.1) with Newton polytopes Q_1 and Q_2, as shown in Fig. 3.1, once again. Recall that

$$\mathcal{M}(Q_1, Q_2) = \text{ area of } (Q_1 + Q_2) - \text{ area of } (Q_1) - \text{ area of } (Q_2).$$

To calculate the area of $Q_1 + Q_2$, we may subdivide the polytope $Q_1 + Q_2$ into convenient pieces, such as squares or triangles, in whichever way as we prefer. Among all the possible subdivisions of $Q_1 + Q_2$, the subdivision I, as shown in Fig. 3.2, exhibits some special properties.

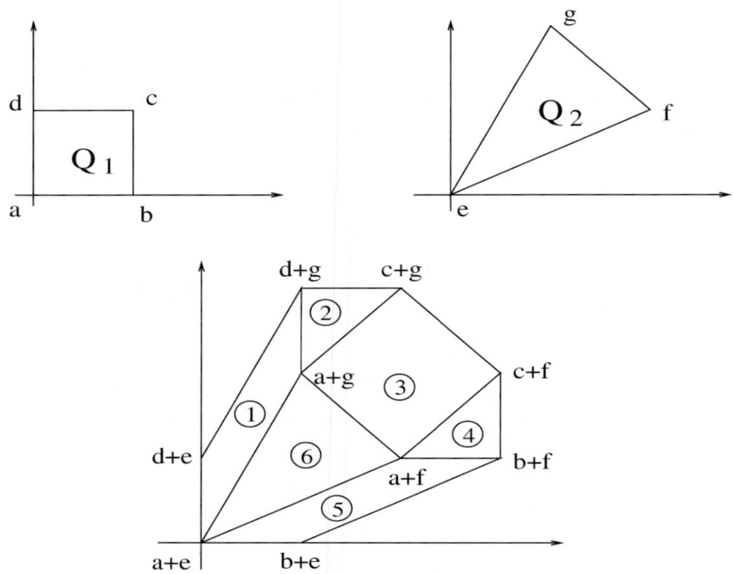

FIG. 3.2. Subdivision I for $Q_1 + Q_2$.

It is easy to verify that all the members, called the *cells*, in this subdivision satisfy the following properties:

(a) Each cell is a Minkowski sum of the convex hulls of $C_1 \subset S_1$ and $C_2 \subset S_2$.
For instance, cell ① $= \mathrm{conv}(\{a,d\}) + \mathrm{conv}(\{e,g\})$.

(b) When $C_1 \subset S_1$ and $C_2 \subset S_2$ form a cell in the subdivision, $\mathrm{conv}(C_i)$ is a simplex of dimension $\#(C_i) - 1$ for $i = 1, 2$. Here, $\#(C_i)$ is the number of points in C_i.
For instance, convex hulls of both $C_1 = \{a,d\}$ and $C_2 = \{e,g\}$ of cell ① are one-dimensional simplices.

(c) Simplices $\mathrm{conv}(C_1)$ and $\mathrm{conv}(C_2)$ are complementary to each other in the sense: $\dim(\mathrm{conv}(C_1)) + \dim(\mathrm{conv}(C_2)) = \dim(\mathrm{conv}(C_1) + \mathrm{conv}(C_2))$.
That is, $v_1 = a - d$ and $v_2 = e - g$ are linearly independent.

In light of properties (a) and (b), each cell $C = \mathrm{conv}(C_1) + \mathrm{conv}(C_2)$ in subdivision I can be characterized as a cell of type (l_1, l_2) where $l_1 = \dim(\mathrm{conv}(C_1))$ and $l_2 = \dim(\mathrm{conv}(C_2))$.

For $j = 1, \ldots, n$, let A_j be a simplex of dimension $k_j \geqslant 0$ in $\mathbb{R}^n$ with vertices $\{q_0^{(j)}, \ldots, q_{k_j}^{(j)}\}$ where $k_1 + \cdots + k_n = n$. Let V be the $n \times n$ matrix whose rows are $q_i^{(j)} - q_0^{(j)}$ for $1 \leqslant j \leqslant n$ and $1 \leqslant i \leqslant k_j$. Notice that any 0-dimensional simplex consists of only one point, and therefore contributes no rows to V. It can be shown that the n-dimensional volume of the Minkowski sum of $A_1, \ldots, A_n$ is equal to

$$\mathrm{Vol}_n(A_1 + \cdots + A_n) = \frac{1}{k_1! \cdots k_n!} |\det V|. \tag{3.10}$$

Applying (3.10) to cell ① ($= \mathrm{conv}(\{a,d\}) + \mathrm{conv}(\{e,g\})$) in subdivision I,

$$\mathrm{Vol}_2(\text{cell ①}) = \left| \det \begin{pmatrix} d - a \\ g - e \end{pmatrix} \right|.$$

When Q_1 and Q_2 are scaled by λ_1 and λ_2 respectively, cell ① becomes a cell ①′ $= \mathrm{conv}(\{\lambda_1 a, \lambda_2 d\}) + \mathrm{conv}(\{\lambda_2 e, \lambda_2 g\})$ in the subdivision I′ of $\lambda_1 Q_1 + \lambda_2 Q_2$ as shown in Fig. 3.3 and its volume becomes

$$\begin{aligned} \left| \det \begin{pmatrix} \lambda_1 d - \lambda_1 a \\ \lambda_2 g - \lambda_2 e \end{pmatrix} \right| &= \left| \det \begin{pmatrix} d - a \\ g - e \end{pmatrix} \right| \times \lambda_1 \lambda_2 \\ &= (\text{volume of cell ①}) \times \lambda_1 \lambda_2. \end{aligned}$$

So, volume of the original cell ① constitutes part of the mixed volume $\mathcal{M}(Q_1, Q_2)$ which, by definition, is the coefficient of $\lambda_1 \lambda_2$ in the homogeneous polynomial $\mathrm{Vol}_2(\lambda_1 Q_1 + \lambda_2 Q_2)$. On the other hand, after scaling, cell ② in subdivision I becomes cell ②′ $= \mathrm{conv}(\{\lambda_1 a, \lambda_1 c, \lambda_1 d\}) + \{\lambda_2 g\}$ in subdivision I′ of $\lambda_1 Q_1 + \lambda_2 Q_2$, and its volume becomes

$$\begin{aligned} \frac{1}{2} \left| \det \begin{pmatrix} \lambda_1 c - \lambda_1 a \\ \lambda_1 d - \lambda_1 a \end{pmatrix} \right| &= \frac{1}{2} \left| \det \begin{pmatrix} c - a \\ d - a \end{pmatrix} \right| \times \lambda_1^2 \\ &= (\text{volume of cell ②}) \times \lambda_1^2. \end{aligned}$$

Thus, volume of the original cell ② plays no part in the mixed volume $\mathcal{M}(Q_1, Q_2)$.

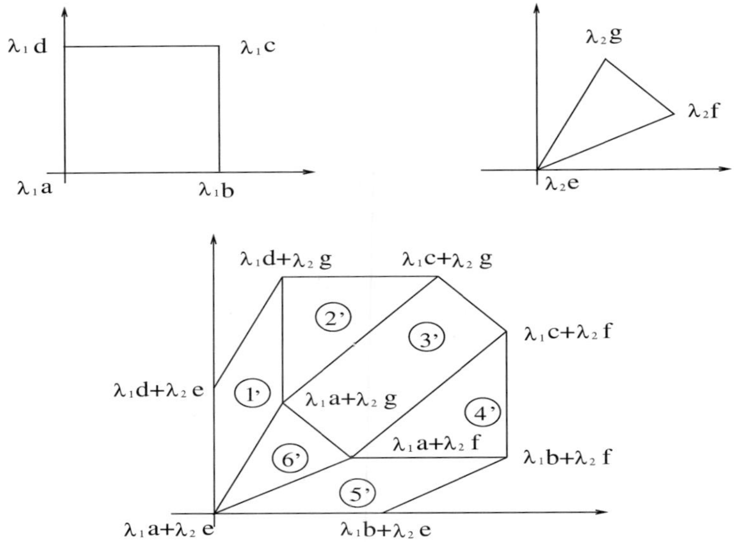

FIG. 3.3. Subdivision I′ for $\lambda_1 Q_1 + \lambda_2 Q_2$.

In general, cells of subdivision I of $Q_1 + Q_2$ become cells of subdivision I′ of $\lambda_1 Q_1 + \lambda_2 Q_2$ with corresponding scalings. Because of properties (a), (b) and (c), volumes of those cells after scaling can all be calculated by (3.10), and only volumes of cells of type $(1, 1)$ in $\lambda_1 Q_1 + \lambda_2 Q_2$ can have the factor $\lambda_1 \lambda_2$ with volumes of the corresponding original cells before scaling as their coefficients. Thus,

$$\begin{aligned}\mathcal{M}(Q_1, Q_2) &= \text{the sum of the volumes of the original cells} \\ &\quad\ \text{of type } (1,1) \text{ in subdivision I before scaling} \\ &= \text{volume of cell ① + volume of cell ③ + volume of cell ⑤} \\ &= 1 + 2 + 1 = 4.\end{aligned}$$

Assembling the mixed volume $\mathcal{M}(Q_1, Q_2)$ in this manner is independent of the scaling factors λ_1 and λ_2, and is valid for any such subdivisions, called the *fine mixed subdivisions*, of $Q_1 + Q_2$ that share the special properties (a)–(c).

To state a formal definition for such subdivisions with less notations, we shall omit those "+" and "conv" in most of the occasions. For instance, rather than formulating the subdivision for $Q_1 + \cdots + Q_n$ $(= \mathrm{conv}(S_1) + \cdots + \mathrm{conv}(S_n))$ we shall deal with the n-tuple $(S_1, \ldots, S_n)$ for short.

Let $S = (S_1, \ldots, S_n)$ be a set of finite subsets of $\mathbb{N}^n$, whose union affinely spans $\mathbb{R}^n$. By a *cell* of $S = (S_1, \ldots, S_n)$ we mean a tuple $C = (C_1, \ldots, C_n)$ where $C_j \subset S_j$ for $j = 1, \ldots, n$. Define

$$\begin{aligned}\mathrm{type}(C) &:= \bigl(\dim\bigl(\mathrm{conv}(C_1)\bigr), \ldots, \dim\bigl(\mathrm{conv}(C_n)\bigr)\bigr), \\ \mathrm{conv}(C) &:= \mathrm{conv}(C_1) + \cdots + \mathrm{conv}(C_n),\end{aligned}$$

and $\mathrm{Vol}(C) := \mathrm{Vol}(\mathrm{conv}(C))$. A *face* of C is a subcell $F = (F_1, \ldots, F_n)$ of C where $F_j \subset C_j$ and some linear functional $\alpha \in (\mathbb{R}^n)^\vee$ attains its minimum over C_j at F_j for $j = 1, \ldots, n$. We call such an α an *inner normal* of F. Clearly, if F is a face of C, then $\mathrm{conv}(F_j)$ is a face of the polytope $\mathrm{conv}(C_j)$ for each $j = 1, \ldots, n$. We call F a *facet* of C if $\mathrm{conv}(F)$ is a co-dimension one face of $\mathrm{conv}(C)$.

DEFINITION 3.1. A mixed subdivision of $S = (S_1, \ldots, S_n)$ is a set of cells $\{C^{(1)}, \ldots, C^{(m)}\}$ such that

(a) For all $i = 1, \ldots, m$, $\dim(\mathrm{conv}(C^{(i)})) = n$,

(b) $\mathrm{conv}(C^{(l)}) \cap \mathrm{conv}(C^{(k)})$ is a proper common face of $\mathrm{conv}(C^{(l)})$ and $\mathrm{conv}(C^{(k)})$ when it is nonempty for $l \neq k$,

(c) $\bigcup_{i=1}^m \mathrm{conv}(C^{(i)}) = \mathrm{conv}(S)$,

(d) For $i = 1, \ldots, m$, write $C^{(i)} = (C_1^{(i)}, \ldots, C_n^{(i)})$, then

$$\dim\bigl(\mathrm{conv}(C_1^{(i)})\bigr) + \cdots + \dim\bigl(\mathrm{conv}(C_n^{(i)})\bigr) = n.$$

This subdivision is called a fine mixed subdivision if in addition we have

(e) Each $\mathrm{conv}(C_j^{(i)})$ is a simplex of dimension $\#C_j^{(i)} - 1$ for $i = 1, \ldots, m$ and $j = 1, \ldots, n$.

Similar to what we have discussed for the system in (3.1), the following important property holds for a polynomial system $P(x) = (p_1(x), \ldots, p_n(x))$ with support $S = (S_1, \ldots, S_n)$, see HUBER and STURMFELS [1995]:

PROPOSITION 3.1. *The mixed volume $\mathcal{M}(S_1, \ldots, S_n)$ of $S = (S_1, \ldots, S_n)$ equals to the sum of the volumes of cells of type $(1, \ldots, 1)$ in a fine mixed subdivision of $S = (S_1, \ldots, S_n)$.*

So, to calculate mixed volume $\mathcal{M}(S_1, \ldots, S_n)$ following the above result, one must find a fine mixed subdivision of $S = (S_1, \ldots, S_n)$ in the first place. This can be accomplished by the standard process: Choose real-valued functions $\omega_j : S_j \to \mathbb{R}$ for each $j = 1, \ldots, n$. We call the n-tuple $\omega = (\omega_1, \ldots, \omega_n)$ a *lifting function* on S, and ω *lifts* S_j to its graph $\widehat{S}_j = \{(a, \omega_j(a)) \colon a \in S_j\} \subset \mathbb{R}^{n+1}$. This notation is extended in the obvious way: $\widehat{S} = (\widehat{S}_1, \ldots, \widehat{S}_n)$, $\widehat{Q}_j = \mathrm{conv}(\widehat{S}_j)$, $\widehat{Q} = \widehat{Q}_1 + \cdots + \widehat{Q}_n$, etc. The lifting function $\omega = (\omega_1, \ldots, \omega_n)$ is known as a *generic lifting* if each ω_j for $j = 1, \ldots, n$ is generic in the sense that its images are chosen at random. Let S_ω be the set of cells $\{C\}$ of S satisfying

(a) $\dim(\mathrm{conv}(\widehat{C})) = n$,

(b) $\widehat{C}$ is a facet of $\widehat{S}$ whose inner normal $\alpha \in (\mathbb{R}^{n+1})^\vee$ has positive last coordinate. (In other words, $\mathrm{conv}(\widehat{C})$ is a facet in the *lower hull* of $\widehat{Q}$.)

It can be shown that (see GEL'FAND, KAPRANOV and ZELEVINSKIĬ [1994], HUBER and STURMFELS [1995] and LEE [1991]):

PROPOSITION 3.2. *When $\omega = (\omega_1, \ldots, \omega_n)$ is a generic lifting, then S_ω provides a fine mixed subdivision of $S = (S_1, \ldots, S_n)$.*

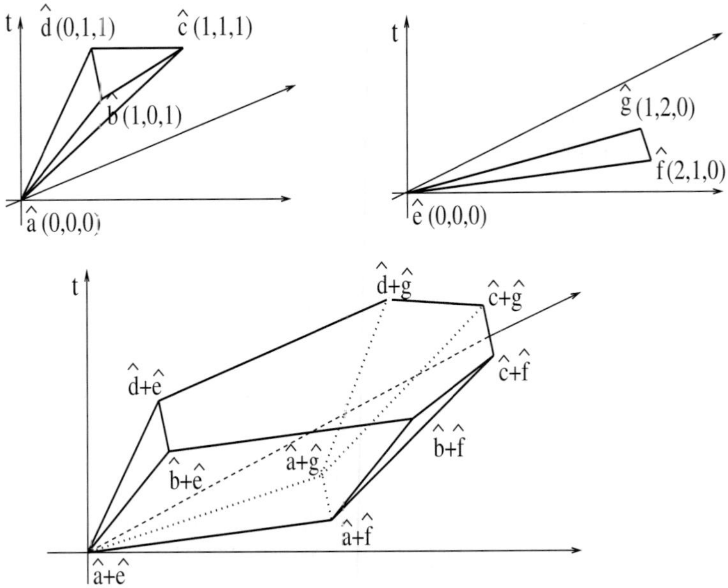

FIG. 3.4.

The subdivision I in Fig. 3.2 for system (3.1) is, in fact, induced by the lifting $\omega = ((0, 1, 1, 1), (0, 0, 0))$, that is

$$\widehat{S} = \left(\left\{(a, 0), (b, 1), (c, 1), (d, 1)\right\}, \left\{(e, 0), (f, 0), (g, 0)\right\}\right)$$

(see Fig. 3.4). While this lifting does not appear to be as generic, it is sufficient to induce a fine mixed subdivision.

4. Polyhedral homotopy method

In light of Theorem 3.2, to find all isolated zeros of a given polynomial system $P(x) = (p_1(x), \ldots, p_n(x))$ in $\mathbb{C}^n$ with support $S = (S_1, \ldots, S_n)$, we first augment the monomial x^0 $(= 1)$ to those p_j's in $P(x)$ in which constant terms are absent. Followed by choosing coefficients of all the monomials generically, a new system $Q(x) = (q_1(x), \ldots, q_n(x))$ with support $S' = (S'_1, \ldots, S'_n)$ where $S'_j = S_j \cup \{0\}$ for $j = 1, \ldots, n$ is produced. We will solve this system first, and consider the linear homotopy

$$H(x, t) = (1 - t)cQ(x) + tP(x) = 0 \quad \text{for generic } c \in \mathbb{C}^* \tag{4.1}$$

afterwards. Recall that, by Theorem 2.3, Properties 1 and 2 (Smoothness and Accessibility) hold for this homotopy, and because all the isolated solutions of $Q(x) = 0$ are known, Property 0 also holds. Therefore, every isolated zero of $P(x)$ lies at the end of a homotopy path of $H(x, t) = 0$, emanating from an isolated solution of $Q(x) = 0$.

4.1. The polyhedral homotopy

To solve $Q(x)=0$, write

$$Q(x)=\begin{cases} q_1(x)=\displaystyle\sum_{a\in S_1'} \bar{c}_{1,a}x^a,\\ \qquad\vdots\\ q_n(x)=\displaystyle\sum_{a\in S_n'} \bar{c}_{n,a}x^a.\end{cases} \tag{4.2}$$

Since all those coefficients $\bar{c}_{j,a}$ for $a\in S_j'$ and $j=1,\ldots,n$ are chosen generically, this system can be considered *in general position*. Namely, there exists a polynomial system

$$G(c)=\bigl(g_1(c),\ldots,g_l(c)\bigr) \tag{4.3}$$

in the variables $c=(c_{j,a})$ of the coefficients in (4.2) for $a\in S_j'$ and $j=1,\ldots,n$ where $G(\bar{c})\neq 0$ for the coefficients $\bar{c}=(\bar{c}_{j,a})$ of $Q(x)$, and all such systems reach the maximum root count $\mathcal{M}(S_1',\ldots,S_n')$ in $\mathbb{C}^n$.

Let t be a new complex variable and consider the polynomial system $\widehat{Q}(x,t)=(\hat{q}_1(x,t),\ldots,\hat{q}_n(x,t))$ in the $n+1$ variables (x,t) given by

$$\widehat{Q}(x,t)=\begin{cases} \hat{q}_1(x,t)=\displaystyle\sum_{a\in S_1'} \bar{c}_{1,a}x^a t^{w_1(a)},\\ \qquad\vdots\\ \hat{q}_n(x,t)=\displaystyle\sum_{a\in S_n'} \bar{c}_{n,a}x^a t^{w_n(a)},\end{cases} \tag{4.4}$$

where each $w_j: S_j'\to\mathbb{R}$ for $j=1,\ldots,n$ is a generic function whose images are chosen at random. For a fixed t_0, we rewrite the system in (4.4) as

$$\widehat{Q}(x,t_0)=\begin{cases} \hat{q}_1(x,t_0)=\displaystyle\sum_{a\in S_1'} \bigl(\bar{c}_{1,a}t_0^{w_1(a)}\bigr)x^a,\\ \qquad\vdots\\ \hat{q}_n(x,t_0)=\displaystyle\sum_{a\in S_n'} \bigl(\bar{c}_{n,a}t_0^{w_n(a)}\bigr)x^a.\end{cases}$$

This system is in general position if for $G(c)$ in (4.3),

$$T(t_0)\equiv G\bigl(\bar{c}_{j,a}t_0^{w_j(a)}\bigr)\neq 0 \quad \text{for } a\in S_j' \text{ and } j=1,\ldots,n.$$

The equation $T(t)=0$ can have at most finitely many solutions, since $T(t)$ is not identically 0 because $T(1)=G(\bar{c}_{j,a})\neq 0$. Let

$$t_1=r_1\mathrm{e}^{\mathrm{i}\theta_1},\quad \ldots,\quad t_k=r_k\mathrm{e}^{\mathrm{i}\theta_k}$$

be the solutions of $T(t) = 0$. Then, for any $\theta \neq \theta_l$ for $l = 1, \ldots, k$, the systems $\overline{Q}(x,t) = (\bar{q}_1(x,t), \ldots, \bar{q}_n(x,t))$ given by

$$\overline{Q}(x,t) = \begin{cases} \bar{q}_1(x,t) = \sum_{a \in S_1'} \left(\bar{c}_{1,a} \mathrm{e}^{\mathrm{i}w_1(a)\theta}\right) x^a t^{w_1(a)}, \\ \quad \vdots \\ \bar{q}_n(x,t) = \sum_{a \in S_n'} \left(\bar{c}_{n,a} \mathrm{e}^{\mathrm{i}w_n(a)\theta}\right) x^a t^{w_n(a)}, \end{cases}$$

are in general position for all $t > 0$ because

$$\bar{c}_{j,a} \mathrm{e}^{\mathrm{i}w_j(a)\theta} t^{w_j(a)} = \bar{c}_{j,a} \left(t\mathrm{e}^{\mathrm{i}\theta}\right)^{w_j(a)}$$

and

$$G\left(\bar{c}_{j,a} \left(t\mathrm{e}^{\mathrm{i}\theta}\right)^{w_j(a)}\right) = T\left(t\mathrm{e}^{\mathrm{i}\theta}\right) \neq 0.$$

Therefore, without loss of generality (choose an angle θ at random and change the coefficients $\bar{c}_{j,a}$ to $\bar{c}_{j,a}\mathrm{e}^{\mathrm{i}w_j(a)\theta}$ if necessary) we may suppose the systems $\widehat{Q}(x,t)$ in (4.4) are in general position for all $t > 0$. By Lemma 3.1, they have the same number of isolated zeros in $(\mathbb{C}^*)^n$ for all $t > 0$, the mixed volume $\mathcal{M}(S_1', \ldots, S_n') := k$.

We now regard $\widehat{Q}(x,t) = 0$ as a homotopy, known as the *polyhedral homotopy*, defined on $(\mathbb{C}^*)^n \times [0,1]$ with target system $\widehat{Q}(x,1) = Q(x)$. The zero set of this homotopy is made up of k homotopy paths $x^{(1)}(t), \ldots, x^{(k)}(t)$. Since each $\hat{q}_j(x,t)$ has nonzero constant term for all $j = 1, \ldots, n$, by a standard application of generalized Sard's theorem (ABRAHAM and ROBBIN [1967]), all those homotopy paths are smooth with no bifurcations. Therefore, both Property 1 (Smoothness) and Property 2 (Accessibility) hold for this homotopy. However, at $t = 0$, $\widehat{Q}(x,0) \equiv 0$, so those homotopy paths cannot get started because their starting points $x^{(1)}(0), \ldots, x^{(k)}(0)$ cannot be identified (see Fig. 4.1). This problem can be resolved by the following design.

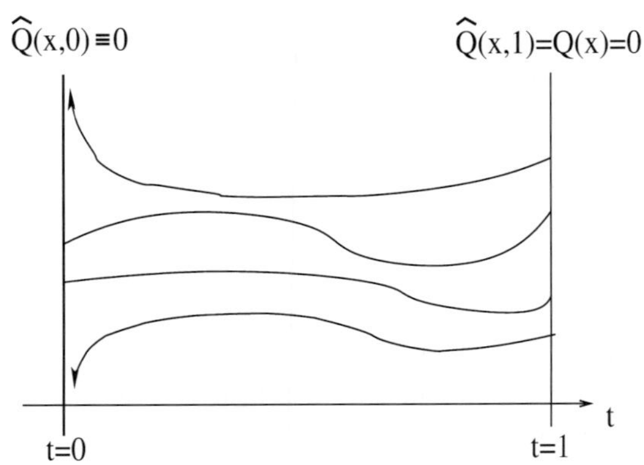

FIG. 4.1.

The function $\omega = (\omega_1, \ldots, \omega_n)$ with $\omega_j : S'_j \to \mathbb{R}$, $j = 1, \ldots, n$, may be considered as a *generic lifting* on the support $S' = (S'_1, \ldots, S'_n)$ of $Q(x)$ which lifts S'_j to its graph

$$\widehat{S}'_j = \{\hat{a} = (a, w_j(a)) \mid a \in S'_j\}, \quad j = 1, \ldots, n.$$

Let $\hat{\alpha} = (\alpha, 1) \in \mathbb{R}^{n+1}$ satisfy the following condition:

There exists a collection of pairs $\{a_1, a'_1\} \subset S'_1, \ldots, \{a_n, a'_n\} \subset S'_n$, where $\{a_1 - a'_1, \ldots, a_n - a'_n\}$ is linearly independent, and

$$\begin{aligned} &\langle \hat{a}_j, \hat{\alpha}\rangle = \langle \hat{a}'_j, \hat{\alpha}\rangle, \\ &\langle \hat{a}, \hat{\alpha}\rangle > \langle \hat{a}_j, \hat{\alpha}\rangle \quad \text{for } a \in S'_j \setminus \{a_j, a'_j\}. \end{aligned} \tag{A}$$

Here, $\langle \cdot, \cdot \rangle$ stands for the usual inner product in the Euclidean space. For such $\hat{\alpha} = (\alpha, 1)$ where $\alpha = (\alpha_1, \ldots, \alpha_n)$, let

$$\begin{aligned} y_1 &= t^{-\alpha_1} x_1, \\ &\vdots \\ y_n &= t^{-\alpha_n} x_n. \end{aligned} \tag{4.5}$$

In short, we write $y = t^{-\alpha} x$ with $y = (y_1, \ldots, y_n)$ and $y t^{\alpha} = x$. With this transformation and $a = (a_1, \ldots, a_n) \in \mathbb{N}^n$,

$$\begin{aligned} x^a &= x_1^{a_1} \cdots x_n^{a_n} \\ &= (y_1 t^{\alpha_1})^{a_1} \cdots (y_n t^{\alpha_n})^{a_n} \\ &= y_1^{a_1} \cdots y_n^{a_n} t^{\alpha_1 a_n + \cdots - \alpha_n a_n} \\ &= y^a t^{\langle a, \alpha \rangle}, \end{aligned} \tag{4.6}$$

and $\hat{q}_j(x, t)$ of $\widehat{Q}(x, t)$ in (4.4) becomes

$$\begin{aligned} \hat{q}_j(y t^{\alpha}, t) &= \sum_{a \in S'_j} \bar{c}_{j,a} y^a t^{\langle a, \alpha \rangle} t^{w_j(a)} \\ &= \sum_{a \in S'_j} \bar{c}_{j,a} y^a t^{\langle (a, w_j(a)), (\alpha, 1) \rangle} \\ &= \sum_{a \in S'_j} \bar{c}_{j,a} y^a t^{\langle \hat{a}, \hat{\alpha} \rangle}, \quad j = 1, \ldots, n. \end{aligned} \tag{4.7}$$

Let

$$\beta_j = \min_{a \in S'_j} \langle \hat{a}, \hat{\alpha} \rangle, \quad j = 1, \ldots, n, \tag{4.8}$$

and consider the homotopy

$$H^{\alpha}(y, t) = (h_1^{\alpha}(y, t), \ldots, h_n^{\alpha}(y, t)) = 0 \tag{4.9}$$

on $(\mathbb{C}^*)^n \times [0,1]$ where for $j = 1, \ldots, n$

$$h_j^\alpha(y,t) = t^{-\beta_j}\hat{q}_j(yt^\alpha, t) = \sum_{a\in S_j'} \bar{c}_{j,a} y^a t^{\langle \hat{a},\hat{\alpha}\rangle - \beta_j}$$

$$= \sum_{\substack{a\in S_j' \\ \langle \hat{a},\hat{\alpha}\rangle = \beta_j}} \bar{c}_{j,a} y^a + \sum_{\substack{a\in S_j' \\ \langle \hat{a},\hat{\alpha}\rangle > \beta_j}} \bar{c}_{j,a} y^a t^{\langle \hat{a},\hat{\alpha}\rangle - \beta_j}. \tag{4.10}$$

This homotopy retains most of the properties of the homotopy $\widehat{Q}(x,t) = 0$; in particular, both Properties 1 (Smoothness) and 2 (Accessibility) remain valid and

$$H^\alpha(y,1) = \widehat{Q}(y,1) = Q(y). \tag{4.11}$$

From condition (A), for each $j = 1, \ldots, n$, $\langle \hat{a}_j, \hat{\alpha}\rangle = \langle \hat{a}_j', \hat{\alpha}\rangle = \beta_j$ and $\langle \hat{a}, \hat{\alpha}\rangle > \beta_j$ for $a \in S_j' \backslash \{a_j, a_j'\}$, hence,

$$H^\alpha(y,0) = \begin{cases} h_1^\alpha(y,0) = \displaystyle\sum_{\substack{a\in S_1' \\ \langle \hat{a},\hat{\alpha}\rangle = \beta_1}} \bar{c}_{1,a} y^a = \bar{c}_{1,a_1} y^{a_1} + c_{1,a_1'} y^{a_1'} = 0, \\ \quad\vdots \\ h_n^\alpha(y,0) = \displaystyle\sum_{\substack{a\in S_n' \\ \langle \hat{a},\hat{\alpha}\rangle = \beta_n}} \bar{c}_{n,a} y^a = \bar{c}_{n,a_n} y^{a_n} + c_{n,a_n'} y^{a_n'} = 0. \end{cases} \tag{4.12}$$

Such system is known as the *binomial system*, and its isolated solutions in $(\mathbb{C}^*)^n$ are constructively available as shown in the next section.

4.2. *Solutions of binomial systems in* $(\mathbb{C}^*)^n$

PROPOSITION 4.1. *The binomial system*

$$\begin{aligned} &\bar{c}_{1,a_1} y^{a_1} + \bar{c}_{1,a_1'} y^{a_1'} = 0, \\ &\qquad\vdots \\ &\bar{c}_{n,a_n} y^{a_n} + \bar{c}_{n,a_n'} y^{a_n'} = 0, \end{aligned} \tag{4.13}$$

has

$$k_\alpha := \left| \det \begin{pmatrix} a_1 - a_1' \\ \vdots \\ a_n - a_n' \end{pmatrix} \right| \tag{4.14}$$

nonsingular isolated solutions in $(\mathbb{C}^*)^n$.

PROOF. For $j = 1, \ldots, n$, let $v_j = a_j - a'_j$. To look for solutions of the system (4.13) in $(\mathbb{C}^*)^n$, we rewrite the system as

$$\begin{aligned} y^{v_1} &= b_1, \\ &\vdots \\ y^{v_n} &= b_n, \end{aligned} \tag{4.15}$$

where $b_j = \bar{c}_{j,a'_j}/\bar{c}_{j,a_j}$ for $j = 1, \ldots, n$. Let

$$V = \begin{bmatrix} v_1 \mid v_2 \mid \cdots \mid v_n \end{bmatrix} \tag{4.16}$$

and $\mathbf{b} = (b_1, \ldots, b_n)$. Then, (4.15) becomes

$$y^V = \mathbf{b}. \tag{4.17}$$

When matrix V is upper triangular, i.e.

$$V = \begin{bmatrix} v_{11} & v_{12} & \cdots & v_{1n} \\ 0 & v_{22} & \cdots & v_{2n} \\ \vdots & \ddots & \ddots & \vdots \\ 0 & \cdots & 0 & v_{nn} \end{bmatrix},$$

then, equations in (4.17) become

$$\begin{aligned} y_1^{v_{11}} &= b_1, \\ y_1^{v_{12}} y_2^{v_{22}} &= b_2, \\ &\vdots \\ y_1^{v_{1n}} y_2^{v_{2n}} \cdots y_n^{v_{nn}} &= b_n. \end{aligned} \tag{4.18}$$

By forward substitutions, all the solutions of the system (4.18) in $(\mathbb{C}^*)^n$ can be found, and the total number of solutions is $|v_{11}| \times \cdots \times |v_{nn}| = |\det V|$.

When V is a general matrix, we may upper triangularize it by the following process. Recall that the greatest common divisor d of two nonzero integers a and b, denoted by $\gcd(a, b)$, can be written as

$$d = \gcd(a, b) = ra + lb$$

for certain nonzero integers r and l. Let

$$M = \begin{bmatrix} r & l \\ -\frac{b}{d} & \frac{a}{d} \end{bmatrix}.$$

Clearly, $\det(M) = 1$, and

$$M \begin{bmatrix} a \\ b \end{bmatrix} = \begin{bmatrix} r & l \\ -\frac{b}{d} & \frac{a}{d} \end{bmatrix} \begin{bmatrix} a \\ b \end{bmatrix} = \begin{bmatrix} d \\ 0 \end{bmatrix}.$$

Similar to using Givens rotation to produce zeros in a matrix for its QR factorization, the matrix M may be used to upper triangularize V as follows. For $v \in \mathbb{Z}^n$, let a and b

be its ith and jth (nonzero) components with $i < j$, that is

$$v = \begin{bmatrix} \vdots \\ a \\ \vdots \\ b \\ \vdots \end{bmatrix} \begin{array}{l} \\ \to i\text{th} \\ \\ \to j\text{th}. \\ \\ \end{array}$$

With $d = \gcd(a, b)$, let

$$U(i, j) := \begin{array}{c} \begin{array}{ccccccccccc} & & & i\text{th} & & & & j\text{th} & & & \end{array} \\ \left[\begin{array}{ccccccccccc} 1 & & & & & & & & & & \\ & \ddots & & & & & & & & & \\ & & 1 & & & & & & & & \\ & & & r & & & & l & & & \\ & & & & 1 & & & & & & \\ & & & & & \ddots & & & & & \\ & & & & & & 1 & & & & \\ & & & -\frac{b}{d} & & & & \frac{a}{d} & & & \\ & & & & & & & & 1 & & \\ & & & & & & & & & \ddots & \\ & & & & & & & & & & 1 \end{array}\right] \begin{array}{l} \\ \\ \\ i\text{th} \\ \\ \\ \\ j\text{th} \\ \\ \\ \\ \end{array} \end{array} \tag{4.19}$$

in which all blank entries are zero. Clearly, $U(i, j)$ is an integer matrix with $\det(U(i, j)) = 1$ and

$$U(i, j)v = \begin{bmatrix} \vdots \\ d \\ \vdots \\ 0 \\ \vdots \end{bmatrix} \begin{array}{l} \\ i\text{th} \\ \\ j\text{th} \\ \\ \end{array}$$

Thus multiplication on the left by a series of matrices in the form of $U(i, j)$ in (4.19) can successively produce zeros in the lower triangular part of the matrix V, resulting in an upper triangular matrix. Let U be the product of all those $U(i, j)$'s. Then $\det U = 1$, and since UV is upper triangular, we may solve the system

$$\left(z^U\right)^V = z^{UV} = \mathbf{b} \tag{4.20}$$

in $(\mathbb{C}^*)^n$ by forward substitutions. And the total number of solutions in $(\mathbb{C}^*)^n$ is

$$\left|\det(UV)\right| = \left|\det(U)\right| \cdot \left|\det(V)\right| = \left|\det(V)\right|.$$

By letting $y = z^U$ for each solution z of (4.20) in $(\mathbb{C}^*)^n$, we obtain $|\det(V)|$ number of solutions of the system (4.13) in $(\mathbb{C}^*)^n$. □

4.3. Polyhedral homotopy procedure

By (4.11), following paths $y(t)$ of the homotopy $H^\alpha(y,t)=0$ in (4.9) that emanate from k_α, as in (4.14), isolated zeros in $(\mathbb{C}^*)^n$ of the binomial start system $H^\alpha(y,0)=0$ in (4.12), yields k_α isolated zeros of the system $Q(x)$ in (4.3) when $t=1$. Moreover, different $\hat{\alpha}=(\alpha,1)\in\mathbb{R}^{n+1}$ that satisfy condition (A) will induce different homotopies $H^\alpha(y,t)=0$ in (4.10) and following corresponding paths of those different homotopies will reach different sets of isolated zeros of $Q(x)$. Those different sets of isolated zeros of $Q(x)$ are actually disjoint from each other, and they therefore provide $\sum_\alpha k_\alpha$ isolated zeros of $Q(x)$ in total. To see they are disjoint, let paths $y^{\alpha^{(1)}}(t)$ of $H^{\alpha^{(1)}}(y,t)=0$ and $y^{\alpha^{(2)}}(t)$ of $H^{\alpha^{(2)}}(y,t)=0$ for $\alpha^{(1)}=(\alpha_1^{(1)},\ldots,\alpha_n^{(1)})$ and $\alpha^{(2)}=(\alpha_1^{(2)},\ldots,\alpha_n^{(2)})\in\mathbb{R}^n$ reach the same point at $t=1$, then their corresponding homotopy paths $x(t)=y(t)t^\alpha$ of $\widehat{Q}(x,t)=0$ are the same since zeros of the system $Q(x)=\widehat{Q}(x,1)$ are isolated and nonsingular. Thus, $x(t)=y^{\alpha^{(1)}}(t)t^{\alpha^{(1)}}=y^{\alpha^{(2)}}(t)t^{\alpha^{(2)}}$ implies

$$1=\lim_{t\to 0}\frac{y_j^{\alpha^{(1)}}(t)}{y_j^{\alpha^{(2)}}(t)}t^{\alpha_j^{(1)}-\alpha_j^{(2)}},\quad \text{for each } j=1,\ldots,n. \tag{4.21}$$

Hence, $\alpha_j^{(1)}=\alpha_j^{(2)}$ for all $j=1,\ldots,n$.

On the other hand, when $\omega=(\omega_1,\ldots,\omega)$ is a generic lifting, it induces, by Proposition 3.2, a fine mixed subdivision S'_ω of $S'=(S'_1,\ldots,S'_n)$. It's easy to see that when the collection of pairs $C^\alpha=(\{a_1,a'_1\},\ldots,\{a_n,a'_n\})$ with $\{a_j,a'_j\}\subset S'_j$, $j=1,\ldots,n$, satisfies condition (A) with $\hat{\alpha}=(\alpha,1)\in\mathbb{R}^{n+1}$, it is a cell of type $(1,\ldots,1)$ in S'_ω with $\hat{\alpha}$ being the inner normal of $(\{\hat{a}_1,\hat{a}'_1\},\ldots,\{\hat{a}_n,\hat{a}'_n\})$ in $(\widehat{S}'_1,\ldots,\widehat{S}'_n)$ and, by (3.10),

$$\mathrm{Vol}_n(C^\alpha)=\left|\det\begin{pmatrix}a_1-a'_1\\ \vdots\\ a_n-a'_n\end{pmatrix}\right|=k_\alpha. \tag{4.22}$$

By Proposition 3.1, the mixed volume $\mathcal{M}(S'_1,\ldots,S'_n)$, the root count of $Q(x)$ in $(\mathbb{C}^*)^n$, is the sum of all the volumes of C^α. That is,

$$\mathcal{M}(S'_1,\ldots,S'_n)=\sum_\alpha k_\alpha.$$

In other words, each isolated zero of $Q(x)$ lies at the end of certain homotopy path of the homotopy $H^\alpha(y,t)=0$ induced by certain $\hat{\alpha}=(\alpha,1)\in\mathbb{R}^{n+1}$ that satisfies condition (A).

A key step in the procedure of solving system $Q(x)$ by the polyhedral homotopy method described above is the search for all those vectors $\hat{\alpha}=(\alpha,1)\in\mathbb{R}^{n+1}$ as well as their associated cells $C^\alpha=(\{a_1,a'_1\},\ldots,\{a_n,a'_n\})$ that satisfy condition (A). This step is actually the main bottleneck of the method in practice. We shall address this important issue in the next section.

In conclusion, we list the polyhedral homotopy procedure.

POLYHEDRAL HOMOTOPY PROCEDURE.

Given polynomial system $P(x) = (p_1(x), \dots, p_n(x))$ with support $S = (S_1, \dots, S_n)$, let $S' = (S'_1, \dots, S'_n)$ with $S'_j = S_j \cup \{0\}$ for $j = 1, \dots, n$.

Step 0: Initialization.

Choose polynomial system $Q(x) = (q_1(x), \dots, q_n(x))$ with support $S' = (S'_1, \dots, S'_n)$ and generically chosen coefficients. Write

$$q_j(x) = \sum_{a \in S'_j} c_{j,a} x^a, \quad j = 1, \dots, n.$$

Step 1: Solve $Q(x) = 0$.

Step 1.1. Choose a set of real valued functions $w_j : S'_j \to \mathbb{R}$, $j = 1, \dots, n$, their images are generic numbers.

Step 1.2. Find all the cells $C^\alpha = (\{a_1, a'_1\}, \dots, \{a_n, a'_n\})$ of type $(1, \dots, 1)$ with $\{a_j, a'_j\} \subset S'_j$, $j = 1, \dots, n$, in the fine mixed subdivision S'_ω of $S' = (S'_1, \dots, S'_n)$ induced by $\omega = (\omega_1, \dots, \omega_n)$ with $\hat{\alpha} = (\alpha, 1) \in \mathbb{R}^{n+1}$ being the inner normal of $(\{\hat{a}_1, \hat{a}'_1\}, \dots, \{\hat{a}_n, \hat{a}'_n\})$ in $(\widehat{S'_1}, \dots, \widehat{S'_n})$. (The algorithm of this part will be given in the next section.)

Step 1.3. For each $\hat{\alpha} = (\alpha, 1) \in \mathbb{R}^{n+1}$ and its associated cell C^α obtained in Step 1.2.

Step 1.3.1. Solve the binomial system

$$c_{j,a_j} y^{a_j} + c_{j,a'_j} y^{a'_j} = 0, \quad j = 1, \dots, n,$$

in $(C^*)^n$. Let the solution set be X^*_α.

Step 1.3.2. Follow homotopy paths $y(t)$ of the homotopy $H^\alpha(y, t) = (h^\alpha_1(y, t), \dots, h^\alpha_n(y, t)) = 0$ with

$$h^\alpha_j(y, t) = \sum_{a \in S'_j} c_{j,a} y^a t^{\langle \hat{a}, \hat{\alpha} \rangle - \beta_j}, \quad j = 1, \dots, n,$$

where $\beta_j = \langle \hat{a}_j, \hat{\alpha} \rangle$, starting from the solutions in X^*_α. Collect all the points of $y(1)$ as a subset of isolated zeros of $Q(x)$.

Step 2: Solve $P(x) = 0$.

Follow homotopy paths of the homotopy

$$H(x, t) = (1 - t)cQ(x) + tP(x) = 0 \quad \text{for generic } c \in \mathbb{C}^*$$

starting from the solutions of $Q(x) = 0$ obtained in Step 1 to reach all isolated solutions of $P(x) = 0$ at $t = 1$.

REMARK 4.1. As we can see in the above procedure, in order to find all isolated zeros of $P(x)$ in $\mathbb{C}^n$, there are $k = \mathcal{M}(S'_1, \dots, S'_n)$ homotopy paths need to be followed in both Step 1.3 and Step 2, hence $2k$ in total. This work may be reduced in half by the following strategy:

For

$$p_j(x) = \sum_{a \in S'_j} \bar{c}_{j,a} x^a, \quad j = 1, \ldots, n,$$

we select the coefficients $c_{j,a}$'s of $q_j(x)$, $j = 1, \ldots, n$, at Step 0 to be $\bar{c}_{j,a} + \varepsilon_{j,a}$, where $\varepsilon_{j,a}$'s are generically chosen small numbers to ensure each $q_j(x)$ is in general position. And at Step 1.3.2, we follow homotopy paths of the homotopy $\overline{H}^\alpha(y,t) = (\bar{h}_j^\alpha(y,t), \ldots, \bar{h}_n^\alpha(y,t)) = 0$, where

$$\bar{h}_j^\alpha(y,t) = \sum_{a \in S'_j} \left[\bar{c}_{j,a} + (1-t)\varepsilon_{j,a}\right] y^a t^{\langle \hat{a}, \hat{\alpha}\rangle - \beta_j}, \quad j = 1, \ldots, n.$$

It can be shown that the starting system $\overline{H}^\alpha(y,0) = 0$ of this homotopy retain the same binomial system as before which was solved at Step 1.3.1 (with different coefficients of course). Most importantly, since $\overline{H}^\alpha(y,1) = \overline{H}^\alpha(x,1) = P(x)$, Step 2 in the above procedure is no longer necessary and we only need to follow k paths.

5. Mixed volume computation

It was mentioned in the last section, a key step in the polyhedral homotopy method for solving polynomial system $P(x) = (p_1(x), \ldots, p_n(x))$ in $\mathbb{C}^n$ with support $S = (S_1, \ldots, S_n)$ is the identification of all the vectors $\hat{\alpha} = (\alpha, 1) \in \mathbb{R}^{n+1}$ and their associate pairs $(\{a_1, a'_1\}, \ldots, \{a_n, a'_n\})$ that satisfy condition (A). We repeat the condition here with the assumption that p_j's in $P(x)$ all have constant terms, namely $S_j = S_j \cup \{0\}$ for $j = 1, \ldots, n$:

For a generic lifting $\omega = (\omega_1, \ldots, \omega_n)$ on $S = (S_1, \ldots, S_n)$ with $w_j : S_j \to \mathbb{R}$ for $j = 1, \ldots, n$, write

$$\widehat{S}_j = \{\hat{a} = (a, \omega_j(a)) \mid a \in S_j\}.$$

Then $\hat{\alpha} = (\alpha, 1) \in \mathbb{R}^{n+1}$ satisfies condition (A) if

> There exists a collection of pairs $\{a_1, a'_1\} \subset S_1, \ldots, \{a_n, a'_n\} \subset S_n$ where $\{a_1 - a'_1, \ldots, a_n - a'_n\}$ is linearly independent and
>
> $$\begin{aligned} &\langle \hat{a}_j, \hat{\alpha}\rangle = \langle \hat{a}'_j, \hat{\alpha}\rangle, \\ &\langle \hat{a}, \hat{\alpha}\rangle > \langle \hat{a}_j, \hat{\alpha}\rangle \quad \text{for } a \in S_j \backslash \{a_j, a'_j\}. \end{aligned} \tag{A}$$

The geometric meaning of this problem, as shown in Fig. 5.1, is that with generic *lifting* ω_j on lattice points $S_j \subset \mathbb{N}^n$ for each $j = 1, \ldots, n$, we look for hyperplanes in the form $\hat{\alpha} = (\alpha, 1) \in \mathbb{R}^{n+1}$ where each hyperplane supports the convex hull of $\widehat{S}_j$ at exactly two points $\{\hat{a}_j, \hat{a}'_j\}$ of $\widehat{S}_j$ for each $j = 1, \ldots, n$.

The collection of pairs $\{a_1, a'_1\}, \ldots, \{a_n, a'_n\}$ in condition (A), denote it by $C^\alpha = (C_1, \ldots, C_n)$ with $C_j = \{a_j, a'_j\} \subset S_j$ for $j = 1, \ldots, n$, is, as we mentioned before, a cell of type $(1, \ldots, 1)$ in the subdivision $\mathcal{S}_w$ of $S = (S_1, \ldots, S_n)$ induced by the lifting

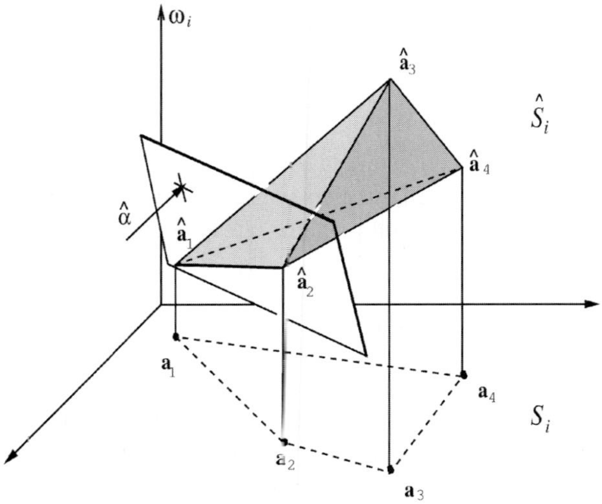

FIG. 5.1. A lifting on lattice points.

$w=(w_1,\dots,w_n)$. We shall call such C^α a *mixed cell* of S_w without specifying its type. By (3.10), the volume of C^α is

$$\mathrm{Vol}_n(C^\alpha)=\left|\det\begin{pmatrix} a_1'-a_1\\ \vdots\\ a_n'-a_n\end{pmatrix}\right|.$$

On the other hand, cells of type $(1,\dots,1)$ in S_w in the form $(\{a_1,a_1'\},\dots,\{a_n,a_n'\})$ where $\{a_j,a_j'\}\subset S_j$ for $j=1,\dots,n$ with $\hat{\alpha}=(\alpha,1)\in\mathbb{R}^{n+1}$ being the inner normal of $(\{\hat{a}_1,\hat{a}_1'\},\dots,\{\hat{a}_n,\hat{a}_n'\})$ in $\widehat{S}=(\widehat{S}_1,\dots,\widehat{S}_n)$ automatically satisfies condition (A). Hence, by Proposition 3.1, the mixed volume $\mathcal{M}(S_1,\dots,S_n)$ is the sum of the volumes of all the mixed cells C^α of S_w. Therefore, when all those mixed cells are identified, the mixed volume $\mathcal{M}(S_1,\dots,S_n)$ can be assembled with little extra computational effort.

In this section, we shall present an algorithm given in LI and LI [2001] for finding all those mixed cells and their associated vectors $\hat{\alpha}=(\alpha,1)\in\mathbb{R}^{n+1}$ following the route described below.

For $1\leqslant j\leqslant n$, $\hat{\mathbf{e}}=\{\hat{a},\hat{a}'\}\subset\widehat{S}_j$ is called a *lower edge* of $\widehat{S}_j$ if there is a vector $\hat{\alpha}=(\alpha,1)\in\mathbb{R}^{n+1}$ for which

$$\begin{aligned}&\langle\hat{a},\hat{\alpha}\rangle=\langle\hat{a}',\hat{\alpha}\rangle,\\&\langle\hat{a},\hat{\alpha}\rangle\leqslant\langle\hat{b},\hat{\alpha}\rangle,\quad\forall\hat{b}\in\widehat{S}_j\backslash\{\hat{a},\hat{a}'\}.\end{aligned}$$

For $1\leqslant k\leqslant n$, $\widehat{E}_k=(\hat{\mathbf{e}}_1,\dots,\hat{\mathbf{e}}_k)$ with $\hat{\mathbf{e}}_j=\{\hat{a}_j,\hat{a}_j'\}\subset\widehat{S}_j$ for $j=1,\dots,k$ is called a *level-k subface* of $\widehat{S}=(\widehat{S}_1,\dots,\widehat{S}_n)$ if there is a vector $\hat{\alpha}=(\alpha,1)\in\mathbb{R}^{n+1}$ such that for all $j=1,\dots,k$,

$$\begin{aligned}&\langle\hat{a}_j,\hat{\alpha}\rangle=\langle\hat{a}_j',\hat{\alpha}\rangle,\\&\langle\hat{a}_j,\hat{\alpha}\rangle\leqslant\langle\hat{a},\hat{\alpha}\rangle,\quad\forall\hat{a}\in\widehat{S}_j\backslash\{\hat{a}_j,\hat{a}_j'\}.\end{aligned}$$

Obviously, a level-1 subface of $\widehat{S}$ is just a lower edge of $\widehat{S}_1$ and a level-n subface of $\widehat{S}$ induces a mixed cell in the subdivision S_w of $S=(S_1,\ldots,S_n)$. Thus, to find the mixed cells of S_w, we may proceed by first finding all the lower edges of $\widehat{S}_j$ for $j=1,\ldots,n$, followed by extending the level-k subfaces of $\widehat{S}$ from $k=1$ to $k=n$.

5.1. A basic linear programming algorithm

In computing mixed cells, we will repeatedly encounter the LP (linear programming) problems of the following type:

$$\begin{aligned}&\text{Minimize } \langle \mathbf{f},\mathbf{z}\rangle\\&\langle \mathbf{c}_j,\mathbf{z}\rangle \leqslant b_j,\quad j=1,\ldots,m\end{aligned}\tag{5.1}$$

where $\{\mathbf{f},\mathbf{c}_j,\mathbf{z}\}\subset\mathbb{R}^n$ and $m>n$. To solve those problems, we will employ the classical simplex algorithm instead of using the faster *interior point method* (YE [1997]) because in the course the main algorithm for finding mixed cells takes a great advantage of the rich information generated by the pivoting process in the simplex method.

When the simplex method is used to solve the LP problem in (5.1), it is customary, for historical as well as practical reasons, to convert the form in (5.1) into the following *standard* form:

$$\begin{aligned}&\text{Minimize } \langle \mathbf{f},\mathbf{y}\rangle\\&A\mathbf{y}=\mathbf{d},\\&\mathbf{y}\geqslant 0,\end{aligned}$$

where $\{\mathbf{f},\mathbf{y}\}\subset\mathbb{R}^n$, $\mathbf{d}\in\mathbb{R}^r$ and A is a $r\times n$ matrix. In fact, many linear programming software packages apply an automatic internal conversion procedure to create such standard-form problems. However, for our needs, it is critically important to solve the problem in the form given in (5.1) directly. The algorithm for this purpose is briefly outlined below, and the details can be found in, e.g., BEST and RITTER [1985].

The feasible region of (5.1), denoted by R, defines a polyhedral set. A *nondegenerate* vertex of R is a feasible point of R with exactly n active constraints. From a feasible point of the problem, or a point in R, one may attain a nondegenerate vertex of R. Let $\mathbf{z}^0$ be a nondegenerate vertex of R and $J=\{j_1,\ldots,j_n\}$ be the set of indices of active constraints at $\mathbf{z}^0$, that is

$$\begin{aligned}&\langle \mathbf{c}_j,\mathbf{z}^0\rangle=b_j,\quad \text{if } j\in J,\\&\langle \mathbf{c}_j,\mathbf{z}^0\rangle<b_j,\quad \text{if } j\notin J.\end{aligned}$$

Since $\mathbf{z}^0$ is nondegenerate, $D^{\mathrm{T}}=[\mathbf{c}_{j_1},\ldots,\mathbf{c}_{j_n}]$ must be nonsingular. Let $D^{-1}=[\mathbf{u}_1,\ldots,\mathbf{u}_n]$. (When $\mathbf{z}^0$ is degenerate where more than n constraints are active, D^{T} is formulated in accordance with the Blend's rule.) The n edges of the feasible region R emanating from $\mathbf{z}^0$ can be represented in the form

$$\mathbf{z}^0-\sigma\mathbf{u}_k,\quad \sigma>0,\ k=1,\ldots,n.$$

Along all those edges, the objective function $\langle \mathbf{f},\mathbf{z}^0-\sigma\mathbf{u}_k\rangle$ decreases as a function of $\sigma>0$ when $\langle \mathbf{f},\mathbf{u}_k\rangle>0$. For such a direction $\mathbf{u}_k$, the largest possible $\sigma>0$ for $\mathbf{z}^0-\sigma\mathbf{u}_k$

to stay feasible is

$$\sigma_0 = \min\left\{ \frac{\langle \mathbf{c}_j, \mathbf{z}^0\rangle - b_j}{\langle \mathbf{c}_j, \mathbf{u}_k\rangle} \;\middle|\; j \notin J \text{ with } \langle \mathbf{c}_j, \mathbf{u}_k\rangle < 0 \right\}$$

and the point $\mathbf{z}^1 = \mathbf{z}^0 - \sigma_0 \mathbf{u}_k$ yields an adjacent vertex of R with reduced objective function value. When $\mathbf{z}^0$ is degenerate, it may occur that $\sigma_0 = 0$. In such cases, one may either choose an alternative $\mathbf{u}_k$ or, if no more $\mathbf{u}_k$ is available, reformulate D^{T} with different set of constraints.

Clearly, $\mathbf{z}^0$ is an optimal solution of (5.1) if $\langle \mathbf{f}, \mathbf{u}_k\rangle \leqslant 0$ for all $k = 1, \ldots, n$. To solve the LP problem in (5.1) directly we may move from one vertex of R to an adjacent one in a direction $\mathbf{u}_k$ where $\langle \mathbf{f}, \mathbf{u}_k\rangle > 0$, forcing the objective function to decrease, until a vertex with $\langle \mathbf{f}, \mathbf{u}_k\rangle \leqslant 0$ for all $k = 1, \ldots, n$ is reached. On the other hand, if for all k where $\langle \mathbf{f}, \mathbf{u}_k\rangle > 0$, $\langle \mathbf{c}_j, \mathbf{u}_k\rangle$ are nonnegative for all j, then the problem is unbounded and the solution does not exist.

Most frequently, the LP problems arise in our algorithm are of the following type:

$$\begin{aligned} &\text{Minimize } \langle \mathbf{f}, \mathbf{z}\rangle \\ &\langle \mathbf{a}_i, \mathbf{z}\rangle = b_i, \quad i \in I_1 = \{1, \ldots, q\}, \\ &\langle \mathbf{c}_j, \mathbf{z}\rangle \leqslant b_j, \quad j \in I_2 = \{q+1, \ldots, m\}, \end{aligned} \tag{5.2}$$

where $\{\mathbf{f}, \mathbf{a}_i, \mathbf{c}_j\} \subset \mathbb{R}^n$, $(b_1, \ldots, b_m) \in \mathbb{R}^m$ and $q < n < m$. Actually, problems of this type can be converted to the LP problem in (5.1) by eliminating the equality constraints by reducing an equal number of variables in $\mathbf{z} = (z_1, \ldots, z_n)$. For instance,

$$\begin{aligned} \langle \mathbf{a}_1, \mathbf{z}\rangle &= b_1, \\ &\vdots \\ \langle \mathbf{a}_q, \mathbf{z}\rangle &= b_q \end{aligned}$$

implies, without loss of generality,

$$\begin{array}{ccc} z_1 & & + a'_{1,q+1} z_{q+1} + \cdots + a'_{1,n} z_n = b_1, \\ & \ddots & \vdots \\ & z_q & + a'_{q,q+1} z_{q+1} + \cdots + a'_{q,n} z_n = b_q. \end{array}$$

Solving $(z_1, \ldots, z_q)$ in terms of $(z_{q+1}, \ldots, z_n)$ in the above and substituting them into the inequality constraints in (5.2), the LP problem in (5.2) becomes

$$\begin{aligned} &\text{Minimize } \langle \mathbf{f}', \mathbf{y}\rangle \\ &\langle \mathbf{c}'_j, \mathbf{y}\rangle \leqslant b'_j, \quad j \in I_2 \end{aligned} \tag{5.3}$$

in the variables $\mathbf{y} = (z_{q+1}, \ldots, z_n)$, where $\{\mathbf{f}', \mathbf{c}'_j\} \subset \mathbb{R}^{n-q}$, and $\mathbf{b} = (b'_q, \ldots, b'_m)^{\mathrm{T}} \in \mathbb{R}^{m-q}$.

5.2. Finding all lower edges of a lifted point set

PROPOSITION 5.1. *For $S = (S_1, \ldots, S_n)$ with $S_j \subset \mathbb{N}^n$ and $Q_j = \mathrm{conv}(S_j)$ for $j = 1, \ldots, n$, the mixed volume $\mathcal{M}(S_1, \ldots, S_n)$ of S equals to*

$$
\begin{aligned}
\mathcal{M}(S_1, \ldots, S_n) = \mathrm{Vol}_n(Q_1 + \cdots + Q_n) + \cdots \\
+ (-1)^{n-2} \sum_{i<j} \mathrm{Vol}_n(Q_j + Q_j) \\
+ (-1)^{n-1} \sum_{j=1}^{n} \mathrm{Vol}_n(Q_j).
\end{aligned}
$$

PROOF. For $\lambda = (\lambda_1, \ldots, \lambda_n)$, let $f(\lambda) = \mathrm{Vol}_n(\lambda_1 Q_1 + \cdots + \lambda_n Q_n)$. Let A be the set of all the monomials $\lambda^{\mathbf{k}}$ in $f(\lambda)$ where $\mathbf{k} = (k_1, \ldots, k_n)$ with $k_1 + \cdots + k_n = n$, and for $j = 1, \ldots, n$, let A_j be the monomials in $f(\lambda)$ in which the variable λ_j is absent. For each subset $I \subset \{1, \ldots, n\}$, let

$$A_I = \bigcap_{j \in I} A_j$$

with $A_\emptyset = A$, and for each subset $A_s \subset A$, let

$$g(A_s) := \sum_{\lambda^{\mathbf{k}} \in A_s} \text{the coefficient of } \lambda^{\mathbf{k}}.$$

It is easy to see that

$$
\begin{aligned}
g(A_I) = f(\lambda_I) \quad \text{where } (\lambda_I)_j &= 0 \quad \text{if } j \in I, \\
(\lambda_I)_j &= 1 \quad \text{if } j \notin I.
\end{aligned}
$$

Let $\bar{A}_s = A \backslash A_s$ and $|I| :=$ the number of elements in I. By the Principle of Inclusion–Exclusion in combinatorics (see, e.g., STANLEY [1997]), $g(\bar{A}_1 \cap \cdots \cap \bar{A}_n)$, the coefficient of $\lambda_1 \cdots \lambda_n$ in $f(\lambda)$, equals

$$
\begin{aligned}
\sum_{I \subset \{1,\ldots,n\}} (-1)^{|I|} g(A_I) = f(1, \ldots, 1) - \sum_{j=1}^{n} f(1, \ldots, 1, \overset{j\text{th}}{0}, 1, \ldots, 1) \\
+ \sum_{i<j} f(1, \ldots, 1, \overset{i\text{th}}{0}, 1, \ldots, 1, \overset{j\text{th}}{0}, 1, \ldots, 1) \\
- \cdots \\
+ (-1)^{n-2} \sum_{i<j} f(0, \ldots, 0, \overset{i\text{th}}{1}, 0, \ldots, 0, \overset{j\text{th}}{1}, 0, \ldots, 0) \\
+ (-1)^{n-1} \sum_{j=1}^{n-1} f(0, \ldots, 0, \overset{j\text{th}}{1}, 0, \ldots, 0)
\end{aligned}
$$

$$
\begin{aligned}
&= \mathrm{Vol}_n(Q_1 + \cdots + Q_n) + \cdots \\
&\quad + (-1)^{n-2} \sum_{i<j} \mathrm{Vol}_n(Q_j + Q_j) \\
&\quad + (-1)^{n-1} \sum_{j=1}^{n} \mathrm{Vol}_n(Q_j). \qquad \square
\end{aligned}
$$

As a consequence of the above, points in $S_j = \{a_{j1}, \ldots, a_{jm_j}\}$ which are not vertices of $Q_j = \mathrm{conv}(S_j)$, called *non-extreme points* of S_j, play no role in the mixed volume $\mathcal{M}(S_1, \ldots, S_n)$. So, when we compute the mixed volume $\mathcal{M}(S_1, \ldots, S_n)$ all those non-extreme points should be eliminated in the first place and we will assume in this section S_j admits only extreme points for each $j = 1, \ldots, n$.

A non-extreme point of S_j is a convex combination of other points of S_j. Namely, if a_{jk} is a non-extreme point of S_j, the following system of equations

$$
\begin{aligned}
&\lambda_1 a_{j1} + \cdots + \lambda_{k-1} a_{jk-1} + \lambda_{k+1} a_{jk+1} + \cdots + \lambda_{m_j} a_{im_j} = a_{jk}, \\
&\lambda_1 + \cdots + \lambda_{k-1} + \lambda_{k+1} + \cdots + \lambda_{m_j} = 1, \\
&\quad \lambda_1, \ldots, \lambda_{k-1}, \lambda_{k+1}, \ldots, \lambda_{m_j} \geqslant 0
\end{aligned}
$$

must have a solution. Testing the existence of solutions of this system by actually finding one is a standard Phase I problem in linear programming, and algorithms for this problem can be found in many standard linear programming books, e.g., PAPADIMITRIOU and STEIGLITZ [1982]. In essence, the existence of solutions of the above system is equivalent to zero optimal value of the optimization problem:

$$
\begin{aligned}
&\text{Minimize } \lambda_k \\
&\lambda_1 a_{j1} + \cdots + \lambda_{m_j} a_{jm_j} = a_{jk}, \\
&\lambda_1 + \cdots + \lambda_{m_j} = 1, \\
&\quad \lambda_i \geqslant 0, \quad i = 1, \ldots, m_j.
\end{aligned}
$$

An obvious feasible point of this LP problem is $\lambda_k = 1$ and $\lambda_i = 0$ for $i \neq k$.

For $w = (w_1, \ldots, w_n)$ with generically chosen $w_j : S_j \to \mathbb{R}$ for $j = 1, \ldots, n$, and

$$
\widehat{S}_j = \big\{\hat{a} = \big(a, w_j(a)\big) \mid a \in S_j\big\},
$$

denote the set of all lower edges of $\widehat{S}_j$ by $\mathcal{L}(\widehat{S}_j)$. To elaborate the algorithm for finding $\mathcal{L}(\widehat{S}_j)$, we let $\mathcal{B} = \{a_1, \ldots, a_m\} \subset \mathbb{N}^n$ represent general S_j's, and $w : \mathcal{B} \to \mathbb{R}$ be a generic lifting on $\mathcal{B}$. For $\widehat{\mathcal{B}} = \{\hat{a} = (a, w(a)) \mid a \in \mathcal{B}\}$, consider the inequality system

$$
\langle \hat{a}_j, \hat{\alpha} \rangle \geqslant \alpha_0, \quad j = 1, \ldots, m \tag{5.4}
$$

in the variables $\hat{\alpha} = (\alpha, 1) \in \mathbb{R}^{n+1}$ and $\alpha_0 \in \mathbb{R}$. Obviously, when this inequality system has a solution $(\alpha_0, \hat{\alpha})$ for which $\langle \hat{a}_i, \hat{\alpha} \rangle = \langle \hat{a}_j, \hat{\alpha} \rangle = \alpha_0$ for $1 \leqslant i, j \leqslant m$, then $\{\hat{a}_i, \hat{a}_j\}$ is a lower edge of $\widehat{\mathcal{B}}$.

With $a_j = (a_{j,1}, \ldots, a_{j,n})$ for $j = 1, \ldots, m$ and $\alpha = (\alpha_1, \ldots, \alpha_n)$, we write system (5.4) explicitly as

$$\begin{pmatrix} 1 & -a_{1,1} & \cdots & -a_{1,n} \\ 1 & -a_{2,1} & \cdots & -a_{2,n} \\ \vdots & \vdots & & \vdots \\ 1 & -a_{m,1} & \cdots & -c_{m,n} \end{pmatrix} \begin{pmatrix} \alpha_0 \\ \alpha_1 \\ \vdots \\ \alpha_n \end{pmatrix} \leqslant \begin{pmatrix} w(a_1) \\ w(a_2) \\ \vdots \\ w(a_m) \end{pmatrix}. \tag{5.5}$$

Letting $\upsilon \leqslant n$ be the rank of the coefficient matrix on the left-hand side of (5.5), we assume without loss that the first υ rows are linearly independent. By Gaussian elimination, an upper triangular nonsingular matrix U exists by which

$$\begin{pmatrix} 1 & -a_{1,1} & \cdots & -c_{1,n} \\ 1 & -a_{2,1} & \cdots & -c_{2,n} \\ \vdots & \vdots & & \vdots \\ 1 & -a_{\upsilon,1} & \cdots & a_{\upsilon,n} \\ \vdots & \vdots & & \vdots \\ 1 & -a_{m,1} & \cdots & -a_{m,n} \end{pmatrix} \cdot U = \begin{pmatrix} c_{1,1} & 0 & \cdots & 0 & 0 & \cdots & 0 \\ c_{2,1} & c_{2,2} & \cdots & 0 & 0 & \cdots & 0 \\ \vdots & \vdots & \ddots & \vdots & \vdots & & \vdots \\ c_{\upsilon,1} & c_{\upsilon,2} & \cdots & c_{\upsilon,\upsilon} & 0 & \cdots & 0 \\ \vdots & \vdots & & \vdots & \vdots & & \vdots \\ c_{m,1} & c_{m,2} & \cdots & c_{m,\upsilon} & 0 & \cdots & 0 \end{pmatrix},$$

where $c_{i,i} \neq 0$ for $i = 1, \ldots, \upsilon$. Replacing $U^{-1}(\alpha_0, \alpha_1, \ldots, \alpha_n)^{\mathrm{T}}$ by $(y_1, \ldots, y_{n+1})^{\mathrm{T}}$ in the following system

$$\begin{pmatrix} 1 & -a_{1,1} & \cdots & -a_{1,n} \\ 1 & -a_{2,1} & \cdots & -a_{2,n} \\ \vdots & \vdots & & \vdots \\ 1 & -a_{\upsilon,1} & \cdots & -a_{\upsilon,n} \\ \vdots & \vdots & & \vdots \\ 1 & -a_{m,1} & \cdots & -a_{m,n} \end{pmatrix} U \cdot U^{-1} \begin{pmatrix} \alpha_0 \\ \alpha_1 \\ \vdots \\ \alpha_n \end{pmatrix} \leqslant \begin{pmatrix} w(a_1) \\ w(a_2) \\ \vdots \\ w(a_\upsilon) \\ \vdots \\ w(a_m) \end{pmatrix} \tag{5.6}$$

yields

$$\begin{pmatrix} c_{1,1} & 0 & \cdots & 0 \\ c_{2,1} & c_{2,2} & \cdots & 0 \\ \vdots & \vdots & \ddots & \vdots \\ c_{\upsilon,1} & c_{\upsilon,2} & \cdots & c_{\upsilon,\upsilon} \\ \vdots & \vdots & & \vdots \\ c_{m,1} & c_{m,2} & \cdots & c_{m,\upsilon} \end{pmatrix} \begin{pmatrix} y_1 \\ y_2 \\ \vdots \\ y_\upsilon \end{pmatrix} \leqslant \begin{pmatrix} b_1 \\ b_2 \\ \vdots \\ b_\upsilon \\ \vdots \\ b_m \end{pmatrix}, \tag{5.7}$$

where $b_j = w(a_j)$, $j = 1, \ldots m$.

The existence of a solution $(\alpha_0, \hat{\alpha})$ of the system in (5.4) satisfying

$$\langle \hat{a}_i, \hat{\alpha} \rangle = \langle \hat{a}_j, \hat{\alpha} \rangle = \alpha_0$$

is now equivalent to the existence of a solution $(y_1, \ldots, y_\upsilon)$ of the system in (5.7) satisfying

$$c_{i,1}y_1 + c_{i,2}y_2 + \cdots + c_{i,\upsilon}y_\upsilon = b_i,$$
$$c_{j,1}y_1 + c_{j,2}y_2 + \cdots + c_{j,\upsilon}y_\upsilon = b_j \quad \text{for } 1 \leqslant i, j \leqslant m.$$

The inequality system (5.7) defines a polyhedron R in $\mathbb{R}^{\upsilon}$, and for any vertex $\mathbf{y}_0$ of R, there are at least υ active constraints, or υ equalities in (5.7). Let $J = \{i_1, \ldots, i_u\}$ with $u \geqslant \upsilon$ be the set of indices of the active constraints at $\mathbf{y}_0$. Clearly, $\{\hat{a}_{i_k}, \hat{a}_{i_l}\}$ is a lower edge of $\widehat{\mathcal{B}}$ for any $i_k, i_l \in J$. On the other hand, if $\{\hat{a}_i, \hat{a}_j\}$ is a lower edge of $\widehat{\mathcal{B}}$, there is a lower facet of $\operatorname{conv}(\widehat{\mathcal{B}})$ with inner normal $\hat{\alpha} = (\alpha, 1) \in \mathbb{R}^{n+1}$ that contains the line segment of $\{\hat{a}_i, \hat{a}_j\}$. Let $\alpha_0 = \langle \hat{a}_i, \hat{\alpha}\rangle = \langle \hat{a}_j, \hat{\alpha}\rangle$ and $\mathbf{y}_0 = (y_1, \ldots, y_\upsilon)$ be the first υ components of $U^{-1}(\alpha_0, \alpha_1, \ldots, \alpha_n)^{\mathrm{T}}$ in (5.6). Then $\mathbf{y}_0$ is a vertex of R.

Therefore, finding all lower edges of $\widehat{\mathcal{B}}$ is equivalent to finding all the vertices of the polyhedron R defined by the inequalities in (5.7). To find all the vertices of R, our strategy is: Find an initial vertex of R in the first place and generate all other vertices of R from this vertex thereafter.

To find an initial vertex of R, we first solve the triangular system

$$\begin{aligned} c_{1,1}y_1 \qquad\qquad\qquad\qquad &= b_1,\\ c_{2,1}y_1 + c_{2,2}y_2 \qquad\qquad\qquad &= b_2,\\ \vdots \qquad\qquad\qquad\qquad\qquad &\ \ \vdots\\ c_{\upsilon,1}y_1 + c_{\upsilon,2}y_2 + \cdots + c_{\upsilon,\upsilon}y_\upsilon &= b_\upsilon \end{aligned}$$

in (5.7) and let the solution be $\mathbf{y}_0 = (y_{01}, y_{02}, \ldots, y_{0\upsilon})$. Let

$$\begin{aligned} &d_j = b_j - (c_{j,1}y_{01} + c_{j,2}y_{02} + \cdots + c_{j,\upsilon}y_{0\upsilon}),\\ &\quad j = \upsilon + 1, \ldots, m. \end{aligned}$$

If $d_l := \min_{\upsilon+1\leqslant j\leqslant m} d_j \geqslant 0$, then $\mathbf{y}_0$ is already a vertex of R. Otherwise we solve the following LP problem in the form given in (5.1):

$$\begin{aligned} &\text{Minimize } \varepsilon\\ &c_{1,1}y_1 &&\leqslant b_1,\\ &c_{2,1}y_1 + c_{2,2}y_2 &&\leqslant b_2,\\ &\quad\vdots &&\ \ \vdots\\ &c_{\upsilon,1}y_1 + c_{\upsilon,2}y_2 + \cdots + c_{\upsilon,\upsilon}y_\upsilon &&\leqslant b_\upsilon,\\ &c_{\upsilon+1,1}y_1 + c_{\upsilon+1,2}y_2 + \cdots + c_{\upsilon+1,\upsilon}y_\upsilon - c_{\upsilon+1,\upsilon+1}\varepsilon &&\leqslant b_{\upsilon+1},\\ &\quad\vdots &&\ \ \vdots\\ &c_{m,1}y_1 + c_{m,2}y_2 + \cdots + c_{m,\upsilon}y_\upsilon - c_{m,\upsilon+1}\varepsilon &&\leqslant b_m,\\ & &-\varepsilon &\leqslant 0,\\ &\text{where } c_{j,\upsilon+1} = \begin{cases} 0, & \text{if } d_j \geqslant 0,\\ 1, & \text{if } d_j < 0, \end{cases} \quad \text{for } \upsilon + 1 \leqslant j \leqslant m, \end{aligned} \tag{5.8}$$

in the variables $(\mathbf{y}, \varepsilon) = (y_1, \ldots, y_\upsilon, \varepsilon)$ with feasible point $(\mathbf{y}, \varepsilon) = (\mathbf{y}_0, -d_l)$ and initial indices of constraints $J = \{1, 2, \ldots, \upsilon, l\}$. Obviously, ε in the optimal solution $(\bar{\mathbf{y}}, \varepsilon)$ of this LP problem must be zero and in such case $\bar{\mathbf{y}}$ becomes a vertex of R.

To generate all other vertices, we first introduce the following LP problem:

TWO-POINT TEST PROBLEM.

$$
\begin{aligned}
&\text{Minimize } -(c_{i_0,1}+c_{j_0,1})y_1-(c_{i_0,2}+c_{j_0,2})y_2-\cdots-(c_{i_0,\upsilon}+c_{j_0,\upsilon})y_\upsilon\\
&c_{1,1}y_1 \leqslant b_1,\\
&c_{2,1}y_1+c_{2,2}y_2 \leqslant b_2,\\
&\quad\vdots\\
&c_{\upsilon,1}y_1+c_{\upsilon,2}y_2+\cdots+c_{\upsilon,\upsilon}y_\upsilon \leqslant b_\upsilon,\\
&\quad\vdots\\
&c_{m,1}y_1+c_{m,2}y_2+\cdots+c_{m,\upsilon}y_\upsilon \leqslant b_m,
\end{aligned}
\tag{5.9}
$$

where $1 \leqslant i_0, j_0 \leqslant m$.

If the optimal value of this problem is $-b_{i_0}-b_{j_0}$ and attained at $(y_1,\ldots,y_\upsilon)$, then

$$
\begin{aligned}
&-(c_{i_0,1}+c_{j_0,1})y_1-(c_{i_0,2}+c_{j_0,2})y_2-\cdots-(c_{i_0,\upsilon}+c_{j_0,\upsilon})y_\upsilon\\
&\quad=(-c_{i_0,1}y_1-c_{i_0,2}y_2-\cdots-c_{i_0,\upsilon}y_\upsilon)+(-c_{j_0,1}y_1-c_{j_0,2}y_2-\cdots-c_{j_0,\upsilon}y_\upsilon)\\
&\quad=-b_{i_0}-b_{j_0}.
\end{aligned}
$$

But $(y_1,\ldots,y_\upsilon)$ also satisfies constraints

$$
\begin{aligned}
&c_{i_0,1}y_1+c_{i_0,2}y_2+\cdots+c_{i_0,\upsilon}y_\upsilon \leqslant b_{i_0},\\
&c_{j_0,1}y_1+c_{j_0,2}y_2+\cdots-c_{j_0,\upsilon}y_\upsilon \leqslant b_{j_0}.
\end{aligned}
$$

Therefore,

$$
\begin{aligned}
&c_{i_0,1}y_1+c_{i_0,2}y_2+\cdots+c_{i_0,\upsilon}y_\upsilon = b_{i_0},\\
&c_{j_0,1}y_1+c_{j_0,2}y_2+\cdots+c_{j_0,\upsilon}y_\upsilon = b_{j_0}.
\end{aligned}
$$

Accordingly, $\{\hat{a}_{i_0}, \hat{a}_{j_0}\}$ is a lower edge of $\widehat{\mathcal{B}}$.

The constraints in (5.9) is the same inequality system in (5.7) which defines polyhedron R. Therefore an initial vertex $\bar{\mathbf{y}}$ of R provides a feasible point of the LP problem in (5.9), and we may solve this problem to determine if $\{\hat{a}_{i_0}, \hat{a}_{j_0}\}$ for given $1 \leqslant i_0, j_0 \leqslant m$ is a lower edge of $\widehat{\mathcal{B}}$. When the simplex method, as outlined in the last section, is used for solving this problem, we pivot from one vertex of R to another vertex in the direction where the objective function decreases. Every time a newly obtained vertex of R in the process will carry a new set of equalities in (5.7), and therefore provides a new collection of lower edges of $\widehat{\mathcal{B}}$. This important feature makes the exhaustive testings on all the possible pairs in $\widehat{\mathcal{B}}$ unnecessary.

The details of the algorithm for finding all lower edges of $\widehat{\mathcal{B}}$ is given in the following

ALGORITHM 5.1. Given $\widehat{\mathcal{B}}=\{\hat{a}_0,\hat{a}_1,\ldots,\hat{a}_m\}$, construct $\mathcal{L}(\widehat{\mathcal{B}})$.

Step 0: Initialization.

Set up inequality system (5.7). Let $\mathcal{P}=\{\{\hat{a}_i,\hat{a}_j\}\mid 1\leqslant i,j\leqslant m\}$ be all the possible pairs of $\widehat{\mathcal{B}}$. If $\upsilon=m$, set $\mathcal{L}(\widehat{\mathcal{B}}):=\mathcal{P}$ and stop. Otherwise, solve the optimization problem (5.8), find an initial vertex $\mathbf{y}_0$ of system (5.7) with the set of indices of active constraints $J=\{i_1,\ldots,i_\upsilon\}$ and $D^{-1}=[\mathbf{u}_1,\ldots,\mathbf{u}_\upsilon]$, where $D^{\mathrm{T}}=[\mathbf{c}_{i_1},\ldots,\mathbf{c}_{i_\upsilon}]$ and

$\mathbf{c}_{i_j} = (c_{i_j,1}, \ldots, c_{i_j,\upsilon})$ for $j = 1, \ldots, \upsilon$. Set $\mathcal{L}(\widehat{\mathcal{B}}) := \{\{\hat{a}_k, \hat{a}_l\} \mid k, l \in J\}$ and $\mathcal{P} := \mathcal{P}\backslash\{\{\hat{a}_k, \hat{a}_l\} \mid k, l \in J\}$, go to Step 1.

Step 1: Setting up objective function for the Two-Point Test.

If $\mathcal{P} = \emptyset$, stop. Otherwise select $\{\hat{a}_{i_0}, \hat{a}_{j_0}\} \in \mathcal{P}$, set $\mathbf{f} := (-c_{i_0,1} - c_{j_0,1}, \ldots, -c_{i_0,\upsilon} - c_{j_0,\upsilon})$, and $\mathcal{P} := \mathcal{P}\backslash\{\{\hat{a}_{i_0}, \hat{a}_{j_0}\}\}$, go to Step 2.

Step 2: Solving the LP problem.

Step 2.1. Determine the smallest index k such that

$$\langle \mathbf{f}, \mathbf{u}_k \rangle = \max\bigl\{\langle \mathbf{f}, \mathbf{u}_j \rangle \mid j = 1, \ldots, \upsilon\bigr\}.$$

If $\langle \mathbf{f}, \mathbf{u}_k \rangle \leqslant 0$, go to Step 1. Otherwise, set $\mathbf{s} = \mathbf{u}_k$, go to Step 2.2.

Step 2.2. Compute the smallest index l and σ such that

$$\sigma = \frac{\langle \mathbf{c}_l, \mathbf{y}_0 \rangle - b_l}{\langle \mathbf{c}_l, \mathbf{s} \rangle} = \min\left\{ \frac{\langle \mathbf{c}_j, \mathbf{y}_0 \rangle - b_j}{\langle \mathbf{c}_j, \mathbf{s} \rangle} \;\middle|\; j \notin J \text{ with } \langle \mathbf{c}_j, \mathbf{s} \rangle < 0 \right\}.$$

Go to Step 2.3.

Step 2.3. Set $\mathbf{y}_0 := \mathbf{y}_0 - \sigma \mathbf{s}$ and update $J = \{i_1, \ldots, i_\upsilon\}$ and D^{-1}. Set $\mathcal{L}(\widehat{\mathcal{B}}) := \mathcal{L}(\widehat{\mathcal{B}}) \cup (\mathcal{P} \cap \{\{\hat{a}_k, \hat{a}_l\} \mid k, l \in J\})$, and $\mathcal{P} := \mathcal{P}\backslash\{\{\hat{a}_k, \hat{a}_l\} \mid k, l \in J\}$.
Go to Step 2.1.

5.3. *Extending level-k subfaces*

For a level-k subface $\widehat{E}_k = (\hat{\mathbf{e}}_1, \ldots, \hat{\mathbf{e}}_k)$ of $\widehat{S} = (\widehat{S}_1, \ldots, \widehat{S}_n)$ with $1 \leqslant k < n$ where $\hat{\mathbf{e}}_j = \{\hat{a}_j, \hat{a}'_j\} \in \mathcal{L}(\widehat{S}_j)$ for $j = 1, \ldots, k$, we say $\hat{\mathbf{e}}_{k+1} = \{\hat{a}_{k+1}, \hat{a}'_{k+1}\} \in \mathcal{L}(\widehat{S}_{k+1})$ *extends* $\widehat{E}_k$ if $\widehat{E}_{k+1} = (\hat{\mathbf{e}}_1, \ldots, \hat{\mathbf{e}}_{k+1})$ is a level-$(k+1)$ subface of $\widehat{S}$. Let

$$\mathcal{E}(\widehat{E}_k) = \bigl\{\{\hat{a}_{k+1}, \hat{a}'_{k+1}\} \in \mathcal{L}(\widehat{S}_{k+1}) \bigm| \{\hat{a}_{k+1}, \hat{a}'_{k+1}\} \text{ extends } \widehat{E}_k\bigr\}.$$

$\widehat{E}_k$ is called *extendible* if $\mathcal{E}(\widehat{E}_k) \neq \emptyset$, it is *nonextendible* otherwise. Obviously, an extendible $\widehat{E}_{n-1}$ yields mixed cells of S_w induced by elements in $\mathcal{E}(\widehat{E}_{n-1})$ (possibly several). So, to find all mixed cells, we may start from $k = 1$ and extend $\widehat{E}_k$ step by step. If $\widehat{E}_k$ is nonextendible, then no mixed cells of S_w whose liftings contain subface $(\{\hat{a}_1, \hat{a}'_1\}, \ldots, \{\hat{a}_k, \hat{a}'_k\})$. Hence, the extension attempt on $\widehat{E}_k$ will only continue when it is extendible.

For a given level-k subface $\widehat{E}_k = (\hat{\mathbf{e}}_1, \ldots, \hat{\mathbf{e}}_k)$ where $\hat{\mathbf{e}}_j = \{\hat{a}_j, \hat{a}'_j\} \subset \widehat{S}_j$ for $j = 1, \ldots, k$, consider the system

$$\begin{aligned} &\langle \hat{a}, \hat{\alpha} \rangle \geqslant \alpha_0, && \hat{a} \in \widehat{S}_{k+1}, \\ &\langle \hat{a}_j, \hat{\alpha} \rangle \leqslant \langle \hat{a}, \hat{\alpha} \rangle, && \text{for } \hat{a} \in \widehat{S}_j \text{ and } j = 1, \ldots, k, \\ &\langle \hat{a}_j, \hat{\alpha} \rangle = \langle \hat{a}'_j, \hat{\alpha} \rangle, && j = 1, \ldots, k, \end{aligned} \tag{5.10}$$

in the variables $\hat{\alpha} = (\alpha, 1) \in \mathbb{R}^{n+1}$ and $\alpha_0 \in \mathbb{R}$. Clearly, if this system has a solution $(\alpha_0, \hat{\alpha})$ satisfying

$$\langle \hat{a}_l, \hat{\alpha} \rangle = \langle \hat{a}_i, \hat{\alpha} \rangle = \alpha_0 \quad \text{for } \hat{a}_l, \hat{a}_i \in \widehat{S}_{k+1}$$

then $\{\hat{a}_l, \hat{a}_i\}$ extends $\widehat{E}_k$.

More explicitly, with $S_j = \{a_{j,1}, \ldots, a_{j,m_j}\}$ for $j = 1, \ldots, n$ and $\alpha = (\alpha_1, \ldots, \alpha_n)$, the above system becomes

$$
\begin{aligned}
&\langle a_{k+1,i}, \alpha\rangle - \alpha_0 \leqslant -w_{k+1}(a_{k+1,i}), && i = 1, \ldots, m_{k+1},\\
&\langle a_j - a_{j,i}, \alpha\rangle \leqslant w_j(a_{j,i}) - w_j(a_j), && i = 1, \ldots, m_j,\ a_{j,i} \in S_j \backslash \{a_j, a_j'\},\\
& && j = 1, \ldots, k,\\
&\langle a_j - a_j', \alpha\rangle = w_j(a_j') - w_j(a_j), && j = 1, \ldots, k.
\end{aligned}
$$

The last k equality constraints can be used to eliminate k variables in $\alpha = (\alpha_1, \ldots, \alpha_n)$, resulting in the following inequality system:

$$
\begin{aligned}
c'_{1,j_1}\alpha_{j_1} + c'_{1,j_2}\alpha_{j_2} + \cdots + c'_{1,j_{\eta'}}\alpha_{j_{\eta'}} &\leqslant b_1,\\
c'_{2,j_1}\alpha_{j_1} + c'_{2,j_2}\alpha_{j_2} + \cdots + c'_{2,j_{\eta'}}\alpha_{j_{\eta'}} &\leqslant b_2,\\
\vdots \qquad\qquad\qquad\qquad & \quad \vdots\\
c'_{\mu,j_1}\alpha_{j_1} + c'_{\mu,j_2}\alpha_{j_2} + \cdots + c'_{\mu,j_{\eta'}}\alpha_{j_{\eta'}} &\leqslant b_\mu,
\end{aligned}
\tag{5.11}
$$

where $\mu = \sum_{j=1}^{k+1} m_j - 2k$ and $\eta' = n - k + 1$. As before, by a coordinate transformation $(y_1, \ldots, y_{\eta'}) = (\alpha_{j_1}, \ldots, \alpha_{j_{\eta'}})U$ where $U \in \mathbb{R}^{\eta' \times \eta'}$ is a nonsingular matrix, the system can further be reduced to

$$
\begin{aligned}
c_{1,1}y_1 \qquad\qquad\qquad\qquad\qquad & \leqslant b_1,\\
c_{2,1}y_1 + c_{2,2}y_2 \qquad\qquad\qquad & \leqslant b_2,\\
\vdots \qquad\qquad \ddots \qquad\qquad & \quad \vdots\\
c_{\eta,1}y_1 + c_{\eta,2}y_2 + \cdots + c_{\eta,\eta}y_\eta & \leqslant b_\eta,\\
\vdots \qquad\qquad\qquad\qquad & \quad \vdots\\
c_{\mu,1}y_1 + c_{\mu,2}y_2 + \cdots + c_{\mu,\eta}y_\eta & \leqslant b_\mu.
\end{aligned}
\tag{5.12}
$$

Consequently, system (5.10) has a solution $(\alpha_0, \hat{\alpha})$ satisfying

$$\langle \hat{a}_{k+1,i}, \hat{\alpha}\rangle = \langle \hat{a}_{k+1,l}, \hat{\alpha}\rangle = \alpha_0 \quad \text{for } 1 \leqslant i, l \leqslant m_{k+1}$$

if and only if system (5.12) has a solution $(y_1, \ldots, y_\eta)$ satisfying

$$
\begin{aligned}
c_{i,1}y_1 + c_{i,2}y_2 + \cdots + c_{i,\eta}y_\eta &= b_i,\\
c_{l,1}y_1 + c_{l,2}y_2 + \cdots + c_{l,\eta}y_\eta &= b_l.
\end{aligned}
$$

Inequalities in (5.12) defines a polyhedron $\overline{R}$ in $\mathbb{R}^\eta$, and for any vertex of $\overline{R}$, there are at least η active constraints. If $\mathbf{z}_0$ is a vertex of $\overline{R}$ and $J = \{i_1, \ldots, i_t\}$ with $t \geqslant \eta$ is the indices of active constraints at $\mathbf{z}_0$, then $\{\hat{a}_{k+1,i_p}, \hat{a}_{k+1,i_q}\}$ extends $\widehat{E}_k$ for any pair $\{i_p, i_q\} \subset J \cap \{1, \ldots, m_{k+1}\}$.

Similar to the discussion given before, for $\{\hat{a}_{k+1,i}, a_{k+1,l}\} \in \mathcal{E}(\widehat{E}_k)$, there exists a corresponding vertex $\mathbf{z}_0$ of $\overline{R}$ whose indices of active constraints includes $\{i, l\} \subset I \equiv \{1, \ldots, m_{k+1}\}$. Hence, to construct $\mathcal{E}(\widehat{E}_k) \subset \mathcal{L}(\widehat{S}_{k+1})$, we may look for all those vertices of $\overline{R}$ whose indices of active constraints contain at least a pair of $\{i, l\}$ in I. To this end, we may apply the Two-Point Test introduced in the last section again and confine the indices of the "two points" to be tested to I. However, most of the $\hat{a}_{k+1,i}$'s that appear in the pairs in $\mathcal{L}(\widehat{S}_{k+1})$ do not exist in any of the pairs in $\mathcal{E}(\widehat{E}_k)$. This never occurs when we compute the set of lower edges $\mathcal{L}(\widehat{\mathcal{B}})$ of $\widehat{\mathcal{B}}$ in the last section since all the points in $\mathcal{B}$ are assumed to be extreme points, and consequently every point of $\widehat{\mathcal{B}}$ appears in certain pairs of $\mathcal{L}(\widehat{\mathcal{B}})$. This important observation stimulates the following One-Point Test to be used in addition to the Two-Point Test in the algorithm.

ONE-POINT TEST PROBLEM.

$$\begin{aligned}
&\text{Minimize } -c_{i_0,1}y_1 - c_{i_0,2}y_2 - \cdots - c_{i_0,\eta}y_\eta\\
&c_{1,1}y_1 \leqslant b_1,\\
&c_{2,1}y_1 + c_{2,2}y_2 \leqslant b_2,\\
&\quad\vdots \qquad\qquad \ddots \qquad\qquad \vdots\\
&c_{\eta,1}y_1 + c_{\eta,2}y_2 + \cdots + c_{\eta,\eta}y_\eta \leqslant b_\eta,\\
&\quad\vdots \qquad\qquad\qquad\qquad \vdots\\
&c_{\mu,1}y_1 + c_{\mu,2}y_2 + \cdots + c_{\mu,\eta}y_\eta \leqslant b_\mu
\end{aligned}\tag{5.13}$$

where $1 \leqslant i_0 < m_{k+1}$.

If the optimal value of this LP problem is greater than $-b_{i_0}$, then $\{\hat{a}_{k+1,i_0}, \hat{a}_{k+1,i}\}$ does not extend $\widehat{E}_k$ for all $i \in \{1,\ldots,m_{k+1}\}\backslash\{i_0\}$. For if there exists $1 \leqslant j_0 \leqslant m_{k+1}$ for which $\{\hat{a}_{k+1,i_0}, \hat{a}_{k+1,j_0}\}$ extends $\widehat{E}_k$, then system (5.12) has a solution $(y_1,\ldots,y_\eta)$ satisfying

$$\begin{aligned}
c_{i_0,1}y_1 + c_{i_0,2}y_2 + \cdots + c_{i_0,\eta}y_\eta &= b_{i_0},\\
c_{j_0,1}y_1 + c_{j_0,2}y_2 + \cdots + c_{j_0,\eta}y_\eta &= b_{j_0},
\end{aligned}$$

and the objective function value at $(y_1,\ldots,y_\eta)$ is then

$$-c_{i_0,1}y_1 - c_{i_0,2}y_2 - \cdots - c_{i_0,\eta}y_\eta = -b_{i_0}.$$

Thus, for the construction of $\mathcal{E}(\widehat{E}_k)$, points which have appeared in the pairs in $\mathcal{L}(\widehat{S}_{k+1})$ will be tested systematically by using the One-Point Test to check for their possible appearance in the pairs in $\mathcal{E}(\widehat{E}_k)$. When the optimal value is not as desired for a particular point $\hat{a}_{k+1,i_0}$, then all the pairs associated with $\hat{a}_{k+1,i_0}$ in $\mathcal{L}(\widehat{S}_{k+1})$ would be deleted from the list of further testing. The constraints in the LP problem in (5.13) is the same inequality system in (5.12) which defines the polyhedron $\overline{R}$ in $\mathbb{R}^\eta$. To find an initial vertex of $\overline{R}$ to start solving the problem, one may employ the same strategy by augmenting a new variable $\varepsilon \geqslant 0$ as in calculating $\mathcal{L}(\widehat{\mathcal{B}})$ of $\widehat{\mathcal{B}}$ in the last section.

In the process of achieving the optimal value, a newly obtained vertex of $\overline{R}$ provides a collection of new pairs of $\mathcal{E}(\widehat{E}_k)$ as long as its active constraints contain a pair of $\{i,l\}$ in $I = \{1,\ldots,m_{k+1}\}$. We will delete those points $\hat{a}_{k+1,i}$ in the pairs in $\mathcal{L}(\widehat{S}_{k+1})$ from the list of testing once their index i have appeared in any of the indices of the active constraints of the vertices of $\overline{R}$ being obtained. After the One-Point Test has exhausted all testings on possible candidates the Two-Point Test will then be used for the remaining pairs in $\mathcal{L}(\widehat{S}_{k+1})$. Empirically, when the One-Point Test is finished, most of the pairs in $\mathcal{E}(\widehat{E}_k)$ have been found and the Two-Point Test only plays a minor role in the process.

REMARK 5.1. Recall that the set of constraints in (5.13) is a modified version of the system in (5.10): the equality constraints in (5.13)

$$\langle \hat{a}_j, \hat{\alpha} \rangle = \langle \hat{a}'_j, \hat{\alpha} \rangle, \quad j = 1, \ldots, k,$$

are used to eliminate equal number of variables in $\alpha = (\alpha_1, \ldots, \alpha_n)$. When we solve the LP problem in (5.13), the inequality constraints

$$\langle \hat{a}_j, \hat{\alpha} \rangle \leqslant \langle \hat{a}, \hat{\alpha} \rangle, \quad \hat{a} \in \widehat{S}_j, \; j = 1, \ldots, k,$$

in (5.10) where either $\{\hat{a}, \hat{a}_j\}$ or $\{\hat{a}, \hat{a}'_j\}$ does not belong to $\mathcal{L}(\widehat{S}_j)$ will obviously never become active in the process. Those constraints should be removed before solving the LP problem in (5.13). In practice, the successive omission of such extraneous constraints in all those LP problems greatly reduces the amount of computation cumulatively.

Combining One-Point Test and Two-Point Test, we list the following algorithm for constructing $\mathcal{E}(\widehat{E}_k)$.

ALGORITHM 5.2. Given $\widehat{E}_k$, construct $\mathcal{E}(\widehat{E}_k)$.

Step 0: Initialization.

Set up inequality system (5.12) after removing extraneous constraints. Start from a vertex $\mathbf{z}_0$ with the set of indices of active constraints $J = \{i_1, \ldots, i_\eta\}$ and $D^{-1} = [\mathbf{u}_1, \ldots, \mathbf{u}_\eta]$ where $D^{\mathrm{T}} = [\mathbf{c}_{i_1}, \ldots, \mathbf{c}_{i_\eta}]$ with $\mathbf{c}_{i_j} = (c_{i_j,1}, \ldots, c_{i_j,\eta})$ and set $\widehat{F}_{k+1} := \mathcal{L}(\widehat{S}_{k+1})$.

Step 1: One-Point Test Problem.

Step 1.0. Set $i_0 := 0$, go to Step 1.1.

Step 1.1. Set up objective function.

Find

$$\tau = \min\bigl\{ j \mid j > i_0, \{\hat{a}_{k+1,j}, \hat{a}_{k+1,j'}\} \subset \widehat{F}_{k+1} \text{ for some } j' \bigr\}.$$

If no such τ exists, go to Step 2. Otherwise set $i_0 := \tau$ and $\mathbf{f} = (-c_{i_0,1}, \ldots, -c_{i_0,\eta})$, go to Step 1.2.

Step 1.2. Determine the smallest index k such that

$$\langle \mathbf{f}, \mathbf{u}_k \rangle = \max\bigl\{ \langle \mathbf{f}, \mathbf{u}_j \rangle \mid j = 1, \ldots, \eta \bigr\}.$$

If $\langle \mathbf{f}, \mathbf{u}_k \rangle \leqslant 0$, go to Step 1.5. Otherwise, set $\mathbf{s} = \mathbf{u}_k$ and go to Step 1.3.

Step 1.3. Compute the smallest index l and σ such that

$$\sigma = \frac{\langle \mathbf{c}_l, \mathbf{z}_0 \rangle - b_l}{\langle \mathbf{c}_l, \mathbf{s} \rangle} = \min\left\{ \frac{\langle \mathbf{c}_j, \mathbf{z}_0 \rangle - b_j}{\langle \mathbf{c}_j, \mathbf{s} \rangle} \;\middle|\; j \notin J, \; \langle \mathbf{c}_j, \mathbf{s} \rangle < 0 \right\}.$$

Go to Step 1.4.

Step 1.4. Set $\mathbf{z}_0 := \mathbf{z}_0 - \sigma \mathbf{s}$ and update $J = \{i_1, \ldots, i_\eta\}$ and D^{-1}. If $l < m_{k+1}$, check if any lower edge $\{\hat{a}_{k+1,l}, \hat{a}_{k+1,j}\}$ in $\widehat{F}_{k+1}$ extends $\widehat{F}_{k+1}$. Collect these lower edges, if they exist, and delete them from $\widehat{F}_{k+1}$.

Go to Step 1.2.

Step 1.5. If the current value of objective function is different from $-b_{i_0}$, delete all lower edges that contain $\hat{a}_{k+1,i_0}$ from $\widehat{F}_{k+1}$.
Go to Step 1.1.

Step 2: Two-Point Test Problems.

Step 2.1. Set up objective function.
If $\widehat{F}_{k+1} = \emptyset$, stop. Otherwise select a lower edge $\{\hat{a}_{k+1,i_0}, \hat{a}_{k+1,j_0}\} \in \widehat{F}_{k+1}$. Set $\mathbf{f} := (-c_{i_0,1} - c_{j_0,1}, \ldots, -c_{i_0,\eta} - c_{j_C,\eta})$, and $\widehat{F}_{k+1} := \widehat{F}_{k+1} \backslash \{\hat{a}_{k+1,i_0}, \hat{a}_{k+1,j_0}\}$, go to Step 2.2.

Step 2.2. Determine the smallest index k such that

$$\langle \mathbf{f}, \mathbf{u}_k \rangle = \max\bigl\{\langle \mathbf{f}, \mathbf{u}_j \rangle \mid j = 1, \ldots, \eta\bigr\}.$$

If $\langle \mathbf{f}, \mathbf{u}_k \rangle \leqslant 0$, go to Step 2.1. Otherwise, set $\mathbf{s} = \mathbf{u}_k$, go to Step 2.3.

Step 2.3. Compute the smallest index l and σ such that

$$\sigma = \frac{\langle \mathbf{c}_l, \mathbf{z}_0 \rangle - b_l}{\langle \mathbf{c}_l, \mathbf{s} \rangle} = \min\left\{ \frac{\langle \mathbf{c}_j, \mathbf{z}_0 \rangle - b_j}{\langle \mathbf{c}_j, \mathbf{s} \rangle} \;\middle|\; j \notin J,\ \langle \mathbf{c}_j, \mathbf{s} \rangle < 0 \right\}.$$

Go to Step 2.4.

Step 2.4. Set $\mathbf{z}_0 := \mathbf{z}_0 - \sigma \mathbf{s}$ and update $J = \{i_1, \ldots, i_\eta\}$ as well as D^{-1}. If $l < m_{k+1}$, check if any lower edge $\{\hat{a}_{k+1,l}, \hat{a}_{k+1,j}\}$ in $\widehat{F}_{k+1}$ extends $\widehat{F}_{k+1}$. Collect those lower edges, if they exist, and delete them from $\widehat{F}_{k+1}$.
Go to Step 2.2.

REMARK 5.2. Setting up inequality system (5.12) can be very time consuming. To be more efficient, one may save all inequality systems at previous levels to help the establishment of the inequality system in the current level.

5.4. Finding all mixed cells

We now insert the algorithms for finding lower edges of $\widehat{S}_j$ and extending subfaces of $\widehat{S}$ in the following procedure for finding all the mixed cells in S_ω induced by a generic lifting $\omega = (\omega_1, \ldots, \omega_n)$ on $S = (S_1, \ldots, S_n)$.

PROCEDURE FOR FINDING ALL MIXED CELLS.
Find all mixed cells in S_ω induced by a generic lifting $\omega = (\omega_1, \ldots, \omega_n)$ on $S = (S_1, \ldots, S_n)$.

Step 0: Initialization.
Find $\mathcal{L}(\widehat{S}_j)$, for all $j = 1, \ldots, n$ by Algorithm 5.1.
Set $\widehat{\mathcal{F}}_1 := \mathcal{L}(\widehat{S}_1)$, $k := 1$.

Step 1: Backtracking.
If $k = 0$ Stop.
If $\widehat{\mathcal{F}}_k = \emptyset$, set $k := k - 1$ and go to Step 1.
Otherwise go to Step 2.

Step 2: Selecting next level-*k* subface to extend.
Select $\hat{\mathbf{e}}_k \in \widehat{\mathcal{F}}_k$, and set $\widehat{\mathcal{F}}_k := \widehat{\mathcal{F}}_k \backslash \{\hat{\mathbf{e}}_k\}$.
Let $\widehat{E}_k = (\hat{\mathbf{e}}_1, \ldots, \hat{\mathbf{e}}_k)$ and go to Step 3.

Step 3: Extending the level-k subface.
Find $\mathcal{E}(\widehat{E}_k)$ by Algorithm 5.2.
If $\mathcal{E}(\widehat{E}_k) = \emptyset$, go to Step 1, otherwise set $\widehat{\mathcal{F}}_{k+1} = \mathcal{E}(\widehat{E}_k)$, $k := k + 1$ then go to Step 4.

Step 4: Collecting mixed cells.
If $k = n$, all $C = (\hat{\mathbf{e}}_1, \ldots, \hat{\mathbf{e}}_{n-1}, \hat{\mathbf{e}})$ with $\hat{\mathbf{e}} \in \widehat{\mathcal{F}}_n$ induce mixed cells of S_ω, pick up all these mixed cells, then set $k := k - 1$, go to Step 1.
Otherwise go to Step 2.

As we suggested earlier, when all the mixed cells are found, the mixed volume $\mathcal{M}(S_1, \ldots, S_n)$ can be computed as the sum of the volumes of all those mixed cells. The algorithm described in this section for calculating the mixed volume by locating all mixed cells in a fine mixed subdivision of the support induced by a generic lifting on the support has been successfully implemented and tested on the supports of a large variety of polynomial systems (LI and LI [2001]). Currently, this algorithm represents the state of the art in computing mixed volumes this way. It leads other existing algorithms (EMIRIS and CANNY [1995], VERSCHELDE [1999], GAO and LI [2000]) in speed as well as storage requirements by a considerable margin. See also TAKEDA, KOJIMA and FUJISAWA [2002].

6. Mixed volume of semi-mixed support

6.1. *Semi-mixed polynomial systems*

A polynomial system $P(x) = (p_1(x), \ldots, p_n(x))$ with support $S = (S_1, \ldots, S_n)$ is called *semi-mixed* of type $(k_1, \ldots, k_r)$ when the supports S_j's are not all distinct, but they are equal within r blocks of sizes $k_1, \ldots, k_r$. To be more precise, there are r finite subsets $S^{(1)}, \ldots, S^{(r)}$ of $\mathbb{N}^n$ such that

$$S^{(i)} = S_{i1} = \cdots = S_{ik_i},$$

where

$$S_{il} \in \{S_1, \ldots, S_n\} \quad \text{for } 1 \leqslant i \leqslant r,\ 1 \leqslant l \leqslant k_i,$$

and $k_1 + \cdots + k_r = n$. The system $P(x)$ is called *unmixed* if $r = 1$ and is *fully mixed* when $r = n$.

For $1 \leqslant i \leqslant r$, let $P^{(i)}(x)$ be the subset of polynomials in $P(x) = (p_1(x), \ldots, p_n(x))$ having support $S^{(i)}$. Thus each polynomial of $P^{(i)}(x)$ can be written as

$$p_{il}(x) = \sum_{a \in S^{(i)}} c_{ila} x^a \quad \text{for } 1 \leqslant i \leqslant r,\ 1 \leqslant l \leqslant k_i. \tag{6.1}$$

We abbreviate $S = (S^{(1)}, \ldots, S^{(r)})$ and $P(x) = (P^{(1)}(x), \ldots, P^{(r)}(x))$.

We may, of course, solve a semi-mixed polynomial system $P(x) = (p_1(x), \ldots, p_n(x))$ in $\mathbb{C}^n$ by following the standard polyhedral homotopy procedure described in Section 4 without paying special attention to the semi-mixed structure of its supports.

However, when this special structure is taken into account, a revised polyhedral homotopy procedure can be developed with a great reduction in the amount of computation, especially when $P(x)$ is unmixed, such as the 9-point problem in mechanism design (WAMPLER, MORGAN and SOMMESE [1992]).

For $P(x) = (p_1(x), \dots, p_n(x))$ with support $S = (S_1, \dots, S_n)$ in general position, we assume as before all the p_j's have constant terms, namely, $S_j = S'_j = S_j \cup \{0\}$ for $j = 1, \dots, n$. Recall that at the beginning of the polyhedral homotopy procedure, we first assign a generic lifting $\omega = (\omega_1, \dots, \omega_n)$ on $S = (S_1, \dots, S_n)$. Now for semi-mixed system $P(x) = (P^{(1)}(x), \dots, P^{(r)}(x))$ of type $(k_1, \dots, k_r)$ with support $S = (S^{(1)}, \dots, S^{(r)})$ and generic coefficients $c_{ila} \in \mathbb{C}^*$ as given in (6.1), we choose generic lifting $\omega = (\omega_1, \dots, \omega_r)$ on $S = (S^{(1)}, \dots, S^{(r)})$ where $\omega_i : S^{(i)} \to \mathbb{R}$ for $i = 1, \dots, r$ and consider the homotopy $Q(x,t) = (Q^{(1)}(x,t), \dots, Q^{(r)}(x,t)) = 0$ where equations in $Q^{(i)}(x,t) = 0$ for $1 \leqslant i \leqslant r$ are

$$q_{il}(x,t) = \sum_{a \in S^{(i)}} c_{ila} x^a t^{\omega_i(a)} = 0, \quad 1 \leqslant l \leqslant k_i. \tag{6.2}$$

Clearly, $Q(x,1) = P(x)$. With $\hat{a} = (a, \omega_i(a))$ for $a \in S^{(i)}$, let $\hat{\alpha} = (\alpha, 1) \in \mathbb{R}^{n+1}$ satisfy the following condition:

$$\begin{array}{l} \text{There exists } C = (C_1, \dots, C_r) \text{ where } C_i = \{a_{i0}, \dots, a_{ik_i}\} \subset S^{(i)} \text{ and} \\ \text{conv}(C_i) \text{ is a simplex of dimension } k_i \text{ for each } i = 1, \dots, r \text{ satisfying} \\ \dim(\text{conv}(C_1) + \dots + \text{conv}(C_r)) = n, \text{ and for } 1 \leqslant i \leqslant r \\ \qquad \langle \hat{a}_{il}, \hat{\alpha} \rangle = \langle \hat{a}_{il'}, \hat{\alpha} \rangle \quad \text{for } 0 \leqslant l, l' \leqslant k_i, \\ \qquad \langle \hat{a}, \hat{\alpha} \rangle > \langle \hat{a}_{il}, \hat{\alpha} \rangle \quad \text{for } 0 \leqslant l \leqslant k_i,\ a \in S^{(i)} \backslash C_i. \end{array} \tag{A$'$}$$

For such $\hat{\alpha} = (\alpha, 1) \in \mathbb{R}^{n+1}$ and $\alpha = (\alpha_1, \dots, \alpha_n)$, the coordinate transformation $y = t^{-\alpha} x$ where $y_j = t^{-\alpha_j} x_j$ for $j = 1, \dots, n$ transforms equations in (4.3) to

$$q_{il}(yt^\alpha, t) = \sum_{a \in S^{(i)}} c_{ila} y^a t^{\langle \hat{a}, \hat{\alpha} \rangle}, \quad 1 \leqslant i \leqslant r,\ 1 \leqslant l \leqslant k_i.$$

Let

$$\beta_i = \min_{a \in S^{(i)}} \langle \hat{a}, \hat{\alpha} \rangle, \quad i = 1, \dots, r,$$

and consider the homotopy

$$H^\alpha(y,t) = \bigl(H_1^\alpha(y,t), \dots, H_r^\alpha(y,t)\bigr) = 0 \tag{6.3}$$

on $(C^*)^n \times [0,1]$ where for $1 \leqslant i \leqslant r$ equations in $H_i^\alpha(y,t) = 0$ are

$$\begin{aligned} h_{il}^\alpha(y,t) &= t^{-\beta_i} q_{il}(yt^\alpha, t) \\ &= \sum_{a \in S^{(i)}} c_{ila} y^a t^{\langle \hat{a}, \hat{\alpha} \rangle - \beta_i} \\ &= \sum_{\substack{a \in S^{(i)} \\ \langle \hat{a}, \hat{\alpha} \rangle = \beta_i}} c_{ila} y^a + \sum_{\substack{a \in S^{(i)} \\ \langle \hat{a}, \hat{\alpha} \rangle > \beta_i}} c_{ila} y^a t^{\langle \hat{a}, \hat{\alpha} \rangle - \beta_i} = 0, \quad 1 \leqslant l \leqslant k_i. \end{aligned}$$

When $t=0$, equations in $\bar{H}_i^\alpha(y,0)=0$ become, by condition (A′),

$$h_{il}^\alpha(y,0)=\sum_{a\in C_i=\{a_{i0},\ldots,a_{ik_i}\}} c_{ila}y^a=0,\quad 1\leqslant l\leqslant k_i. \tag{6.4}$$

For each $1\leqslant i\leqslant r$, the above system consists of k_i equations, each one has the same k_i+1 monomials $\{y^{a_{i0}},\ldots,y^{a_{ik_i}}\}$. By applying Gaussian elimination to its $k_i\times(k_i+1)$-coefficient matrix (c_{ila}), we can replace $H_i^\alpha(y,0)=0$ by an equivalent binomial system

$$\begin{aligned} c'_{i11}y^{a_{i1}}+c'_{i10}y^{a_{i0}}&=0,\\ &\vdots\\ c'_{ik_i1}y^{a_{ik_i}}+c'_{ik_i0}y^{a_{i0}}&=0.\end{aligned}$$

If we repeat this process for each $H_i^\alpha(y,0)=0$, $i=1,\ldots,r$, and collect all the resulting binomial equations, a system of $k_1+\cdots+k_n=n$ binomial equations in n variables is attained. This binomial system is equivalent to the start system $H^\alpha(y,0)=0$ of the homotopy $H^\alpha(y,t)=0$ in (5.3), which, as shown in Proposition 4.1, admits $|\det(V^{(\alpha)})|$ nonsingular isolated zeros in $(\mathbb{C}^*)^n$ where

$$V^{(\alpha)}=\begin{pmatrix} a_{11}-a_{10}\\ \vdots\\ a_{1k_1}-a_{10}\\ \vdots\\ a_{r1}-a_{r0}\\ \vdots\\ a_{rk_r}-a_{r0}\end{pmatrix}.$$

Following homotopy paths of $H^\alpha(y,t)=0$ emanating from those isolated zeros, we will reach $|\det(V^{(\alpha)})|$ isolated zeros of $P(x)$.

It was shown in HUBER and STURMFELS [1995], the root count of $P(x)$ in $(\mathbb{C}^*)^n$, or its mixed volume, is equal to

$$\mathcal{M}(\overbrace{S^{(1)},\ldots,S^{(1)}}^{k_1},\ldots,\overbrace{S^{(r)},\ldots,S^{(r)}}^{k_r})=\sum_\alpha\left|\det(V^{(\alpha)})\right|,$$

where the summation is taken over those α where $(\alpha,1)\in\mathbb{R}^{n+1}$ satisfy condition (A′). Therefore, to find all isolated zeros of $P(x)$ we may repeat this process for all such $(\alpha,1)\in\mathbb{R}^{n+1}$ along with their associate cells $C^\alpha=(C_1^\alpha,\ldots,C_r^\alpha)$ where $C_i^\alpha=\{a_{i0},\ldots,a_{ik_i}\}\subset S^{(i)}$ for $i=1,\ldots,r$.

To identify all those α's and their associate cells $C^\alpha=(C_1^\alpha,\ldots,C_r^\alpha)$, one may follow the same route for finding all $(\alpha,1)\in\mathbb{R}^{n+1}$ that satisfy condition (A) in the last section with certain modifications. First of all, the definition of *fine mixed subdivision* of $S=(S^{(1)},\ldots,S^{(r)})$ for $S^{(i)}\subset\mathbb{N}^n$, $i=1,\ldots,r$, can be repeated word for word in Definition 3.1, but replacing $S=(S_1,\ldots,S_n)$ by $S=(S^{(1)},\ldots,S^{(r)})$ and cell $C^{(i)}=(C_1^{(i)},\ldots,C_n^{(i)})$ by $C^{(i)}=(C_1^{(i)},\ldots,C_r^{(i)})$. Similarly, generic lifting $\omega=(\omega_1,\ldots,\omega_r)$

on $S=(S^{(1)},\dots,S^{(r)})$ will induce a fine mixed subdivision S_ω of $S=(S^{(1)},\dots,S^{(r)})$. With $\hat{a}=(a,\omega_i(a))$ for $a\in S^{(i)}$ and $\widehat{C}_i=\{\hat{a}\mid a\in C_i\}$ for $C_i\subset S^{(i)}$, it is easy to see that for $\alpha\in\mathbb{R}^n$ along with its associated cell $C^\alpha=(C_1^\alpha,\dots,C_r^\alpha)$ satisfies condition (A′) if and only if $C^\alpha=(C_1^\alpha,\dots,C_r^\alpha)$ is a cell of type $(k_1,\dots,k_r)$ in the fine mixed subdivision S_ω of $S=(S^{(1)},\dots,S^{(r)})$ induced by $\omega=(\omega_1,\dots,\omega_r)$ where $\hat{\alpha}=(\alpha,1)$ is the inner normal of $\widehat{C}^\alpha=(\widehat{C}_1^\alpha,\dots,\widehat{C}_r^\alpha)$ in $\widehat{S}=(\widehat{S}^{(1)},\dots,\widehat{S}^{(r)})$. The procedure we described in the last section to identify mixed cells, cells of type $(1,\dots,1)$, in S_ω is no longer effective for identifying cells of type $(k_1,\dots,k_r)$ in S_ω, because a straightforward generalization of the Two-Point test in the algorithm to k_i-Point test would enormously increase the number of candidates that need to be tested. In the next few subsections, we shall present a revised procedure given in GAO and LI [2003] for finding cells of type $(k_1,\dots,k_r)$ in which the Two-Point test is replaced by successive One-Point tests. It is therefore applicable in general to test k_i points consecutively.

We again assume that $S^{(i)}$ admits only extreme points for each $i=1,\dots,r$. Meanwhile, when LP problems arise, for the sake of simplicity, we will not deal with the details of the possible degeneracy of the constraints as in the last section. We therefore assume the matrix that represents the set of constraints of any LP problem is always of full rank.

6.2. *The relation table*

For generic $\omega_i: S^{(i)}\to\mathbb{R}$ and $\widehat{S}^{(i)}=\{\hat{\mathbf{a}}=(\mathbf{a},\omega_i(\mathbf{a}))\mid \mathbf{a}\in S^{(i)}\}$ for each $i=1,\dots,r$, an important primary step of our algorithm for finding cells of type $(k_1,\dots,k_r)$ in S_ω is to complete the *relation table* consisting of pairwise relation subtables $\mathrm{T}(i,j)$ between $\widehat{S}^{(i)}$ and $\widehat{S}^{(j)}$ for all $1\leqslant i\leqslant j\leqslant r$ as shown in Table 6.1. For $\widehat{S}^{(i)}=\{\mathbf{a}_1^{(i)},\dots,\mathbf{a}_{s_i}^{(i)}\}$, $i=1,\dots,r$, Table $\mathrm{T}(i,j)$ displays the relationships between elements of $\widehat{S}^{(i)}$ and $\widehat{S}^{(j)}$ in the following sense:

> Given elements $\hat{\mathbf{a}}_l^{(i)}\in\widehat{S}^{(i)}$ and $\hat{\mathbf{a}}_m^{(j)}\in\widehat{S}^{(j)}$ where $l\neq m$ when $i=j$, does there exist an $\hat{\alpha}=(\alpha,1)\in\mathbb{R}^{n+1}$ such that
>
> $$\begin{aligned}&\langle\hat{\mathbf{a}}_l^{(i)},\hat{\alpha}\rangle\leqslant\langle\hat{\mathbf{a}}_k^{(i)},\hat{\alpha}\rangle,\quad \forall\hat{\mathbf{a}}_k^{(i)}\in\widehat{S}^{(i)},\\ &\langle\hat{\mathbf{a}}_m^{(j)},\hat{\alpha}\rangle\leqslant\langle\hat{\mathbf{a}}_k^{(j)},\hat{\alpha}\rangle,\quad \forall\hat{\mathbf{a}}_k^{(j)}\in\widehat{S}^{(j)}?\end{aligned}\tag{6.5}$$

TABLE $\mathrm{T}(i,i)$

$\widehat{S}^{(i)}$	$\hat{\mathbf{a}}_2^{(i)}$	$\hat{\mathbf{a}}_3^{(i)}$	$\cdots$	$\hat{\mathbf{a}}_{s_i-1}^{(i)}$	$\hat{\mathbf{a}}_{s_i}^{(i)}$
$\hat{\mathbf{a}}_1^{(i)}$	$[\hat{\mathbf{a}}_1^{(i)},\hat{\mathbf{a}}_2^{(i)}]$	$[\hat{\mathbf{a}}_1^{(i)},\hat{\mathbf{a}}_3^{(i)}]$	$\cdots$	$[\hat{\mathbf{a}}_1^{(i)},\hat{\mathbf{a}}_{s_i-1}^{(i)}]$	$[\hat{\mathbf{a}}_1^{(i)},\hat{\mathbf{a}}_{s_i}^{(i)}]$
	$\hat{\mathbf{a}}_2^{(i)}$	$[\hat{\mathbf{a}}_2^{(i)},\hat{\mathbf{a}}_3^{(i)}]$	$\cdots$	$[\hat{\mathbf{a}}_2^{(i)},\hat{\mathbf{a}}_{s_i-1}^{(i)}]$	$[\hat{\mathbf{a}}_2^{(i)},\hat{\mathbf{a}}_{s_i}^{(i)}]$
			$\ddots$	$\vdots$	$\vdots$
				$\hat{\mathbf{a}}_{s_i-1}^{(i)}$	$[\hat{\mathbf{a}}_{s_i-1}^{(i)},\hat{\mathbf{a}}_{s_i}^{(i)}]$
					$\hat{\mathbf{a}}_{s_i}^{(i)}$

TABLE T(i, j)

		$\widehat{S}^{(j)}$				
		$\hat{\mathbf{a}}_1^{(j)}$	$\hat{\mathbf{a}}_2^{(j)}$	$\hat{\mathbf{a}}_3^{(j)}$	$\cdots$	$\hat{\mathbf{a}}_{s_j}^{(j)}$
$\widehat{S}^{(i)}$	$\hat{\mathbf{a}}_1^{(i)}$	$[\hat{\mathbf{a}}_1^{(i)}, \hat{\mathbf{a}}_1^{(j)}]$	$[\hat{\mathbf{a}}_1^{(i)}, \hat{\mathbf{a}}_2^{(j)}]$	$[\hat{\mathbf{a}}_1^{(i)}, \hat{\mathbf{a}}_3^{(j)}]$	$\cdots$	$[\hat{\mathbf{a}}_1^{(i)}, \hat{\mathbf{a}}_{s_j}^{(j)}]$
	$\hat{\mathbf{a}}_2^{(i)}$	$[\hat{\mathbf{a}}_2^{(i)}, \hat{\mathbf{a}}_1^{(j)}]$	$[\hat{\mathbf{a}}_2^{(i)}, \hat{\mathbf{a}}_2^{(j)}]$	$[\hat{\mathbf{a}}_2^{(i)}, \hat{\mathbf{a}}_3^{(j)}]$	$\cdots$	$[\hat{\mathbf{a}}_2^{(i)}, \hat{\mathbf{a}}_{s_j}^{(j)}]$
	$\vdots$	$\vdots$	$\vdots$	$\vdots$	$\cdots$	$\vdots$
	$\hat{\mathbf{a}}_{s_i}^{(i)}$	$[\hat{\mathbf{a}}_{s_i}^{(i)}, \hat{\mathbf{a}}_1^{(j)}]$	$[\hat{\mathbf{a}}_{s_i}^{(i)}, \hat{\mathbf{a}}_2^{(j)}]$	$[\hat{\mathbf{a}}_{s_i}^{(i)}, \hat{\mathbf{a}}_3^{(j)}]$	$\cdots$	$[\hat{\mathbf{a}}_{s_i}^{(i)}, \hat{\mathbf{a}}_{s_j}^{(j)}]$

TABLE 6.1

The Relation Table.

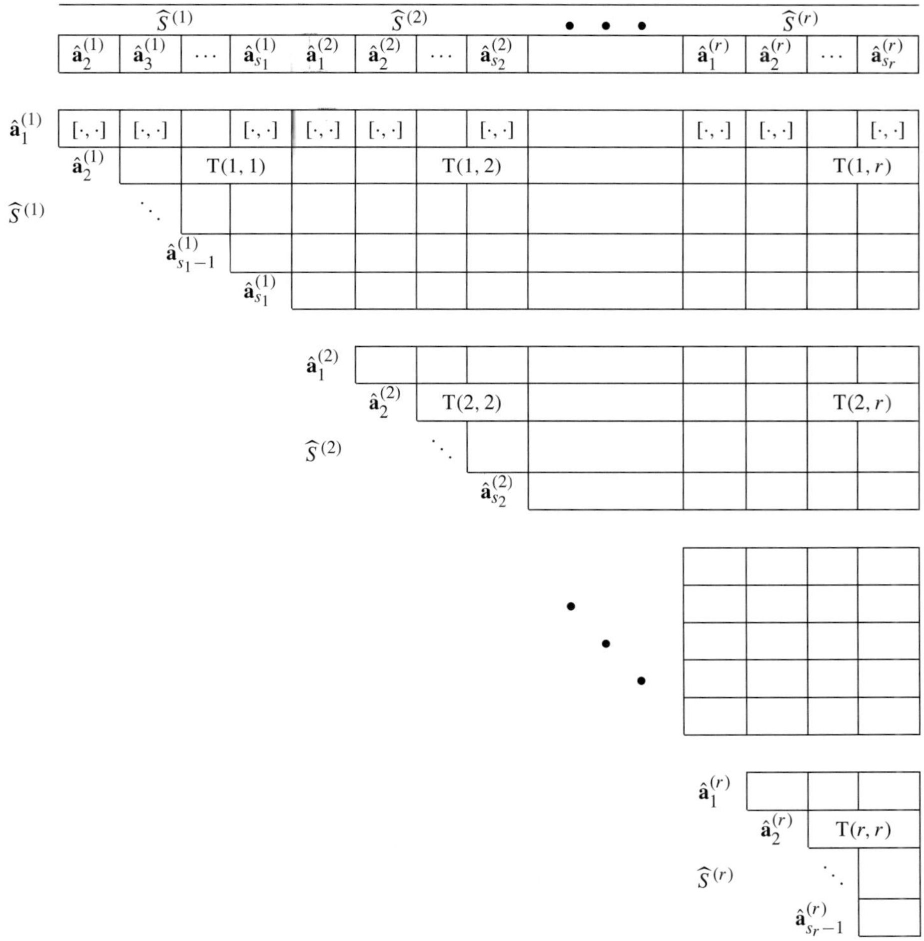

Denote the entry on Table T(i, j) located at the intersection of the row containing $\hat{\mathbf{a}}_l^{(i)}$ and the column containing $\hat{\mathbf{a}}_m^{(j)}$ by $[\hat{\mathbf{a}}_l^{(i)}, \hat{\mathbf{a}}_m^{(j)}]$. We set $[\hat{\mathbf{a}}_l^{(i)}, \hat{\mathbf{a}}_m^{(j)}] = 1$ when the answer of problem (6.5) for $\hat{\mathbf{a}}_l^{(i)}$ and $\hat{\mathbf{a}}_m^{(j)}$ is positive and $[\hat{\mathbf{a}}_l^{(i)}, \hat{\mathbf{a}}_m^{(j)}] = 0$ otherwise. When $i = j$, $[\hat{\mathbf{a}}_l^{(i)}, \hat{\mathbf{a}}_m^{(i)}] = [\hat{\mathbf{a}}_m^{(i)}, \hat{\mathbf{a}}_l^{(i)}]$ for $l \neq m$, therefore we always assume $l < m$ in such cases.

To fill out the relation table, Table 6.1, we first *fix* $\hat{\mathbf{a}}_1^{(1)}$ on the first row

$$\hat{\mathbf{a}}_1^{(1)}:\ \overbrace{[\hat{\mathbf{a}}_1^{(1)}, \hat{\mathbf{a}}_2^{(1)}], \ldots, [\hat{\mathbf{a}}_1^{(1)}, \hat{\mathbf{a}}_{s_1}^{(1)}]}^{\mathrm{T}(1,1)}, \ldots, \overbrace{[\hat{\mathbf{a}}_1^{(1)}, \hat{\mathbf{a}}_1^{(r)}], \ldots, [\hat{\mathbf{a}}_1^{(1)}, \hat{\mathbf{a}}_{s_r}^{(r)}]}^{\mathrm{T}(1,r)}. \tag{6.6}$$

To determine $[\hat{\mathbf{a}}_1^{(1)}, \hat{\mathbf{a}}_2^{(1)}]$, we consider the LP problem,

$$\begin{aligned}
&\text{Minimize } \langle \hat{\mathbf{a}}_2^{(1)}, \hat{\alpha} \rangle - \alpha_0 \\
&\alpha_0 = \langle \hat{\mathbf{a}}_1^{(1)}, \hat{\alpha} \rangle, \\
&\alpha_0 \leqslant \langle \hat{\mathbf{a}}_k^{(1)}, \hat{\alpha} \rangle, \quad \forall k \in \{2, \ldots, s_1\},
\end{aligned} \tag{6.7}$$

in the variables $\hat{\alpha} = (\alpha, 1) \in \mathbb{R}^{n+1}$ and $\alpha_0 \in \mathbb{R}$. More explicitly, we may write this problem in the form (5.1) in Section 5.1:

$$\begin{aligned}
&\text{Minimize } \langle \mathbf{a}_2^{(1)} - \mathbf{a}_1^{(1)}, \alpha \rangle + \omega_1(\mathbf{a}_2^{(1)}) - \omega_1(\mathbf{a}_1^{(1)}) \\
&\langle \mathbf{a}_1^{(1)} - \mathbf{a}_k^{(1)}, \alpha \rangle \leqslant \omega_1(\mathbf{a}_k^{(1)}) - \omega_1(\mathbf{a}_1^{(1)}), \quad \forall k \in \{2, \ldots, s_1\}.
\end{aligned} \tag{6.8}$$

Since $\mathbf{a}_1^{(1)}$ is a vertex point of $Q_1 = \operatorname{conv}(S^{(1)})$, $\hat{\mathbf{a}}_1^{(1)}$ must be in the lower hull of $\widehat{Q}_1 = \operatorname{conv}(\widehat{S}^{(1)})$, and any hyperplane in the form $\hat{\alpha} = (\alpha, 1) \in \mathbb{R}^{n+1}$ that supports $\hat{\mathbf{a}}_1^{(1)}$ in $\widehat{Q}_1$ decides a feasible point of the constraints in (6.7). Such feasible point can be obtained by solving a standard Phase I problem for the constraints in (6.8):

$$\begin{aligned}
&\text{Minimize } \varepsilon \\
&\langle \mathbf{a}_1^{(1)} - \mathbf{a}_k^{(1)}, \alpha \rangle - \varepsilon \leqslant \omega_1(\mathbf{a}_k^{(1)}) - \omega_1(\mathbf{a}_1^{(1)}), \quad \forall k \in \{2, \ldots, s_1\}, \\
&-\varepsilon \leqslant 0
\end{aligned}$$

in the variables $\alpha \in \mathbb{R}^n$ and $\varepsilon \geqslant 0$ with feasible point $\alpha = 0$ along with large enough $\varepsilon > 0$.

If the optimal value of the LP problem (6.7) is zero, then at the optimal solution $(\hat{\alpha}, \alpha_0)$ we have

$$\langle \hat{\mathbf{a}}_1^{(1)}, \hat{\alpha} \rangle = \langle \hat{\mathbf{a}}_2^{(1)}, \hat{\alpha} \rangle \leqslant \langle \hat{\mathbf{a}}_k^{(1)}, \hat{\alpha} \rangle, \quad \forall k \in \{3, \ldots, s_1\}. \tag{6.9}$$

This makes $[\hat{\mathbf{a}}_1^{(1)}, \hat{\mathbf{a}}_2^{(1)}] = 1$. Otherwise, $[\hat{\mathbf{a}}_1^{(1)}, \hat{\mathbf{a}}_2^{(1)}]$ must be zero, for if there exists $\hat{\alpha}' = (\alpha, 1) \in \mathbb{R}^{n+1}$ for which the inequalities in (6.9) hold, then this $\hat{\alpha}'$ together with $\alpha_0' = \langle \hat{\mathbf{a}}_1^{(1)}, \hat{\alpha}' \rangle$ yields a feasible point of (6.7) at which the objective function value is zero.

This process essentially uses the *One-Point Test* introduced in the last section to test if $\{\hat{\mathbf{a}}_1^{(1)}, \hat{\mathbf{a}}_2^{(1)}\}$ is a lower edge of $\widehat{S}^{(1)}$. When the simplex method is used, the pivoting process in the algorithm generates rich information on other entries of the relation table. Since the images of $\omega_1 : S^{(1)} \to \mathbb{R}$ is generically chosen, we assume without loss that there are exactly $n+1$ active constraints at any stage of the pivoting process, say,

$$\begin{aligned}
&\alpha_0 = \langle \hat{\mathbf{a}}_1^{(1)}, \hat{\alpha} \rangle,\\
&\alpha_0 = \langle \hat{\mathbf{a}}_{l_1}^{(1)}, \hat{\alpha} \rangle,\\
&\quad \vdots\\
&\alpha_0 = \langle \hat{\mathbf{a}}_{l_n}^{(1)}, \hat{\alpha} \rangle,\\
&\alpha_0 < \langle \hat{\mathbf{a}}_k^{(1)}, \hat{\alpha} \rangle, \quad \forall k \in \{2, 3, \ldots, s_1\} \backslash \{l_1, \ldots, l_n\},
\end{aligned}$$

then $[\hat{\mathbf{a}}_{j_1}^{(1)}, \hat{\mathbf{a}}_{j_2}^{(1)}] = 1$ for all $j_1, j_2 \in \{1, l_1, \ldots, l_n\}$ with $j_1 < j_2$. This important feature considerably reduces the number of One-Point tests needed for completely determining the entries of the relation table.

To determine the rest of the unknown entries in the first row of the table in (6.6) from left to right: for $[\hat{\mathbf{a}}_1^{(1)}, \hat{\mathbf{a}}_j^{(1)}]$ for $j > 2$, we apply the One-Point test on $\hat{\mathbf{a}}_j^{(1)}$, or solve the LP problem,

$$\begin{aligned}
&\text{Minimize } \langle \hat{\mathbf{a}}_j^{(1)}, \hat{\alpha} \rangle - \alpha_0\\
&\alpha_0 = \langle \hat{\mathbf{a}}_1^{(1)}, \hat{\alpha} \rangle,\\
&\alpha_0 \leqslant \langle \hat{\mathbf{a}}_l^{(1)}, \hat{\alpha} \rangle, \quad \forall l \in \{2, \ldots, s_1\},
\end{aligned} \tag{6.10}$$

and for $[\hat{\mathbf{a}}_1^{(1)}, \hat{\mathbf{a}}_j^{(i)}]$ for $i > 1$, $j \in \{1, \ldots, s_i\}$, solve the LP problem

$$\begin{aligned}
&\text{Minimize } \langle \hat{\mathbf{a}}_j^{(i)}, \hat{\alpha} \rangle - \alpha_0\\
&\langle \hat{\mathbf{a}}_1^{(1)}, \hat{\alpha} \rangle \leqslant \langle \hat{\mathbf{a}}_l^{(1)}, \hat{\alpha} \rangle, \quad \forall l \in \{2, \ldots, s_1\}\\
&\alpha_0 \leqslant \langle \hat{\mathbf{a}}_m^{(i)}, \hat{\alpha} \rangle, \qquad \forall m \in \{1, 2, \ldots, s_i\}.
\end{aligned} \tag{6.11}$$

If the corresponding optimal values are zero, then $[\hat{\mathbf{a}}_1^{(1)}, \hat{\mathbf{a}}_j^{(1)}] = 1$, or $[\hat{\mathbf{a}}_1^{(1)}, \hat{\mathbf{a}}_j^{(i)}] = 1$. They are zero otherwise.

Feasible points of the above LP problems are always available, there is no need to solve the sometimes costly Phase I problem here. Because when we determine $[\hat{\mathbf{a}}_1^{(1)}, \hat{\mathbf{a}}_2^{(1)}]$, there exists $\hat{\alpha} = (\alpha, 1)$ for which

$$\langle \hat{\mathbf{a}}_1^{(1)}, \hat{\alpha} \rangle \leqslant \langle \hat{\mathbf{a}}_l^{(1)}, \hat{\alpha} \rangle, \quad \forall l \in \{2, \ldots, s_1\}.$$

This $\hat{\alpha}$ together with $\alpha_0 = \langle \hat{\mathbf{a}}_1^{(1)}, \hat{\alpha} \rangle$ for (6.10) or

$$\alpha_0 = \min\{ \langle \hat{\mathbf{a}}_m^{(1)}, \hat{\alpha} \rangle \mid m = 1, \ldots, s_i \}$$

for (6.11) provides feasible points for the constraints of the respective LP problems.

An important remark here is, a substantial number of constraints in both (6.10) and (6.11) can be removed. For instance, if we have known $[\hat{\mathbf{a}}_1^{(1)}, \hat{\mathbf{a}}_\mu^{(i)}] = 0$ for certain $\mu \in \{1, \ldots, s_i\}$, $i \geqslant 1$, before solving the LP problems in (6.10) or (6.11), then its corresponding constraint

$$\alpha_0 \leqslant \langle \hat{\mathbf{a}}_\mu^{(i)}, \hat{\alpha} \rangle$$

should be removed, because this constraint will never become active (otherwise, $[\hat{\mathbf{a}}_1^{(1)}, \hat{\mathbf{a}}_\mu^{(i)}] = 1$) during the process of solving the LP problems. In fact, numerous extraneous constraints of this sort appear in all the LP problems below. The successive omission of those extraneous constraints yields a considerable reduction in the amount of computation and plays a crucially important role in the efficiency of the algorithm. We will not elaborate the details of the omission here, they can be found in GAO and LI [2003].

Similarly, when we determine the entries of the row

$$\hat{\mathbf{a}}_\mu^{(\nu)}: \overbrace{[\hat{\mathbf{a}}_\mu^{(\nu)}, \hat{\mathbf{a}}_{\mu+1}^{(\nu)}], \ldots, [\hat{\mathbf{a}}_\mu^{(\nu)}, \hat{\mathbf{a}}_{s_\nu}^{(\nu)}]}^{\mathrm{T}(\nu,\nu)}, \ldots, \overbrace{[\hat{\mathbf{a}}_\mu^{(\nu)}, \hat{\mathbf{a}}_1^{(r)}], \ldots, [\hat{\mathbf{a}}_\mu^{(\nu)}, \hat{\mathbf{a}}_{s_r}^{(r)}]}^{\mathrm{T}(\nu,r)} \tag{6.12}$$

on the relation table assuming all the entries in the previous rows have all been determined, for the unknown entries $[\hat{\mathbf{a}}_\mu^{(\nu)}, \hat{\mathbf{a}}_j^{(\nu)}]$ for $j > \mu$, we solve the LP problem,

$$\begin{aligned}
&\text{Minimize } \langle \hat{\mathbf{a}}_j^{(\nu)}, \hat{\alpha} \rangle - \alpha_0 \\
&\alpha_0 = \langle \hat{\mathbf{a}}_\mu^{(\nu)}, \hat{\alpha} \rangle, \\
&\alpha_0 \leqslant \langle \hat{\mathbf{a}}_l^{(\nu)}, \hat{\alpha} \rangle, \quad \forall l \in \{1, \ldots, s_\nu\} \backslash \{\mu\},
\end{aligned} \tag{6.13}$$

and for $[\hat{\mathbf{a}}_\mu^{(\nu)}, \hat{\mathbf{a}}_j^{(i)}]$ for $j \in \{1, \ldots, s_i\}$ and $\nu < i \leqslant r$, solve the LP problem

$$\begin{aligned}
&\text{Minimize } \langle \hat{\mathbf{a}}_j^{(i)}, \hat{\alpha} \rangle - \alpha_0 \\
&\langle \hat{\mathbf{a}}_\mu^{(\nu)}, \hat{\alpha} \rangle \leqslant \langle \hat{\mathbf{a}}_l^{(\nu)}, \hat{\alpha} \rangle, \quad \forall l \in \{1, \ldots, s_1\} \backslash \{\mu\}, \\
&\alpha_0 \leqslant \langle \hat{\mathbf{a}}_m^{(i)}, \hat{\alpha} \rangle, \quad \forall m \in \{1, \ldots, s_i\}.
\end{aligned} \tag{6.14}$$

When the LP problem in (6.14) is solved by the simplex method, information on other unknown entries of the table provided by the pivoting process becomes particularly fruitful. We assume without loss that there are exactly $n + 1$ active constraints at any stage of the pivoting process, say

$$\begin{aligned}
\langle \hat{\mathbf{a}}_\mu^{(\nu)}, \hat{\alpha} \rangle &= \langle \hat{\mathbf{a}}_{l_1}^{(\nu)}, \hat{\alpha} \rangle, \\
&\vdots \\
\langle \hat{\mathbf{a}}_\mu^{(\nu)}, \hat{\alpha} \rangle &= \langle \hat{\mathbf{a}}_{l_s}^{(\nu)}, \hat{\alpha} \rangle, \quad \text{and} \\
\alpha_0 &= \langle \hat{\mathbf{a}}_{m_1}^{(i)}, \hat{\alpha} \rangle, \\
&\vdots \\
\alpha_0 &= \langle \hat{\mathbf{a}}_{m_t}^{(i)}, \hat{\alpha} \rangle,
\end{aligned}$$

where $s + t = n + 1$. Then for $l', l'' \in \{l_1, \ldots, l_s\}$ with $l' < l''$ and $m', m'' \in \{m_1, \ldots, m_t\}$ with $m' < m''$, we have

$$[\hat{\mathbf{a}}_{l'}^{(\nu)}, \hat{\mathbf{a}}_{l''}^{(\nu)}] = 1, \qquad [\hat{\mathbf{a}}_{m'}^{(i)}, \hat{\mathbf{a}}_{m''}^{(i)}] = 1.$$

And, for $l_0 \in \{l_1, \ldots, l_s\}$ and $m_0 \in \{m_1, \ldots, m_t\}$, $[\hat{\mathbf{a}}_{l_0}^{(\nu)}, \hat{\mathbf{a}}_{m_0}^{(i)}] = 1$.

6.3. Level-ξ subfaces and their extensions

For $1 \leqslant \xi \leqslant r$ and $\widehat{F}_i \subset \widehat{S}^{(i)}$ with $\dim(\widehat{F}_i) = d_i$ for $i = 1, \ldots, \xi$, $(\widehat{F}_1, \ldots, \widehat{F}_\xi)$ is called a *level-ξ subface* of $\widehat{S} = (\widehat{S}^{(1)}, \ldots, \widehat{S}^{(r)})$ of type $(d_1, \ldots, d_\xi)$ if there exists $\hat{\alpha} = (\alpha, 1) \in \mathbb{R}^{n+1}$ such that for each $i = 1, \ldots, \xi$,

$$\langle \hat{\mathbf{a}}^{(i)}, \hat{\alpha} \rangle = \langle \hat{\mathbf{a}}^{(i)\prime}, \hat{\alpha} \rangle \quad \forall \hat{\mathbf{a}}^{(i)}, \hat{\mathbf{a}}^{(i)\prime} \in \widehat{F}_i,$$
$$\langle \hat{\mathbf{a}}^{(i)}, \hat{\alpha} \rangle \leqslant \langle \hat{\mathbf{a}}^{(i)\prime\prime}, \hat{\alpha} \rangle \quad \forall \hat{\mathbf{a}}^{(i)} \in \widehat{F}_i \text{ and } \hat{\mathbf{a}}^{(i)\prime\prime} \in \widehat{S}^{(i)} \backslash \widehat{F}_i.$$

Equivalently, $\widehat{F}_i$ is a lower face of $\widehat{S}^{(i)}$ of dimension d_i for each $i = 1, \ldots, \xi$. A level-ξ subface $(\widehat{F}_1, \ldots, \widehat{F}_\xi)$ is said to be *extendible* if there is a lower face $\widehat{F}_{\xi+1}$ of $\widehat{S}^{(\xi+1)}$ which makes $(\widehat{F}_1, \ldots, \widehat{F}_\xi, \widehat{F}_{\xi+1})$ a level-$(\xi + 1)$ subface. It is *nonextendible* otherwise.

A level-r subface $(\widehat{F}_1, \ldots, \widehat{F}_r)$ of $\widehat{S} = (\widehat{S}^{(1)}, \ldots, \widehat{S}^{(r)})$ of type $(k_1, \ldots, k_r)$ is a lower facet of $\widehat{S}$ of type $(k_1, \ldots, k_r)$ when

$$\begin{aligned}\dim(\widehat{F}_1 + \cdots + \widehat{F}_r) &= \dim(\widehat{F}_1) + \cdots + \dim(\widehat{F}_r)\\ &= k_1 + \cdots + k_r = n.\end{aligned}$$

In such case, $(F_1, \ldots, F_r)$ becomes a cell of type $(k_1, \ldots, k_r)$ in S_ω. To find all such lower facets of $\widehat{S}$ of type $(k_1, \ldots, k_r)$ for the purpose of finding all cells of type $(k_1, \ldots, k_r)$ in S_ω, we first find all level-1 subfaces of $\widehat{S} = (\widehat{S}^{(1)}, \ldots, \widehat{S}^{(r)})$ of type (k_1), followed by extending each such subface step by step from $\xi = 1$ to $\xi = r$ to reach a level-r subface of $\widehat{S}$ of type $(k_1, \ldots, k_r)$.

6.3.1. Level-1 subfaces of $\widehat{S} = (\widehat{S}^{(1)}, \ldots, \widehat{S}^{(r)})$

Clearly, level-1 subfaces of $\widehat{S} = (\widehat{S}^{(1)}, \ldots, \widehat{S}^{(r)})$ of type (k_1) are faces of dimension k_1 in the lower hull of $\widehat{S}^{(1)} = \{\hat{\mathbf{a}}_1^{(1)}, \ldots, \hat{\mathbf{a}}_{s_1}^{(1)}\}$, they are faces of dimension k_1 of $\widehat{S}^{(1)}$ having inner normal of type $\hat{\alpha} = (\alpha, 1) \in \mathbb{R}^{n+1}$. When $k_1 = 1$, such subfaces are the pairs of points $\{\mathbf{a}_{l_0}^{(1)}, \mathbf{a}_{l_1}^{(1)}\}$ on the Relation Table T(1, 1) with $[\mathbf{a}_{l_0}^{(1)}, \mathbf{a}_{l_1}^{(1)}] = 1$, $1 \leqslant l_0 < l_1 \leqslant s_1$. So only the case $k_1 > 1$ will be discussed here, and we will find all those faces by extending each lower face of $\widehat{S}^{(1)}$ of dimension one, a lower *edge* of $\widehat{S}^{(1)}$, step by step. More precisely, for lower edge $\{\hat{\mathbf{a}}_{l_0}^{(1)}, \hat{\mathbf{a}}_{l_1}^{(1)}\}$ of $\widehat{S}^{(1)}$ with $l_0 < l_1$, we look for all possible points $\hat{\mathbf{a}}_l^{(1)}$ in $\widehat{S}^{(1)}$ with $l > l_1$ for which $\{\hat{\mathbf{a}}_{l_0}^{(1)}, \hat{\mathbf{a}}_{l_1}^{(1)}, \hat{\mathbf{a}}_l^{(1)}\}$ is a lower face of $\widehat{S}^{(1)}$ of dimension two. And inductively, for known face $\{\hat{\mathbf{a}}_{l_0}^{(1)}, \hat{\mathbf{a}}_{l_1}^{(1)}, \ldots, \hat{\mathbf{a}}_{l_j}^{(1)}\}$ of $\widehat{S}^{(1)}$ of dimension j with $j < k_1$ and $l_0 < l_1 < \cdots < l_j$, we look for all possible points $\hat{\mathbf{a}}_l^{(1)}$ in $\widehat{S}^{(1)}$ with $l > l_j$ for which $\{\hat{\mathbf{a}}_{l_0}^{(1)}, \hat{\mathbf{a}}_{l_1}^{(1)}, \ldots, \hat{\mathbf{a}}_{l_j}^{(1)}, \hat{\mathbf{a}}_l^{(1)}\}$ is a lower face of $\widehat{S}^{(1)}$ of dimension $j + 1$. Lower face $\{\hat{\mathbf{a}}_{l_0}^{(1)}, \hat{\mathbf{a}}_{l_1}^{(1)}, \ldots, \hat{\mathbf{a}}_{l_j}^{(1)}\}$ is called *extendible* if such point exists. This task of extension can be carried out systematically by employing the One-Point test successively.

We will extend lower edges of $\widehat{S}^{(1)}$ one by one in the order from left to right and top to bottom of their corresponding entries on the Relation Table T(1, 1).

For $[\hat{\mathbf{a}}_{l_0}^{(1)}, \hat{\mathbf{a}}_{l_1}^{(1)}] = 1$ where $1 \leqslant l_0 < l_1 < s_1$, we first identify on Table T(1, 1) the set

$$\mathcal{C}^{(1)} = \{1 \leqslant l \leqslant s_1 \mid \hat{\mathbf{a}}_l^{(1)} \text{ has positive relations with both } \hat{\mathbf{a}}_{l_0}^{(1)} \text{ and } \hat{\mathbf{a}}_{l_1}^{(1)}\},$$

TABLE T(1, 1)

	$\widehat{S}^{(1)}$				
	$\hat{\mathbf{a}}_2^{(1)}$	$\hat{\mathbf{a}}_3^{(1)}$	$\cdots$	$\hat{\mathbf{a}}_{s_1-1}^{(1)}$	$\hat{\mathbf{a}}_{s_1}^{(1)}$
$\hat{\mathbf{a}}_1^{(1)}$	$[\hat{\mathbf{a}}_1^{(1)}, \hat{\mathbf{a}}_2^{(1)}]$	$[\hat{\mathbf{a}}_1^{(1)}, \hat{\mathbf{a}}_3^{(1)}]$	$\cdots$	$[\hat{\mathbf{a}}_1^{(1)}, \hat{\mathbf{a}}_{s_1-1}^{(1)}]$	$[\hat{\mathbf{a}}_1^{(1)}, \hat{\mathbf{a}}_{s_1}^{(1)}]$
	$\hat{\mathbf{a}}_2^{(1)}$	$[\hat{\mathbf{a}}_2^{(1)}, \hat{\mathbf{a}}_3^{(1)}]$	$\cdots$	$[\hat{\mathbf{a}}_2^{(1)}, \hat{\mathbf{a}}_{s_1-1}^{(1)}]$	$[\hat{\mathbf{a}}_2^{(1)}, \hat{\mathbf{a}}_{s_1}^{(1)}]$
$\widehat{S}^{(1)}$		$\hat{\mathbf{a}}_3^{(1)}$	$\cdots$	$[\hat{\mathbf{a}}_3^{(1)}, \hat{\mathbf{a}}_{s_1-1}^{(1)}]$	$[\hat{\mathbf{a}}_3^{(1)}, \hat{\mathbf{a}}_{s_1}^{(1)}]$
			$\ddots$	$\vdots$	$\vdots$
				$\hat{\mathbf{a}}_{s_1-1}^{(1)}$	$[\hat{\mathbf{a}}_{s_1-1}^{(1)}, \hat{\mathbf{a}}_{s_1}^{(1)}]$

and let $\mathcal{T}^{(1)}$ be the elements in $\mathcal{C}^{(1)}$ which are bigger than l_1, i.e.

$$\mathcal{T}^{(1)} = \{l \in \mathcal{C}^{(1)},\ l > l_1 \mid [\hat{\mathbf{a}}_{l_0}^{(1)}, \hat{\mathbf{a}}_l^{(1)}] = [\hat{\mathbf{a}}_{l_1}^{(1)}, \hat{\mathbf{a}}_l^{(1)}] = 1\}.$$

Clearly, the set

$$\mathcal{P}^{(1)} = \{\hat{\mathbf{a}}_l^{(1)} \mid l \in \mathcal{T}^{(1)}\}$$

contains all the possible points which may subsequently extend $\{\hat{\mathbf{a}}_{l_0}^{(1)}, \hat{\mathbf{a}}_{l_1}^{(1)}\}$ to a k_1-dimensional lower face $\{\hat{\mathbf{a}}_{l_0}^{(1)}, \ldots, \hat{\mathbf{a}}_{l_{k_1}}^{(1)}\}$ of $\widehat{S}^{(1)}$ with $l_0 < l_1 < \cdots < l_{k_1}$. Hence all subsequential extension attempts on $\{\hat{\mathbf{a}}_{l_0}^{(1)}, \hat{\mathbf{a}}_{l_1}^{(1)}\}$ will be restricted to this set. To extend $\{\hat{\mathbf{a}}_{l_0}^{(1)}, \hat{\mathbf{a}}_{l_1}^{(1)}\}$ we apply the One-Point test on points $\hat{\mathbf{a}}_\tau^{(1)}$ in $\mathcal{P}^{(1)}$ by solving the LP problem

$$\begin{aligned}
&\text{Minimize } \langle \hat{\mathbf{a}}_\tau^{(1)}, \hat{\alpha}\rangle - \alpha_0 \\
&\alpha_0 = \langle \hat{\mathbf{a}}_{l_0}^{(1)}, \hat{\alpha}\rangle = \langle \hat{\mathbf{a}}_{l_1}^{(1)}, \hat{\alpha}\rangle, \\
&\alpha_0 \leqslant \langle \hat{\mathbf{a}}_k^{(1)}, \hat{\alpha}\rangle, \quad \forall k \in \mathcal{C}^{(1)},
\end{aligned} \tag{6.15}$$

in the variables $\hat{\alpha} = (\alpha, 1) \in \mathbb{R}^{n+1}$ and $\alpha_0 \in \mathbb{R}$.

Since $[\hat{\mathbf{a}}_{l_0}^{(1)}, \hat{\mathbf{a}}_{l_1}^{(1)}] = 1$ implies the existence of $\hat{\alpha} \in \mathbb{R}^{n+1}$ for which

$$\begin{aligned}
&\langle \hat{\mathbf{a}}_{l_0}^{(1)}, \hat{\alpha}\rangle = \langle \hat{\mathbf{a}}_{l_1}^{(1)}, \hat{\alpha}\rangle, \\
&\langle \hat{\mathbf{a}}_{l_0}^{(1)}, \hat{\alpha}\rangle \leqslant \langle \hat{\mathbf{a}}_k^{(1)}, \hat{\alpha}\rangle, \quad \forall k \in \{1, \ldots, s_1\} \backslash \{l_0, l_1\},
\end{aligned}$$

this $\hat{\alpha}$ along with

$$\alpha_0 = \min_{k \in \mathcal{C}^{(1)}} \langle \hat{\mathbf{a}}_k^{(1)}, \hat{\alpha}\rangle$$

yields a feasible point of the constraints in (6.15). Clearly, zero optimal value of this LP problem makes $\{\hat{\mathbf{a}}_{l_0}^{(1)}, \hat{\mathbf{a}}_{l_1}^{(1)}, \hat{\mathbf{a}}_\tau^{(1)}\}$ a lower face of $\widehat{S}^{(1)}$ of dimension two, and the point $\hat{\mathbf{a}}_\tau^{(1)}$ will be retained for further extension considerations. It will be deleted otherwise.

Again, the pivoting process in solving the LP problem in (6.15) by the simplex method provides abundant information on the extendibility of $\{\hat{\mathbf{a}}_{l_0}^{(1)}, \hat{\mathbf{a}}_{l_1}^{(1)}\}$ by other points in $\mathcal{P}^{(1)}$. For instance, at any stage of the pivoting process, when the set of active constraints contains

$$\alpha_0 = \langle \hat{\mathbf{a}}_l^{(1)}, \hat{\alpha} \rangle$$

for any $l \in \mathcal{T}^{(1)} \backslash \{\tau\}$, then $\hat{\mathbf{a}}_l^{(1)}$ extends $\{\hat{\mathbf{a}}_{l_0}^{(1)}, \hat{\mathbf{a}}_{l_1}^{(1)}\}$ and can be omitted from the list of further testings.

When the testing on points of $\mathcal{P}^{(1)}$ is completed, we have extended $\{\hat{\mathbf{a}}_{l_0}^{(1)}, \hat{\mathbf{a}}_{l_1}^{(1)}\}$ to all possible 2-dimensional lower faces. This process may be repeated along the same line as we extend a j-dimensional lower face $\{\hat{\mathbf{a}}_{l_0}^{(1)}, \dots, \hat{\mathbf{a}}_{l_j}^{(1)}\}$ for $j < k_1$ to $(j+1)$-dimensional lower faces. In the end, all k_1-dimensional lower faces $\{\hat{\mathbf{a}}_{l_0}^{(1)}, \hat{\mathbf{a}}_{l_1}^{(1)}, \dots, \hat{\mathbf{a}}_{l_{k_1}}^{(1)}\}$ of $\widehat{S}^{(1)}$ with $l_0 < l_1 < \dots < l_{k_1}$ can be obtained if they exist.

6.3.2. The extension of level-ξ subfaces

Let $\widehat{E}_\xi = (\widehat{F}_1, \dots, \widehat{F}_\xi)$ be a level-ξ subface of $\widehat{S} = (\widehat{S}^{(1)}, \dots, \widehat{S}^{(r)})$ of type $(k_1, \dots, k_\xi)$ with $\xi < r$ where $\widehat{F}_i \subset \widehat{S}^{(i)} = \{\hat{\mathbf{a}}_1^{(i)}, \dots, \hat{\mathbf{a}}_{s_i}^{(i)}\}$ for $i = 1, \dots, \xi$. To continue the extension of $\widehat{E}_\xi$, we look for lower faces $\{\widehat{F}_{\xi+1}\}$ of $\widehat{S}^{(\xi+1)} = \{\hat{\mathbf{a}}_1^{(\xi+1)}, \dots, \hat{\mathbf{a}}_{s_{\xi+1}}^{(\xi+1)}\}$ of dimension $k_{\xi+1}$ so that $\widehat{E}_{\xi+1} = (\widehat{F}_1, \dots, \widehat{F}_{\xi+1})$ is a level-$(\xi+1)$ subface of $\widehat{S}$ of type $(k_1, \dots, k_{\xi+1})$. To find all such lower faces, we first find all the vertices $\hat{a}_l^{(\xi+1)}$ in the lower hull of $\widehat{S}^{(\xi+1)}$ for which $(\widehat{F}_1, \dots, \widehat{F}_\xi, \{\hat{a}_l^{(\xi+1)}\})$ is a level-$(\xi+1)$ subface of $\widehat{S}$ of type $(k_1, \dots, k_\xi, 0)$, followed by extending each such vertex of $\widehat{S}^{(\xi+1)}$ to lower faces $\widehat{F}_{\xi+1}^j$ of $\widehat{S}^{(\xi+1)}$ of dimension j for $j = 1, \dots, k_{\xi+1}$ consecutively, where for each j, $\widehat{F}_{\xi+1}^j \subset \widehat{F}_{\xi+1}^{j+1}$ and $(\widehat{F}_1, \dots, \widehat{F}_\xi \widehat{F}_{\xi+1}^j)$ is a level-$(\xi+1)$ subface of $\widehat{S}$ of type $(k_1, \dots, k_\xi, j)$.

For each $i = 1, \dots, \xi$, since $\dim(\widehat{F}_i) = k_i$, let

$$\widehat{F}_i = \{\hat{\mathbf{a}}_{l_0}^{(i)}, \dots, \hat{\mathbf{a}}_{l_{k_i}}^{(i)}\}.$$

To extend $\widehat{E}_\xi$, we begin by collecting on Table $\mathrm{T}(i, \xi+1)$ for $i = 1, \dots, \xi$ all the points $\hat{\mathbf{a}}_l^{(\xi+1)}$ in $\widehat{S}^{(\xi+1)}$ where $[\hat{\mathbf{a}}_{l_j}^{(i)}, \hat{\mathbf{a}}_l^{(\xi+1)}] = 1$ for all $j = 0, \dots, k_i$ and $i = 1, \dots, \xi$, and denote this set by $\mathcal{P}^{(\xi+1)}$. This set clearly contains all the vertex points of any lower face of $\widehat{S}^{(\xi+1)}$ of dimension $k_{\xi+1}$ that extends $\widehat{E}_\xi$. To examine points $\hat{\mathbf{a}}_\tau^{(\xi+1)}$ in $\mathcal{P}^{(\xi+1)}$ for its possibility to extend $\widehat{E}_\xi$, we apply the One-Point test on $\hat{\mathbf{a}}_\tau^{(\xi+1)}$:

$$\begin{aligned}
&\text{Minimize } \langle \hat{\mathbf{a}}_\tau^{(\xi+1)}, \hat{\alpha} \rangle - \alpha_0 \\
&\left.\begin{aligned}
&\langle \hat{\mathbf{a}}_{l_0}^{(i)}, \hat{\alpha} \rangle = \dots = \langle \hat{\mathbf{a}}_{l_{k_i}}^{(i)}, \hat{\alpha} \rangle, \\
&\langle \hat{\mathbf{a}}_{l_0}^{(i)}, \hat{\alpha} \rangle \leqslant \langle \hat{\mathbf{a}}_l^{(i)}, \hat{\alpha} \rangle, \quad \forall l \in \mathcal{C}^{(i)}
\end{aligned}\right\} \quad i = 1, \dots, \xi \\
&\alpha_0 \leqslant \langle \hat{\mathbf{a}}_k^{(\xi+1)}, \hat{\alpha} \rangle, \qquad \forall k \in \mathcal{C}^{(\xi+1)},
\end{aligned} \tag{6.16}$$

in the variables $\hat{\alpha} = (\alpha, 1) \in \mathbb{R}^{n+1}$ and $\alpha_0 \in \mathbb{R}$, where for $i = 1, \dots, \xi$, $\mathcal{C}^{(i)}$ is the set of indices of points $\hat{\mathbf{a}}_l^{(i)}$ in $\widehat{S}^{(i)}$ with $[\hat{\mathbf{a}}_{l_j}^{(i)}, \hat{\mathbf{a}}_l^{(i)}] = 1$ for all $j = 0, \dots, k_i$, and $\mathcal{C}^{(\xi+1)}$

TABLE T$(i, \xi+1)$

		$\widehat{S}^{(\xi+1)}$				
		$\hat{\mathbf{a}}_1^{(\xi+1)}$	$\hat{\mathbf{a}}_2^{(\xi+1)}$	$\hat{\mathbf{a}}_3^{(\xi+1)}$	$\cdots$	$\hat{\mathbf{a}}_{s_{\xi+1}}^{(\xi+1)}$
	$\hat{\mathbf{a}}_1^{(i)}$	$[\hat{\mathbf{a}}_1^{(i)}, \hat{\mathbf{a}}_1^{(\xi+1)}]$	$[\hat{\mathbf{a}}_1^{(i)}, \hat{\mathbf{a}}_2^{(\xi+1)}]$	$[\hat{\mathbf{a}}_1^{(i)}, \hat{\mathbf{a}}_3^{(\xi+1)}]$	$\cdots$	$[\hat{\mathbf{a}}_1^{(i)}, \hat{\mathbf{a}}_{s_{\xi+1}}^{(\xi+1)}]$
$\widehat{S}^{(i)}$	$\hat{\mathbf{a}}_2^{(i)}$	$[\hat{\mathbf{a}}_2^{(i)}, \hat{\mathbf{a}}_1^{(\xi+1)}]$	$[\hat{\mathbf{a}}_2^{(i)}, \hat{\mathbf{a}}_2^{(\xi+1)}]$	$[\hat{\mathbf{a}}_2^{(i)}, \hat{\mathbf{a}}_3^{(\xi+1)}]$	$\cdots$	$[\hat{\mathbf{a}}_2^{(i)}, \hat{\mathbf{a}}_{s_{\xi+1}}^{(\xi+1)}]$
	$\vdots$	$\vdots$	$\vdots$	$\vdots$	$\cdots$	$\vdots$
	$\hat{\mathbf{a}}_{s_i}^{(i)}$	$[\hat{\mathbf{a}}_{s_i}^{(i)}, \hat{\mathbf{a}}_1^{(\xi+1)}]$	$[\hat{\mathbf{a}}_{s_i}^{(i)}, \hat{\mathbf{a}}_2^{(\xi+1)}]$	$[\hat{\mathbf{a}}_{s_i}^{(i)}, \hat{\mathbf{a}}_3^{(\xi+1)}]$	$\cdots$	$[\hat{\mathbf{a}}_{s_i}^{(i)}, \hat{\mathbf{a}}_{s_{\xi+1}}^{(\xi+1)}]$

contains the indices of the points in $\mathcal{P}^{(\xi+1)}$. When the optimal value is zero, point $\hat{\mathbf{a}}_\tau^{(\xi+1)}$ will be retained for further extension considerations, otherwise it would be deleted. As before, beneficial information provided by the pivoting in the simplex method averts the necessity to apply One-Point test on all the points in $\mathcal{P}^{(\xi+1)}$.

When the examination on the points in $\mathcal{P}^{(\xi+1)}$ for the extension of $\widehat{E}_\xi$ is completed, let $\mathcal{E}^{(\xi+1)}$ be the set of points in $\mathcal{P}^{(\xi+1)}$ which are capable of extending $\widehat{E}_\xi$; namely, for each such point $\hat{\mathbf{a}}_l^{(\xi+1)}$,

$$\left(\widehat{F}_1, \ldots, \widehat{F}_\xi, \{\hat{\mathbf{a}}_l^{(\xi+1)}\}\right)$$

is a level-$(\xi+1)$ subface of $\widehat{S}$ of type $(k_1, \ldots, k_\xi, 0)$. Let the indices of its points be

$$\tau_1 < \tau_2 < \cdots < \tau_t$$

and, in this order, we continue our attempt to extend

$$\left(\widehat{F}_1, \ldots, \widehat{F}_\xi, \{\hat{\mathbf{a}}_{\tau_j}^{(\xi+1)}\}\right)$$

for $j = 1, \ldots, t$ by examining points in $\{\hat{\mathbf{a}}_{\tau_l}^{(\xi+1)}\}_{l>j} \subset \mathcal{E}^{(\xi+1)}$.

This procedure may be continued along the same line as we extend the lower faces of $\widehat{S}_1$ until subface

$$\widehat{F}_{\xi+1} = \left\{\hat{\mathbf{a}}_{l_0}^{(\xi+1)}, \ldots, \hat{\mathbf{a}}_{l_{k_{\xi+1}}}^{(\xi+1)}\right\}$$

of $\widehat{S}^{(\xi+1)}$ of dimension $k_{\xi+1}$ for which

$$\widehat{E}_{\xi+1} := \left(\widehat{F}_1, \ldots, \widehat{F}_\xi, \widehat{F}_{\xi+1}\right)$$

is a level-$(\xi+1)$ subface of $\widehat{S}$ of type $(k_1, \ldots, k_\xi, k_{\xi+1})$ are all obtained.

6.4. *Finding all cells of type* $(k_1, \ldots, k_r)$

In summary, we list the procedure for finding all cells of type $(k_1, \ldots, k_r)$ in S_ω induced by a generic lifting $\omega = (\omega_1, \ldots, \omega_r)$ on $S = (S_1^{(1)}, \ldots, S^{(r)})$:

PROCEDURE FOR FINDING ALL CELLS OF TYPE $(k_1, \ldots, k_r)$.

(1) With $\widehat{S}^{(i)} = \{\hat{\mathbf{a}} = (\mathbf{a}, \omega_i(\mathbf{a})) \mid \mathbf{a} \in S^{(i)}\}$ for $i = 1, \ldots, r$, fill out the relation table, Table 6.1, consisting of Tables $\mathrm{T}(i, j)$ between $\widehat{S}^{(i)}$ and $\widehat{S}^{(j)}$ for all $1 \leqslant i, j \leqslant r$.

(2) Find all k_1 dimensional faces $\widehat{F}_1$ in the lower hull of $\widehat{S}^{(1)}$.

(3) For $1 \leqslant \xi < r$, extend each level-ξ surface $(\widehat{F}_1, \ldots, \widehat{F}_\xi)$ of $\widehat{S} = (\widehat{S}^{(1)}, \ldots, \widehat{S}^{(r)})$ of type $(k_1, \ldots, k_\xi)$ to level-$(\xi + 1)$ subfaces of $\widehat{S}$ of type $(k_1, \ldots, k_{\xi+1})$.

(4) When $\xi + 1 = r$, $(F_1, \ldots, F_r)$ is a cell of type $(k_1, \ldots, k_r)$ in S_ω.

The above procedure has been successfully implemented in GAO and LI [2003] to calculate the mixed volume for semi-mixed polynomial systems. The algorithm achieves a dramatical speed-up, especially when the systems are unmixed, such as the 9-point problem in WAMPLER, MORGAN and SOMMESE [1992].

7. Finding isolated zeros in $\mathbb{C}^n$ via stable cells

7.1. Stable mixed volume

As remarked in the end of Section 3.1, in order to find all isolated zeros of a polynomial system $P(x) = (p_1(x), \ldots, p_n(x))$ with support $S = (S_1, \ldots, S_n)$ in $\mathbb{C}^n$, we need to follow $\mathcal{M}(S'_1, \ldots, S'_n)$ homotopy paths, where $S'_j = S_j \cup \{0\}$, $j = 1, \ldots, n$. By Theorem 3.2, $\mathcal{M}(S'_1, \ldots, S'_n)$ provides an upper bound for the root count of the system $P(x)$ in $\mathbb{C}^n$. However, as the following example shows, this bound may not be exact:

EXAMPLE 7.1 (HUBER and STURMFELS [1997]). Using linear homotopy with start system such as (1.3), one finds six isolated zeros in $\mathbb{C}^2$ of the system

$$P(x_1, x_2) = \begin{cases} p_1(x_1, x_2) = ax_2 + bx_2^2 + cx_1x_2^3, \\ p_2(x_1, x_2) = dx_1 + ex_1^2 + fx_1^3x_2 \end{cases}$$

for generic coefficients $\{a, b, c, d, e, f\}$. But for its augmented system

$$Q(x_1, x_2) = \begin{cases} q_1(x_1, x_2) = \varepsilon_1 + ax_2 + bx_2^2 + cx_1x_2^3, \\ q_2(x_1, x_2) = \varepsilon_2 + dx_1 + ex_1^2 + fx_1^3x_2, \end{cases}$$

the mixed volume $\mathcal{M}(S_1 \cup \{0\}, S_2 \cup \{0\})$, easily calculable by hand, is eight. In this case, eight homotopy paths need to be followed to obtain all six isolated zeros of $P(x_1, x_2)$ in $\mathbb{C}^2$, hence two of them are extraneous.

In HUBER and STURMFELS [1997], a tighter upper bound for the root count in $\mathbb{C}^n$ of the system $P(x) = (p_1(x), \ldots, p_n(x))$ was given. Based on this root count, one may employ alternative algorithms, which we will describe in this section, to approximate all isolated zeros of $P(x)$ in $\mathbb{C}^n$ by following fewer homotopy paths.

As before, for a given lifting $\omega = (\omega_1, \ldots, \omega_n)$ on $S' = (S'_1, \ldots, S'_n)$, we write $\hat{a} = (a, \omega_j(a))$ for $a \in S'_j$ and $\widehat{C}_j = \{\hat{a} \mid a \in C_j\}$ for $C_j \subset S'_j$. Cell $\widehat{C} = (\widehat{C}_1, \ldots, \widehat{C}_n)$ where $C_j \subset S'_j$ for $j = 1, \ldots, n$ is a *lower facet* of $\widehat{S}' = (\widehat{S}'_1, \ldots, \widehat{S}'_n)$ if $\dim(\mathrm{conv}(\widehat{C})) = n$ and there exists $\hat{\alpha} = (\alpha, 1) \in \mathbb{R}^{n+1}$ satisfying, for $j = 1, \dots, n$,

$$\langle \hat{a}, \hat{\alpha} \rangle = \langle \hat{b}, \hat{\alpha} \rangle \quad \text{for all } a, b \in C_j,$$
$$\langle \hat{a}, \hat{\alpha} \rangle < \langle \hat{d}, \hat{\alpha} \rangle \quad \text{for } a \in C_j \text{ and } d \in S'_j \setminus C_j.$$

We will refer to the vector $\alpha \in \mathbb{R}^n$ as the *inner normal* of $C = (C_1, \dots, C_n)$ with respect to the lifting ω, and denote such $C = (C_1, \dots, C_n)$ by C^α. When α is nonnegative, i.e. $\alpha_j \geqslant 0$ for all $j = 1, \dots, n$, we call C^α a *stable cell* of $S = (S_1, \dots, S_n)$ *with respect to the lifting* ω. The term *stable cell* alone, without specification of its corresponding lifting, will be reserved for stable cells with respect to the particular lifting $\omega_0^1 = (\omega_1^{01}, \dots, \omega_n^{01})$ where $\omega_j^{01} : S'_j \to \mathbb{R}$ for $j = 1, \dots, n$ is defined as:

$$\omega_j^{01}(0) = 1 \quad \text{if } 0 \notin S_j,$$
$$\omega_j^{01}(a) = 0 \quad \text{for } a \in S_j.$$

Obviously, $S = (S_1, \dots, S_n)$ itself is a stable cell with inner normal $\alpha = (0, \dots, 0)$ with respect to this particular lifting.

DEFINITION 7.1. The stable mixed volume of $S = (S_1, \dots, S_n)$, denoted by $\mathcal{SM}(S_1, \dots, S_n)$, is the sum of mixed volumes of all stable cells of $S = (S_1, \dots, S_n)$.

With this definition, a tighter bound for the root count of $P(x)$ in $\mathbb{C}^n$ is given in the following

THEOREM 7.1 (HUBER and STURMFELS [1997]). *For polynomial system $P(x) = (p_1(x), \dots, p_n(x))$ with support $S = (S_1, \dots, S_n)$, the stable mixed volume $\mathcal{SM}(S_1, \dots, S_n)$ satisfies*:

$$\mathcal{M}(S_1, \dots, S_n) \leqslant \mathcal{SM}(S_1, \dots, S_n) \leqslant \mathcal{M}\big(S_1 \cup \{0\}, \dots, S_n \cup \{0\}\big). \tag{7.1}$$

Moreover, it provides an upper bound for the root count of $P(x)$ in $\mathbb{C}^n$.

For the system $P(x_1, x_2)$ in Example 7.1,

$$S_1 = \big\{(0, 1), (0, 2), (1, 1)\big\}, \qquad S_2 = \big\{(1, 0), (2, 0), (3, 1)\big\}$$

and

$$S'_1 = S_1 \cup \big\{(0, 0)\big\}, \qquad S'_2 = S_2 \cup \big\{(0, 0)\big\},$$

as shown in Fig. 7.1. With lifting ω_0^1, there are eight lower facets of $\widehat{S}' = (\widehat{S}'_1, \widehat{S}'_2)$. Among them, four stable cells of the system are induced:

(1) $C^{\alpha^{(1)}} = (\{(0, 0), (0, 1)\}, \{(1, 0), (2, 0)\})$ with $\alpha^{(1)} = (0, 1)$,
(2) $C^{\alpha^{(2)}} = (\{(0, 1), (0, 2)\}, \{(0, 0), (1, 0)\})$ with $\alpha^{(2)} = (1, 0)$,
(3) $C^{\alpha^{(3)}} = (\{(0, 0), (0, 1)\}, \{(0, 0), (1, 0)\})$ with $\alpha^{(3)} = (1, 1)$,
(4) $C^{\alpha^{(4)}} = (S_1, S_2)$ with $\alpha^{(4)} = (0, 0)$.

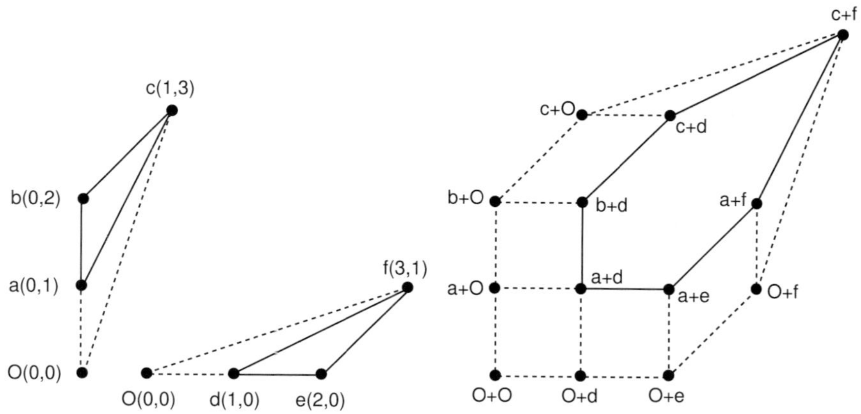

FIG. 7.1.

Clearly, the mixed volumes of $C^{\alpha^{(1)}}$, $C^{\alpha^{(2)}}$, and $C^{\alpha^{(3)}}$ are just their volumes, they all equal to 1. The mixed volume of $C^{\alpha^{(4)}} = (S_1, S_2)$ is

$$\mathcal{M}(S_1, S_2) = \mathrm{Vol}_2(S_1, S_2) - \mathrm{Vol}_2(S_1) - \mathrm{Vol}_2(S_2) = 4 - 0.5 - 0.5 = 3.$$

Therefore, the stable mixed volume $\mathcal{SM}(S_1, S_2) = 6$ for the system, while the mixed volume of its augment system $Q(x_1, x_2)$ with support $(S_1 \cup \{0\}, S_2 \cup \{0\})$ is

$$\begin{aligned}&\mathcal{M}(S_1 \cup \{0\}, S_2 \cup \{0\})\\&= \mathrm{Vol}_2(S_1 \cup \{0\}, S_2 \cup \{0\}) - \mathrm{Vol}_2(S_1 \cup \{0\}) - \mathrm{Vol}_2(S_2 \cup \{0\})\\&= 10 - 1 - 1 = 8.\end{aligned}$$

Hence,

$$\mathcal{M}(S_1, S_2) < \mathcal{SM}(S_1, S_2) < \mathcal{M}(S_1 \cup \{0\}, S_2 \cup \{0\})$$

for the system and the inequalities in (7.1) are strict in this case.

7.2. *An alternative algorithm*

Based on the derivation of Theorem 7.1, it was suggested in HUBER and STURMFELS [1997] that one may find all isolated zeros of polynomial system $P(x) = (p_1(x), \ldots, p_n(x))$ in $\mathbb{C}^n$ with support $S = (S_1, \ldots, S_n)$ where

$$p_j(x) = \sum_{a \in S_j} c_{j,a} x^a, \quad j = 1, \ldots, n,$$

by the following procedure:

Step 1: Identify all stable cells $C^\alpha = (C_1, \ldots, C_n)$ of $S = (S_1, \ldots, S_n)$.
Step 2: For each stable cell $C^\alpha = (C_1, \ldots, C_n)$ with inner normal $\alpha = (\alpha_1, \ldots, \alpha_n) \geqslant 0$, find all isolated zeros in $(\mathbb{C}^*)^n$ of the *support* system $P^\alpha(x) = (p_1^\alpha(x), \ldots, p_n^\alpha(x))$

where for $j = 1, \ldots, n$,

$$p_j^{\alpha}(x) = \sum_{a \in C_j \cap S_j} c_{j,a} x^a + \varepsilon_j \tag{7.2}$$

with $\varepsilon_j = 0$ if $0 \in S_j$, an arbitrary nonzero number otherwise.

Step 3: For each isolated zero $z = (z_1, \ldots, z_n)$ of $P^{\alpha}(x)$ in $(\mathbb{C}^*)^n$, let

$$\bar{z}_j = z_j \quad \text{if } \alpha_j = 0,$$
$$\bar{z}_j = 0 \quad \text{if } \alpha_j \neq 0.$$

Then $\bar{z} = (\bar{z}_1, \ldots, \bar{z}_n)$ is an isolated zero of $P(x) = (p_1(x), \ldots, p_n(x))$.

Inevitably, zeros $z = (z_1, \ldots, z_n)$ of $P^{\alpha}(x)$ will depend on $\varepsilon = (\varepsilon_1, \ldots, \varepsilon_n)$. However, it can be shown that the transition from z to $\bar{z}$ given above actually eliminates this dependency as the following example shows:

EXAMPLE 7.2. For the system $P(x_1, x_2) = (p_1(x_1, x_2), p_2(x_1, x_2))$ where

$$p_1(x_1, x_2) = x_1 x_2,$$
$$p_2(x_1, x_2) = x_1 + x_2 - 1,$$

$S_1 = \{(1,1)\}$, $S_2 = \{(1,0), (0,1), (0,0)\}$ and $S_1' = \{(1,1), (0,0)\}$, $S_2' = S_2$.

With lifting ω_0^1, there are two stable cells with nonzero mixed volumes:

(1) $C^{\alpha^{(1)}} = (\{(1,1), (0,0)\}, \{(1,0), (0,0)\})$ with $\alpha^{(1)} = (0,1)$,

(2) $C^{\alpha^{(2)}} = (\{(1,1), (0,0)\}, \{(0,1), (0,0)\})$ with $\alpha^{(2)} = (1,0)$.

The support system of $C^{\alpha^{(1)}}$ is

$$p_1^{\alpha^{(1)}}(x_1, x_2) = x_1 x_2 + \varepsilon = 0, \quad \varepsilon \neq 0,$$
$$p_2^{\alpha^{(1)}}(x_1, x_2) = x_1 - 1 = 0,$$

with isolated zero $(1, -\varepsilon)$ in $(\mathbb{C}^*)^2$. This leads to an isolated zero $(1,0)$ of $P(x_1, x_2)$ in $\mathbb{C}^2$ since the second component of $\alpha^{(1)}$ is positive. Similarly, isolated zero $(-\varepsilon, 1)$ of the support system of $C^{\alpha^{(2)}}$,

$$p_1^{\alpha^{(2)}}(x_1, x_2) = x_1 x_2 + \varepsilon = 0, \quad \varepsilon \neq 0,$$
$$p_2^{\alpha^{(2)}}(x_1, x_2) = x_2 - 1 = 0,$$

in $(\mathbb{C}^*)^2$ gives an isolated zero $(0,1)$ of $P(x_1, x_2)$ since the first component of $\alpha^{(2)}$ is positive.

In Step 2 above, when polyhedral homotopy is used to find all isolated zeros in $(\mathbb{C}^*)^n$ of the support system $P^{\alpha}(x)$ corresponding to the stable cell $C^{\alpha} = (C_1, \ldots, C_n)$, one follows $\mathcal{M}(C_1, \ldots, C_n)$ homotopy paths. Accordingly, the total number of homotopy paths one needs to follow to reach all isolated zeros of $P(x)$ in $\mathbb{C}^n$ equals to the stable mixed volume $\mathcal{SM}(S_1, \ldots, S_n)$, which is strictly fewer than $\mathcal{M}(S_1 \cup \{0\}, \ldots, S_n \cup \{0\})$ in general, therefore admitting less extraneous paths.

However, there are difficulties to implement this procedure efficiently. First of all, types of those stable cells are undetermined in general. They may not be mixed cells, cells of type $(1,\ldots,1)$, with respect to the lifting $\omega_0^1 = (\omega_1^{01},\ldots,\omega_n^{01})$, which invalidates the algorithm we developed in Section 5 for finding mixed cells. This makes the identification of all the stable cells in Step 1 rather difficult. Secondly, when polyhedral homotopy is used in Step 2 to solve $P^\alpha(x)$ in $(\mathbb{C}^*)^n$, one must find all mixed cells of a subdivision of $C^\alpha = (C_1,\ldots,C_n)$ induced by a further generic lifting on C^α in the first place. This accumulated work for all the stable cells can be very costly, which may not be more favorable compared to solving $P(x)$ in $\mathbb{C}^n$ by simply following the polyhedral homotopy procedure given in Section 4 directly with a generic lifting on $S' = (S_1',\ldots,S_n')$ permitting some of the homotopy paths to be extraneous.

7.3. A revision

We will elaborate in this section a revision given in GAO, LI and WANG [1999] for the procedure suggested by HUBER and STURMFELS [1997] in the last section. To begin, for $k \geqslant 0$, let $\omega_0^k = (\omega_1^{0k},\ldots,\omega_n^{0k})$ be the lifting on $S' = (S_1',\ldots,S_n')$ where for $j = 1,\ldots,n$,

$$\begin{aligned} \omega_j^{0k}(0) &= k \quad \text{if } 0 \notin S_j, \\ \omega_j^{0k}(a) &= 0 \quad \text{for } a \in S_j. \end{aligned} \tag{7.3}$$

Clearly, the set of stable cells with respect to ω_0^k remains invariant for different k's. For instance, if $C = (C_1,\ldots,C_n)$ is a stable cell with respect to the lifting $\omega_0^{k_1}$ with inner normal $\alpha \geqslant 0$, then $C = (C_1,\ldots,C_n)$ is also a stable cell with respect to the lifting $\omega_0^{k_2}$ with inner normal $k_2/k_1 \cdot \alpha \geqslant 0$. Denote this set of stable cells by $\mathcal{T}$. Let $\omega = (\omega_1,\ldots,\omega_n)$ be a generic lifting on $S' = (S_1',\ldots,S_n')$ where for $j = 1,\ldots,n$,

$$\begin{aligned} \omega_j(0) &= k & &\text{for } 0 \notin S_j, \\ \omega_j(a) &= \text{a generic number in } (0,1) & &\text{for } a \in S_j. \end{aligned} \tag{7.4}$$

For a cell $C = (C_1,\ldots,C_n)$ in the subdivision of $S' = (S_1',\ldots,S_n')$ induced by the lifting $\omega_0^k = (\omega_1^{0k},\ldots,\omega_n^{0k})$, let ω^C be the restriction of ω on C, which can, of course, be regarded as a generic lifting on C. It was shown in GAO, LI and WANG [1999] that if k is sufficiently large, mixed cell $D = (D_1,\ldots,D_n)$ of subdivision S_ω of $S' = (S_1',\ldots,S_n')$ induced by the lifting ω is also a mixed cell of subdivision S_{ω^C} induced by the lifting ω^C of certain cell $C = (C_1,\ldots,C_n)$ of S' with respect to the lifting ω_0^k. Accordingly, stable cells $C = (C_1,\ldots,C_n)$ in $\mathcal{T}$ can be assembled by grouping a collection of proper cells in S_ω, and consequently, mixed cells in this collection provides all the mixed cells of subdivision S_{ω^C} of $C = (C_1,\ldots,C_n)$. More precisely, when $k \geqslant n(n+1)d^n$ (see GAO, LI and WANG [1999]) where $d = \max_{1\leqslant j\leqslant n} \deg p_j(x)$, any mixed cell $D = (D_1,\ldots,D_n)$ in the subdivision S_ω induced by the lifting $\omega = (\omega_1,\ldots,\omega_n)$ on $S' = (S_1',\ldots,S_n')$ given in (7.4) is a mixed cell of subdivision S_{ω^C} induced by the lifting ω^C of certain cell $C = (C_1,\ldots,C_n)$ in the subdivision $S_{\omega_0^k}$ induced by the lifting $\omega_0^k = (\omega_1^{0k},\ldots,\omega_n^{0k})$ on $S' = (S_1',\ldots,S_n')$ given in (7.3).

Let $D^* = (D_1, \ldots, D_n)$ be any cell in the subdivision S_w which may or may not be of type $(1, \ldots, 1)$. Let

$$D_j = \{a_{j0}, \ldots, a_{jk_j}\}, \quad j = 1, \ldots, n,$$

where $k_1 + \cdots + k_n = n$. For $j = 1, \ldots, n$ and $a \in S'_j$, write $\hat{a}(k) = (a, \omega_j^{0k}(a))$ and $\widehat{D}_j(k) = \{\hat{a}(k) \mid a \in D_j\}$. Let $\widehat{D}^*(k) = (\widehat{D}_1(k), \ldots, \widehat{D}_n(k))$. Clearly, the $n \times (n+1)$ matrix

$$V(\widehat{D}^*(k)) = \begin{pmatrix} \hat{a}_{11}(k) - \hat{a}_{10}(k) \\ \vdots \\ \hat{a}_{1k_1}(k) - \hat{a}_{10}(k) \\ \vdots \\ \hat{a}_{n1}(k) - \hat{a}_{n0}(k) \\ \vdots \\ \hat{a}_{nk_n}(k) - \hat{a}_{n0}(k) \end{pmatrix}$$

is of rank n. Let $\alpha \in \mathbb{R}^n$ be the unique vector where $\hat{\alpha} = (\alpha, 1)$ is in the kernel of $V(\widehat{D}^*(k))$. This α is the inner normal of D^* with respect to ω_0^k. Let $\mathcal{T}(\alpha)$ be the collection of all mixed cells in S_ω with the same nonnegative inner normal α with respect to ω_0^k and let $D = (\{a_{10}, a_{11}\}, \ldots, \{a_{n0}, a_{n1}\})$ where $\{a_{j0}, a_{j1}\} \subset S'_j$ for $j = 1, \ldots, n$ be any mixed cell in $\mathcal{T}(\alpha)$. Let $C = (C_1, \ldots, C_n)$ where

$$C_j = \{a \in S'_j \mid \langle \hat{a}(k), \hat{\alpha} \rangle = \langle \hat{a}_{j0}(k), \hat{\alpha} \rangle\}, \quad j = 1, \ldots, n.$$

This cell satisfies, for $j = 1, \ldots, n$,

$$\begin{aligned} \langle \hat{a}(k), \hat{\alpha} \rangle &= \langle \hat{b}(k), \hat{\alpha} \rangle \quad \text{for } a, b \in C_j, \\ \langle \hat{a}(k), \hat{\alpha} \rangle &< \langle \hat{d}(k), \hat{\alpha} \rangle \quad \text{for } a \in C_j,\ d \in S'_j \backslash C_j. \end{aligned}$$

It is therefore a stable cell with respect to ω_0^k with inner normal α, which, as mentioned above, is also a stable cell with respect to ω_0^1 with inner normal $\frac{1}{k}\alpha$. In the meantime, the cells in the collection $\mathcal{T}(\alpha)$ gives *all* the mixed cells in the subdivision S_{ω^C} of $C = (C_1, \ldots, C_n)$ induced by the lifting ω^C.

From what we have discussed above, the previously listed procedure for solving system $P(x) = (p_1(x), \ldots, p_n(x))$ with $S = (S_1, \ldots, S_n)$ in $\mathbb{C}^n$ suggested in HUBER and STURMFELS [1997] may now be revised as follows:

FINDING ISOLATED ZEROS IN $\mathbb{C}^n$ VIA STABLE CELLS.

Step 0: Let $d = \max_{1 \leqslant i \leqslant n} \deg p_i(x)$. Choose a real number $k > n(n+1)d^n$ at random.

Step 1: Lift the support $S' = (S'_1, \ldots, S'_n)$ by a random lifting $\omega = (\omega_1, \ldots, \omega_n)$ as defined in (7.4) where for $j = 1, \ldots, n$,

$$\begin{aligned} &\omega_j(0) = k && \text{if } 0 \notin S_j, \\ &\omega_j(a) = \text{a randomly chosen number in } (0, 1) && \text{if } a \in S_j. \end{aligned}$$

Find cells of type $(1, \ldots, 1)$ in the induced fine mixed subdivision S_ω of $S' = (S'_1, \ldots, S'_n)$.

Step 2: For cell $D = (\{a_{10}, a_{11}\}, \dots, \{a_{n0}, a_{n1}\})$ of type $(1, \dots, 1)$ in S_ω, let $\hat{a}_{ji}(k) = (a_{ji}, l)$ where for $j = 1, \dots, n$, and $i = 0, 1$,

$$l = k \quad \text{if } a_{ji} = 0 \notin S_j,$$
$$l = 0 \quad \text{if } a_{ji} \in S_j.$$

Form the $n \times (n+1)$ matrix

$$V = \begin{pmatrix} \hat{a}_{11}(k) - \hat{a}_{10}(k) \\ \vdots \\ \hat{a}_{n1}(k) - \hat{a}_{n0}(k) \end{pmatrix},$$

and find the unique vector $\alpha = (\alpha_1, \dots, \alpha_n)$ where $\hat{\alpha} = (\alpha, 1)$ is in the kernel of V. This α is the inner normal of D with respect to ω_0^k. Let $\mathcal{T}(\alpha)$ be the collection of all cells of type $(1, \dots, 1)$ in S_ω with the same nonnegative inner normal $\alpha = (\alpha_1, \dots, \alpha_n)$ with respect to ω_0^k.

Step 3: (a) Choose any mixed cell $D = (\{a_{10}, a_{11}\}, \dots, \{a_{n0}, a_{n1}\})$ from $\mathcal{T}(\alpha)$, let

$$C_j = \{a \in S_j' \mid \langle \hat{a}(k), \hat{\alpha} \rangle = \langle \hat{a}_{j0}(k), \hat{\alpha} \rangle\}, \quad j = 1, \dots, n,$$

where $\hat{a}(k) = (a, l)$ with

$$l = k \quad \text{if } a = 0 \notin S_j,$$
$$l = 0 \quad \text{if } a \in S_j.$$

Then $C = (C_1, \dots, C_n)$ is a stable cell of $S = (S_1, \dots, S_n)$ with respect to the inner normal α in $S_{\omega_0^k}$. Notice that

$$S_{\omega^C} = \{(D_1, \dots, D_r) \in S_\omega \mid D_j \subseteq C_j \text{ for all } 1 \leqslant j \leqslant n\}$$

is the fine mixed subdivision of C induced by ω^C, the restriction of ω on C, and $\mathcal{T}(\alpha)$ gives all the mixed cells, cells of type $(1, \dots, 1)$, in S_{ω^C}.

(b) Find all the isolated zeros of the system

$$P^\alpha(x) = (p_1^\alpha(x), \dots, p_n^\alpha(x)) \tag{7.5}$$

where

$$p_j^\alpha(x) = \sum_{\alpha \in C_j \cap S_j} c_{j,\alpha} x^a + \varepsilon_j, \quad j = 1, \dots, n,$$

and

$$\varepsilon_j = 0 \quad \text{if } 0 \in S_j,$$
$$\varepsilon_j = 1 \quad \text{if } 0 \notin S_j$$

in $(\mathbb{C}^*)^n$ by employing the polyhedral homotopy procedure developed in Section 4 with lifting ω^C.

(c) For zeros $\mathbf{e} = (e_1, \dots, e_n)$ of $P^\alpha(x)$ found in (b), let

$$\bar{e}_j = e_j \quad \text{if } \alpha_j = 0,$$
$$\bar{e}_j = 0 \quad \text{if } \alpha_j \neq 0.$$

Then $\bar{\mathbf{e}} = (\bar{e}_1, \ldots, \bar{e}_n)$ is a zero of $P(x)$ in $\mathbb{C}^n$.
Step 4: Repeat Step 3 for all $\mathcal{T}(\alpha)$ with $\alpha \geqslant 0$.

REMARK 7.1. For $d_j = \deg p_j(x)$, $j = 1, \ldots, n$, we assume without loss $d_1 \leqslant d_2 \leqslant \cdots \leqslant d_n$. It was mentioned in GAO, LI and WANG [1999], in Step 0 of the above procedure, d may be replaced by $d_2 \times \cdots \times d_n \times d_n$ which usually gives a much smaller number.

REMARK 7.2. It is commonly known that when the polyhedral homotopy method is used to solve polynomial systems, large differences between the powers of parameter t in the polyhedral homotopies may cause computational instability when homotopy curves are followed. In the above algorithm, the point 0 often receives very large lifting value k, compared to the rest of the lifting values in $(0, 1)$. It was shown in GAO, LI and WANG [1999] that the stability of the algorithm is independent of the large lifting value k when polyhedral homotopies are used in Step 3(b).

The revised procedure listed above has been successfully implemented in GAO, LI and WANG [1999] with remarkable numerical results.

8. Solutions of positive dimension

8.1. Solutions of positive dimension

For polynomial system $P(x) = (p_1(x), \ldots, p_n(x))$, positive dimensional components of the solution set of $P(x) = 0$ are a common occurrence. Sometimes they are an unpleasant side show (SOMMESE and WAMPLER [1996]) that happens with a system generated using a model for which only the isolated nonsingular solutions are of interest; and sometimes, the positive dimensional solution components are of primary interest. In either case, dealing with positive dimensional solution components, is usually computationally difficult.

Putting aside formal definition with technical terms, by a *generic point* of an irreducible component X of the solution set of $P(x) = 0$, it usually means a point of X which has no special properties not possessed by the whole component X. Numerically, it is modeled by a point in X with random coordinates. In SOMMESE and WAMPLER [1996], a procedure consisted in slicing X with linear subspaces in general position to obtain generic point of X as the isolated solutions of an auxiliary system was presented. By Noether's normalization theorem combined with Bertini's theorem, it can be shown that if X is of k-dimensional, then a general affine linear subspace $\mathbb{C}^{n-k}$ meets X at isolated points. Those are generic points of X. A generic linear subspace $\mathbb{C}^{n-k}$ can be given by

$$\begin{aligned}
\lambda_{11}x_1 + \cdots + \lambda_{1n}x_n &= \lambda_1, \\
&\ \ \vdots \\
\lambda_{k1}x_1 + \cdots + \lambda_{kn}x_n &= \lambda_k
\end{aligned}$$

with all the λ's being random numbers. Thus, the existence of isolated solutions of the system

$$\begin{aligned} p_1(x_1,\ldots,x_n) &= 0,\\ &\vdots\\ p_n(x_1,\ldots,x_n) &= 0,\\ \lambda_{11}x_1+\cdots+\lambda_{1n}x_n &= \lambda_1,\\ &\vdots\\ \lambda_{k1}x_1+\cdots+\lambda_{kn}x_n &= \lambda_k \end{aligned} \tag{8.1}$$

warrants the existence of k-dimensional components of the solution set of the original system $P(x) = (p_1(x),\ldots,p_n(x)) = 0$. Furthermore, the set of isolated solutions of (8.1) contains at least one generic point of each irreducible component of dimension k of the solution set of $P(x) = 0$.

System (8.1) is overdetermined and the procedure suggested in SOMMESE and WAMPLER [1996] for solving its isolated solutions is quite computationally expensive. A more efficient method which we shall describe below is given in SOMMESE and VERSCHELDE [2000], SOMMESE, VERSCHELDE and WAMPLER [2001]. This method can determine the existence of components of various dimensions, including isolated solutions, of the solution set of $P(x) = 0$.

With extra parameters $z_1,\ldots,z_n$, consider, for $j = 1,\ldots,n$, the embeddings:

$$\mathcal{E}_j(x,z_1,\ldots,z_j) = \begin{cases} p_1(x)+\lambda_{11}z_1+\cdots+\lambda_{1j}z_j,\\ \quad\vdots\\ p_n(x)+\lambda_{n1}z_1+\cdots+\lambda_{nj}z_j,\\ c_1+a_{11}x_1+\cdots+a_{1n}x_n+z_1,\\ \quad\vdots\\ c_j+a_{j1}x_1+\cdots+a_{jn}x_n+z_j. \end{cases}$$

For $j = 0$, we let $\mathcal{E}_0(x) = P(x)$. Let Y denote the space $\mathbb{C}^{n\times(n+1)}\times\mathbb{C}^{n\times n}$ of parameters

$$\begin{bmatrix} a_1 & a_{11} & \cdots & a_{1n} & \lambda_{11} & \cdots & \lambda_{n1}\\ \vdots & \vdots & \ddots & \vdots & \vdots & \ddots & \vdots\\ a_n & a_{n1} & \cdots & a_{nn} & \lambda_{1n} & \cdots & \lambda_{nn} \end{bmatrix} \in \mathbb{C}^{n\times(n+1)}\times\mathbb{C}^{n\times n}.$$

LEMMA 8.1 (SOMMESE and VERSCHELDE [2000]). *There is an open dense set U of full measure of points*

$$\begin{bmatrix} a_1 & a_{11} & \cdots & a_{1n} & \lambda_{11} & \cdots & \lambda_{n1}\\ \vdots & \vdots & \ddots & \vdots & \vdots & \ddots & \vdots\\ a_n & a_{n1} & \cdots & a_{nn} & \lambda_{1n} & \cdots & \lambda_{nn} \end{bmatrix} \in \mathbb{C}^{n\times(n+1)}\times\mathbb{C}^{n\times n} \tag{8.2}$$

such that for each $j = 1,\ldots,n$,

(1) *The solution set of* $\mathcal{E}_j(x, z_1, \ldots, z_j) = 0$ *with* $(z_1, \ldots, z_j) \neq 0$ *are isolated and nonsingular*;
(2) *Given any irreducible component* X *of the solution set of* $P(x) = 0$ *of dimension* j, *the set of isolated solutions of* $\mathcal{E}_j(x, z_1, \ldots, z_j) = 0$ *with* $(z_1, \ldots, z_j) = 0$ *contains as many generic points as the degree of* X;
(3) *The solutions of* $\mathcal{E}_j(x, z_1, \ldots, z_j) = 0$ *with* $(z_1, \ldots, z_j) \neq 0$ *are the same as the solutions of* $\mathcal{E}_j(x, z_1, \ldots, z_j) = 0$ *with* $z_j \neq 0$.

By the third assertion of the above lemma, the solutions of $\mathcal{E}_j(x, z_1, \ldots, z_j) = 0$ are either $z_j \neq 0$ or $(z_1, \ldots, z_j) = 0$, namely, $z_j = 0$ implies $z_1 = z_2 = \cdots = z_{j-1} = 0$ for any solutions. Moreover, when all those parameters a's and λ's in (8.2) are chosen generically, the existence of the solutions of $\mathcal{E}_j(x, z_1, \ldots, z_j) = 0$ with $(z_1, \ldots, z_j) = 0$ reveals the possibilities of the existence of components of dimension j of the solution set of $P(x) = 0$. While those solutions might lie in components of dimension higher then j, it is clear that if j is the largest integer for which the solution set of $\mathcal{E}_j(x, z_1, \ldots, z_j) = 0$ is nonempty then the dimension of the solution set V of $P(x) = 0$, defined to be the largest dimension of all the irreducible components of V, must be j.

For j from n to 1 and $t \in [0, 1]$, consider homotopies H_j defined by

$$H_j(x, z_1, \ldots, z_j, t) = (1-t)\mathcal{E}_j(x, z_1, \ldots, z_j) + t\begin{pmatrix} \mathcal{E}_{j-1}(x, z_1, \ldots, z_{j-1}) \\ z_j \end{pmatrix}$$

$$= \begin{cases} p_1(x) + \sum_{i=1}^{j-1} \lambda_{1,i} z_i + (1-t)\lambda_{1,j} z_j, \\ \vdots \\ p_n(x) + \sum_{i=1}^{j-1} \lambda_{n,i} z_i + (1-t)\lambda_{n,j} z_j, \\ a_1 + a_{11}x_1 + \cdots + a_{1n}x_n + z_1, \\ \vdots \\ a_{j-1} + a_{j-1,1}x_1 + \cdots + a_{j-1,n}x_n + z_{j-1}, \\ (1-t)(a_j + a_{j,1}x_1 + \cdots + a_{j,n}x_n) + z_j. \end{cases}$$

Let Z_j denote the solutions of $\mathcal{E}_j(x, z_1, \ldots, z_j) = 0$ with $z_j \neq 0$. Following the paths of the homotopy $H_j(x, z_1, \ldots, z_j, t) = 0$ starting at points of Z_j will produce a set of solutions of $\mathcal{E}_{j-1}(x, z_1, \ldots, z_{j-1}) = 0$ at $t = 1$. Among them, let W_{j-1} denote the set of points with $z_{j-1} = 0$. By convention, when $j = 1$ the set W_0 is empty. As discussed above, we have

THEOREM 8.1 (SOMMESE and VERSCHELDE [2000]). *If* j *is the largest integer for which* $W_j \neq \emptyset$, *then the dimension of the solution set* V *of* $P(x) = 0$ *is* j. *Moreover, given any irreducible component* X *of* V *of dimension* $k \leqslant j$, *then the finite set* W_k *contains generic points of* X.

To find the sets W_k for $k = 1, \ldots, n-1$, we may follow the following procedure:

CASCADE OF HOMOTOPIES BETWEEN EMBEDDED SYSTEMS.

Step 0: Initialize the embedding sequences

$$\mathcal{E}_n(x, z_1, \dots, z_n), \quad \mathcal{E}_{n-1}(x, z_1, \dots, z_{n-1}), \quad \dots, \quad \mathcal{E}_1(x, z_1).$$

Step 1: Solve $Z_n :=$ all isolated zeros of $\mathcal{E}_n(x, z_1, \dots, z_n) = 0$ with $z_n \neq 0$.

Step 2: For $j = n - 1$ down to 0, follow paths of the homotopy

$$H_{j+1} = (1-t)\mathcal{E}_{j+1} + t\begin{pmatrix} \mathcal{E}_j \\ z_{j+1} \end{pmatrix} = 0$$

from $t = 0 \to 1$ with starting points

$$Z_{j+1} := \text{solutions of } \mathcal{E}_{j+1}(x, z_1, \dots, z_{j+1}) = 0 \quad \text{with } z_{j+1} \neq 0.$$

Let $W_j :=$ limits of solutions of $H_{j+1} = 0$ as $t \to 1$ with $z_j = 0$.

In the above procedure, by default, Step 1 begins by solving Z_n, the set of all isolated zeros of $\mathcal{E}_n(x, z_1, \dots, z_n) = 0$ with $z_n \neq 0$. But for efficiency reasons, if it can be determined by some other means that there are no components of the solution set V having dimension greater than k, one may start Step 1 with $k < n$.

When we proceed the cascade and identify the largest integer j_0 with $W_{j_0} \neq \emptyset$, then the dimension of the solution set V of $P(x) = 0$ is j_0. We may continue the procedure by following the paths of $H_{j_0}(x, z_1, \dots, z_{j_0}, t) = 0$ with starting points in Z_{j_0} to identify the lower dimensional components of V. However, as we mentioned earlier, the existence of nonempty W_j for $j < j_0$ may no longer warrant the existence of irreducible components of V of dimension j, because points in W_j may all land in components of V with higher dimensions. In SOMMESE, VERSCHELDE and WAMPLER [2001], an algorithm, called *WitnessClassify*, is developed to clarify this problem:

Let the decomposition of the entire solution set V of $P(x) = 0$ be the nested union:

$$V := \bigcup_{j=0}^{j_0} V_j := \bigcup_{j=0}^{j_0} \bigcup_{i \in I_j} V_{ji},$$

where V_j is the union of all j-dimensional components, the V_{ji} are the irreducible components of dimension j, and the index sets I_j are finite and possibly empty. Let

$$W = \bigcup_{j=0}^{j_0} W_j.$$

The *WitnessClassify* algorithm proceeds to classify the points in each of the nonempty W_j for $j < j_0$ by first separating out the points on higher-dimensional components. This can be accomplished by the construction of the *filtering polynomials*, p_{ji}, each of which vanishes on the entire irreducible component V_{ji} but which is nonzero, with probability 1, for any point in W that is not on V_{ji}. The geometric concept of the filtering polynomials is as follows. For an irreducible component V_{ji}, which by definition has dimension j, pick a generic linear subspace of directions having dimension $n - j - 1$ and define a new set constructed by replacing each point in V_{ji} with its expansion along the chosen subspace directions. The result is an $(n-1)$-dimensional hypersurface. The filtering polynomial is the unique polynomial that vanishes on this hypersurface.

When filtering polynomials for all components of dimension greater than j are available, those points J_j in W_j that lie on the higher-dimensional components can be identified. Let $\widehat{W}_j = W_j \backslash J_j$. The next task is to sort the points in $\widehat{W}_j$ by their membership in the irreducible component V_{ji}. For some arbitrary point $w \in \widehat{W}_j$, we move the slicing planes that pick w out of the underlying irreducible components V_{ji}, using continuation. In this manner, an arbitrary number of new points can be generated on V_{ji}. After picking a generic linear subspace of directions to expand V_{ji} into a hypersurface, one can find the lowest-degree polynomial that interpolates the samples, always taking extra samples as necessary to ensure the true hypersurface has been correctly determined. This is the filtering polynomial p_{ji}, which can then be used to find any additional points in $\widehat{W}_j$ that lies on V_{ji}. We then choose a new point from those in $\widehat{W}_j$ that are not sorted, and repeat the process until all points are sorted. With all the filtering polynomials p_{ji} in hand, we proceed to dimension $j-1$ and apply the same method.

Instead of giving the details, we shall illustrate the above decomposition procedure by presenting the following example given in SOMMESE, VERSCHELDE and WAMPLER [2001]:

EXAMPLE 8.1. Consider the polynomial system $P(x_1,x_2,x_3) = (p_1(x_1,x_2,x_3), p_2(x_1,x_2,x_3), p_3(x_1,x_2,x_3))$ where

$$\begin{aligned}
p_1(x_1,x_2,x_3) &= \qquad\qquad (x_2 - x_1^2)(x_1^2+x_2^2+x_3^2-1)(x_1-0.5),\\
p_2(x_1,x_2,x_3) &= \qquad\qquad (x_3 - x_1^3)(x_1^2+x_2^2+x_3^2-1)(x_2-0.5),\\
p_3(x_1,x_2,x_3) &= (x_2 - x_1^2)(x_3 - x_1^3)(x_1^2+x_2^2+x_3^2-1)(x_3-0.5).
\end{aligned}$$

The decomposition of the solution set V of $P(x_1,x_2,x_3)=0$ into its irreducible components is obvious:

$$V = V_2 \cup V_1 \cup V_0 = \{V_{21}\} \cup \{V_{11}\} \cup \{V_{12}\} \cup \{V_{13}\} \cup \{V_{14}\} \cup \{V_{01}\},$$

where

(1) V_{21} is the sphere $x_1^2+x_2^2+x_3^2=1$,
(2) V_{11} is the line $(x_1=0.5, x_3=(0.5)^3)$,
(3) V_{12} is the line $(x_1=\sqrt{0.5}, x_2=0.5)$,
(4) V_{13} is the line $(x_1=-\sqrt{0.5}, x_2=0.5)$,
(5) V_{14} is the twisted cubic $(x_2=x_1^2, x_3=x_1^3)$,
(6) V_{01} is the point $(x_1=0.5, x_2=0.5, x_3=0.5)$.

Solving $\mathcal{E}_3(x_1,x_2,x_3,z_1,z_2,z_3)=0$ in Step 1 of the cascade procedure by the polyhedral homotopy method given in Section 4, yields 139 isolated zeros which constitute Z_3. By following 139 paths of

$$H_3 = (1-t)\mathcal{E}_3 + t\begin{pmatrix}\mathcal{E}_2\\ z_3\end{pmatrix} = 0$$

starting from points in Z_3 obtained in Step 2, we reach 2 solutions in W_2 consisting of solutions with $z_1=z_2=z_3=0$, and 38 solutions in Z_2 consisting of solutions with

$z_1 z_2 \neq 0$. The rest of the paths, 99 of them, all went to infinity. Samples by continuation from the first point in W_2 were found to be interpolated by a quadratic surface (the sphere) and the second point was found to also fall on the sphere. Thus, component V_{21} is determined to be a second-degree variety. The sphere equation is appended to the filter, and the algorithm proceeds to the next level.

	Z_3	Z_2	W_2
H_3	139	38	2

W_2	Sphere
2	2

By following 38 paths of the homotopy

$$H_2 = (1-t)\mathcal{E}_2 + t \begin{pmatrix} \mathcal{E}_1 \\ z_2 \end{pmatrix} = 0$$

starting from points in Z_2 we obtain 14 solutions in W_1 and 20 solutions in Z_1. Among 14 solutions in W_1, 8 of them are found to lie on the sphere and are discarded as J_1. Using the sample and interpolate procedures, the remaining 6 are classified as 3 falling on 3 lines, one on each, and 3 on a cubic. A filtering polynomial for each of these is appended to the filter and the algorithm proceeds to the last level.

	Z_2	Z_1	W_1
H_2	38	20	14

W_1	J_1	Line 1	Line 2	Line 3	Cubic
14	8	1	1	1	3

By following 20 paths of the homotopy

$$H_1 = (1-t)\mathcal{E}_1 + t \begin{pmatrix} \mathcal{E}_0 = P(x_1, x_2, x_3) \\ z_1 \end{pmatrix} = 0$$

starting from points in Z_1 yields 19 solutions in W_0. Among them, 13 lie on the sphere, 2 on line 1, 2 on line 2 and 1 on line 3, leaving a single isolated point as W_{01}.

	Z_1	Z_0	W_0
H_1	20	0	19

W_0	J_0				W_{10}
	Sphere	Line 1	Line 2	Line 3	
19	13	2	2	1	1

9. Numerical considerations

9.1. *Fundamental procedure for following paths*

There are well established numerical techniques to track homotopy paths of a homotopy $H(x,t) = 0$, see ALLGOWER and GEORG [1990], ALLGOWER and GEORG [1993], ALLGOWER and GEORG [1997], in which homotopy paths are usually parametrized by the arc length. However, in the content of solving polynomial systems in $\mathbb{C}^n$ where homotopies $H(x,t)$ are defined on $\mathbb{C}^n \times [0,1]$, we will show below that for any point on the smooth homotopy path $(x(s), t(s))$ of $H(x,t) = 0$, parametrized by the arc length s, dt/ds is always nonzero, and therefore $dt/ds > 0$. Meaning: those paths do not "turn back in t". In other words, they extend across the interval $0 \leqslant t < 1$ and can always be parametrized by t. Accordingly, standard procedures in tracing general homotopy paths need to be properly adjusted to capitalize this special feature as we will elaborate in this section.

LEMMA 9.1 (CHOW, MALLET-PARET and YORKE [1979]). *Regard the $n \times n$ complex matrix M as a linear transformation of complex variables $(x_1, \ldots, x_n)$ in $\mathbb{C}^n$ into itself. If this transformation is regarded as one on the space $\mathbb{R}^{2n}$ of real variables $(u_1, v_1, \ldots, u_n, v_n)$ where $x_j = u_j + \mathrm{i}v_j$, $j = 1, \ldots, n$, and is represented by the $2n \times 2n$ real matrix N then*

$$\det N = \left|\det M\right|^2 \geqslant 0$$

and

$$\dim_R(\text{kernel } N) = 2 \times \dim_C(\text{kernel } M).$$

Here, $\dim_R$ *and* $\dim_{\mathbb{C}}$ *refer to real and complex dimension.*

PROOF. The relation between M and N is the following: if the (j,k)-entry of M is the complex number $m_{jk} = \xi_{jk} + \mathrm{i}\eta_{jk}$, and N is written in block form as an $n \times n$ array of 2×2 blocks, then the (j,k)-block of N is the real matrix

$$\begin{pmatrix} \xi_{jk} & -\eta_{jk} \\ \eta_{jk} & \xi_{jk} \end{pmatrix}.$$

Denoting this relation by $\alpha(M) = N$, we have $\alpha(AB) = \alpha(A)\alpha(B)$ for complex matrices A and B, and $\alpha(A^{-1}) = \alpha(A)^{-1}$.

Now when M is upper triangular, the assertion is clear. For general M, there exists complex nonsingular matrix A for which $A^{-1}MA$ is upper triangular. Because

$$\alpha(A^{-1}MA) = \alpha(A^{-1})\alpha(M)\alpha(A) = \alpha(A)^{-1}N\alpha(A),$$

it follows that

$$\det\bigl(\alpha(A^{-1}MA)\bigr) = \det\bigl(\alpha(A)\bigr)^{-1} \times \det N \times \det\bigl(\alpha(A)\bigr) = \det N.$$

The assertion holds, since

$$\det\bigl(\alpha(A^{-1}MA)\bigr) = \left|\det(A^{-1}MA)\right|^2 = \left|\det M\right|^2. \qquad \square$$

PROPOSITION 9.1. *If (x_0, t_0) is a point on any smooth homotopy paths $(x(s), t(s))$ of the homotopy $H(x,t) = 0$ defined on $\mathbb{C}^n \times [0,1]$ with $t_0 \in [0,1)$, then $H_x(x_0, t_0)$ is nonsingular. Hence, $dt/ds \neq 0$ at (x_0, t_0).*

PROOF. Regard H as a map from $\mathbb{R}^{2n} \times \mathbb{R}$ to $\mathbb{R}^{2n}$. Since the $2n \times (2n+1)$ Jacobian matrix $DH = [H_x, H_t]$ must be of full rank at (x_0, t_0) (otherwise it would be a bifurcation point (ALLGOWER [1984])), its kernel is at most one-dimensional. By the above lemma, the matrix H_x must have zero kernel, so it is nonsingular. Hence, $dt/ds \neq 0$ at

(x_0, t_0), because

$$H_x \frac{dx}{ds} + H_t \frac{dt}{ds} = 0. \qquad \square$$

Algorithms for following homotopy paths vary but are typically of the *predictor–corrector* form, in which the next point on the path is "predicted" by some easy but relatively inaccurate means, and then a series of Newton-like "corrector" iterations is employed to return approximately to the path.

Since homotopy paths of the homotopy $H(x,t) = 0$ in $\mathbb{C}^n \times [0,1]$ can always be parametrized by t, let $x(t)$ be a path in $\mathbb{C}^n$ satisfying the homotopy equation $H(x,t) = 0$, that is,

$$H\bigl(x(t), t\bigr) = 0, \quad 0 \leqslant t \leqslant 1. \tag{9.1}$$

From here on we shall denote dx/dt by $x'(t)$. Now, differentiating (9.1) with respect to t, yields

$$H_x x'(t) + H_t = 0, \quad 0 \leqslant t \leqslant 1,$$

or

$$x'(t) = -H_x^{-1} H_t, \quad 0 \leqslant t \leqslant 1. \tag{9.2}$$

For a fixed $0 \leqslant t_0 \leqslant 1$, to proceed from the point $x(t_0)$ already attained on $x(t)$, one takes the following fundamental steps:

Step 1: Euler Prediction:

For an adaptive step size $\delta > 0$, let $t_1 = t_0 + \delta < 1$ and

$$\tilde{x}(t_1) = x(t_0) + \delta x'(t_0). \tag{9.3}$$

Step 2: Newton's Correction:

For fixed t_1, $H(x, t_1) = 0$ becomes a system of n equations in n unknowns. So, Newton's iteration can be employed to solve the solution of $H(x, t_1) = 0$ with starting point $\tilde{x}(t_1)$, i.e.

$$x^{(m+1)} = x^{(m)} - \bigl[H_x\bigl(x^{(m)}, t_1\bigr)\bigr]^{-1} H\bigl(x^{(m)}, t_1\bigr), \quad m = 0, 1, \ldots, \tag{9.4}$$

with $x^{(0)} = \tilde{x}(t_1)$. When the iteration fails to converge, Step 1 will be repeat with $\delta \leftarrow \frac{\delta}{2}$. Eventually, an approximate value of $x(t_1)$ can be obtained.

For the efficiency of the algorithm, one seldom stays with a predetermined and fixed step size in practice. Based on the smoothness of the path $x(t)$ around t_0, there are different strategies of choosing step size $\delta > 0$ in Step 1, and of course the smoother the path is, the larger the step size can be adapted. An effective tool to measure the smoothness of $x(t)$ at t_0 is the angle between two consecutive tangent vectors $x'(t_{-1})$ and $x'(t_0)$, where $x(t_{-1})$ is the previous point on the path. For the prediction in Step 1, there are several alternatives for the Euler prediction in (9.3), which are more efficient empirically:

- *The cubic Hermite interpolation*
For $x(t) = (x_1(t), \dots, x_n(t))$ and $j = 1, \dots, n$, let $\tilde{x}_j(t)$ be the cubic polynomial which interpolates $x_j(t)$ and $x_j'(t)$ at $t = t_{-1}$ and $t = t_0$. Namely,

$$\tilde{x}_j(t_{-1}) = x_j(t_{-1}), \quad \tilde{x}_j(t_0) = x_j(t_0)$$

and

$$\tilde{x}_j'(t_{-1}) = x_j'(t_{-1}), \quad \tilde{x}_j'(t_0) = x_j'(t_0).$$

Writing $\tilde{x}(t) = (\tilde{x}_1(t), \dots, \tilde{x}_n(t))$, the value $\tilde{x}(t_1)$ will be taken as the prediction of $x(t)$ at $t = t_1$.
This method usually provides more accurate prediction for $x(t_1)$ with no extra computational cost.
- *The cubic interpolation*
Let $x(t_{-3}), x(t_{-2}), x(t_{-1})$ be three consecutive points immediately previous to $x(t_0)$ on $x(t)$. For $j = 1, \dots, n$, let $\tilde{x}_j(t)$ be the cubic polynomial which interpolates $x_j(t)$ at $t = t_{-3}, t_{-2}, t_{-1}$ and t_0. With $\tilde{x}(t) = (\tilde{x}_1(t), \dots, \tilde{x}_n(t))$, naturally we let $\tilde{x}(t_1)$ be the predicted value of $x(t)$ at $t = t_1$.
To start this method of prediction, one may use some other means to find the beginning four points on $x(t)$, starting at $t = 0$, such as the Euler method in (9.3). The most important advantage of this interpolation is the absence of the derivative computation in the prediction steps all along the path following (except, perhaps, at the first few points), which may sometimes be very costly.

9.2. *Projective coordinates and the projective Newton method*

Solution paths of $H(x, t) = 0$ that do not proceed to a solution of the target polynomial equation $P(x) = 0$ in $\mathbb{C}^n$ diverge to infinity: a very poor state of affairs for numerical methods. However, there is a simple idea from classical mathematics which improves the situation. If the system $H(x, t)$ is viewed in the projective space $\mathbb{P}^n$, the diverging paths are simply proceeding to a "point at infinity". Since projective space is compact, we can force all paths, including the extraneous ones, to have finite length by using projective coordinates.

For $P(x) = (p_1(x), \dots, p_n(x)) = 0$ and $x = (x_1, \dots, x_n) \in \mathbb{C}^n$, we first homogenize the $p_j(x)$'s, that is, for $j = 1, \dots, n$,

$$\tilde{p}_j(x_0, \dots, x_n) = x_0^{d_j} p_j\left(\frac{x_1}{x_0}, \dots, \frac{x_n}{x_0}\right), \quad d_j = \text{degree of } p_j(x).$$

Then consider the system of $n + 1$ equations in $n + 1$ unknowns

$$\widetilde{P}: \begin{cases} \tilde{p}_1(x_0, \dots, x_n) & = 0, \\ & \vdots \\ \tilde{p}_n(x_0, \dots, x_n) & = 0, \\ a_0 x_0 + \dots + a_n x_n - 1 & = 0, \end{cases}$$

where $a_0, \ldots, a_n$ are generically chosen complex numbers. In short, we augment the homogenization of $P(x)$ with a generic hyperplane

$$a_0x_0 + \cdots + a_nx_n - 1 = 0.$$

When a start system

$$Q(x) = \big(q_1(x), \ldots, q_n(x)\big) = 0$$

is chosen, we augment its homogenization with the same hyperplane,

$$\widetilde{Q}\colon \begin{cases} \tilde{q}_1(x_0, \ldots, x_n) & = 0, \\ & \vdots \\ \tilde{q}_n(x_0, \ldots, x_n) & = 0, \\ a_0x_0 + \cdots + a_nx_n - 1 & = 0. \end{cases}$$

When the classical linear homotopy continuation procedure is used to follow all the solution paths of the homotopy

$$\widetilde{H}(x_0, x, t) = (1-t)c\widetilde{Q}(x_0, x) + t\widetilde{P}(x_0, x) = 0, \quad c \in \mathbb{C}^* \text{ is generic},$$

the paths stay in $\mathbb{C}^{n+1}$ for almost all choices of $a_0, \ldots, a_n$. It only remains to ignore solutions with $x_0 = 0$ when t reaches 1. For the remaining solutions with $x_0 \neq 0$, $x = (x_1/x_0, \ldots, x_n/x_0)$ are the corresponding solutions of $P(x) = 0$ in $\mathbb{C}^n$.

A similar technique, called *projective transformation* is described in MORGAN and SOMMESE [1987a]. It differs from the above in the following way. Instead of increasing the size of the problem from $n \times n$ to $(n+1) \times (n+1)$, they implicitly consider solving the last equation for x_0 and substituting it in the other equations, essentially retaining n equations in n unknowns. Then the chain rule is used for the Jacobian calculations needed for path following.

A more advanced technique, called the *projective Newton method*, was suggested in BLUM, CUCKER, SHUB and SMALE [1998] and SHUB and SMALE [1993]. For fixed $0 < t_0 < 1$, the homogenization of the homotopy equation $H(x, t) = 0$ becomes $\widetilde{H}(\tilde{x}, t_0) = 0$ with $\tilde{x} = (x_0, x_1, \ldots, x_n)$. It is a system of n equations in $n+1$ variables. Instead of following solution paths of $\widetilde{H}(\tilde{x}, t) = 0$ in $\mathbb{C}^{n+1}$ with an additional *fixed* hyperplane $a_0x_0 + \cdots + a_nx_n - 1 = 0$ as described above, the new strategy follows the paths in $\mathbb{C}^{n+1}$ with variant hyperplanes. Without hyperplane $a_0x_0 + \cdots + a_nx_n - 1 = 0$, Newton's iteration is unsuitable for approximating the solution of an $n \times (n+1)$ system $\widetilde{H}(\tilde{x}, t_0) = 0$ in the fundamental correction step after the prediction. However, for any nonzero constant $c \in \mathbb{C}$, $\tilde{x}$ and $c\tilde{x}$ in $\mathbb{C}^{n+1}$ are considered to be equal in projective space $\mathbb{P}^n$, whence the magnitude of $\tilde{x}$ in $\mathbb{C}^{n+1}$ is no longer significant in $\mathbb{P}^n$. Therefore it is reasonable to *project* every step of Newton's iteration onto the hyperplane that is perpendicular to the current point in $\mathbb{C}^{n+1}$. More precisely, to approximate the solutions of $\widetilde{H}(\tilde{x}, t_0) = 0$ for fixed $0 < t_0 < 1$, at a point $\tilde{x}^{(m)} \in \mathbb{C}^{n+1}$ during the correction process, we augment the equation $\widetilde{H}(\tilde{x}, t_0) = 0$ with the hyperplane $(\tilde{x} - \tilde{x}^{(m)}) \cdot \tilde{x}^{(m)} = 0$ to form an $(n+1) \times (n+1)$ square system

$$\overline{H}(\tilde{x}, t_0) = \begin{cases} \widetilde{H}(\tilde{x}, t_0) = 0, \\ (\tilde{x} - \tilde{x}^{(m)}) \cdot \tilde{x}^{(m)} = 0. \end{cases} \tag{9.5}$$

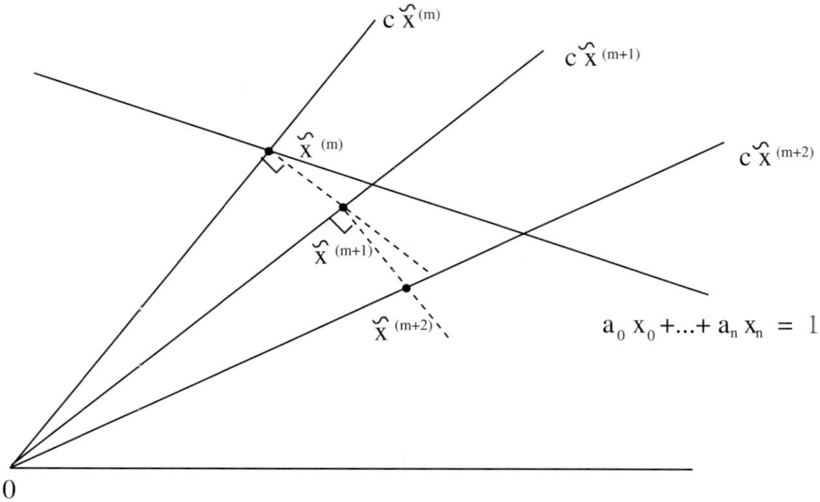

FIG. 9.1.

One step of Newton's iteration applied to this system at $\tilde{x}^{(m)}$ yields

$$\tilde{x}^{(m+1)} = \tilde{x}^{(m)} - \left[\overline{H}_{\tilde{x}}(\tilde{x}^{(m)}, t_0)\right]^{-1} \overline{H}\left(\tilde{x}^{(m)}, t_0\right). \tag{9.6}$$

The iteration is continued by replacing $\tilde{x}^{(m)}$ in (9.5) with $\tilde{x}^{(m+1)}$ obtained in (9.6) until $\widetilde{H}(\tilde{x}^{(m)}, t_0)$ becomes sufficiently small.

The efficiency of this strategy when applied to following the homotopy paths in $\mathbb{P}^n$, is intuitively clear. See Fig. 9.1. It frequently allows a bigger step size at the prediction step.

9.3. Balancing the lifting values in polyhedral homotopies

Polyhedral homotopies are nonlinear in the continuation parameter t. Powers of the continuation parameter too close to zero can be scaled away from zero by a suitable scalar multiplication. After scaling, if very high powers exist in a polyhedral homotopy, small step sizes must be taken in order to successfully trace a solution path of the polyhedral homotopy. Although the end of the solution path can be reached as long as the required step size is not smaller than the available machine precision, the efficiency of the path-tracing is greatly reduced. A more serious problem occurs when the continuation parameter is not yet close enough to 1, some terms of the polyhedral homotopy with high powers of t have values smaller than the machine precision and some solution curves may come close to "valleys" where the values of the homotopy are numerically zero, but no solution paths exist inside the "valleys". This situation can easily cause the path-tracings to be trapped in those "valleys" with no chance of reaching the ends of solution paths unless the paths are retraced with smaller step sizes.

For instance, consider the polynomial system $P(x) = (p_1(x), p_2(x)) = 0$ where $x = (x_1, x_2)$, and

$$p_1(x) = c_1x_1^2 + c_2x_2^2 + c_3 = 0,$$
$$p_2(x) = c_4x_1^3x_2^3 + c_5x_1 + c_6x_2 = 0$$

with support $S_1 = \{(2,0),(0,2),(0,0)\}$ and $S_2 = \{(3,3),(1,0),(0,1)\}$. By a simple calculation, the mixed volume $\mathcal{M}(S_1, S_2)$ of this system is 12. We choose a generic lifting $\omega = (\omega_1, \omega_2)$ where $\omega_1 : S_1 \to \mathbb{R}$ and $\omega_2 : S_2 \to \mathbb{R}$ with

$$\omega_1(2,0) = 0.655416, \quad \omega_1(0,2) = 0.200995, \quad \omega_1(0,0) = 0.893622$$

and

$$\omega_2(3,3) = 0.281886, \quad \omega_2(1,0) = 0.525000, \quad \omega_2(0,1) = 0.314127.$$

Then the mixed-cell configuration $\mathcal{M}_\omega$ in the fine mixed subdivision S_ω of S induced by ω consists of two mixed cells:

$$C^{(1)} = \big(\{(2,0),(0,2)\},\{(3,3),(1,0)\}\big)$$

and

$$C^{(2)} = \big(\{(0,2),(0,0)\},\{(1,0),(0,1)\}\big).$$

To construct a system $G(x) = (g_1(x), g_2(x)) = 0$ in general position with the same support $S = (S_1, S_2)$, we choose the following set of randomly generated coefficients,

$$c_1' = -0.434847 - 0.169859\mathrm{i}, \qquad c_2' = 0.505911 + 0.405431\mathrm{i},$$
$$c_3' = 0.0738596 + 0.177798\mathrm{i}, \qquad c_4' = -0.0906755 + 0.208825\mathrm{i},$$
$$c_5' = 0.175313 - 0.163549\mathrm{i}, \qquad c_6' = 0.527922 - 0.364841\mathrm{i}.$$

The polyhedral homotopy induced by the cell $C^{(1)}$, following the procedure given in Section 4, is $\overline{G}(y,t) = (\bar{g}_1(y,t), \bar{g}_2(y,t)) = 0$, where

$$\bar{g}_1(y,t) = c_1'y_1^2 + c_2'y_2^2 + c_3't^{50.63523},$$
$$\bar{g}_2(y,t) = c_4'y_1^3y_2^3 + c_5'y_1 + c_6'y_2t^2.$$

It is easy to see that $\mathrm{Vol}_2(C^{(1)}) = 10$. Therefore, there are ten solution paths of $\overline{G}(y,t) = 0$ emanating from the ten solutions of $\overline{G}(y,0) = 0$. At $t = 0.65$, five of those ten paths have phase space tangent vectors $(dy_1/dt, dy_2/dt)$ all pointing closely to $y = (0,0)$ at the points on the curves. For the standard prediction–correction method, starting at these points, the prediction step with step size 0.025 yield the predicted points close to $(y,t) = (0,0,0.675)$ for all those five paths. Since $t^{50.63523}$ is about 10^{-9} for $t = 0.675$, the function values of $\overline{G}(y,t)$ in a very small neighborhood of $(0,0,0.675)$, the "valley", are almost zero. Starting from these predicted points at $t = 0.675$, Newton's iterations for the correction step will converge to $(0,0)$, center of the "valley", rather than the points on the paths at $t = 0.675$. But there are no solution paths of $\overline{G}(y,t) = 0$ passing through the "valley". This shows that generic liftings may induce highly nonlinear polyhedral homotopies which may produce numerical instabilities.

Two known geometric approaches to control the numerical stability of polyhedral homotopy continuation methods are recursive liftings as in VERSCHELDE, VERLINDEN and COOLS [1994] and dynamic liftings in VERSCHELDE, GATERMANN and COOLS [1996] and VERSCHELDE [1996]. However, because of using multiple liftings or flattenings, these approaches both require more expensive construction of subdivisions and create more homotopy curves which need to be traced than a random floating-point lifting.

To minimize the height of powers of the continuation parameter t in polyhedral homotopies, we search in the cone of all lifting vectors that induce the same mixed-cell configuration to obtain better-balanced powers of the continuation parameter of polyhedral homotopies. As given in Section 4, the first step of the polyhedral homotopy procedure to find all isolated zeros of a polynomial system $P(x) = (p_1(x), \ldots, p_n(x))$ with $x = (x_1, \ldots, x_n) \in \mathbb{C}^n$ and support $S = (S_1, \ldots, S_n)$ where

$$p_j(x) = \sum_{a \in S_j} c_{j,a} x^a, \quad j = 1, \ldots, n,$$

is to solve a generic polynomial system $G(x) = (g_1(x), \ldots, g_n(x)) = 0$ with support $S' = (S'_1, \ldots, S'_n)$ where $S'_j = S_j \cup \{0\}$ for $j = 1, \ldots, n$, but with randomly chosen coefficients $c'_{j,a}$, namely,

$$g_j(x) = \sum_{a \in S'_j} c'_{j,a} x^a, \quad j = 1, \ldots, n.$$

Secondly, the system $G(x) = 0$ is used as the start system in the linear homotopy

$$H(x,t) = (1-t)cG(x) + tP(x) = 0, \quad c \in \mathbb{C}^* \text{ is generic},$$

to solve $P(x) = 0$.

To find all isolated zeros of $G(x) = (g_1(x), \ldots, g_n(x))$, we first construct a nonlinear homotopy $\widehat{G}(x,t) = (\hat{g}_1(x,t), \ldots, \hat{g}_n(x,t)) = 0$ where

$$\hat{g}_j(x,t) = \sum_{a \in S'_j} c'_{j,a} x^a t^{\omega_j(a)}, \quad j = 1, \ldots, n, \tag{9.7}$$

and the powers $\omega_j(a)$ of t are determined by the generic *liftings* $\omega_j : S'_j \to \mathbb{R}$ on the support S'_j, $j = 1, \ldots, n$, followed by calculating all mixed cells of the fine mixed subdivision S_ω induced by the lifting $\omega = (\omega_1, \ldots, \omega_n)$. Let $\mathcal{M}_\omega$ denote the set of all those mixed cells and call it the *mixed-cell configuration* of the support $S' = (S'_1, \ldots, S'_n)$. For a mixed cell $(\{a_1, a'_1\}, \ldots, \{a_n, a'_n\}) \in \mathcal{M}_\omega$ with inner normal $\alpha = (\alpha_1, \ldots, \alpha_n) \in \mathbb{R}^n$, substituting the coordinate transformation $y = xt^\alpha$ where $y_j = x_j t^{\alpha_j}$, $j = 1, \ldots, n$ into system (9.7), and factoring out the lowest powers of t, we obtain the homotopy $H^\alpha(y,t) = (h_1^\alpha(y,t), \ldots, h_n^\alpha(y,t)) = 0$ defined on $(\mathbb{C}^*)^n \times [0,1]$ where

$$h_j^\alpha(y,t) = c'_{j,a_j} y^{a_j} + c'_{j,a'_j} y^{a'_j} + \sum_{a \in S'_j \backslash \{a_j, a'_j\}} c'_{j,a} y^a t^{e_a^\omega}, \quad j = 1, \ldots, n,$$

and e_a^ω, the powers of t, are given by

$$e_a^\omega := \langle a, \alpha\rangle - \langle a_j, \alpha\rangle + \omega_j(a) - \omega_j(a_j), \quad \forall a \in S'_j \backslash \{a_j, a'_j\}.$$

To reduce the power of t for numerical stability, the strategy suggested in GAO, LI, VERSCHELDE and WU [2000] is to find a new lifting function $\nu = (\nu_1, \ldots, \nu_n)$ on $S' = (S'_1, \ldots, S'_n)$ based on the *already computed* mixed-cell configuration $\mathcal{M}_\omega$. The mixed-cell configuration $\mathcal{M}_\nu$ induced by the new lifting ν will be the same as $\mathcal{M}_\omega$, but the highest power of the continuation parameter t in the polyhedral homotopies induced by $\nu = (\nu_1, \ldots, \nu_n)$ is the smallest among all those polyhedral homotopies induced by the liftings having the *same* mixed-cell configuration $\mathcal{M}_\omega$. In this way, very time-consuming procedure for re-identifying the mixed-cell configuration $\mathcal{M}_\nu$ becomes unnecessary.

Let $C = (C_1, \ldots, C_n) \in \mathcal{M}_\omega$ where $C_j = \{a_j, a'_j\} \subset S_j$ for $j = 1, \ldots, n$. To keep $\mathcal{M}_\omega$ invariant, we impose on any new lifting function $\nu = (\nu_1, \ldots, \nu_n)$ the conditions:

$$\begin{aligned}&\langle (a_j, \nu_j(a_j)), (\gamma, 1)\rangle = \langle (a'_j, \nu_j(a'_j)), (\gamma, 1)\rangle,\\ &\langle (a_j, \nu_j(a_j)), (\gamma, 1)\rangle < \langle (a, \nu_j(a)), (\gamma, 1)\rangle, \quad \forall a \in S'_j \backslash \{a_j, a'_j\},\ j = 1, \ldots, n,\end{aligned}$$

or,

$$\langle a_j, \gamma\rangle + \nu_j(a_j) = \langle a'_j, \gamma\rangle + \nu_j(a'_j), \tag{9.8}$$

$$\langle a_j, \gamma\rangle + \nu_j(a_j) < \langle a, \nu\rangle + \nu_j(a), \quad \forall a \in S_j \backslash \{a_i, a'_j\},\ j = 1, \ldots, n, \tag{9.9}$$

where γ is the inner normal of C in $\mathcal{M}_\nu$. Since $C = (C_1, \ldots, C_n)$ is a mixed cell in the fine mixed subdivision S_ω, the edges spanned by $\{a_j, a'_j\}$ determine a full-dimensional parallelepiped in $\mathbb{R}^n$, that is, the matrix

$$A = \begin{bmatrix} a_1 - a'_1 \\ \vdots \\ a_n - a'_n \end{bmatrix}$$

is nonsingular. So, γ can be expressed uniquely as a linear combination of the lifting values $\nu_j(a_j)$ and $\nu_j(a'_j)$ for $j = 1, \ldots, n$. More explicitly, from (9.8), we have

$$\langle a_j - a'_j, \gamma\rangle = \nu_j(a'_j) - \nu_j(a_j), \quad j = 1, \ldots, n,$$

therefore

$$\gamma^{\mathrm{T}} = A^{-1} \begin{bmatrix} \nu_1(a'_1) - \nu_1(a_1) \\ \vdots \\ \nu_n(a'_n) - \nu_n(a_n) \end{bmatrix}. \tag{9.10}$$

The polyhedral homotopy induced by the lifting $\nu = (\nu_1, \ldots, \nu_n)$ and mixed cell $C = (\{a_1, a'_1\}, \ldots, \{a_n, a'_n\})$ with inner normal γ is

$$\bar{h}_j^\gamma(y, t) = c'_{j,a_j} y^{a_j} + c'_{j,a'_j} y^{a'_j} + \sum_{a \in S'_j \backslash \{a_j, a'_j\}} c'_{j,a} y^a t^{e_a^\gamma}, \quad j = 1, \ldots, n, \tag{9.11}$$

where e_a^γ, the powers of t, are given by

$$e_a^\gamma := \langle a, \gamma\rangle - \langle a_j, \gamma\rangle + \nu_j(a) - \nu_j(a_j), \quad \forall a \in S_j'\backslash\{a_j, a_j'\}.$$

They are always positive by (9.9). By (9.10), γ may be removed from e_a^γ. We denote the resulting expressions by e_a. Explicitly, $\forall a \in S_j'\backslash\{a_j, a_j'\}$,

$$e_a := (a - a_j)A^{-1}\begin{bmatrix} \nu_1(a_1') - \nu_1(a_1) \\ \vdots \\ \nu_n(a_n') - \nu_n(a_n)\end{bmatrix} + \nu_j(a) - \nu_j(a_j'). \tag{9.12}$$

To avoid the powers e_a being too large or too small, it is natural to consider the following minimization problem:

$$\begin{aligned} &\text{Minimize } M \\ &1 \leqslant e_a \leqslant M, \ e_a \text{ is given in (9.12)} \\ &\forall C = \big(\{a_1, a_1'\}, \dots, \{a_n, a_n'\}\big) \in \mathcal{M}_\omega, \quad \text{and} \\ &\forall a \in S_j'\backslash\{a_j, a_j'\}, \quad j = 1, \dots, n. \end{aligned} \tag{9.13}$$

Clearly, the lifting function $\nu = (\nu_1, \dots, \nu_n)$ having values obtained by the solution of this minimization problem satisfies (9.8) and (9.9) with γ defined in (9.10). Therefore, the mixed-cell configuration $\mathcal{M}_\nu$ induced by ν coincides with $\mathcal{M}_\omega$. Moreover, the powers of the continuation parameter t in the polyhedral homotopies induced by ν are much better balanced.

The minimization problem (9.13) has $1 + \sum_{j=1}^n \#S_j'$ unknowns; they are: M as well as $\nu_j(a)$ for $a \in S_j'$, $j = 1, \dots, n$. It has $2(\#\mathcal{M}_\omega)\sum_{j=1}^n(\#S_j' - 2)$ inequalities. For practical considerations, we wish to reduce both the number of unknowns and the number of inequalities. We will show below that for a fixed mixed cell $\overline{C} = (\{\bar a_1, \bar a_1'\}, \dots, \{\bar a_n, \bar a_n'\})$, where $\{\bar a_j, \bar a_i'\} \subset S_j'$, $j = 1, \dots, n$, in the mixed-cell configuration $\mathcal{M}_\omega$ with inner normal β, we may set $\nu_j(\bar a_j)$ and $\nu_j(\bar a_j)$ to be zero for $j = 1, \dots, n$ in (9.13), so the number of unknowns is reduced by $2n$. But the lifting function $\nu' = (\nu_1', \dots, \nu_n')$ defined by the solution of the new minimization problem still induces the same mixed-cell configuration $\mathcal{M}_\omega$.

For a fixed mixed cell $\overline{C} = (\{\bar a_1, \bar a_1'\}, \dots, \{\bar a_n, \bar a_n'\}) \in \mathcal{M}_\omega$ with inner normal β, we define a lifting function $\omega' = (\omega_1', \dots, \omega_n')$ as follows: for $j = 1, \dots, n$ and $a \in S_j'$,

$$\omega_j'(a) := \langle a, \beta\rangle - \langle \bar a_j, \beta\rangle + \omega_j(a) - \omega_j(\bar a_j). \tag{9.14}$$

Then ω_j' vanishes at both $\bar a_j$ and $\bar a_j'$. Let $\mathcal{M}_{\omega'}$ be the mixed-cell configuration in the subdivision $S_{\omega'}$ of $S' = (S_1', \dots, S_n')$ induced by ω'.

LEMMA 9.2. *$\mathcal{M}_{\omega'} = \mathcal{M}_\omega$. More precisely, $C = (C_1, \dots, C_n) \in \mathcal{M}_\omega$ with inner normal α with respect to ω if and only if $C = (C_1, \dots, C_n) \in \mathcal{M}_{\omega'}$ with inner normal $\alpha - \beta$ with respect to ω'.*

PROOF. Let $C_j = \{a_j, a_j'\} \subset S_j'$ for $j = 1, \ldots, n$. Then, $C \in \mathcal{M}_\omega$ with inner normal

$$\alpha \quad \Leftrightarrow \quad \begin{cases} \langle (a_j, \omega_j(a_j)), (\alpha, 1) \rangle = \langle (a_j', \omega_j(a_j')), (\alpha, 1) \rangle, \\ \langle (a_j, \omega_j(a_j)), (\alpha, 1) \rangle < \langle (a, \omega_j(a)), (\alpha, 1) \rangle, \\ \quad \forall a \in S_j \backslash \{a_j, a_j'\}, \ j = 1, \ldots, n. \end{cases}$$

Or,

$$\begin{cases} \langle a_j - a_j', \alpha \rangle = \omega_j(a_j') - \omega_j(a_j), \\ \langle a - a_j, \alpha \rangle > \omega_j(a_j) - \omega_j(a), \quad \forall a \in S_j' \backslash \{a_j, a_j'\}, \ j = 1, \ldots, n. \end{cases}$$

On the other hand, $C \in \mathcal{M}_{\omega'}$ with inner normal

$$\alpha - \beta \quad \Leftrightarrow \quad \begin{cases} \langle (a_j, \omega_j'(a_j)), (\alpha - \beta, 1) \rangle = \langle (a_j', \omega_j'(a_j')), (\alpha - \beta, 1) \rangle, \\ \langle (a_j, \omega_j'(a_j)), (\alpha - \beta, 1) \rangle < \langle (a, \omega_j'(a)), (\alpha - \beta, 1) \rangle, \\ \quad \forall a \in S_j \backslash \{a_j, a_j'\}, \ j = 1, \ldots, n. \end{cases}$$

By (9.14),

$$\begin{aligned} \langle a_j - a_j', \alpha - \beta \rangle &= \omega_j'(a_j') - \omega_j'(a_j) \\ &= \langle a_j', \beta \rangle - \langle \bar{a}_j, \beta \rangle + \omega_j(a_j') - \omega_j(\bar{a}_j) \\ &\quad - \big(\langle a_j, \beta \rangle - \langle \bar{a}_j, \beta \rangle + \omega_j(a_j) - \omega_j(\bar{a}_j) \big) \\ &= \langle a_j - a_j', -\beta \rangle + \omega_j(a_j') - \omega_j(a_j), \quad \text{i.e.} \\ \langle a_j - a_j', \alpha \rangle &= \omega_j(a_j') - \omega_j(a_j), \quad j = 1, \ldots, n. \end{aligned}$$

And, for $a \in S_j' \backslash \{a_j, a_j'\}$,

$$\begin{aligned} \langle a - a_j, \alpha - \beta \rangle &> \omega_j'(a_j) - \omega_j'(a) \\ &= \langle a_j, \beta \rangle - \langle \bar{a}_j, \beta \rangle + \omega_j(a_j) - \omega_j(\bar{a}_j) \\ &\quad - \big(\langle a, \beta \rangle - \langle \bar{a}_j, \beta \rangle + \omega_j(a) - \omega_j(\bar{a}_j) \big) \\ &= \langle a - a_j, -\beta \rangle + \omega_j(a_j) - \omega_j(a), \quad \text{i.e.} \\ \langle a - a_j, \alpha \rangle &> \omega_j(a_j) - \omega_j(a), \quad j = 1, \ldots, n. \end{aligned}$$

The proof is completed. □

Most importantly, a straightforward calculation shows that the polyhedral homotopy, as in (9.11), induced by the cell $C = (C_1, \ldots, C_n)$ in $\mathcal{M}_\omega$ with inner normal α is exactly the same as the one induced by cell $C = (C_1, \ldots, C_n)$ in $\mathcal{M}_{\omega'}$ with inner normal $\alpha - \beta$. So, we may solve the generic polynomial system $G(x) = 0$ by using the polyhedral homotopies induced by mixed cells in $\mathcal{M}_{\omega'}$ along with their corresponding inner normals.

Now, with lifting ω', we consider the minimization problem

$$\begin{aligned}&\text{Minimize } M\\&1 \leqslant \langle a - a_j, \gamma\rangle + \nu_j(a) - \nu_j(a_j) \leqslant M,\\&\forall C = (\{a_1, a_1'\}, \ldots, \{a_n, a_n'\}) \in \mathcal{M}_{\omega'},\ \forall a \in S_j \backslash \{a_j, a_j'\},\\&\nu_j(\bar{a}_j) = \nu_j(\bar{a}_j') = 0, \quad j = 1, \ldots, n.\end{aligned} \tag{9.15}$$

Here, γ can be expressed, as in (9.12), as a linear combination of the values of ν_j's:

$$\gamma^{\mathrm{T}} = \begin{bmatrix} a_1 - a_1' \\ \vdots \\ a_n - a_n' \end{bmatrix}^{-1} \begin{bmatrix} \nu_1(a_1') - \nu_1(a_1) \\ \vdots \\ \nu_n(a_n') - \nu_n(a_n) \end{bmatrix}.$$

This problem has $2n$ fewer unknowns than the original problem in (9.13), and its solution yields a lifting function $\nu' = (\nu_1', \ldots, \nu_n')$, and the mixed-cell configuration $\mathcal{M}_{\nu'}$ induced by which is the same as $\mathcal{M}_{\omega'}$.

For the feasibility of this LP problem, the values of the lifting function $\omega' = (\omega_1', \ldots, \omega_n')$ clearly satisfy

$$0 < \langle \alpha - a_j, \alpha\rangle + \omega_j'(a) - \omega_j'(a_j), \quad j = 1, \ldots, n, \tag{9.16}$$

for all $C = (\{a_1, a_1'\}, \ldots, \{a_n, a_n'\}) \in \mathcal{M}_{\omega'}$ with corresponding inner normal α and $a \in S_j' \backslash \{a_j, a_j'\}$, where

$$\alpha^{\mathrm{T}} = \begin{bmatrix} a_1 - a_1' \\ \vdots \\ a_n - a_n' \end{bmatrix}^{-1} \begin{bmatrix} \omega_1'(a_1') - \omega_1'(a_1) \\ \vdots \\ \omega_n'(a_n') - \omega_n'(a_n) \end{bmatrix}.$$

Therefore the function values of $\nu^{(l)} := l\omega'$ for $l > 0$ also satisfy (9.16) with α replaced by $l\alpha$. We may choose $l_0 > 0$ for which

$$1 \leqslant \langle a - a_j, l_0\alpha\rangle + \nu_j^{(l_0)}(a) - \nu_j^{(l_0)}(a_j), \quad j = 1, \ldots, n, \tag{9.17}$$

for all $C = (\{a_1, a_1'\}, \ldots, \{a_n, a_n'\}) \in \mathcal{M}_{\omega'}$ and $a \in S_j' \backslash \{a_j, a_j'\}$. This makes $(\nu^{(l_0)}, M_0)$, where M_0 is the maximum of the right-hand side of (9.17), a feasible solution of the constraints in (9.15).

The number of variables in each inequality in the constraints of (9.15) is no greater than $2n + 2$ which is usually much smaller than the total number of variables in (9.15). This sparsity in the constraints can be exploited in the algorithm implementation and results in a remarkable speed-up. Furthermore, a substantial amount of the inequalities in (9.15) are exactly the same, and they can easily be detected by comparisons and deleted when the constraints are being generated.

The algorithm in balancing the powers of t by solving the LP problem (9.15) has been successfully implemented (GAO, LI, VERSCHELDE and WU [2000]). The numerical results of applying the algorithm to several well-known polynomial systems are listed in Tables 9.1 and 9.2.

TABLE 9.1
Sizes of the LP problems. Here, n is the number of variables and $\#\mathcal{M}_\omega$ is the number of mixed cells in the mixed-cell configuration $\mathcal{M}_\omega$.

Polynomial system	n	$\#\mathcal{M}_\omega$	Size of the LP in (9.13)		Size of the LP in (9.15)	
			Number of variables	Number of constraints	Number of variables	Number of constraints
Cohn-2	4	17	31	748	23	690
Cassou–Noguès	4	3	28	114	20	106
Planar 4-bar	4	4	33	192	25	168
Cyclic-6	6	25	33	1000	21	692
Cyclic-7	7	126	45	7560	31	4982
Cyclic-8	8	297	59	24948	43	16118

TABLE 9.2
Height of powers and CPU time in seconds.

Polynomial system	Average highest power of t		Average CPU time	
	Before balancing	After balancing	Finding mixed cells	Balancing method
Cohn-2	1391	85	0.21	0.19
Cassou–Noguès	251	11	0.05	0.03
Planar 4-bar	429	8	0.17	0.08
Cyclic-6	425	31	0.46	0.17
Cyclic-7	3152	139	7.1	1.9
Cyclic-8	10281	398	81	16.6

The data in Table 9.1 are generated by the program with one random lifting function ω for each of the polynomial systems Cohn-2 (MOURRAIN [1996], Cassou–Noguès (TRAVERSO [1997]), Planar 4-bar (MORGAN and WAMPLER [1990]), and cyclic-6, -7, -8 problems (BJORCK and FROBERG [1991]). The fourth and fifth columns give the size of the LP problem in (9.13). The last two columns are the size of the LP problem in (9.15) after all repeated constraints are deleted. For cyclic-n polynomial systems, about $1/3$ of the constraints are deleted.

For the data in Table 9.2, the algorithm was run with ten different real random liftings for each polynomial system. The powers of t was first scaled in the polyhedral homotopies before balancing where the lowest power of t in the homotopies is one, and the average of the highest powers of t in the polyhedral homotopies for the ten random liftings are listed in the second column. The third column lists the average of the highest powers of t in the polyhedral homotopies induced by the ten liftings obtained from the optimal solutions of the corresponding LP problems (9.15). The fourth column gives the average time elapsed for finding all mixed cells. The last column is the average time elapsed for finding the optimal lifting functions ν', including the constructing and solving of the LP problems (9.15). From these results, we see that the highest powers of t in the polyhedral homotopies are substantially reduced. The overall reduced powers of

t in the polyhedral homotopies greatly limit the chance of running into "valleys" which may cause the failure of path-tracings.

9.4. *The end game*

When we approximate all isolated zeros of the polynomial system $P(x)$ in $\mathbb{C}^n$ by following homotopy paths of a homotopy $H(x,t)=0$ on $\mathbb{C}^n \times [0,1]$, every isolated zero of $P(x)$ lies at the end of some path $x(t)$. However, as we mentioned before, there may be many other paths which do not lead to finite solutions. They are divergent in the sense that some coordinates will become arbitrarily large, leaving us with the problem of deciding if a path is indeed diverging or if it is just converging to a solution with large coordinates.

Let $H:\mathbb{C}^n \times [0,1] \to \mathbb{C}^n$ be a homotopy with $H(x,1)=P(x)$ and $H^{-1}(0)$ consisting of finite many smooth paths $x(t)=(x_1(t),\dots,x_n(t))$. It was shown in MORGAN, SOMMESE and WAMPLER [1992b] that, in the neighborhood of $t=1$, each path can be written in the form

$$x_j(t)=a_j(1-t)^{w_j/m}\left(1+\sum_{i=1}^{\infty}a_{ij}(1-t)^{i/m}\right), \quad j=1,\dots,n, \tag{9.18}$$

where m, called the *cyclic number*, is a positive integer and $w=(w_1,\dots,w_n)\in\mathbb{Z}^n$. Evidently, path $x(t)$ diverges to infinity when $w_j<0$ for some j, and $x(t)\in\mathbb{C}^n$ when $t\to 1$ if $w_j\geqslant 0$ for all $j=1,\dots,n$.

REMARK 9.1. In MORGAN, SOMMESE and WAMPLER [1992b], only expansions in the form (9.18) of those paths which lead to finite solutions in $\mathbb{C}^n$ were discussed. Of course, $w_j\geqslant 0$ for all $j=1,\dots,n$ in all those expansions. However, the theory established in MORGAN, SOMMESE and WAMPLER [1992b] can easily be extended to cover the case for diverging paths with $w_j<0$ for some j in those expansions.

To decide if $x(t)=(x_1(t),\dots,x_n(t))$ leads to a solution of $P(x)=0$ at infinity, one must distinguish the signs of w_j/m for all $j=1,\dots,n$. If none of them are negative, then $x(t)$ will converge to a finite solution of $P(x)=0$. Those (w_j/m)'s can be estimated as follows. Taking the logarithm of the absolute value of the expression in (9.18) yields, for $j=1,\dots,n$,

$$\log\left|x_j(t)\right|=\log|a_j|+\frac{w_j}{m}\log(1-t)+\sum_{i=1}^{\infty}c_{ij}(1-t)^i, \tag{9.19}$$

where $\sum_{i=1}^{\infty}c_{ij}(1-t)^i$ is the Taylor expansion of $\log(1+\sum_{i=1}^{\infty}a_{ij}(1-t)^{i/m})$. During the continuation process, a sequence of points $x(t_k)$, for $k=0,1,\dots$ with $t_0<t_1<\cdots<1$ were generated. For two consecutive points t_k and t_{k+1} very close to one, computing their differences of (9.19) yields,

$$\frac{\log|x_j(t_k)|-\log|x_j(t_{k+1})|}{\log(1-t_k)-\log(1-t_{k+1})}=\frac{w_j}{m}+0(1-t_k). \tag{9.20}$$

We may therefore estimate w_j/m by the value on the left hand side of the above equation. While this estimation is only of order 1, this will not cause difficulties in practice. Because, in theory (MORGAN, SOMMESE and WAMPLER [1992b]), m is not a very big number in general, therefore even lower order estimation is capable of distinguishing w_j/m from 0, especially when t_k and t_{k+1} are very close to 1. Nevertheless, higher order approximation of w_j/m can be found in HUBER and VERSCHELDE [1998].

9.5. *Softwares*

Industrial-Quality software for solving polynomial systems by homotopy continuation methods was first established by MORGAN [1983]. Later, it appeared in HOMPACK by L.T. Watson et al. (WATSON, BILLUPS and MORGAN [1987], MORGAN, SOMMESE and WATSON [1989], WATSON, SOSONKINA, MELVILLE, MORGAN and WALKER [1997]), in which the polyhedral homotopy methods, emerged in the middle of 90's and important in practice, were not implemented. Polyhedral homotopies exist in the package PHC (VERSCHELDE [1999]) written in *Ada* by J. Verschelde, the code HOM4PS written in *Fortran* developed by T. Gao and T.Y. Li (available at: http://www.math.msu.edu/~li/software) and PHoM written in C++ developed by Gunji, Kim, Kojima, Takeda, Fujisawa and Mizutani (available at: http://www.is.titech.ac.jp/~kojima/polynomials/). The excellent performance of these codes on a large collection of polynomial systems coming from a wide variety of application fields provides practical evidence that the homotopy algorithms constitute a powerful general purpose solver for polynomial equations.

Modern scientific computing is marked by the advent of vector and parallel computers and the search for algorithms that are to a large extent parallel in nature. A great advantage of the homotopy continuation algorithm for solving polynomial systems is that it is to a large degree parallel, in the sense that each isolated zero can be computed independently. In this respect, it stands in contrast to the highly serial algebraic elimination methods, which use resultants or Gröbner bases. Excellent speed-ups of parallel algorithms for symmetric eigenvalue problems, considered as polynomial systems, were reported in HUANG and LI [1995], LI and ZOU [1999] and TREFFTZ, MCKINLEY, LI and ZENG [1995]. The performance of the homotopy algorithms for solving general polynomial systems on multi-processor machines with shared or distributed memory is currently under active investigation (ALLISON, CHAKRABORTY and WATSON [1989], HARIMOTO and WATSON [1989]). One may expect a very high level of speed-up on different types of architectures of those algorithms.

References

ABRAHAM, R., ROBBIN, J. (1967). *Transversal Mapping and Flows* (W.A. Benjamin, New York, Amsterdam).

ALLGOWER, E.L. (1984). Bifurcation arising in the calculation of critical points via homotopy methods. In: Kupper, T., Mittelman, H.D., Weber, H. (eds.), *Numerical Methods for Bifurcation Problems* (Birkhäuser-Verlag, Basel), pp. 15–28.

ALLGOWER, E.L., GEORG, K. (1990). *Numerical Continuation Methods, an Introduction*, Springer Series in Comput. Math. **13** (Springer-Verlag, Berlin).

ALLGOWER, E.L., GEORG, K. (1993). Continuation and path following. *Acta Numerica*, 1–64.

ALLGOWER, E.L., GEORG, K. (1997). Numerical path following. In: Ciarlet, P.G., Lions, J.L. (eds.). In: Handbook of Numerical Analysis **5** (North-Holland, Amsterdam), pp. 3–203.

ALLISON, D.C.S., CHAKRABORTY, A., WATSON, L.T. (1989). Granularity issues for solving polynomial systems via globally convergent algorithms on a hypercube. *J. Supercomput.* **3**, 5–20.

BERNSHTEÍN, D.N. (1975). The number of roots of a system of equations. *Functional Anal. Appl.* **9** (3), 183–185; Transl. from: *Funktsional. Anal. i Prilozhen.* **9** (3) (1975), 1–4.

BEST, M.J., RITTER, K. (1985). *Linear Programming: Active Set Analysis and Computer Programs* (Prentice-Hall, Englewood Cliffs, NJ).

BJORCK, G., FROBERG, R. (1991). A faster way to count the solutions of inhomogeneous systems of algebraic equations, with applications to cyclic n-roots. *J. Symbolic Comput.* **12**, 329–336.

BLUM, L., CUCKER, F., SHUB, M., SMALE, S. (1998). *Complexity and Real Computation* (Springer-Verlag, New York).

BUCHBERGER, B. (1985). Gröbner basis: An algorithmic method in polynomial ideal theory. In: Bose, N.K. (ed.), *Multidimensional System Theory* (D. Reidel, Dordrecht), pp. 184–232.

CANNY, J., ROJAS, J.M. (1991). An optimal condition for determining the exact number of roots of a polynomial system. In: *Proceedings of the 1991 International Symposium on Symbolic and Algebraic Computation* (ACM, New York), pp. 96–101.

CHOW, S.N., MALLET-PARET, J., YORKE, J.A. (1979). Homotopy method for locating all zeros of a system of polynomials. In: Peitgen, H.O., Walther, H.O. (eds.), *Functional Differential Equations and Approximation of Fixed Points*. In: Lecture Notes in Math. **730** (Springer-Verlag, Berlin), pp. 77–88.

DREXLER, F.J. (1977). Eine Methode zur Berechnung sämtlicher Lösungen von Polynomgleichungssystemen. *Numer. Math.* **29**, 45–58.

EMIRIS, I.Z., CANNY, J. (1995). Efficient incremental algorithms for the sparse resultant and the mixed volume. *J. Symbolic Comput.* **20**, 117–149.

FULTON, W. (1984). *Intersection Theory* (Springer-Verlag, New York).

GAO, T., LI, T.Y. (2000). Mixed volume computation via linear programming. *Taiwanese J. Math.* **4** (4), 599–619.

GAO, T., LI, T.Y. (2003). Mixed volume computation for semi-mixed systems. *Discrete Comput. Geom.*, to appear.

GAO, T., LI, T.Y., VERSCHELDE, J., WU, M. (2000). Balancing the lifting values to improve the numerical stability of polyhedral homotopy continuation methods. *Appl. Math. Comput.* **114**, 233–247.

GAO, T., LI, T.Y., WANG, X. (1999). Finding all isolated zeros of polynomial systems in $\mathbb{C}^n$ via stable mixed volumes. *J. Symbolic Comput.* **28**, 187–211.

GARCIA, C.B., ZANGWILL, W.I. (1979). Finding all solutions to polynomial systems and other systems of equations. *Math. Programm.* **16**, 159–176.

GEL'FAND, I.M., KAPRANOV, M.M., ZELEVINSKIĬ, A.V. (1994). *Discriminants, Resultants and Multidimensional Determinants* (Birkhäuser, Boston).

HARIMOTO, S., WATSON, L.T. (1989). The granularity of homotopy algorithms for polynomial systems of equations. In: Rodrigue, G. (ed.), *Parallel Processing for Scientific Computing* (SIAM, Philadelphia, PA), pp. 115–120.

HUANG, L., LI, T.Y. (1995). Parallel homotopy algorithm for symmetric large sparse eigenproblems. *J. Comput. Appl. Math.* **60**, 77–100.

HUBER, B. (1996). Solving sparse polynomial systems. Ph.D. thesis, Department of Mathematics, Cornell University.

HUBER, B., STURMFELS, B. (1995). A polyhedral method for solving sparse polynomial systems. *Math. Comp.* **64**, 1541–1555.

HUBER, B., STURMFELS, B. (1997). Bernstein's theorem in affine space. *Discrete Comput. Geom.* **7** (2), 137–141.

HUBER, B., VERSCHELDE, J. (1998). Polyhedral end games for polynomial continuation. *Numer. Algorithms* **18** (1), 91–108.

KHOVANSKIĬ, A.G. (1978). Newton polyhedra and the genus of complete intersections. *Functional Anal. Appl.* **12** (1), 38–46; Transl. from: *Funktsional. Anal. i Prilozhen.* **12** (1) (1978), 51–61.

KUSHNIRENKO, A.G. (1976). Newton polytopes and the Bézout theorem. *Functional Anal. Appl.* **10** (3), 233–235; Transl. from: *Funktsional. Anal. i Prilozhen.* **10** (3) (1976), 82–83.

LEE, C.W. (1991). Regular triangulations of convex polytopes. In: Gritzmann, P., Sturmfels, B. (eds.), *Applied Geometry and Discrete Mathematics – The Victor Klee Festschrift*. In: DIMACS Series **4** (AMS, Providence, RI), pp. 443–456.

LI, T.Y. (1983). On Chow, Mallet-Paret and Yorke homotopy for solving systems of polynomials. *Bull. Inst. Math. Acad. Sinica* **11**, 433–437.

LI, T.Y., LI, X. (2001). Finding mixed cells in the mixed volume computation. *Found. Comput. Math.* **1**, 161–181.

LI, T.Y., SAUER, T. (1987). Regularity results for solving systems of polynomials by homotopy method. *Numer. Math.* **50**, 283–289.

LI, T.Y., SAUER, T. (1989). A simple homotopy for solving deficient polynomial systems. *Japan J. Appl. Math.* **6**, 409–419.

LI, T.Y., SAUER, T., YORKE, J.A. (1987a). Numerical solution of a class of deficient polynomial systems. *SIAM J. Numer. Anal.* **24**, 435–451.

LI, T.Y., SAUER, T., YORKE, J.A. (1987b). The random product homotopy and deficient polynomial systems. *Numer. Math.* **51**, 481–500

LI, T.Y., SAUER, T., YORKE, J.A. (1988). Numerically determining solutions of systems of polynomial equations. *Bull. Amer. Math. Soc.* **18**, 173–177.

LI, T.Y., SAUER, T., YORKE, J.A. (1989). The cheater's homotopy: an efficient procedure for solving systems of polynomial equations. *SIAM J. Numer. Anal.* **26**, 1241–1251.

LI, T.Y., WANG, X. (1990). A homotopy for solving the kinematics of the most general six- and five-degree of freedom manipulators. In: *Proc. of ASME Conference on Mechanisms* **25**, pp. 249–252.

LI, T.Y., WANG, X. (1991). Solving deficient polynomial systems with homotopies which keep the subschemes at infinity invariant. *Math. Comp.* **56**, 693–710.

LI, T.Y., WANG, X. (1992). Nonlinear homotopies for solving deficient polynomial systems with parameters. *SIAM J. Numer. Anal.* **29**, 1104–1118.

LI, T.Y., WANG, X. (1997). The BKK root count in $\mathbb{C}^n$. *Math. Comp.* **65**, 1477–1484.

LI, T.Y., ZOU, X. (1999). Implementing the parallel quasi-Laguerre's algorithm for symmetric tridiagonal eigenproblems. *SIAM J. Sci. Comput.* **20** (6), 1954–1963.

LIU, C., LI T., BAI F. Generalized Bézout number computation for sparse polynomial systems. Preprint.

MORGAN, A.P. (1983). A method for computing all solutions to systems of polynomial equations. *ACM Trans. Math. Software* **9** (1), 1–17

MORGAN, A.P. (1986). A homotopy for solving polynomial systems. *Appl. Math. Comput.* **18**, 173–177.

MORGAN, A.P. (1987). *Solving Polynomial Systems Using Continuation for Engineering and Scientific Problems* (Prentice-Hall, Englewood Cliffs, NJ).

MORGAN, A.P., SOMMESE, A.J. (1987a). Computing all solutions to polynomial systems using homotopy continuation. *Appl. Math. Comput.* **24**, 115–138.

MORGAN, A.P., SOMMESE, A.J. (1987b). A homotopy for solving general polynomial systems that respect m-homogeneous structures. *Appl. Math. Comput.* **24** (2), 101–113.

MORGAN, A.P., SOMMESE, A.J. (1989). Coefficient-parameter polynomial continuation. *Appl. Math. Comput.* **29** (2), 123–160; Errata. *Appl. Math. Comput.* **51** (1992), 207.

MORGAN, A.P., SOMMESE, A.J., WAMPLER, C.W. (1992b). A power series method for computing singular solutions to nonlinear analytic systems. *Numer. Math.* **63** (3), 391–409.

MORGAN, A.P., SOMMESE, A.J., WATSON, L.T. (1989). Finding all isolated solutions to polynomial systems using HOMPACK. *ACM Trans. Math. Software* **15**, 93–122.

MORGAN, A.P., WAMPLER, C.W. (1990). Solving a planar four-bar design problem using continuation. *ASME J. Mechanical Design* **112**, 544–550.

MOURRAIN, B. (1996). *The Handbook of Polynomial Systems*, available at http://www.inria.fr/safir/POL/index.html.

PAPADIMITRIOU, C.H., STEIGLITZ, K. (1982). *Combinatorial Optimization: Algorithms and Complexity* (Prentice-Hall, Englewood Cliffs, NJ).

ROJAS, J.M. (1994). A convex geometric approach to counting the roots of a polynomial system. *Theoret. Comput. Sci.* **133**, 105–140.

ROJAS, J.M., WANG, X. (1996). Counting affine roots of polynomial systems via pointed Newton polytopes. *J. Complexity* **12** (2), 116–133.

SHAFAREVICH, I.R. (1977). *Basic Algebraic Geometry* (Springer-Verlag, New York).

SELBY, S.M. (ed.) (1971). *CRC Standard Mathematical Tables* (The Chemical Rubber Company, Cleveland, OH).

SHUB, M., SMALE, S. (1993). Complexity of Bézout's theorem. I: Geometric aspects. *J. Amer. Math Soc.* **6**, 459–501.

SOMMESE, A.J., VERSCHELDE, J. (2000). Numerical homotopies to compute points on positive dimensional algebraic sets. *J. Complexity* **16** (3), 572–602.

SOMMESE, A.J., VERSCHELDE, J., WAMPLER, C.W. (2001). Numerical decomposition of the solution sets of polynomial systems into irreducible components. *SIAM J. Numer. Anal.* **38** (6), 2022–2046.

SOMMESE, A.J., WAMPLER, C.W. (1996). Numerical algebraic geometry. In: Renegar, J., Shub, M., Smale, S. (eds.), *The Mathematics of Numerical Analysis*, Proceedings of the 1995 AMS–SIAM Summer Seminar in Applied Mathematics, Park City, UT. In: Lectures in Appl. Math. **32**, pp. 749–763.

STANLEY, R.P. (1997). *Enumerative Combinatorics (I)* (Cambridge University Press, Cambridge).

TAKEDA, A., KOJIMA, M., FUJISAWA, K. (2002). Enumeration of all solutions of a combinatorial linear inequality system arising from the polyhedral homotopy continuation method. *J. Oper. Soc. Japan* **45**, 64–82.

TRAVERSO, C. (1997). The PoSSo test suite examples. Available at: http://www.inria.fr/safir/POL/index.html.

TREFFTZ, C., MCKINLEY, P., LI, T.Y., ZENG, Z. (1995). A scalable eigenvalue solver for symmetric tridiagonal matrices. *Parallel Comput.* **21**, 1213–1240.

TSAI, L.W., MORGAN, A.P. (1985). Solving the kinematics of the most general six- and five-degree-of-freedom manipulators by continuation methods. *ASME J. of Mechanics, Transmissions, and Automation in Design* **107**, 189–200.

VERSCHELDE, J. (1996). Homotopy continuation methods for solving polynomial systems. Ph.D. thesis, Department of Computer Science, Katholieke Universiteit Leuven, Leuven, Belgium.

VERSCHELDE, J. (1999). PHCPACK: A general-purpose solver for polynomial systems by homotopy continuation. *ACM Trans. Math. Software* **25**, 251–276.

VERSCHELDE, J., GATERMANN, K., COOLS, R. (1996). Mixed-volume computation by dynamic lifting applied to polynomial system solving. *Discrete Comput. Geom.* **16**, 69–112.

VERSCHELDE, J., HAEGEMANS, A. (1993). The *GBQ*-algorithm for constructing start systems of homotopies for polynomial systems. *SIAM J. Numer. Anal.* **30** (2), 583–594.

VERSCHELDE, J., VERLINDEN, P., COOLS, R. (1994). Homotopies exploiting Newton polytopes for solving sparse polynomial systems. *SIAM J. Numer. Anal.* **31**, 915–930.

WAMPLER, C.W. (1992). Bézout number calculations for multi-homogeneous polynomial systems. *Appl. Math. Comput.* **51**, 143–157.

WAMPLER, C.W. (1994). An efficient start system for multi-homogeneous polynomial continuation. *Numer. Math.* **66**, 517–523.

WAMPLER, C.W., MORGAN, A.P., SOMMESE, A.J. (1992). Complete solution of the nine-point path synthesis problem for four-bar linkages. *ASME J. Mechanical Design* **114**, 153–159.

WATSON, L.T., BILLUPS, S.C., MORGAN, A.P. (1987). Algorithm 652: HOMPACK: a suite of codes for globally convergent homotopy algorithms. *ACM Trans. Math. Software* **13** (3), 281–310.

WATSON, L.T., SOSONKINA, M., MELVILLE, R.C., MORGAN, A.P., WALKER, H.F. (1997). HOMPACK90: A suite of Fortran 90 codes for globally convergent homotopy algorithms. *ACM Trans. Math. Software* **23** (4), 514–549. Available at http://www.cs.vt.edu/~ltw/.

WRIGHT, A.H. (1985). Finding all solutions to a system of polynomial equations. *Math. Comp.* **44**, 125–133.

YE, Y. (1997). *Interior Point Algorithm: Theory and Analysis* (John Wiley & Sons, New York).

ZULENER, W. (1988). A simple homotopy method for determining all isolated solutions to polynomial systems. *Math. Comp.* **50**, 167–177.

Further reading

BRUNOVSKÝ, P., MERAVÝ, P. (1984). Solving systems of polynomial equations by bounded and real homotopy. *Numer. Math.* **43**, 397–418.

DAI, Y., KIM, S., KOJIMA, M. (2001). Computing all nonsingular solutions of cyclic-n polynomial using polyhedral homotopy continuation methods, Technical Report B-373, available at http://www.is.titech.ac.jp/~kojima/sdp.html, to appear: J. Comput. Appl. Math.

EMIRIS, I.Z. (1994). Sparse elimination and applications in kinematics. Ph.D. thesis, Computer Science Division, Dept. of Electrical Engineering and Computer Science, University of California, Berkeley.

EMIRIS, I.Z. (1996). On the complexity of sparse elimination. *J. Complexity* **12** (2), 134–166.

EMIRIS, I.Z., VERSCHELDE, J. (1999). How to count efficiently all affine roots of a polynomial system. *Discrete Appl. Math.* **93** (1), 21–32.

GARCIA, C.B., LI, T.Y. (1980). On the number of solutions to polynomial systems of equations. *SIAM J. Numer. Anal.* **17**, 540–546.

HENDERSON, M.E., KELLER, H.B. (1990). Complex bifurcation from real paths. *SIAM J. Appl. Math.* **50**, 460–482.

LI, T.Y. (1987). Solving polynomial systems. *Math. Intelligencer* **9** (3), 33–39.

LI, T.Y. (1997). Numerical solution of multivariate polynomial systems by homotopy continuation methods. *Acta Numerica*, 399–436.

LI, T.Y. (1999). Solving polynomial systems by polyhedral homotopies. *Taiwanese J. Math.* **3** (3), 251–279.

LI, T.Y., WANG, T., WANG, X. (1996). Random product homotopy with minimal *BKK* bound. In: Renegar, J., Shub, M., Smale, S. (eds.), *The Mathematics of Numerical Analysis, Proceedings of the 1995 AMS–SIAM Summer Seminar in Applied Mathematics, Park City, UT*, pp. 503–512.

LI, T.Y., WANG, X. (1993). Solving real polynomial systems with real homotopies. *Math. Comp.* **60**, 669–680.

LI, T.Y., WANG, X. (1994). Higher order turning points. *Appl. Math. Comput.* **64**, 155–166.

MORGAN, A.P., SOMMESE, A.J., WAMPLER, C.W. (1991). Computing singular solutions to nonlinear analytic systems. *Numer. Math.* **58** (7), 669–684.

MORGAN, A.P., SOMMESE, A.J., WAMPLER, C.W. (1992a). Computing singular solutions to polynomial systems. *Adv. Appl. Math.* **13** (3), 305–327.

MORGAN, A.P., SOMMESE, A.J., WAMPLER, C.W. (1995). A product-decomposition theorem for bounding Bèzout numbers. *SIAM J. Numer. Anal.* **32** (4), 1308–1325.

ROJAS, J.M. (1999). Toric intersection theory for affine root counting. *J. Pure Appl. Algebra* **136**, 67–100.

VERSCHELDE, J. (1995). PHC and MVC: two programs for solving polynomial systems by homotopy continuation. In: Faugère, J.C., Marchand, J., Rioboo, R. (eds.), *Proceedings of the PoSSo Workshop on Software*, pp. 165–175.

VERSCHELDE, J., COOLS, R. (1993). Symbolic homotopy construction. *Appl. Algebra Engrg. Comm. Comput.* **4**, 169–183.

VERSCHELDE, J., COOLS, R. (1994). Symmetric homotopy construction. *J. Comput. Appl. Math.* **50**, 575–592.

VERSCHELDE, J., GATERMANN, K. (1995). Symmetric Newton polytopes for solving sparse polynomial systems. *Adv. Appl. Math.* **16**, 95–127.

Chaos in Finite Difference Schemes

Masaya Yamaguti

Ryukoku University, Fac. of Science and Technology, Dept. Applied Maths & Informatics, Yokotani, Ohe-cho, Otsu-shi, 520-21 Japan

Yoichi Maeda

University of Tokai, Department of Mathematics, 1117 Kitakaname, Hiratsuka, Kanagawa, 259-2192 Japan

Preface

Modern dynamical systems theory has its origin in the study of the three-body problem by H. Poincaré. Applying the techniques of topological geometry, he analyzed the global properties of solutions of nonlinear differential equations. He showed that simple deterministic systems do not always produce simple results, but may yield very complicated ones. In this sense, he is a predictor of chaos.

The word "chaos" appeared for the first time in the field of mathematics in an article of LI and YORKE [1975] entitled "Period Three Implies Chaos". This short and elegant paper caused a great sensation in the world of mathematical physics. The year before, MAY [1974] had gotten a remarkable numerical result which showed that even some simple discrete dynamical systems with small number of freedom can produce very complicated behaviors of orbits.

After these two works, one could no longer believe the well-known dichotomy between the deterministic phenomena and the stochastic ones in the world of mathematics. Although the solutions of the discrete dynamical systems are all deterministic, they can produce very complicated, nearly stochastic behaviors. We, the present authors, were fascinated by these results. We generalized mathematically May's result and proved that some very familiar discretizations of some ordinary differential equations (O.D.E.) give the same results for a certain mesh size (time step). Moreover, some very accurate

Foundations of Computational Mathematics
Special Volume (F. Cucker, Guest Editor) of
HANDBOOK OF NUMERICAL ANALYSIS, VOL. XI
P.G. Ciarlet (Editor)

discretizations of O.D.E. may cause the same phenomena for any small mesh size. We will discuss these results in Section 6.

This paper is mainly concentrated on the discretizations of O.D.E. from the viewpoint of chaos. Depending on the types of discretization, continuous dynamical systems may change into complicated discrete dynamical systems, specifically chaos.

In the first section, we present an interesting example of population dynamics as an introduction. The famous generalized logistic map is naturally deduced.

In Section 2, we review one of the criteria of one-dimensional chaos as defined by Li and Yorke. The snap-back repeller is a generalization of Li–Yorke chaos to higher dimensional dynamics elaborated by MAROTTO [1978].

Section 3 describes the Euler discretization of O.D.E. and the fundamental theorem proposed by YAMAGUTI and MATANO [1979]. This theorem is used in the succeeding sections.

In Section 4, we browse the present state of mathematical economics, illustrating the importance of the discretization of O.D.E. This section owes a great deal to suggestions of T. Kaizoji, a leading economist.

In Section 5, we develop the discretization of O.D.E. treated in Section 3. This time, however, we assume that the original differential equation has only one asymptotically stable equilibrium point in the case of large mesh size discretization.

In Section 6, however, we study the cases of sufficiently small mesh size discretization. Even in these cases, we may get the same chaotic dynamical systems by applying the Euler discretization. We provide one interesting example for which every discretization causes chaotic phenomenon.

Section 7 treats the relation between the symmetric O.D.E. and its Euler discretization. We show that in a sense chaos demands asymmetry.

Section 8 deals with a chaotic dynamical system produced by a modified Euler scheme. The discretization by this scheme shows again chaotic phenomena for certain large mesh size.

Section 9 is a study of the central difference scheme for special O.D.E. It also includes the study of multi-dimensional chaos.

In Section 10, we examine mathematical sociology through a model of a particular fashion with two thresholds which occasionally behaves chaotically.

In the last section, we consider some singular functions which are obtained as a generating function of some chaotic dynamical system. We mention that these fractal singular functions can be considered as the solutions of some Dirichlet problem. There is a striking relation between two fractal singular functions, that is, the Lebesgue's singular function and the Takagi function.

Masaya Yamaguti, one of the authors, passed away on December 24, 1998, at the age of 73. His interests ranged over many fields: Mathematics, physics, technology, biology, sociology, psychology, literature, religion... He was deeply involved in the chaotic and the fractal phenomena, endlessly contemplating the philosophical and religious meanings behind them.

In preparing material for this article the authors were assisted greatly by T. Kaizoji. And thanks are especially due to M. Kishine for proofreading the entire work.

1. Some simple discrete models in ecology and fluid dynamics

1.1. Pioneers in chaotic dynamical system

MAY [1974] illustrated a well-known example of chaotic dynamical system. This simple example came from a discretization of an ordinary differential equation called logistic equation.

Let us explain first the logistic equation. This equation had gotten by VERHULST in his article [1838] ("Notice sur la Loi que la Population Suit dans son Accroissement"). This equation is defined as follows:

$$\frac{\mathrm{d}u}{\mathrm{d}t} = u(\varepsilon - u), \tag{1.1}$$

where ε is a positive constant. The exact solution $u(t)$ of (1.1) with initial data u_0 is given as:

$$u(t) = \frac{u_0 \mathrm{e}^{\varepsilon t}}{1 + u_0/\varepsilon \cdot (\mathrm{e}^{\varepsilon t} - 1)}. \tag{1.2}$$

We can easily integrate (1.1) by the method of separation variables and obtain the exact solution above (1.2).

This model was noted for the simulation of the population growth of some insects which have overlapping generation, i.e. the life span of parents and their children are not disjoined. Indeed, from biological point of view, the derivation of this equation assumed that these insects have overlapping generation. But there are some species of insects which have no such properties which are called insects of nonoverlapping generation. This kind of insects die just after they give birth to their children, cicada (locust) as an example.

It was a Japanese entomologist UTIDA [1941] from the Kyoto school who first hit upon to describe the population growth of the insects with nonoverlapping generation in 1941. He practiced many experiments of a kind of insects (callosobruchus chinensis) in his laboratory. He remarked first that to explain the oscillatory growth of the population of these insects, the logistic equation (1.1) is not available, for the solution (1.2) is never oscillatory. He elaborated then another approach through a model as follows:

$$N_{n+1} = \left(\frac{1}{b + cN_n} - \sigma\right) N_n, \tag{1.3}$$

where N_n means the population of nth generation, and

$$b = \frac{1}{\mathrm{e}^{\varepsilon \Delta t}}, \qquad c = \frac{\mathrm{e}^{\varepsilon \Delta t} - 1}{\varepsilon \mathrm{e}^{\varepsilon \Delta t}},$$

and σ is a positive constant. Here we should remark that in the case of $\sigma = 0$ in (1.3), then we obtain precisely $N_n = u(n\Delta t)$ where u is the exact solution (1.2) for the logistic equation (1.1). Thus FUJITA and UTIDA [1953] succeeded in explaining the oscillatory phenomena observed in the population growth of these insects. The positive constant σ

is essential in his model. If we express (1.3) by

$$N_{n+1} = F_\sigma(N_n),$$

then we can distinguish clearly the difference between the case $\sigma = 0$ and the case $\sigma > 0$. F_σ is monotonous in the former case whereas in the latter F_σ is not monotonous which may cause some oscillatory phenomena of N_n.

Another pioneering work done in the field of fluid dynamics was that of LORENZ [1963]. He studied the equations for a rotating water-filled vessel which is circularly symmetric about its vertical axis. The water is heated around the rim of the vessel and is cooled in the center. When the vessel is annular in shape and the rotation rate is high, waves develop and alter their shape irregularly. He explained these phenomena using a quite simple set of equations solved numerically.

He let X_n be in essence the maximum kinetic energy of successive waves. Plotting (X_{n+1}, X_n) for each n and connecting the points successively, we obtain the following graph (see Fig. 1.1).

As you see, the above two researches which are from completely different fields coincide in so-called discrete dynamical systems expressed in the following form:

$$U_{n+1} = F(U_n). \tag{1.4}$$

Moreover, F is nonmonotonous in both cases.

Besides these two researches, there were many independent works in some mathematical discipline, say, MYRBERG's work [1962] on complex dynamical system, SHARKOVSKII's work [1964] about (1.4) under the simple assumption that F is real

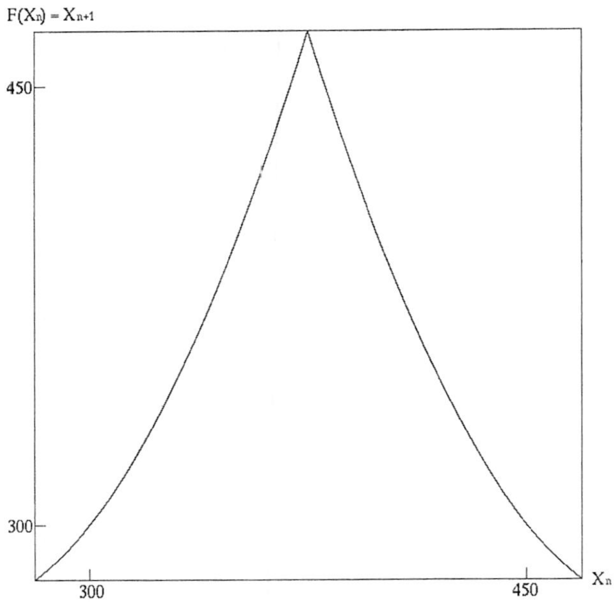

FIG. 1.1.

and continuous. Despite these many works, we may say that those works done in the 1950's and the 1960's remained independent in the sense that each discipline did not recognize the existence of the others who managed the same problem in different context.

Now, we have arrived at the explanation of MAY's work [1974]. May also applied the logistic equation (1.1), but he adopted a method of discretization in more ordinary form than that of Utida which was more sophisticated. That method is so-called the Euler method which is written as below:

$$\frac{U_{n+1} - U_n}{\Delta t} = U_n(\varepsilon - U_n). \tag{1.5}$$

Remark that if we rewrite Utida's scheme ($\sigma = 0$) like the expression (1.5) then we have the equation as

$$\frac{U_{n+1} - U_n}{e^{\varepsilon \Delta t} - 1} = \frac{1}{\varepsilon}(\varepsilon - U_{n+1})U_n.$$

A little complicated as this equation is, it is really surprising that this U_n is precisely equal to $U(n\Delta t)$, where $U(t)$ is the exact solution (1.2) for any Δt. This was shown by another biologist Morishita from the Kyoto school.

This discretization (1.5) is a well-known scheme for numerical solution. It can be proved very easily that, if $\Delta t \to 0$, then $U_n \to U(t)$. Here $U(t)$ is the exact solution with initial data U_0. What happens, however, when we fix Δt and make n increase to $+\infty$? This is a quite different question. May discussed it in the following way: First, deform (1.5) as

$$U_{n+1} = \{(1 + \varepsilon \Delta t) - \Delta t U_n\} U_n.$$

Then change the variable U_n to x_n:

$$x_n = \frac{\Delta t U_n}{1 + \varepsilon \Delta t} \quad (\forall n \in \mathbb{N}),$$

and finally put $1 + \varepsilon \Delta t = a$, then he got

$$x_{n+1} = a(1 - x_n)x_n. \tag{1.6}$$

Naming (1.6) as generalized logistic map, May computed its orbits for many values of a $(0 \leqslant a \leqslant 4)$.

1.2. *From order to chaos*

We can see easily that if a satisfies $0 \leqslant a \leqslant 4$, then $0 \leqslant x_0 \leqslant 1$ implies always $0 \leqslant x_n \leqslant 1$ for all $n = 0, 1, 2, \ldots$. That is, all orbits which start at the initial point x_0 in the interval $[0, 1]$ are confined in the same interval. We call this system as a discrete dynamical system on $[0, 1]$. Although the following results discussed by May are well known, we review them for the preparation of the following sections.

(i) $0 \leqslant a < 1$. As you see in Fig. 1.2, for any x_0 $(0 \leqslant x_0 < 1)$, x_n tends to zero monotonously as n increases to $+\infty$ (see also Fig. 1.3).

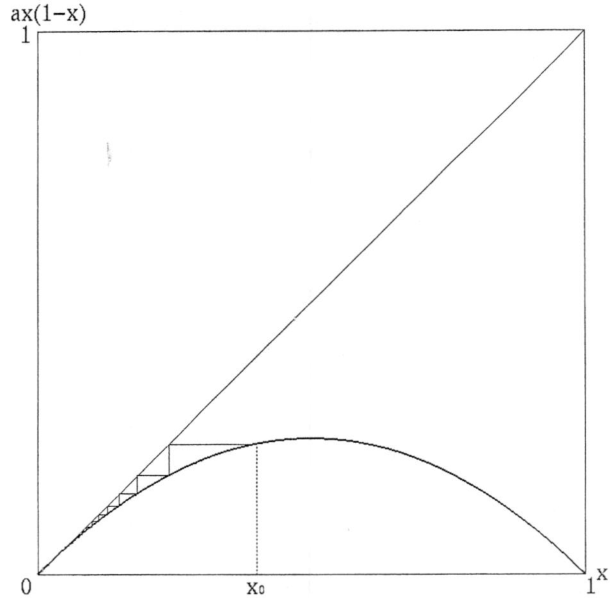

FIG. 1.2.

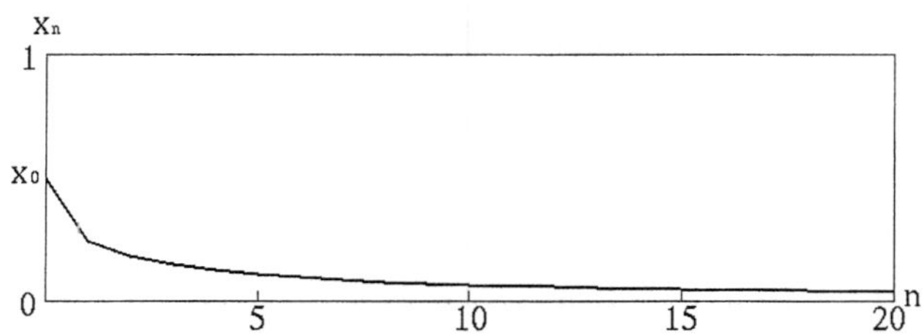

FIG. 1.3.

(ii) $1 \leqslant a \leqslant 2$. For any x_0 $(0 \leqslant x_0 \leqslant 1 - 1/a)$, x_n increases to $1 - 1/a$ as n increases to $+\infty$ as is observed in Fig. 1.4. For any x_0 $(1 - 1/a < x_0 < 1/a)$, x_n also tends to $1 - 1/a$ as n increases to $+\infty$. This time, however, x_n decreases. Fig. 1.5 shows the graph of these two orbits. And as for x_0 satisfies $1/a < x_0 < 1$, $x_1 = ax_0(1 - x_0)$ satisfies $0 < x_1 < 1 - 1/a$, which means x_n $(n \geqslant 2)$ increases monotonously. Consequently, we can summarize that in this interval of a, for any x_0 $(0 < x_0 < 1)$, x_n is monotonously tends to $1 - 1/a$.

(iii) $2 < a \leqslant 3$. Fig. 1.6 shows that for any x_0 such that $0 < x_0 < 1$, x_n tends to $1 - 1/a$ with oscillation. The graph of its orbit is shown in Fig. 1.7.

(iv) $3 < a \leqslant 1 + \sqrt{6} = 3.44949\ldots$. Here we have no convergence of x_n to the fixed point $1 - 1/a$. Instead, for any x_0 in $[0, 1]$, x_n behaves as asymptotically period 2 oscillation, that is, x_n tends to an oscillation with period 2. Period 2 means $x_n = x_{n+2}$ and $x_n \neq x_{n+1}$ for any n.

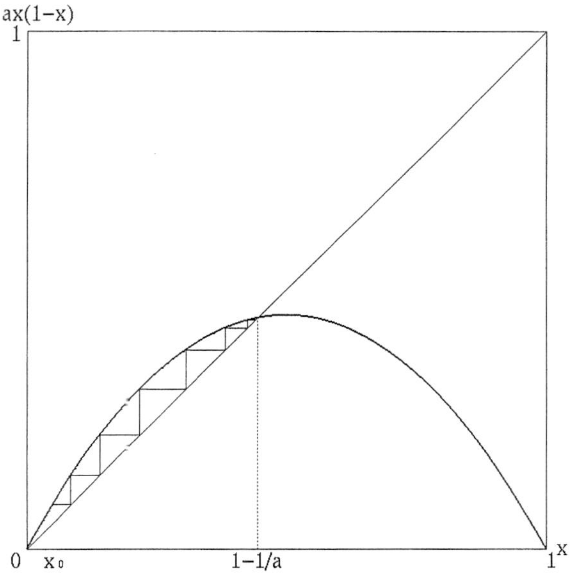

FIG. 1.4.

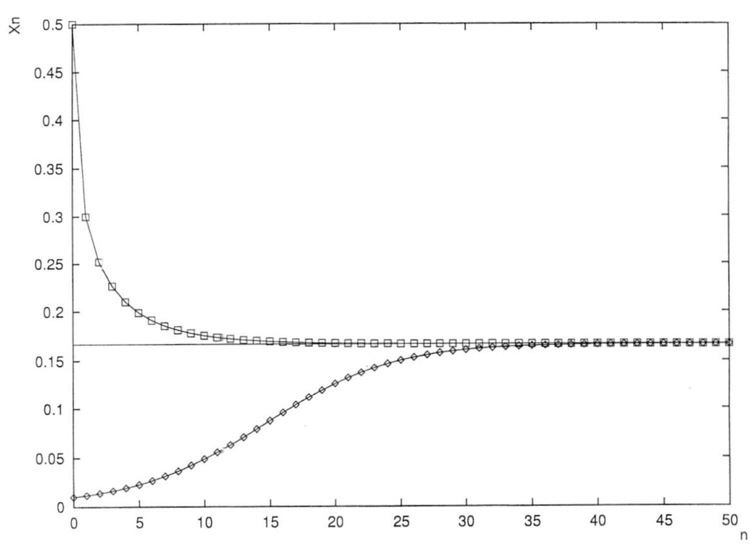

FIG. 1.5.

In the following, we have a monotonously increasing sequence of a_n $(n = 1, 2, \ldots)$ tending a_c $(= 3.5699456\ldots)$.

(v) $1 + \sqrt{6} < a < a_1$. Here we have asymptotically period 4 oscillation of all orbits. Period 4 means $x_n = x_{n+4}$ but $x_n \neq x_{n+i}$ $(1 \leqslant i \leqslant 3)$.

(vi) $a_1 < a < a_2$. Here we have asymptotically period 8 oscillation of all orbits.

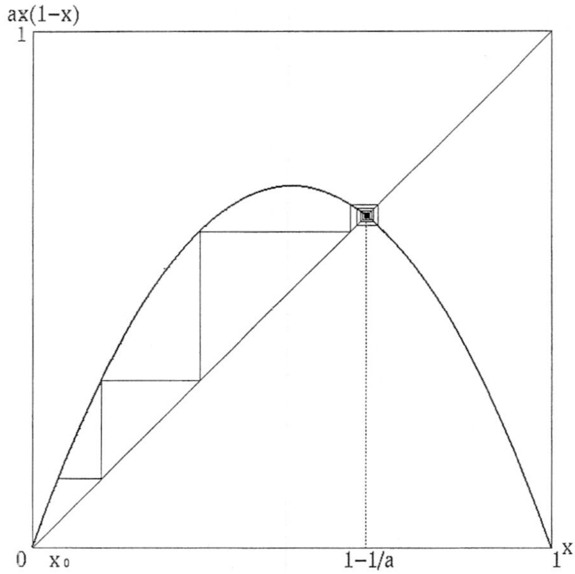

FIG. 1.6.

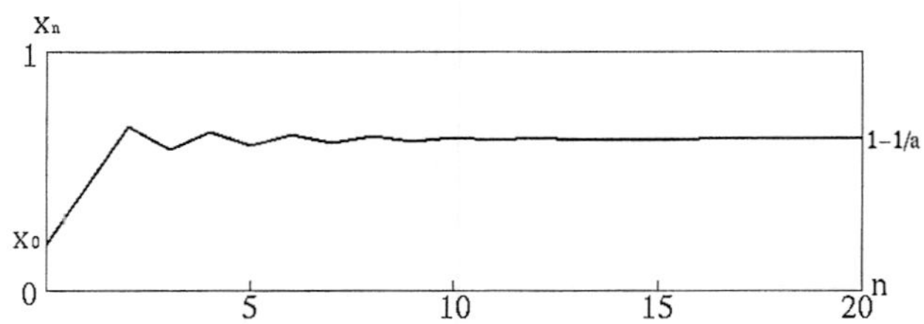

FIG. 1.7.

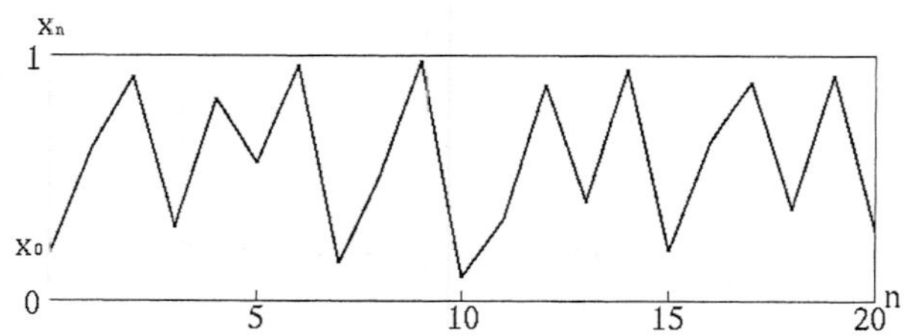

FIG. 1.8.

Of course, the following was a conjecture given by May's numerical experiments. We can guess that in the case of

(vii) $a_{m-1} < a < a_m$, all orbits behave as asymptotically period 2^{m+1} oscillation. Period 2^{m+1} oscillation means $x_n = x_{n+2^{m+1}}$ and $x_n \neq x_{n+i}$ $(1 \leqslant i \leqslant 2^{m+1} - 1)$.

Finally we obtain the important result that, in the case of

(viii) a_c ($= 3.5699456\ldots$) $< a \leqslant 4$, orbits behave in really complicated ways and depend very sensitively on the initial data. In Fig. 1.8, you can observe the graph of orbit for $a = 3.9$, as an example of this type.

2. Li–Yorke theorem and its generalization

2.1. Li–Yorke theorem

LI and YORKE [1975] made their contribution at the same moment that May tried the numerical experiments stated in the previous section. They were fascinated greatly by the paper of LORENZ [1963] and elaborated the following general discrete dynamical system:

$$x_{n+1} = F(x_n). \tag{2.1}$$

Here, they only supposed that $F(x)$ is a continuous function defined on $\mathbb{R}^1$ and its values are also in $\mathbb{R}^1$. They proved the following theorem.

THEOREM 2.1. *If $F(x)$ maps the interval $[0, 1]$ to $[0, 1]$ and satisfies the following condition: There exist a, b, c, d in this interval such that $F(a) = b$, $F(b) = c$, $F(c) = d$ and*

$$d \leqslant a < b < c,$$

then the discrete dynamical system (2.1) *has the following three properties*:

(i) *for any given positive integer p, there exists at least one orbit of* (2.1) *which is p-periodic.*

(ii) *there exists an uncountable set S called "scrambled set" in $[0, 1]$ such that for any points $x_0 \in S$, $y_0 \in S$, the two orbits of* (2.1) $x_n = F^n(x_0)$, $y_n = F^n(y_0)$ *satisfy the followings*:

$$\begin{aligned} &\limsup_{n\to+\infty} |x_n - y_n| > 0, \\ &\liminf_{n\to+\infty} |x_n - y_n| = 0. \end{aligned} \tag{2.2}$$

(iii) *for any periodic point x_0 and for any $y_0 \in S$, the inequality* (2.2) *is true.*

REMARK 2.1. We define x_0 is a p-periodic orbit (or a p-periodic point) if and only if for a $x_0 \in [0, 1]$, $F^p(x_0) = x_0$ and $F^q(x_0) \neq x_0$ for $1 \leqslant q \leqslant p - 1$.

REMARK 2.2. Li and Yorke referred in their appendix the easy extension of the result above to the whole straight line $(-\infty, +\infty)$ instead of the interval $[0, 1]$.

Here we leave the proof of Theorem 2.1 to their simple and elegant original paper: "Period Three Implies Chaos", American Mathematical Monthly **82** (1975) 985–992. Readers are recommended to read it. Instead, we give here a rigorous proof for a simple fact in a dynamical system which is deduced as a special case of this theorem. This

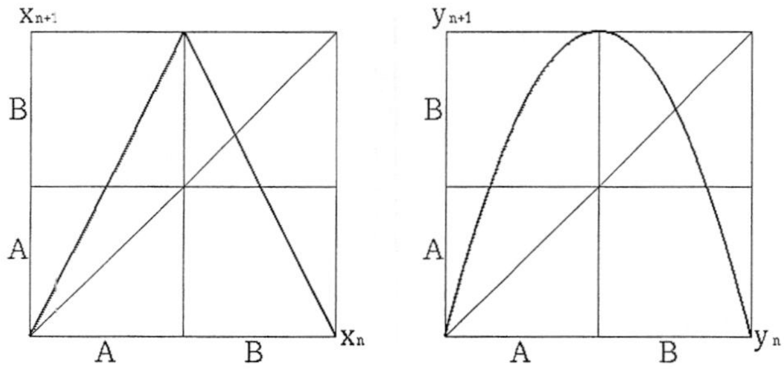

FIG. 2.1.

corresponds also to the case of $a = 4$ in the generalized logistic map as the May's experiment stated in the previous section.

Let us consider a dynamical system:

$$x_{n+1} = \varphi(x_n), \quad x_0 \in [0, 1], \tag{2.3}$$

where $\varphi(x)$ is defined as follows:

$$\varphi(x) = \begin{cases} 2x & (0 \leqslant x \leqslant 1/2), \\ 2(1-x) & (1/2 \leqslant x \leqslant 1). \end{cases}$$

We also consider another dynamical system which is expressed as

$$y_{n+1} = \psi(y_n), \quad y_0 \in [0, 1], \tag{2.4}$$

where $\psi(x)$ is given by

$$\psi(x) = 4x(1-x), \quad x \in [0, 1],$$

the case $a = 4$ of (1.6) stated in the previous section.

We notice that these two systems (2.3) and (2.4) are mutually transformable by the following variable transformation:

$$y_n = \sin^2 \frac{x_n}{2}, \qquad x_n = \arcsin\sqrt{\frac{y_n}{2}}.$$

We show here the graph of φ and ψ in Fig. 2.1. Let us denote $A = [0, 1/2]$, $B = [1/2, 1]$ as in Fig. 2.1. These two dynamical systems share the properties:

$$\begin{aligned} &\varphi(A) \supset A \cup B, \qquad \varphi(B) \supset A \cup B, \\ &\psi(A) \supset A \cup B, \qquad \psi(B) \supset A \cup B. \end{aligned} \tag{2.5}$$

Now we are ready to prove the following important properties of φ and ψ, useful in the rest sections.

THEOREM 2.2. *For any sequence of symbols A and B:*

$$w_0, w_1, w_2, \ldots, w_n, \ldots = \{w_n\}_{n=0}^{+\infty},$$

where $w_i = A$ or B, we can find one x_0 in $[0, 1]$ which insures that

$$x_n \in w_n,$$

for any integer n.

PROOF. First we remark that by the finite intersection property about a sequence of closed set $A_n = \bigcap_{i=0}^{n} \varphi^{-i}(w_i)$, it suffices to prove that for any integer n,

$$\bigcap_{i=0}^{n} \varphi^{-i}(w_i) \neq \emptyset, \tag{2.6}$$

where $\varphi^{-i}(w_i)$ is the inverse image of w_i by $\varphi^i(x)$.

We prove this by induction. For $n = 0$, it is evident that $\varphi^0(w_i) = w_i \neq \emptyset$: Taking A or B, we see that property (2.5) insures this. We assume that (2.6) is true for n, i.e. for any sequence $w_0, w_1, \ldots, w_n$. Now we proceed to prove the case of $n + 1$.

Suppose the contrary, let for some $w_0, w_1, \ldots, w_{n+1}$,

$$\bigcap_{i=0}^{n+1} \varphi^{-i}(w_i) = \emptyset. \tag{2.7}$$

By the induction hypothesis, we have

$$w' = w_1 \cap \varphi^{-1}(w_2) \cap \cdots \cap \varphi^{-n}(w_{n+1}) \neq \emptyset,$$

then (2.7) means $w_0 \cap \varphi^{-1}(w') = \emptyset$. This means $\varphi(x) \notin w'$ for all $x \in w_0$, that is, $\varphi(w_0) \subset (w')^c$ which is a contradiction to (2.5), because

$$w' \cup (w')^c = I \subset \varphi(w_0) \subset (w')^c. \qquad \square$$

2.2. *Snap-back repeller*

The straightforward generalization of Li–Yorke theorem for multi-dimensional case was done by MAROTTO [1978]. Before explaining his results, it is better to introduce a notion called "snap-back repeller".

Let us come back to the example of May:

$$x_{n+1} = ax_n(1 - x_n) = f(x_n). \tag{2.8}$$

We remark that (2.8) has a fixed point $1 - 1/a$, and the differential coefficient of $f(x)$ at this point is

$$f'\left(1 - \frac{1}{a}\right) = -a + 2.$$

It is clear that for the instability of this fixed point (i.e. $|f'(1 - 1/a)| > 1$), a is necessary greater than 3. As we saw in the numerical experiments by May, however, this value of a is not sufficiently large for the occurrence of certain chaotic orbits. What is then the condition for the existence of dynamical system whose one orbit is chaotic?

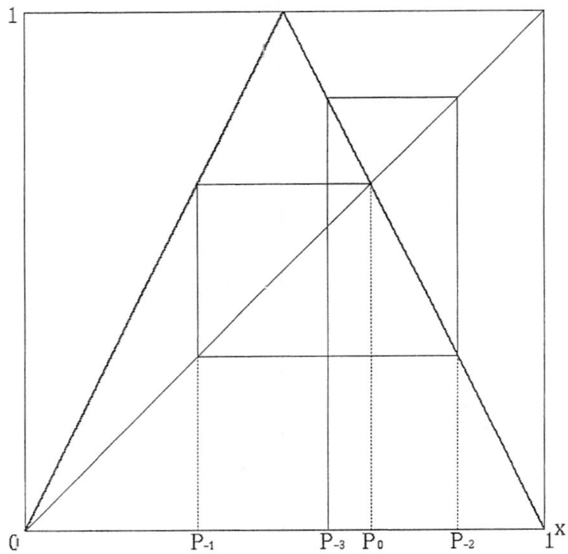

FIG. 2.2.

For an illustration of this, we again examine the simplest dynamical system (2.3), see Fig. 2.2.

First, the non-zero fixed point of (2.3) is $p_0 = 2/3$, and $|\varphi'(p_0)| = 2 > 1$. We can find out such sequence of points as:

$$p_{-1} < p_{-3} < p_{-5} < \cdots < p_0 < \cdots < p_{-4} < p_{-2},$$

where $\varphi(p_{-n}) = p_{-n+1}$, and $p_{-n} \to p_0$ (as $n \to +\infty$). We call this sequence of points a homoclinic orbit of (2.3).

For the proof of the Li–Yorke theorem, they used a lemma (given by the intermediate value theorem) that if a continuous map $f(x):\mathbb{R}^1 \to \mathbb{R}^1$ has an interval I such that $f(I) \supset I$, then there exists a fixed point p of $f(x)$ in I. This lemma is valid only for one-dimensional case, not for multi-dimensional case, and we can easily show that this lemma is no more true even for the two-dimensional dynamical systems. The following is a counterexample: Consider a continuous mapping g which maps two boxes in Fig. 2.3 to a twisted house shoe. Here, we have no fixed point in $A \cup B$. But

$$g(A \cup B) \supset A \cup B.$$

Marotto had generalized the notion of homoclinic orbit to the notion of "snap-back repeller" for one generalization of Li–Yorke theorem in multi-dimensional dynamical system.

To describe the theorem of Marotto, we consider a mapping $F:\mathbb{R}^n \to \mathbb{R}^n$ in C^1-class,

$$X_{n+1} = F(X_n). \tag{2.9}$$

To explain the definition of snap-back repeller, we need first the notion of expanding fixed point. Let $B_r(\bar{u})$ be a closed ball in $\mathbb{R}^n$ of radius r centered at the point $\bar{u}$. The

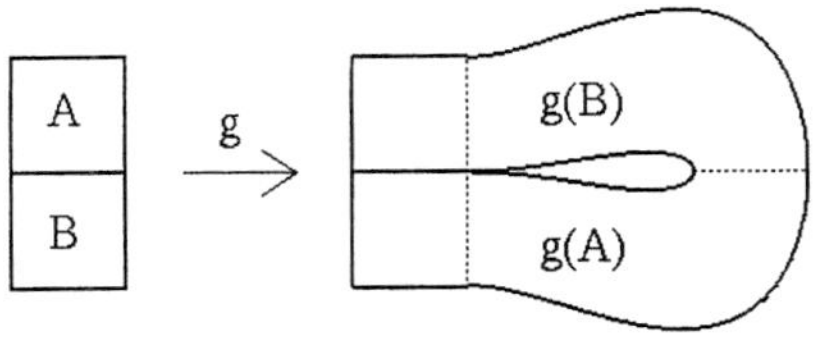

FIG. 2.3.

point $\bar{u} \in \mathbb{R}^n$ is an *expanding fixed point* of F in $B_r(\bar{u})$, if $F(\bar{u}) = \bar{u}$ and all eigenvalues of $\partial F(X)^* \partial F(X)$ exceed one in norm for all $X \in B_r(\bar{u})$ where A^* is an adjoint matrix of A.

A *snap-back repeller* is an expanding fixed point $\bar{u}$ of (2.9) which satisfies the following condition: There exists a point Z ($\neq \bar{u}$) in the neighborhood $B_r(\bar{u})$ such that for some positive integer M, $F^M(Z) = \bar{u}$ and $\det \partial F^M(Z) \neq 0$. Note that we can find a homoclinic orbit $\{\ldots, p_{-3}, p_{-2}, p_{-1}, p_0\ (= Z), p_1, p_2, \ldots, p_M\ (= \bar{u})\}$ such that there exists a neighborhood $V(Z)$ of Z satisfying

$$F^M\bigl(V(Z)\bigr) \subset B_r(\bar{u}), \quad \text{and} \quad p_{-n} \in F^{-n}\bigl(V(Z)\bigr) \to \bar{u} \quad (n \to +\infty),$$

especially, $F^{-n}(V(Z)) \subset F^M(V(Z))$ for some n.

With this definition, Marotto obtained the next theorem.

THEOREM 2.3. *If* (2.9) *has a snap-back repeller, then F has the following properties, similar to those found in the Li–Yorke theorem*:

(i) *There is a positive integer N such that for any integer $p \geqslant N$, F has a p-periodic orbit.*

(ii) *There exists an uncountable set S containing no periodic points of F such that*:

  (a) $F(S) \subset S$,

  (b) *for every $X, Y \in S$ with $X \neq Y$,*

$$\limsup_{n\to+\infty} \bigl\| F^n(X) - F^n(Y) \bigr\| > 0,$$

  (c) *for every $X \in S$ and any periodic point Y of F,*

$$\limsup_{n\to+\infty} \bigl\| F^n(X) - F^n(Y) \bigr\| > 0.$$

(iii) *There is an uncountable subset S_0 of S such that for every $X, Y \in S_0$,*

$$\liminf_{n\to+\infty} \bigl\| F^n(X) - F^n(Y) \bigr\| = 0.$$

3. Discretization of ordinary differential equation

3.1. *Euler's difference scheme*

An ordinary differential equation (logistic equation) can be converted to a chaotic discrete dynamical system, as we have seen through May's example in Section 1. In this section we shall mention that these phenomena can be easily generalized to some class of autonomous differential equations. Namely, certain suitably-settled equilibrium points are actually apt to become sources of "chaos" in the discretized equations of original ordinary differential equations.

Let us consider an autonomous scalar differential equation of the form:

$$\frac{\mathrm{d}u}{\mathrm{d}t} = f(u), \tag{3.1}$$

where f is a continuous function. We assume that (3.1) has at least two equilibrium points, one of which is asymptotically stable.

This assumption reduces (if necessary, after a linear transformation of the unknown variable) to the following conditions:

$$\begin{cases} f(0) = f(\bar{u}) = 0 & \text{for some } \bar{u} > 0, \\ f(u) > 0 & (0 < u < \bar{u}), \\ f(u) < 0 & (\bar{u} < u < K). \end{cases} \tag{3.2}$$

Here, the positive constant K is possibly $+\infty$.

Euler's difference scheme for (3.1) is:

$$x_{n+1} = x_n + \Delta t \cdot f(x_n). \tag{3.3}$$

We adopt the notation $F_{\Delta t}(x) = x + \Delta t f(x)$ in the following section.

3.2. *Yamaguti–Matano theorem*

Yamaguti–Matano theorem [1979] is stated as follows:

THEOREM 3.1.

(i) *Let* (3.2) *hold. Then there exists a positive constant* c_1 *such that for any* $\Delta t > c_1$ *the difference equation* (3.3) *is chaotic in the sense of Li–Yorke.*

(ii) *Suppose in addition that* $K = +\infty$; *then there exists another constant* c_2, $0 < c_1 < c_2$, *such that for any* $0 \leqslant \Delta t \leqslant c_2$, *the map* $F_{\Delta t}$ *has an invariant finite interval* $[0, \alpha_{\Delta t}]$ (*i.e.* $F_{\Delta t}$ *maps* $[0, \alpha_{\Delta t}]$ *into* $[0, \alpha_{\Delta t}]$) *with* $\alpha_{\Delta t} > \bar{u}$. *Moreover, when* $c_1 < \Delta t \leqslant c_2$, *the above-mentioned chaotic phenomenon occurs in this interval.*

REMARK 3.1. If f is analytic and (3.1) has no asymptotically stable equilibrium point, then (3.3) can never be chaotic for any positive value of Δt, because $F_{\Delta t}(x) = x + \Delta t f(x)$ defines a one-to-one correspondence under this assumption.

PROOF. For each $\Delta t \geqslant 0$, we define the following three functions of Δt, $M(\Delta t)$, $R(\Delta t)$ and $r(\Delta t)$:

$$M(\Delta t) = \max_{0 \leqslant x \leqslant \bar{u}} F_{\Delta t}(x),$$

$$R(\Delta t) = \sup\Big\{x \geqslant \bar{u} \;\Big|\; \min_{\bar{u} \leqslant z \leqslant x} F_{\Delta t}(z) \geqslant 0\Big\},$$

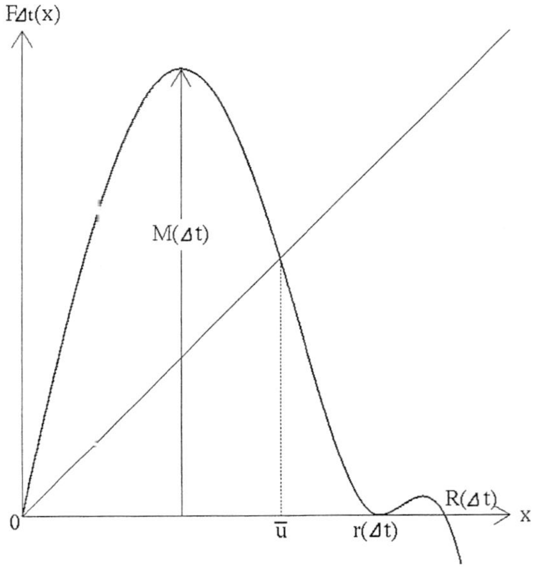

FIG. 3.1.

$$r(\Delta t) = \sup\Big\{x \geqslant \bar{u} \;\Big|\; \min_{\bar{u} \leqslant z \leqslant x} F_{\Delta t}(z) > 0\Big\}.$$

In brief, $r(\Delta t)$ is the first positive zero-point of $F_{\Delta t}(x)$, and $R(\Delta t)$ is the first point where $F_{\Delta t}(x)$ changes its sign. See Fig. 3.1. As Δt varies, $M(\Delta t)$ ranges over the interval $[\bar{u}, +\infty)$, while $R(\Delta t)$ and $r(\Delta t)$ range over $(\bar{u}, +\infty]$.

Since $F_{\Delta t}(x)$ is a linear function of Δt, the function M has the following two properties:

$$\begin{cases} \text{(i)} & M(s) \text{ is monotone increasing and continuous in } s, \\ \text{(ii)} & M(0) = \bar{u} \text{ and } \lim_{s \to +\infty} M(s) = +\infty. \end{cases}$$

And the function R satisfies:

$$\begin{cases} \text{(i)} & R(s) \text{ is monotone decreasing and left continuous in } s \\ & (\text{i.e. } R(s) = R(s-0)), \\ \text{(ii)} & R(0) = +\infty \text{ and } \lim_{s \to +\infty} R(s) = \bar{u}. \end{cases} \tag{3.4}$$

Indeed, the monotonicity is an easy consequence of the fact that $F_{\Delta t}(x) = x + \Delta t f(x)$ is monotone decreasing with respect to Δt for $x \geqslant \bar{u}$. The left continuity can be shown by using a contradiction: Suppose $R(s-0) = R(s) + k$ $(k > 0)$. Then $R(s-\varepsilon) = R(s) + k + g(\varepsilon)$ (where $g(\varepsilon) \searrow 0$ for $\varepsilon \searrow 0$). Therefore, $F_{s-\varepsilon}(R(s)+k) \geqslant 0$ by the definition of R. Then,

$$0 > F_s\big(R(s)+k\big) = F_{s-\varepsilon}\big(R(s)+k\big) + \varepsilon f\big(R(s)+k\big),$$

hence,

$$0 > F_s\big(R(s)+k\big) \geqslant 0 \quad (\varepsilon \to 0).$$

This is a contradiction. Part (ii) of (3.4) is easier to prove.

We need to show

$$r(s) = R(s+0). \tag{3.5}$$

It suffices to prove $r(s-0) = R(s)$ for $s > 0$. By our definition,

$$r(s-\varepsilon) = \sup\Big\{x \geqslant \bar{u} \mid \min_{\bar{u} \leqslant z \leqslant x} F_{s-\varepsilon}(z) > 0\Big\},$$

$$R(s-\varepsilon) = \sup\Big\{x \geqslant \bar{u} \mid \min_{\bar{u} \leqslant z \leqslant x} F_{s-\varepsilon}(z) \geqslant 0\Big\}.$$

First, let us prove $r(s-\varepsilon) \geqslant R(s)$ for small ε. It is easy to see that

$$F_s(z) - \varepsilon f(z) > 0 \quad \text{for } \bar{u} < z \leqslant R(s),$$

and $F_s(\bar{u}) - \varepsilon f(\bar{u}) > 0$. Then this implies

$$\min_{\bar{u} \leqslant z \leqslant R(s)} F_{s-\varepsilon}(z) > 0.$$

This proves $r(s-\varepsilon) \geqslant R(s)$ for all $\varepsilon > 0$, hence, $r(s-0) \geqslant R(s)$.

On the other hand, $R(s-\varepsilon) \geqslant r(s-0)$ $(\forall \varepsilon > 0)$, therefore, $R(s) \geqslant r(s-0)$, i.e. (3.5) is obtained.

From the equality $r(s) = R(s+0)$, it follows that $r(s)$ is right continuous and that $\bar{u} < r(s) \leqslant R(s) \leqslant +\infty$. The strict inequality $r(s) < R(s)$ holds on their discontinuity points. Note that $r(0) = R(0) = +\infty$ so that $\lim_{s\to+0} R(s) = +\infty = R(0)$.

We also need to show that, if $r(\Delta t) \leqslant M(\Delta t)$, then there exists a point $x^* \in [0, \bar{u}]$ satisfying

$$0 < F^3_{\Delta t}(x^*) < x^* < F_{\Delta t}(x^*) < F^2_{\Delta t}(x^*), \tag{3.6}$$

where $F^n_{\Delta t}$ denotes the nth iteration of the map $F_{\Delta t}$.

Take ξ to be one of points at which $F_{\Delta t}(\xi) = r(\Delta t)$. Then we determine a positive x^* to be a point in $[0, \bar{u}]$ which satisfies $F_{\Delta t}(x^*) = \xi$. Then, we obtain the results of iterations for x^* satisfying the conditions (3.6) except $0 < F^3_{\Delta t}(x^*)$:

$$x^*, \quad F_{\Delta t}(x^*) = \xi, \quad F_{\Delta t}(\xi) = r(\Delta t) = F^2_{\Delta t}(x^*), \quad \text{and}$$
$$F^3_{\Delta t}(x^*) = F^2_{\Delta t}(\xi) = F_{\Delta t}\big(r(\Delta t)\big) = 0.$$

Furthermore, by taking $F_{\Delta t}(\xi)$ near $r(\Delta t)$, we can modify very easily this x^* to satisfy all conditions in (3.6).

We are now ready to prove the theorem. Put

$$c_2 = \sup\big\{s \geqslant 0 \mid M(s) - R(s) \leqslant 0\big\}.$$

Then c_2 is positive, because $M(s) - R(s)$ is continuous at $s = 0$, and $R(s) = +\infty$.

For the left continuity of $M(s) - R(s)$, we have

$$R(s) \geqslant M(s) \quad (0 \leqslant s \leqslant c_2), \tag{3.7}$$

$$R(s) < M(s) \quad (s > c_2). \tag{3.8}$$

Combining (3.8) and (3.5), we get

$$r(s) \leqslant M(s) \quad (s \geqslant c_2).$$

This means that when we are given any $\Delta t \geqslant c_2$, there exists a point $x^* \in [0, \bar{u}]$ satisfying the condition (3.6), which makes (3.3) chaotic in the sense of Li and Yorke (see their original paper including their last remark). Since this condition is a stable property under a small perturbation of $F_{\Delta t}$, we can find a constant c_1, $0 < c_1 < c_2$, such that (3.3) is chaotic for any $\Delta t > c_1$. Thus the first statement of the theorem is established.

Suppose now that $K = -\infty$, and let $0 \leqslant \Delta t \leqslant c_2$. Then from (3.7), it immediately follows that when we consider any α with $M(\Delta t) \leqslant \alpha \leqslant R(\Delta t)$, the interval $[0, \alpha]$ is invariant under $F_{\Delta t}$. It is also clear that, when $c_1 < \Delta t \leqslant c_2$, the restriction of $F_{\Delta t}$ to $[0, \alpha]$ possesses a point x^* satisfying (3.6). Hence the second statement follows. This completes the proof of the theorem. □

4. Mathematical model for economics

4.1. Walrasian general equilibrium theory

When we discretize a differential equation, the solutions may behave quite differently from those of original differential equation in the sense that they may be chaotic even though the original differential equation has no chaotic solution, as we have seen in the previous section. In this section, we examine the meanings of this fact in the field of economics.

Most of the traditional economic theories assume that demand and supply are balanced by the price setting. The basic idea behind these theories is that prices adjust fast enough so that demand and supply are equalized in the market.

In this tradition, these prices are therefore supposed effective indices for the efficient allocation of resources in a decentralized economy. Walrasian general equilibrium theory (WALRAS [1926]), which explains how the price mechanism functions in a complex economy with many interrelated markets, can be presented as the most elaborated model. In this theory, the market process is regarded as the results of the price adjustment of a commodity where the price is increased according to its positive excess demand and vice versa. This price-adjustment process, so-called *tatonnement* was first formulated mathematically by HICKS [1946], and has been reformulated more rigorously in recent decades. Studies on the stability of the tatonnement process are mainly modeled by differential equations, as seen in the article by NEGISHI [1962].

The recent trend of theoretical economics is moving to consider the difference equations instead of the differential equations. For instance, SAARI [1985] argued that the correct dynamical process associated with the tatonnement process is an iterative one. One of the reasons for this change is that in the real market the time lag in the price adjustment process does exist. This time lag is caused as follows: A price adjustment process is based upon a reaction of the people in the market as reflected by the disequilibrium of supply and demand. That is, the market maker needs to obtain information on excess demand, and this may be more or less out of date when records are in. Decisions are to be made after that, so the effect of the adjustment process requires some delay.

Therefore, the continuous-time price adjustment process can be viewed as serving as a limiting approximation for the discrete-time one.

The main object of this section is to compare the continuous-time price adjustment process with the discrete-time one, and to study the difference between their dynamical properties. In a Walrasian exchange economy with two commodities we will show that if a continuous-time price adjustment process has two or more equilibrium prices and at least one of those equilibrium prices is asymptotically stable, then the corresponding discrete-time price adjustment process becomes chaotic for appropriate rates of price-adjustment.

4.2. *A mathematical model of an exchange economy*

Let us consider a Walrasian exchange economy with n commodities and m participants. Commodities are denoted by $i = 1, 2, \ldots, n$, while individual participants are represented by $r = 1, 2, \ldots, m$. At the beginning of the market day, each individual has certain amounts of commodities, and the exchange of commodities between the individuals take place. Let the amount of commodity i initially held by individual r be y_i^r, $i = 1, 2, \ldots, n$; $r = 1, 2, \ldots, m$. Using vector notation, we may say that the vector of the initial holdings of individual r is

$$y^r = (y_1^r, y_2^r, \ldots, y_n^r), \quad r = 1, 2, \ldots, m.$$

It is assumed that each individual has a definite demand schedule when a market price vector and his income are given. Let the demand function of individual r be

$$x^r(p, M^r) = [x_1^r(p, M^r), \ldots, x_n^r(p, M^r)],$$

where $p = (p_1, p_2, \ldots, p_n)$ is a market price vector and M^r is the income of individual r.

Each demand function $x^r(p, M^r)$ is assumed to satisfy the budget equation:

$$\sum_{i=1}^{n} p_i x_i^r(p, M^r) = M^r.$$

Once a price vector $p = (p_1, p_2, \ldots, p_n)$ has been announced to prevail in the whole market, the income $M^r(p)$ of individual r is expressed by

$$M^r(p) = \sum_{i=1}^{n} p_i y_i^r, \quad r = 1, 2, \ldots, m.$$

Hence the demand function of individual r becomes

$$x^r[p, M^r(p)], \quad r = 1, 2, \ldots, m.$$

At this stage, the aggregate demand function is defined as

$$x_i(p) = \sum_{r=1}^{m} x_i^r[p, M^r(p)], \quad i = 1, 2, \ldots, n,$$

and the aggregate excess demand function is given as

$$z_i(p) = x_i(p) - y_i,$$

where

$$y_i = \sum_{r=1}^{m} y_i^r, \quad i = 1, 2, \ldots, n.$$

Note that Walras law holds:

$$\sum_{i=1}^{n} p_i z_i(p) = 0, \quad \text{for any price vector } p = (p_1, p_2, \ldots, p_n).$$

A price vector $p^* = (p_1^*, p_2^*, \ldots, p_n^*)$ is defined to be an equilibrium price vector if $z_i(p) = 0$, $i = 1, 2, \ldots, n$.

The main problem in the theory of exchange of commodities is now to investigate whether or not there exists an equilibrium, and whether or not the price adjustment process converges to the equilibrium price vector.

Conventionally, the price adjustment process is formulated by the following ordinary differential equations

$$\frac{\mathrm{d}p_i}{\mathrm{d}t} = z_i(p), \quad i = 1, 2, \ldots, n. \tag{4.1}$$

Thus the price adjustment process is assumed to be successful.

On the other hand, the discrete-time price adjustment process is

$$p_i(t+1) - p_i(t) = \alpha z_i\big[p_i(t)\big], \quad i = 1, 2, \ldots, n, \tag{4.2}$$

where $p_i(t)$ stands for the value of the price of this commodity at the t-step of time, and α is the speed of the determination of prices. Remark that Eq. (4.2) is exactly the Euler discretization of (4.1).

4.3. *Chaotic tatonnement*

Let us now discuss the simplest case: An exchange economy with two commodities and two participants. Before describing the price-adjustment process we shall first derive the demand of an individual for the commodities. In the classical theory of demand for commodities the central assumption on the individual's behavior is that he or she chooses the quantity of the commodities which maximizes the utility function. The individual's demand for a commodity is then the quantity chosen as a result of this utility maximization. Mathematically, this means we want to solve the constrained maximization problem:

$$\max U^r\left(x_1^r, x_2^r\right) \quad \text{subject to} \quad p_1 x_1^r + p_2 x_2^r = p_1 y_1^r + p_2 y_2^r,$$

where x_1^r, x_2^r represent commodity 1 and commodity 2 which individual r actually consumes, and p_1, p_2 are the prices of commodity 1 and commodity 2. Finally, $U^r(x_1^r, x_2^r)$ represents the individual's own subjective evaluation of the satisfaction or utility derived from consuming those commodities. The second equation is the budget constraint that the individual faces. Let us consider the following typical utility function by KEMENY and SNELL [1963],

$$U^r\left(x_1^r, x_2^r\right) = -\alpha_1^r \exp\left(-\beta_1^r\left(x_1^r - y_1^r\right)\right) - \alpha_2^r \exp\left(-\beta_2^r\left(x_2^r - y_2^r\right)\right).$$

The excess demand functions of the individuals are

$$z_1^r(p) = \frac{p \log \theta^r p}{\beta_2^r + \beta_1^r p}, \qquad z_2^r(p) = -\frac{\log \theta^r p}{\beta_2^r + \beta_1^r p},$$

where

$$\theta^r = \frac{\alpha_1^r \beta_1^r}{\alpha_2^r \beta_2^r}, \qquad p = \frac{p_1}{p_2}.$$

We assume here that the two individuals have the different preference ordering over the commodities. We select the parameters for the utility functions of individual 1 and individual 2 as follows:

$$\theta^1 = \mathrm{e}^5, \quad \beta_1^1 = 3, \quad \beta_2^1 = 1, \quad \theta^2 = \mathrm{e}^{-5}, \quad \beta_1^2 = 1, \quad \beta_2^2 = 3.$$

Solving the utility maximization problem, the aggregate excess demand functions are

$$z_1(p) = \frac{p(5 + \log p)}{1 + 3p} + \frac{p(-5 + \log p)}{3 + p},$$

$$z_2(p) = -\frac{(5 + \log p)}{1 + 3p} - \frac{(-5 + \log p)}{3 + p}.$$

In this case there exists the three equilibrium prices, $p^* = 0.17$, $p^{**} = 1$, and $p^{***} = 5.89$. Fig. 4.1 shows the excess demand for commodity 2 and the equilibrium prices, where $p^* = A$, $p^{**} = B$, and $p^{***} = C$. Furthermore, it is easy to confirm that the Walras law holds at the equilibrium prices. That is,

$$z_1(p) + p z_2(p) = 0.$$

The market price is equal to an equilibrium price provided that the excess demand for commodity 2, $z_2(p)$ is equal to zero. Therefore, the continuous-time price-adjustment mechanism is formulated by

$$\frac{\mathrm{d}p}{\mathrm{d}t} = z_2(p).$$

It follows from Fig. 4.1 that the equilibrium prices p^* and p^{***} are asymptotically stable, and the equilibrium price p^{**} is asymptotically unstable.

On the other hand, the corresponding discrete-time price-adjustment process is

$$p(t+1) = p(t) + \alpha z_2\big(p(t)\big). \tag{4.3}$$

Now consider the largest and second largest of the equilibrium prices, p^{**} and p^{***}. Fig. 4.2 shows the map for (4.3). In Fig. 4.2, $p^* = A$, $p^{**} = B$ and $p^{***} = C$.

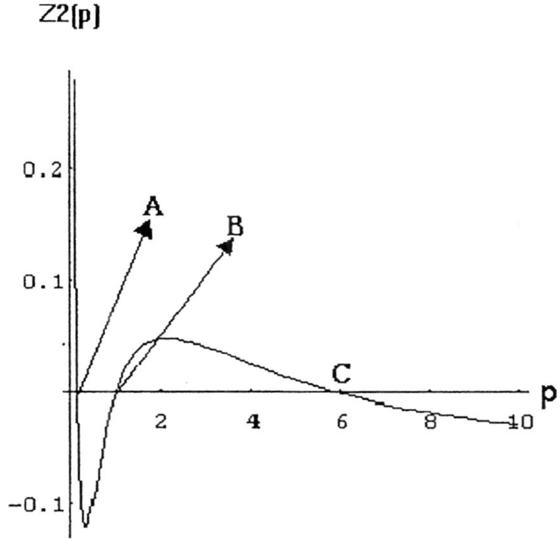

FIG. 4.1.

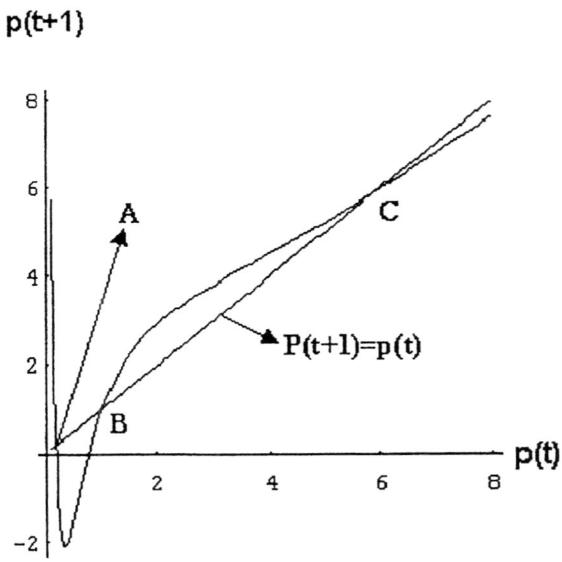

FIG. 4.2.

If follows from Fig. 4.2 that the aggregate excess demand function for commodity 2 satisfies the following conditions:

$$z_2(p^{**}) = z_2(p^{***}) = 0,$$
$$z_2(p) > 0 \quad (p^{**} < p < p^{***}),$$
$$z_2(p) < 0 \quad (p^{***} < p < +\infty).$$

Under these conditions, we can apply the Yamaguti–Matano theorem to the price adjustment process (4.3) and, from a result of KAIZOUJI [1994], we can formulate the following:

THEOREM 4.1. *Let the above conditions hold. Then there exists a positive constant* α_1 *and* α_2 $(0 < \alpha_1 < \alpha_2)$ *such that for any* $\alpha_1 \leqslant \alpha \leqslant \alpha_2$ *the discrete-time price-adjustment process* (4.3) *is chaotic in the sense of Li–Yorke.*

4.4. Future work

In this section the price-adjustment process in the exchange economy with two commodities is investigated making use of Yamaguti–Matano theorem. It is likely that the result that is demonstrated above can be developed in two directions at least. First, it is proved that the same result is true in the case which the exchange economy has only one equilibrium price and the continuous-time price-adjustment process is globally asymptotically stable. The fact is demonstrated by MAEDA [1995], which are mentioned in the following section. Secondly, Theorem 4.1 is able to be generalized in the case of the exchange economy with a large number of commodities and a large number of participants. HATA [1982] has contributed in this generalization of Yamaguti–Matano theorem, which is also discussed in Section 9.3.

Although numerous studies have been made on the existence of equilibria and the local stability for the equilibria in the Walrasian economy, little attention has been given to the global properties of the price-adjustment processes in the markets (see ARROW and HAHN [1971]). In this sense, the recent developments on nonlinear dynamics have thrown a new light over this unsettled question.

5. Discretization with large time steps and chaos

5.1. O.D.E. with globally asymptotical stability

Some types of scalar ordinary differential equations are apt to turn into chaos in its Euler discretization as we have seen in Section 3. There, we discussed this fact under two principal assumptions, viz., that the differential equation has at least two equilibrium points, and that one of them is asymptotically stable. From the present section until Section 7, we relax these conditions by assuming that there is only one equilibrium point (see MAEDA [1995]).

Here is a simple example:

$$\frac{\mathrm{d}u}{\mathrm{d}t} = 1 - \mathrm{e}^{u}.$$

The equilibrium point is $u = 0$ and it is globally asymptotically stable, namely, the orbit starting from any initial point goes toward $u = 0$ as t increases. What then happens in its discretization $F_{\Delta t}(x)$? Fig. 5.1 shows that $F_{\Delta t}(x)$ is chaotic in the sense of Li–Yorke with sufficiently large time step Δt, because there is a periodic orbit with period 3.

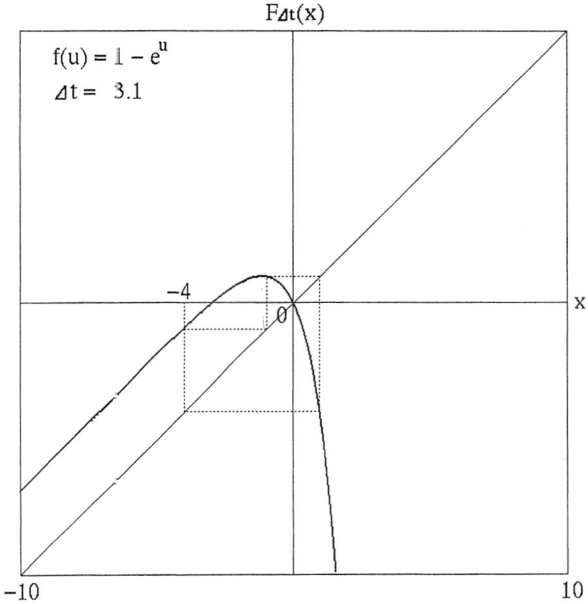

FIG. 5.1.

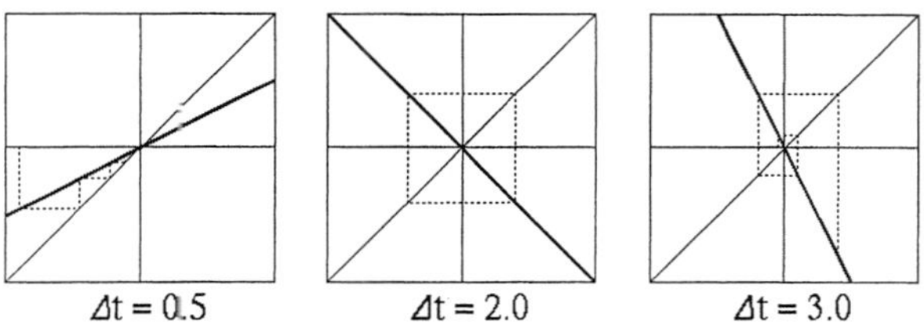

FIG. 5.2.

But any differential equation whose equilibrium point is globally asymptotically stable does not always turn into chaos; see, e.g.,

$$\frac{\mathrm{d}u}{\mathrm{d}t} = -u.$$

This linear differential equation is discretized as

$$F_{\Delta t}(x) = x - \Delta t \cdot x = (1 - \Delta t)x.$$

If Δt is less than 2, any orbit tends to $x = 0$. If $\Delta t = 2$, then any orbit is periodic with period 2 except for the fixed point $x = 0$, and in the case of $\Delta t > 2$, any orbit starting with $x_0 \neq 0$ diverges with oscillations (see Fig. 5.2).

Motivated by the simple observation as above, our aim is to obtain sufficient conditions under which a globally asymptotically stable equilibrium point of a differential equation induces chaos for sufficiently large time step. From now on, we consider a scalar ordinary differential equation

$$\frac{\mathrm{d}u}{\mathrm{d}t} = f(u)$$

with conditions:

$$\begin{cases} f(u) \text{ is continuous in } \mathbb{R}^1, \\ f(0) = 0, \\ f(u) > 0 \quad (u < 0), \\ f(u) < 0 \quad (u > 0). \end{cases} \tag{5.1}$$

We will investigate three types of the differential equation according as f has an upper bound, f has a power root order, and f is piecewise linear. One of them has a remarkable property: Chaos occurs with any small time step Δt (this phenomenon will be explained in Section 6). Recall that the function f given by $f(u) = 1 - \mathrm{e}^u$ (as mentioned above) is bounded for $u < 0$.

5.2. *Three types of O.D.E.'s and chaos with large time steps*

Let us first investigate the bounded case.

THEOREM 5.1 (bounded case). *If f satisfies the following conditions, in addition to* (5.1):

(i) *there exists a positive constant M such that $f(u) \leqslant M$ $(u < 0)$,*

(ii) *there exists a positive c_0 for which $f(c_0) = -2M$,*

then, for sufficiently large Δt, there is an invariant finite interval $[K_1, K_2]$ on which $F_{\Delta t}$ is chaotic in the sense of Li–Yorke.

PROOF. As we have seen in Section 2, to prove chaos in the sense of Li–Yorke, it is enough to establish the existence of a, $b = f(a)$, $c = f(b)$, $d = f(c)$ that satisfy $d \leqslant a < b < c$. Here, we define ΔT as

$$\Delta T = \min_{x<0} \frac{c_0 - x}{f(x)}.$$

Note that

$$\lim_{x \to -0} \frac{c_0 - x}{f(x)} = +\infty, \qquad \lim_{x \to -\infty} \frac{c_0 - x}{f(x)} = +\infty,$$

and $\Delta T \geqslant c_0/M$. Now, let us show that for any $\Delta t > \Delta T$, $F_{\Delta t}$ is chaotic.

First, take a negative b which satisfies $(c_0 - b)/f(b) = \Delta t$. Then,

$$c = F_{\Delta t}(b) = b + \Delta t \cdot f(b) = c_0 \ (> 0 > b),$$

and

$$d = F_{\Delta t}(c) = c_0 + \Delta t \cdot f(c_0) = c_0 - 2\Delta t \cdot M \ (< -c).$$

By the continuity of the function $F_{\Delta t}$, we can confirm the existence of a negative $a \in [b - \Delta t \cdot M, b)$ that satisfies $F_{\Delta t}(a) = b$. Indeed,

$$\begin{aligned} F_{\Delta t}(b - \Delta t \cdot M) &= b - \Delta t \cdot M + \Delta t \cdot f(b - \Delta t \cdot M) \\ &= b + \Delta t\big(f(b - \Delta t \cdot M) - M\big) \leqslant b, \end{aligned}$$

and

$$F_{\Delta t}(b) = b + \Delta t \cdot f(b) > b.$$

Hence, there exists a which satisfies $a < b$.

Finally, $d \leqslant a$ since

$$\begin{aligned} a - d &\geqslant b - \Delta t \cdot M - (c_0 - 2\Delta t \cdot M) \\ &= b - \Delta t \cdot M - \big(b + \Delta t \cdot f(b) - 2\Delta t \cdot M\big) \\ &= \Delta t\big(M - f(b)\big) \geqslant 0. \end{aligned}$$

As to the invariant finite interval $[K_1, K_2]$, it is defined as follows:

$$K_2 = \max_{x \leqslant 0} F_{\Delta t}(x) \quad \text{and} \quad K_1 = \min_{0 \leqslant x \leqslant K_2} F_{\Delta t}(x).$$

Note that $-\infty < K_1 < d$, and $c \leqslant K_2 < +\infty$. □

In the following argument, an invariant finite interval will be defined in the same way. The following theorem is a simple generalization of Theorem 5.1, which is applicable to the case of $f(u) = -u/(u^2 + 1)$ (see Fig. 5.3).

THEOREM 5.2. *If f satisfies, in addition to* (5.1):

(i) *there exists a positive constant m such that*

$$\limsup_{u \to -\infty} f(u) = m,$$

(ii) *there exists a positive c_0 such that $f(c_0) < -2m$,*

then, for sufficiently large Δt, $F_{\Delta t}$ is chaotic in an invariant finite interval.

PROOF. Define a small positive number ε as

$$\varepsilon = -\frac{2m + f(c_0)}{2}.$$

Then, there exists a negative number K such that $f(x) < m + \varepsilon$ $(x \in (-\infty, K))$. Now choose ΔT as

$$\Delta T = \inf_{x < K} \frac{c_0 - x}{f(x)} \left(\geqslant \frac{c_0 - K}{m + \varepsilon} \right).$$

Noticing that

$$\lim_{x \to -\infty} \frac{c_0 - x}{f(x)} = +\infty,$$

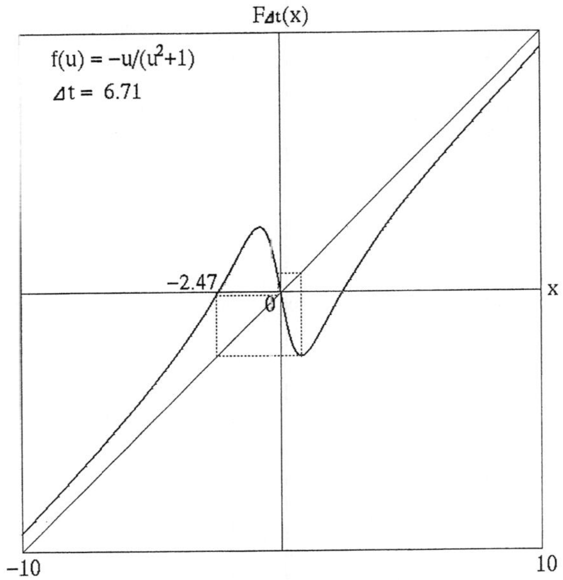

FIG. 5.3.

for any Δt larger than ΔT, we can confirm that there exists b less than K which satisfies $F_{\Delta t}(b) = c_0$. As in the proof of Theorem 5.1, we can find $a \in (b - \Delta t(m+\varepsilon), b)$ such that $F_{\Delta t}(a) = b$.

Finally,

$$\begin{aligned} a - d &> \big(b - \Delta t(m+\varepsilon)\big) - \big(c_0 + \Delta t \cdot f(c_0)\big) \\ &= \Delta t\big(m + \varepsilon - f(b)\big) > 0. \end{aligned} \qquad \square$$

In particular, if $m = 0$, the theorem above is also a generalization of the Yamaguti–Matano theorem [1979]. Next, let us consider whether the boundedness of f is needed for chaos. The answer is no: The next theorems indicate that chaos may occur even in an unbounded case.

THEOREM 5.3 (power root order case). *If f satisfies, in addition to* (5.1):

(i) $\limsup_{u\to-\infty} f(u) = \mathrm{O}((-u)^\alpha)$ $(0 < \alpha < 1)$,

(ii) $\limsup_{u\to+\infty} f(u) = -\mathrm{O}(u^\beta)$ $(1/(1-\alpha) < 1+\beta)$,

then, for sufficiently large Δt, $F_{\Delta t}$ is chaotic in an invariant finite interval.

PROOF. Take any negative number b and fix it. For $\Delta t > -b/f(b)$,

$$c = b + \Delta t \cdot f(b) > 0,$$

$$d = c + \Delta t \cdot f(c) = b + \Delta t\big(f(b) + f\big(b + \Delta t \cdot f(b)\big)\big) = -\mathrm{O}(\Delta t^{1+\beta}).$$

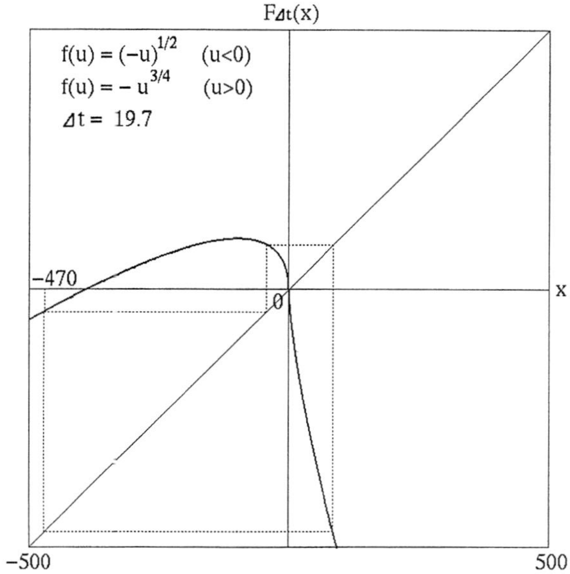

FIG. 5.4.

Since

$$\lim_{x\to-\infty} F_{\Delta t}(x) = x^1 + \mathrm{O}\big((-x)^\alpha\big) = -\infty,$$

for any b and any Δt, there exists a negative number a less than b such that $F_{\Delta t}(a) = b$, i.e. $a + \Delta t \cdot f(a) = b$.

Noticing that

$$\lim_{\Delta t\to+\infty} a = -\infty,$$

let us show that we can find m that satisfies $a = -\mathrm{O}(\Delta t^m)$. If this is the case, then $f(a) = \mathrm{O}(\Delta t^{m\alpha})$ and

$$\Delta t \cdot f(a) = b - a,$$

so that

$$\mathrm{O}(\Delta t^{m\alpha+1}) = \mathrm{O}(\Delta t^m)$$

follows. Thus, we can choose $m = 1/(1-\alpha)$, that is, $a = -\mathrm{O}(\Delta t^{1/(1-\alpha)})$. Hence, $1 + \beta > 1/(1-\alpha)$ implies $d \leqslant a$ for sufficiently large Δt. □

For a specific case, we can weaken the condition on α and β in Theorem 5.3.

THEOREM 5.4. *Assume that the function f is defined as follows:*

$$f(u) = \begin{cases} (-u)^\alpha & (u < 0)\ (0 < \alpha < 1), \\ -u^\beta & (u \geqslant 0) \end{cases}$$

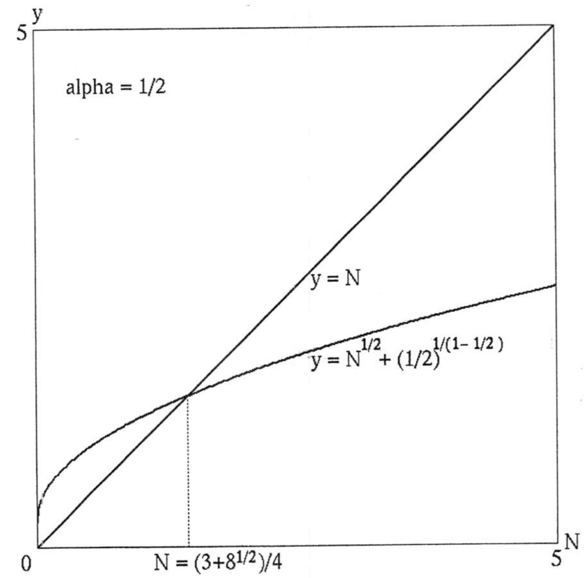

FIG. 5.5.

(see Fig. 5.4). If $\alpha < \beta$, then, for sufficiently large Δt, $F_{\Delta t}$ is chaotic in an invariant finite interval.

PROOF. Take a negative b such that $F'_{\Delta t}(b) = 0$, i.e.

$$b = -(\alpha \Delta t)^{1/(1-\alpha)}.$$

Note that $F_{\Delta t}(x)$ is monotone increasing in $x \in (-\infty, b)$ and that

$$\lim_{x \to -\infty} F_{\Delta t}(x) = -\infty.$$

Hence, we can confirm that there exists a unique a such that $F_{\Delta t}(a) = b$. In fact, a is given as follows:

$$a = -N \Delta t^{1/(1-\alpha)},$$

where N is a unique positive solution of $N = N^{\alpha} + \alpha^{1/(1-\alpha)}$ (see Fig. 5.5). Then

$$c = F_{\Delta t}(b) = \left(\alpha^{\alpha/(1-\alpha)} - \alpha^{1/(1-\alpha)}\right) \Delta t^{1/(1-\alpha)} = A \Delta t^{1/(1-\alpha)},$$

where $A = \alpha^{\alpha/(1-\alpha)} - \alpha^{1/(1-\alpha)}$ and $A \in (0, 1)$; hence c is positive. Then d is given as

$$d = F_{\Delta t}(c) = A \Delta t^{1/(1-\alpha)} - A^{\beta} \Delta t^{(1+\beta/(1-\alpha))}.$$

It suffices for $F_{\Delta t}$ to be chaotic that $1/(1-\alpha) < 1 + \beta/(1-\alpha)$, i.e. $\alpha < \beta$. □

We can also point out that chaos may occur in the case where

$$\lim_{u \to -\infty} f(u) = \mathrm{O}\left((-u)^1\right).$$

Consider a piecewise linear function. The convexity of the function is important.

THEOREM 5.5. *Assume that f is a piecewise linear function defined as*

$$f(u) = \begin{cases} -\alpha(u-b) - b & (u < b), \\ -\beta u & (u \geqslant b), \end{cases}$$

where α and β are positive numbers and b is negative. If $\alpha/\beta < (2-\sqrt{3})/2$, then there exists an interval $J = [\Delta t_1, \Delta t_2]$ such that for any $\Delta t \in J$, $F_{\Delta t}$ is chaotic in an invariant finite interval.

PROOF. By rescaling Δt, we can choose f as follows:

$$f(u) = \begin{cases} -m(u-b) - b & (u < b), \\ -u & (u \geqslant b), \end{cases}$$

where $0 < m < (2-\sqrt{3})/2$. Then,

$$F_{\Delta t}(x) = \begin{cases} (1 - m\Delta t)x + b(m-1)\Delta t & (x < b), \\ (1-\Delta t)x & (x \geqslant b). \end{cases}$$

Note that $0 < m < (2-\sqrt{3})/2$ implies $0 < m < 1/2$. From now on, assume $2 < \Delta t < 1/m$. Since

$$c = F_{\Delta t}(b) = (1-\Delta t)b,$$
$$d = F_{\Delta t}(c) = (1-\Delta t)^2 b,$$

the condition $\Delta t > 2$ yields that $c > 0$ and $d < b$. On the other hand, if Δt is less than $1/m$, there exists a such that,

$$a = \left(1 + \frac{\Delta t}{1 - m\Delta t}\right) b \ (< b).$$

Therefore,

$$a - d = \frac{-b\Delta t}{1 - m\Delta t}\left(-m\Delta t^2 + (2m+1)\Delta t - 3\right).$$

The inequalities $0 < m < (2-\sqrt{3})/2$ imply $-m\Delta t^2 + (2m+1)\Delta t - 3 > 0$ for some Δt. The numbers Δt_1 and Δt_2 are defined as follows:

$$\Delta t_1 = \frac{1}{2m}\left(2m + 1 - \sqrt{4m^2 - 8m + 1}\right),$$
$$\Delta t_2 = \frac{1}{2m}\left(2m + 1 + \sqrt{4m^2 - 8m + 1}\right).$$

Then, $2 < \Delta t_1 < \Delta t_2 < 1/m$, and $d < a$ for any $\Delta t \in [\Delta t_1, \Delta t_2]$. □

Fig. 5.6 shows an example where Theorem 5.5 holds. Finally, let us mention a simple generalization of Theorem 5.5.

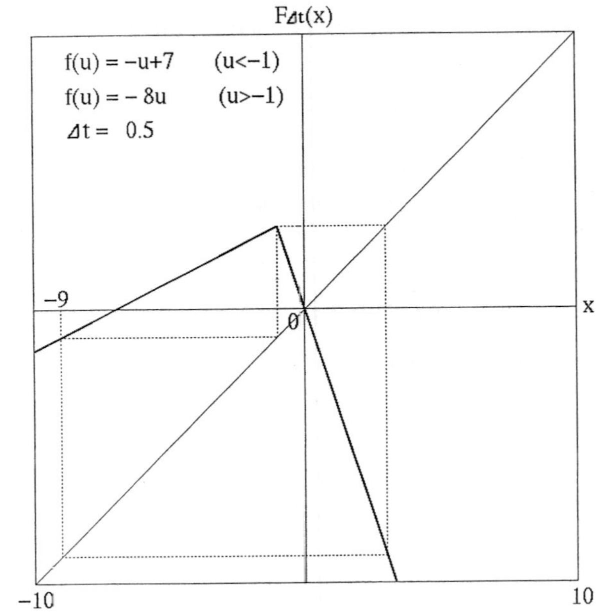

FIG. 5.6.

THEOREM 5.6. *Let f be of class C^1. Assume that f is upward convex, monotone decreasing, and that $f(0) = 0$. If there exists a negative u_0 that satisfies*

$$f(u_0) > \left(4 + 2 + \sqrt{3}\right)u_0 \cdot f'(u_0),$$

then there exists an interval $J = [\Delta t_1, \Delta t_2]$ such that for any $\Delta t \in J$, $F_{\Delta t}$ is chaotic in an invariant finite interval.

PROOF. Define $m = u_0 \cdot f'(u_0)/f(u_0)$; then $0 < m < (2 - \sqrt{3})/2$. Here we consider a function g defined as follows:

$$g(u) = \begin{cases} -m(u - u_0) - u_0 & (u < u_0), \\ -u & (u \geqslant u_0). \end{cases}$$

Theorem 5.5 shows that the discretization of g is chaotic for Δt in some interval. Note that

$$\frac{f(u_0)}{-u_0} \cdot g(u) \geqslant f(u) \quad (u \in (-\infty, u_0] \cup [0, +\infty)).$$

Therefore, the discretization of f is also chaotic. □

For example, functions f given by

$$f(u) = \begin{cases} -u + 7 & (u < -1), \\ -7u^2 - 15u & (u \geqslant -1), \end{cases}$$

or by $f(u) = 1 - e^u$ satisfy the assumptions of Theorem 5.6.

6. Discretization with small time steps and chaos

6.1. Three types of O.D.E.'s and chaos with small time step

In Section 5, we have investigated the discretization of differential equations with sufficiently large time steps Δt. To the contrary, we focus in this section on the discretization with small time steps. This study is motivated by the interesting example following:

$$f(u) = \begin{cases} \sqrt{-u} & (u < 0), \\ -3.2\sqrt{u} & (u \geqslant 0), \end{cases}$$

whose discretization is chaotic for any time step Δt (see Fig. 6.1). After showing several other examples, we elaborate necessary conditions under which chaos occurs for sufficiently small Δt.

First of all, let us show that the example above induces chaos for any time step.

THEOREM 6.1 (MAEDA [1995]). *Suppose that $f(u)$ is given as*

$$f(u) = \begin{cases} (-u)^{\alpha} & (u < 0), \\ -Lu^{\alpha} & (u \geqslant 0), \end{cases} \tag{6.1}$$

where $\alpha \in (0, 1)$ and L is positive. If L is sufficiently large, then $F_{\Delta t}$ is chaotic for any time step Δt.

REMARK 6.1. Solutions of this differential equation become extinct, that is, any trajectory will arrive at the equilibrium point in a finite time and stay there eternally. Note the loss of uniqueness in the backward direction of time, because f is not Lipschitz-continuous at the equilibrium point.

REMARK 6.2. With the central difference scheme, the discretization $F_{\Delta t}(x)$ of (6.1) diverges in an oscillatory manner, while $F_{\Delta t}(x)$ converges with the backward difference scheme for any time step.

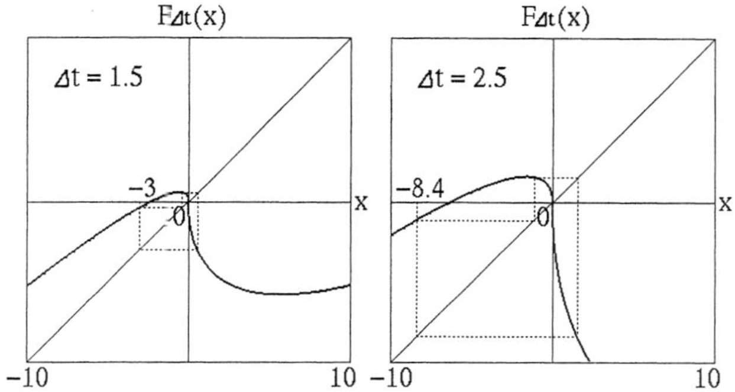

FIG. 6.1.

PROOF. The proof is much the same as that of Theorem 5.4. Take a negative b such that $F'_{\Delta t}(b) = 0$, i.e.

$$b = -(\alpha \Delta t)^{1/(1-\alpha)}.$$

Note that $F_{\Delta t}(x)$ is monotone increasing in $x \in (-\infty, b)$ and that

$$\lim_{x \to -\infty} F_{\Delta t}(x) = -\infty.$$

Therefore, there exists a unique a such that $F_{\Delta t}(a) = b$, given by

$$a = -N \Delta t^{1/(1-\alpha)},$$

where N is a unique positive solution of $N = N^\alpha + \alpha^{1/(1-\alpha)}$. It follows that

$$c = F_{\Delta t}(b) = \left(\alpha^{\alpha/(1-\alpha)} - \alpha^{1/(1-\alpha)}\right) \Delta t^{1/(1-\alpha)} = A \Delta t^{1/(1-\alpha)},$$

where $A = \alpha^{\alpha/(1-\alpha)} - \alpha^{1/(1-\alpha)}$ and $A \in (0, 1)$. Therefore c is positive. Finally,

$$d = F_{\Delta t}(c) = (A - LA^\alpha) \Delta t^{1/(1-\alpha)}.$$

If L is sufficiently large so that L satisfies the inequality $A - LA^\alpha \leqslant -N$, then $d \leqslant a$, which implies that $F_{\Delta t}$ is chaotic for any Δt. □

In the case where $\alpha = 1/2$, we can take $A = 1/4$, $N = (3 + 2\sqrt{2})/4$ and $L \geqslant 2 + \sqrt{2}$ for obtaining chaos. The following two theorems are generalizations of the result above.

THEOREM 6.2 (YAMAGUTI and MAEDA [1996]). *Suppose that f is a continuous function which satisfies the following three conditions*:

(i) $0 \leqslant f(u) \leqslant (-u)^\alpha$ $(u < 0)$,

(ii) $f(u) \leqslant -Lu^\alpha$ $(u > 0)$,

(iii) $\liminf_{h \to +0} (f(-h)/h^\alpha) = \delta > \alpha$,

where $\alpha \in (0, 1)$, and A and N are positive constants as defined in the proof of Theorem 6.1 (*see Fig.* 6.2). *If L satisfies*

$$A + N < L\alpha^{\alpha/(1-\alpha)} \left(\frac{\delta - \alpha}{\alpha}\right)^\alpha, \tag{6.2}$$

then there exists a positive ΔT such that for any $\Delta t < \Delta T$, $F_{\Delta t}$ is chaotic in the sense of Li–Yorke.

PROOF. We now construct a', b', c', and d' correspond to a, b, c, and d in the proof of Theorem 6.1. First, we take $b' = -(\alpha \Delta t)^{1/(1-\alpha)}$ equal to b. Next, we determine a' satisfying

$$a' + \Delta t \cdot f(a') = b' = -(\alpha \Delta t)^{1/(1-\alpha)},$$

then we have $a' \geqslant a = -N \Delta t^{1/(1-\alpha)}$ from condition (i).

On the other hand, from condition (iii), for sufficiently small positive ε $(< \delta - \alpha)$ and taking ΔT small, we get the following inequalities for any $\Delta t < \Delta T$,

$$f\left(-(\alpha \Delta t)^{1/(1-\alpha)}\right) \geqslant (\delta - \varepsilon)(\alpha \Delta t)^{\alpha/(1-\alpha)}, \tag{6.3}$$

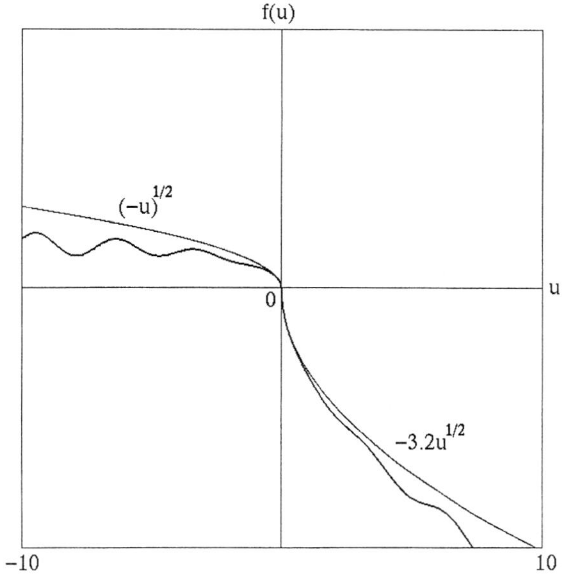

FIG. 6.2.

and also, using (6.2),

$$A + N \leqslant L\alpha^{\alpha/(1-\alpha)}\left(\frac{\delta-\varepsilon-\alpha}{\alpha}\right)^{\alpha}. \tag{6.4}$$

Then $c' > 0$ since

$$\begin{aligned} c' &= b' + \Delta t \cdot f(b') \\ &\geqslant -(\alpha\Delta t)^{1/(1-\alpha)} + \Delta t(\delta-\varepsilon)(\alpha\Delta t)^{\alpha/(1-\alpha)} \qquad \text{(from (6.3))} \\ &= \left(\frac{\delta-\varepsilon-\alpha}{\alpha}\right)\alpha^{1/(1-\alpha)}\Delta t^{1/(1-\alpha)} \\ &> 0. \end{aligned} \tag{6.5}$$

Furthermore,

$$\begin{aligned} d' &= c' + \Delta t \cdot f(c') \\ &= -(\alpha\Delta t)^{1/(1-\alpha)} + \Delta t \cdot f\big(-(\alpha\Delta t)^{1/(1-\alpha)}\big) + \Delta t \cdot f(c') \\ &\leqslant -(\alpha\Delta t)^{1/(1-\alpha)} + \Delta t(\alpha\Delta t)^{\alpha/(1-\alpha)} + \Delta t \cdot f(c') \qquad \text{(from (i))} \\ &\leqslant A\Delta t^{1/(1-\alpha)} - \Delta t \cdot Lc'^{\alpha} \qquad \text{(from (ii))} \\ &\leqslant A\Delta t^{1/(1-\alpha)} - \Delta t \cdot L\left(\left(\frac{\delta-\varepsilon-\alpha}{\alpha}\right)\alpha^{1/(1-\alpha)}\Delta t^{1/(1-\alpha)}\right)^{\alpha} \qquad \text{(from (6.5))} \\ &= A\Delta t^{1/(1-\alpha)} - L\left(\alpha^{1/(1-\alpha)}\left(\frac{\delta-\varepsilon-\alpha}{\alpha}\right)\right)^{\alpha}\Delta t^{1/(1-\alpha)} \end{aligned}$$

$$\leqslant -N\Delta t^{1/(1-\alpha)} \quad \text{(from (6.4))}$$
$$= a \leqslant a'. \qquad \square$$

THEOREM 6.3 (MAEDA and YAMAGUTI [1996]). *Let f be a continuous function which satisfies the following two conditions*:
(i) $f(u) = \mathrm{O}((-u)^\alpha)$ $(u \to -0)$,
(ii) $\lim_{u\to+0} f(u)/u^\alpha = -\infty$,
where $\alpha \in (0,1)$. Then there exists a positive ΔT such that for any $\Delta t < \Delta T$, $F_{\Delta t}$ is chaotic in the sense of Li–Yorke.

PROOF. From condition (i), there exist positive K, L_1, and L_2 $(L_1 > L_2)$ such that

$$L_2(-x)^\alpha \leqslant f(x) \leqslant L_1(-x)^\alpha \quad (x \in (-K, 0)).$$

First, we take $b = -(L_2\alpha\Delta t)^{1/(1-\alpha)}$. Second, we define M as the unique positive solution of

$$M = L_1 M^\alpha + (L_2\alpha)^{1/(1-\alpha)},$$

and we assume that Δt is sufficiently small such that

$$M\Delta t^{1/(1-\alpha)} < K.$$

Then we can find $a \in [-M\Delta t^{1/(1-\alpha)}, b)$ that satisfies $F_{\Delta t}(a) = b$, as follows:

$$\begin{aligned}
F_{\Delta t}\bigl(-M\Delta t^{1/(1-\alpha)}\bigr) &= -M\Delta t^{1/(1-\alpha)} + \Delta t \cdot f\bigl(-M\Delta t^{1/(1-\alpha)}\bigr)\\
&\leqslant -M\Delta t^{1/(1-\alpha)} + \Delta t L_1\bigl(M\Delta t^{1/(1-\alpha)}\bigr)^\alpha\\
&= \bigl(L_1 M^\alpha - M\bigr)\Delta t^{1/(1-\alpha)}\\
&= (L_2\alpha)^{1/(1-\alpha)}\Delta t^{1/(1-\alpha)}\\
&= b,
\end{aligned}$$

while

$$F_{\Delta t}(b) = b + \Delta t \cdot f(b) > b.$$

Since $F_{\Delta t}$ is continuous, there exists a in the interval $[-M\Delta t^{1/(1-\alpha)}, b)$.

So far, we have the following inequalities:

$$-K < -M\Delta t^{1/(1-\alpha)} \leqslant a < b < 0.$$

Next, let us estimate c chosen as $c = F_{\Delta t}(b)$:

$$\begin{aligned}
&b + \Delta t L_2(-b)^\alpha \leqslant c \leqslant b + \Delta L_1(-b)^\alpha,\\
&-(L_2\alpha\Delta t)^{1/(1-\alpha)} + \Delta t L_2(L_2\alpha\Delta t)^{\alpha/(1-\alpha)}\\
&\quad \leqslant c \leqslant -(L_2\alpha\Delta t)^{1/(1-\alpha)} + \Delta t L_1(L_2\alpha\Delta t)^{\alpha/(1-\alpha)},\\
&L_2^{1/(1-\alpha)}\bigl(\alpha^{\alpha/(1-\alpha)} - \alpha^{1/(1-\alpha)}\bigr)\Delta t^{1/(1-\alpha)}\\
&\quad \leqslant c \leqslant \bigl(L_1(L_2\alpha)^{\alpha/(1-\alpha)} - (L_2\alpha)^{1/(1-\alpha)}\bigr)\Delta t^{1/(1-\alpha)},\\
&C_2\Delta t^{1/(1-\alpha)} \leqslant c \leqslant C_1\Delta t^{1/(1-\alpha)},
\end{aligned}$$

where C_1 and C_2 are two constants which do not depend on Δt:

$$C_1 = L_1(L_2\alpha)^{\alpha/(1-\alpha)} - (L_2\alpha)^{1/(1-\alpha)},$$
$$C_2 = L_2^{1/(1-\alpha)}\big(\alpha^{\alpha/(1-\alpha)} - \alpha^{1/(1-\alpha)}\big).$$

Note that c is positive because $\alpha^{\alpha/(1-\alpha)} - \alpha^{1/(1-\alpha)} > 0$. Finally, we compare a with d:

$$\begin{aligned} a - d &\geqslant -M\Delta t^{1/(1-\alpha)} - \big(c + \Delta t \cdot f(c)\big) \\ &\geqslant -M\Delta t^{1/(1-\alpha)} - C_1\Delta t^{1/(1-\alpha)} - \Delta t \cdot f(c) \\ &= \left(-(M + C_1) - \frac{f(c)}{\Delta t^{\alpha/(1-\alpha)}}\right)\Delta t^{1/(1-\alpha)} \\ &= \left(-(M + C_1) + \left(\frac{c}{\Delta t^{1/(1-\alpha)}}\right)^{\alpha}\left(\frac{-f(c)}{c^{\alpha}}\right)\right)\Delta t^{1/(1-\alpha)} \\ &\geqslant \left(-(M + C_1) + C_2^{\alpha}\left(\frac{-f(c)}{c^{\alpha}}\right)\right)\Delta t^{1/(1-\alpha)}. \end{aligned}$$

Since $\lim_{\Delta t \to +0} c = +0$ and condition (ii) holds, we have $d \leqslant a$ for sufficiently small Δt. □

A simple example that satisfies Theorem 6.3 is given by:

$$f(u) = \begin{cases} (-u)^{\alpha} & (u \leqslant 0), \\ -u^{\beta} & (u \geqslant 0), \end{cases}$$

where $\alpha, \beta \in (0, 1)$ and $\alpha \neq \beta$.

6.2. *Lipschitz continuity at the equilibrium point*

We have investigated three examples where the Euler discretization $F_{\Delta t}$ is chaotic for sufficiently small time step Δt. Recall that every example satisfies the condition

$$\lim_{u \to 0} \frac{f(u)}{u} = -\infty. \tag{6.6}$$

In this section, we discuss how condition (6.6) plays an important role for chaos.

As was already suggested by BOOLE [1860], there is a remarkable difference between differential calculus and difference calculus. Theorem 6.1 exemplifies this fact: The exact solution of the differential equation of Theorem 6.1 when $u(0) = u_0 > 0$ is

$$u(t) = \begin{cases} \big(u_0^{1-\alpha} - (1-\alpha)Lt\big)^{1/(1-\alpha)} & (t \leqslant u_0^{1-\alpha}/L(1-\alpha)), \\ 0 & (t \geqslant u_0^{1-\alpha}/L(1-\alpha)), \end{cases}$$

which indicates that the equilibrium point O is so stable that $u(t)$ become extinct in a finite time. However, we also showed that $F_{\Delta t}$ is chaotic for any Δt. While the equilibrium point O is super-stable (extinct at O) in the differential equation, in the difference equation O is super-unstable (infinite derivative at O), which convinces us of the importance of investigating condition (6.6).

Recall that we are discussing the discretization of the following scalar autonomous ordinary differential equation:

$$\frac{du}{dt} = f(u), \quad u \in \mathbb{R}^1,$$

under the conditions:

$$\begin{cases} f(u) \text{ is continuous in } \mathbb{R}^1, \\ f(u) > 0 \quad (u < 0), \\ f(0) = 0, \\ f(u) < 0 \quad (u > 0). \end{cases} \tag{6.7}$$

THEOREM 6.4 (left Lipschitz-continuity). *Assume that f satisfies the following condition, in addition to* (6.7):

$$-\frac{f(u)}{u} < M_0 \quad (u < 0),$$

where M_0 is a positive constant (*see Fig.* 6.3). *Then there exists a positive ΔT such that for any Δt less than ΔT, $F_{\Delta t}$ has no periodic orbit except for the fixed point O. Furthermore, for any initial point x_0, $x_n = F_{\Delta t}^n(x_0)$ converges to the fixed point.*

PROOF. Define four subsets of $\mathbb{R}^2$ as follows:

$$D_- = \{(x, y) \mid x < y < 0\},$$
$$D_+ = \{(x, y) \mid 0 < y < x\},$$

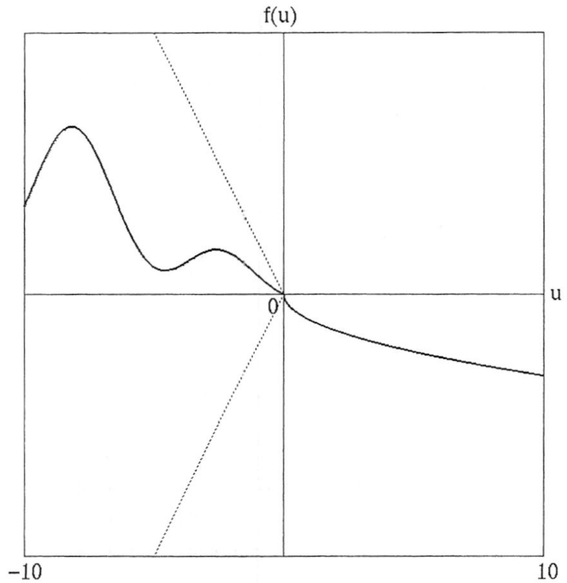

FIG. 6.3.

$$D_0 = \{(x, y) \mid 0 < x,\ y = 0\},$$
$$D' = \{(x, y) \mid y < 0 < x\}.$$

Take $\Delta T = 1/M_0$, and let Δt be less than ΔT. If x is negative,

$$\begin{aligned} F_{\Delta t}(x) &= x + \Delta t \cdot f(x) < x + \Delta T \cdot f(x) \\ &= x + \Delta T \cdot (-M_0 x) = x(1 - M_0 \cdot \Delta T) \\ &= 0. \end{aligned}$$

On the other hand,

$$F_{\Delta t}(x) = x + \Delta t \cdot f(x) > x.$$

Therefore, $x < 0$ implies $(x, F_{\Delta t}(x)) \in D_-$.

The behaviors of x_n are classified into the following four cases. In each case we can show that x_n converges to the fixed point O:

Case (i): $x_0 < 0$. For any $n \geqslant 0$, $(x_n, x_{n+1}) \in D_-$, that is, x_n is monotone increasing to O.

Case (ii): $x_0 > 0$, and for any $n \geqslant 0$, $(x_n, x_{n+1}) \in D_+$. In this case, x_n is monotone decreasing to O.

Case (iii): There exists $N \geqslant 0$ such that $(x_N, x_{N+1}) \in D_0$. In this case, $x_{N+1} = x_{N+2} = \cdots = 0$, i.e. $(x_n, x_{n+1}) = O$ for any n greater than N.

Case (iv): There exists $N \geqslant 0$ such that $(x_N, x_{N+1}) \in D'$. Since $x_{N+1} < 0$, this case is reduced to case (i). □

Clearly Theorem 6.4 also holds under the following condition, in addition to (6.7):

$$\frac{-f(u)}{u} < M_0 \quad (u > 0).$$

Furthermore, if we want to show the stability of $F_{\Delta t}$ only in a neighborhood of the fixed point O, the condition can be relaxed as

$$\frac{f(u)}{-u} < M_0 \quad (u \in [K, 0)),$$

where K is some constant. Therefore, if neither the right nor the left upper limit of $-f(u)/u$ is $+\infty$, then $F^n_{\Delta t}(x)$ converges to O in a neighborhood of O for sufficiently small Δt. By contrast, if both the left and right limits of $f(u)/u$ are $-\infty$, then $F'_{\Delta t}(0) = -\infty$, namely, the fixed point O is super-unstable.

THEOREM 6.5.

(i) *$F^n_{\Delta t}(x)$ converges to the fixed point O in a neighborhood of O for sufficiently small Δt* if

$$\limsup_{u\to -0} \frac{f(u)}{-u} < +\infty \quad \textit{or} \quad \limsup_{u\to +0} \frac{-f(u)}{u} < +\infty.$$

(ii) *If $F^n_{\Delta t}(x)$ never converges to the fixed point for any Δt,* then

$$\limsup_{u\to 0} \frac{f(u)}{u} = -\infty.$$

We find here a necessary condition for chaos, that is,

$$\limsup_{u\to 0} \frac{-f(u)}{u} = +\infty.$$

At the end of this section, we show that there is a periodic orbit with period 2 under the condition $f'(0) = -\infty$.

THEOREM 6.6. *If $f(u)$ satisfies*

$$\lim_{u\to 0} \frac{f(u)}{u} = -\infty,$$

then $F_{\Delta t}$ has a periodic orbit with period 2 for sufficiently small Δt.

PROOF. Assume that Δt is less than

$$\sup_{x\neq 0} \frac{-x}{f(x)} \quad (\in (0, +\infty]).$$

Then, there exists $x_0 \neq 0$ such that $-x_0/f(x_0) = \Delta t$. Without loss of generality, we can assume that x_0 is negative. Since

$$F_{\Delta t}(x_0) = x_0 + \Delta t \cdot f(x_0) = 0,$$
$$F^2_{\Delta t}(x_0) = F_{\Delta t}(0) = 0,$$

we have $F^2_{\Delta t}(x_0) > x_0$. It then suffices to find $x_1 \in (x_0, 0)$ which satisfies $F^2_{\Delta t}(x_1) > x_1$ (then the continuity of the function $F_{\Delta t}$ leads to the existence of 2-periodic point in (x_0, x_1)). For this purpose, choose two numbers K_1 and K_2 such that

$$F_{\Delta t}(x) > -x \quad (x \in (K_1, 0)),$$
$$F_{\Delta t}(x) < -x \quad (x \in (0, K_2)),$$

whose existence is assured by the assumption of this theorem. Note that K_1 is greater than x_0. From $\lim_{x\to -0} F_{\Delta t}(x) = 0$, there exists $x_1 \in (K_1, 0)$ which satisfies $F_{\Delta t}(x_1) < K_2$. Then $x_1 \in (K_1, 0)$ implies $0 < F_{\Delta t}(x_1) < K_2$, and $F^2_{\Delta t}(x_1) < -F_{\Delta t}(x_1) < x_1$. This completes the proof. □

7. Odd symmetry and chaos

7.1. Necessary condition for chaos

In the previous section, we have considered small time steps Δt and investigated Euler's finite difference scheme $F_{\Delta t}$. We already established the following two facts:

(1) $\lim_{u \to 0} f(u)/u = -\infty$ is the necessary condition under which $F_{\Delta t}$ is chaotic for sufficiently small time step Δt.

(2) Under the condition above, $F_{\Delta t}$ always has 2-periodic orbits for sufficiently small Δt (3-periodic orbit implies chaos).

In this section, we look for another necessary condition for chaos. Theorems 6.1, 6.2, and 6.3 suggest that asymmetry of the function f could be necessary for chaos. To confirm this indication, we investigate symmetric differential equations and conclude that chaotic phenomena never occur in their discretization (MAEDA [1996]).

We consider a scalar autonomous O.D.E.

$$\frac{\mathrm{d}u}{\mathrm{d}t} = f(u), \quad u \in \mathbb{R}^1,$$

under the following conditions:

$$\begin{cases} \text{(i)} & f \text{ is continuous and strictly decreasing,} \\ \text{(ii)} & f(0) = 0, \\ \text{(iii)} & f \text{ is of class } C^1 \text{ except for } u = 0, \\ \text{(iv)} & \lim_{u \to 0} f(u)/u = -\infty, \\ \text{(v)} & f(-u) = -f(u). \end{cases} \tag{7.1}$$

Conditions (i) and (ii) imply that $f(u) > 0$ in $u < 0$ and $f(u) < 0$ in $u > 0$. In other words, this differential equation has only one equilibrium point which is asymptotically stable. That f strictly decreases is natural since we consider for sufficiently small time steps. Condition (v) means that $f(u)$ is an odd function, i.e. it has odd symmetry at the equilibrium point (see Fig. 7.1). In the following discussion, we get very simple results for its Euler discretization $F_{\Delta t}$, which will be stated in Theorem 7.3.

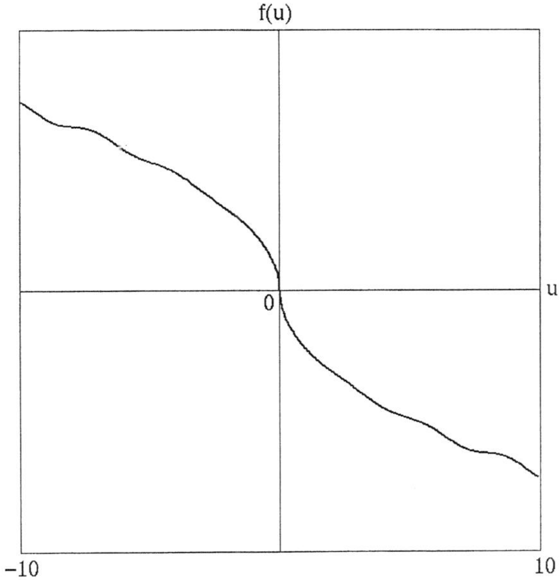

FIG. 7.1.

First of all, it is easy to check that $F_{\Delta t}$ is also an odd function and that $F'_{\Delta t}(x) < 1$ for any Δt. These two facts are important properties.

In the following two preparatory theorems $x_n = F^n_{\Delta t}(x_0)$.

THEOREM 7.1. *If $x_0 x_1 > 0$, then $|x_n| < |x_0|$ for any positive integer n.*

PROOF. Let us prove this theorem by induction of n. Without loss of generality, let both x_0 and x_1 be positive. In the case $n = 1$,

$$x_1 = F_{\Delta t}(x_0) = x_0 + \Delta t f(x_0) < x_0.$$

Therefore, the inequality $0 < x_1 < x_0$ implies $|x_1| < x_0$.

Next, assume that $|x_k| < x_0$ for some positive integer k. There are following three cases.

Case (a): $x_k > 0$. In this case, $x_{k+1} = x_k + \Delta t f(x_k) < x_k$. From the assumption on x_k, we have $x_{k+1} < x_0$. Then, unless the slope of the line AB (which passes two points $A = (x_0, x_1)$ and $B = (x_k, x_{k+1})$) is less than 1, we can find $x_c \in (x_k, x_0)$ which satisfies $F'_{\Delta t}(x_c) \geqslant 1$ from the mean value theorem. It contradicts $F'_{\Delta t}(x) < 1$ for any x in $\mathbb{R}$. Hence we get the following inequality:

$$\frac{x_{k+1} - x_1}{x_k - x_0} < 1.$$

The assumption $x_k < x_0$ leads to $x_{k+1} + x_0 > x_k + x_1 > 0$. Thus we conclude that $|x_{k+1}| < x_0$.

Case (b): $x_k < 0$. This case is easily reduced to the case above. Define $A' = (-x_0, -x_1)$. Then A' is also on the graph of $y = F_{\Delta t}(x)$ because $F_{\Delta t}$ is an odd function.

Case (c): $x_k = 0$. It implies $x_{k+1} = 0$ and of course $|x_{k+1}| < x_0$.

In any case, we have $|x_{k+1}| < x_0$. □

Using this theorem, we can prove the next theorem.

THEOREM 7.2. *$F_{\Delta t}$ does not have any odd periodic orbit except for the fixed point.*

PROOF. Let $\{x_0, x_1, \ldots, x_{n-1}\}$ be a periodic orbit with period $n \geqslant 2$. Note that x_n is equal to x_0. Assume that x_0 is positive. If x_1 is positive, $|x_n| < |x_0|$ from Theorem 7.1 and it is an obvious contradiction. Therefore, x_1 is negative (if x_1 is equal to 0, this periodic orbit is the fixed point). In the same way, x_2 is positive, and x_3 is negative, and so on. If n is odd, x_n is negative, i.e. x_0 is negative because of $x_n = x_0$. It is also a contradiction. The same argument is valid in the case of x_0 is negative. □

Now the main result of this section is as follows:

THEOREM 7.3. *Assume that f satisfies conditions* (7.1). *Then,*

(i) *For any $\Delta t > 0$, $F_{\Delta t}$ does not have any periodic orbit with period n which is greater than 2.*

(ii) *There exists $\Delta T > 0$ such that, for any Δt less than ΔT, there is a periodic orbit with period 2, and any 2-periodic point x_0 satisfies $F_{\Delta t}(x_0) = -x_0$.*

PROOF. Let us prove (i) first. From Theorem 7.2, it is enough to show that $F_{\Delta t}$ does not have any even periodic orbit with period n greater than 2. Because $F_{\Delta t}$ is an odd function, we can assume without loss of generality, that x_0 is positive and satisfies

$$x_0 = \min_{0 \leqslant i \leqslant n-1} |x_i|.$$

First, let us compare x_0 with $-x_{-1}$. From Theorem 7.1, x_{-1} is negative, and the assumption above leads to $x_0 < -x_{-1}$ (note that $x_0 \neq -x_{-1}$, since $x_0 = -x_{-1}$ implies period 2).

Secondly, let us show that $x_0 < -x_1 < -x_{-1}$. In the same way as in the case of x_{-1}, the inequalities $x_0 x_1 < 0$, $x_0 \leqslant |x_1|$, and $n > 2$ imply $x_0 < -x_1$. The slope of the line AB ($A = (-x_{-1}, -x_0)$ and $B = (x_0, x_1)$, which are on the graph $F_{\Delta t}$), is less than 1, i.e.

$$\frac{x_1 + x_0}{x_0 + x_{-1}} < 1,$$

thus we have $x_{-1} < x_1$ since $x_0 + x_{-1} < 0$.

Finally, we can prove $x_0 \leqslant |x_k| < -x_{-1}$ $(1 \leqslant \forall k \leqslant n-1)$ by induction. In the case $k = 1$, it is true by the argument above. Assume that $x_0 \leqslant |x_m| < -x_{-1}$ for some $m \geqslant 1$. The left inequality is trivial because of the assumption on x_0. There are two cases about x_m.

Case (a): x_m is positive. In this case, x_{m+1} is negative. The two points $A = (-x_{-1}, -x_0)$ and $C = (x_m, x_{m+1})$ are different points because $x_m < -x_{-1}$. The slope of the line AC is also less than 1, i.e.

$$\frac{x_{m+1} + x_0}{x_m + x_{-1}} < 1.$$

Then $x_m + x_{-1} < 0$ implies that

$$\begin{aligned} -x_{m+1} &< x_0 - (x_m + x_{-1}) = -x_{-1} - (x_m - x_0) \\ &\leqslant -x_{-1}. \end{aligned}$$

Here we used $x_m - x_0 \geqslant 0$. Therefore, $|x_{m+1}| < -x_{-1}$ is proved.

Case (b): x_m is negative. Then the same argument as in case (a) leads to $|x_{m+1}| < -x_{-1}$ (take $A = (x_{-1}, x_0)$).

Consequently, we have proved that $x_0 \leqslant |x_k| < -x_{-1}$ for any $k \geqslant 1$. In particular, take k as $n-1$, $|x_{n-1}| < -x_{-1}$. It contradicts $x_{n-1} = x_{-1}$.

Now let us prove (ii) of Theorem 7.3. As for the existence of periodic orbit with period 2, it was already proved in Theorem 6.6. The latter part of the statement is also proved by contradiction. Let $\{x_0, x_1\}$ be a 2-periodic orbit which satisfies $x_0 + x_1 \neq 0$. Then $F_{\Delta t}(x_0) = x_1$ and $F_{\Delta t}(x_1) = x_0$. Compare $A(x_0, x_1)$ with $B(-x_1, -x_0)$. Then $A \neq B$ and $F_{\Delta t}(x) < 1$ imply that

$$\frac{x_1 + x_0}{x_0 + x_1} < 1,$$

a contradiction. Thus $x_0 + x_1 = 0$, i.e. $F_{\Delta t}(x_0) = -x_0$. □

7.2. Existence of stable periodic orbits

Before closing this section, let us investigate about the existence of an invariant finite interval and the stability of 2-periodic orbits. Recall that the definition of the invariant finite interval I is as follows: For any initial point x_0 in I, the orbits x_n belong to I for all n.

THEOREM 7.4. *Assume that f satisfies conditions* (7.1). *Then $F_{\Delta t}$ has an invariant finite interval for sufficiently small Δt.*

PROOF. Set

$$\Delta T = \sup_{x<0} \frac{-x}{f(x)} \quad (0 < \Delta T \leqslant +\infty).$$

For any Δt less than ΔT, there exists a negative x_0 such that $F_{\Delta t}(x_0) = 0$. Indeed, an interval $[x_0, -x_0]$ is invariant: From the symmetric property of $F_{\Delta t}$, it is enough to show that $F_{\Delta t}([x_0, 0]) \subset [x_0, -x_0]$. For any $x \in [x_0, 0]$,

$$F_{\Delta t}(x) = x + \Delta t f(x) \geqslant x_0 + \Delta t f(x) > x_0.$$

On the other hand, for $x \in (x_0, 0]$,

$$\frac{F_{\Delta t}(x) - F_{\Delta t}(x_0)}{x - x_0} < 1.$$

This inequality implies that $F_{\Delta t}(x) < x - x_0 \leqslant -x_0$. Of course, $F_{\Delta t}(x_0) = 0 \in [x_0, -x_0]$, hence $F_{\Delta t}([x_0, 0]) \subset [x_0, -x_0]$ is proved. □

THEOREM 7.5. *Assume that f satisfies conditions* (7.1) *and a set of* 2*-periodic points is written as*

$$C = \bigcup_{i=1}^{n} I_i,$$

with $I_i = [a_i, b_i]$, $a_i \leqslant b_i$ $(i = 1, 2, \ldots, n)$, and $b_i < a_{i+1}$ $(i = 1, 2, \ldots, n-1)$. Then there exists at least one stable 2*-periodic orbit.*

PROOF. Note that $F^2_{\Delta t}(x_0) > x_0$. From assumption (iv) of (7.1),

$$\lim_{x\to 0} \frac{F^2_{\Delta t}(x)}{x} = +\infty.$$

Hence there exists $x_1 \in (x_0, 0)$ which satisfies $F^2_{\Delta t}(x_1) < x_1$. Restrict C to $[x_0, x_1]$ and rewrite $C = \bigcup_{i=1}^{n} I_i$ (i.e. $x_0 < a_1$ and $b_n < x_1$). Let $J_1, J_2, \ldots, J_{n+1}$ be a sequence of open intervals such that

$$J_1 = (x_0, a_1), \quad J_2 = (b_1, a_2), \quad \ldots, \quad J_n = (b_{n-1}, a_n), \quad J_{n+1} = (b_n, x_1).$$

The function g defined by $g(x) = F^2_{\Delta t}(x) - x$ vanishes for any $x \in C$. Furthermore, $g(x)$ is positive for $x \in J_1$ and negative for $x \in J_{n+1}$. Thus we can find an integer k

$(2 \leqslant k \leqslant n)$ such that $g(x) > 0$ for $x \in J_k$ and $g(x) < 0$ for $x \in J_{k+1}$. Let us look at a neighborhood of the interval $I_k = [a_k, b_k]$. For any $x \in I_k$,

$$\{F_{\Delta t}^2(x)\}' = F'_{\Delta t}(F_{\Delta t}(x)) \cdot F'_{\Delta t}(x) = F'_{\Delta t}(-x) \cdot F'_{\Delta t}(x) = \{F'_{\Delta t}(x)\}^2 \geqslant 0.$$

Hence, there is $\delta_1 > 0$ such that for $x \in (a_k - \delta_1, a_k)$,

$$\frac{F_{\Delta t}^2(x) - a_k}{x - a_k} > -1.$$

Note that $g(x)$ is positive for $x \in J_k$. This implies that there is $\delta_2 > 0$ such that for $x \in (a_k - \delta_2, a_k)$,

$$\frac{F_{\Delta t}^2(x) - a_k}{x - a_k} < 1.$$

For this reason, for any x in the left side of the neighborhood of I_k, $F_{\Delta t}^n(x)$ converges to a 2-periodic orbit. In the same way, the orbit starting from the right side of the neighborhood of I_k also converges. □

8. Modified Euler's finite difference scheme as a discretization of O.D.E.'s

8.1. Preliminary study

In this section, we study again the problem of the discretization of an O.D.E. We consider the following equation:

$$\frac{du}{dt} = f(u). \tag{8.1}$$

Here, we assume that $f(u)$ has two equilibrium points. We start with the most well-known difference scheme called Euler scheme:

$$x_{n+1} - x_n = \Delta t \cdot f(x_n). \tag{8.2}$$

Needless to say, numerical analysis usually treats (8.2) for a finite time interval. This means for one fixed time T, the solution $\{x_k\}_{k=1,\dots,n}$, with $n\Delta t = T$, converges to the solution $u(t)$ of (8.1) with the same initial data as Δt tends to zero (n tends to $+\infty$). In real numerical computations, however, we cannot expect that the mesh length Δt reduces to zero. Furthermore, in some special cases, we have to compute the solution of (8.2) for very long time interval T. This is the reason why we study the dynamical system (8.2) for fixed Δt, and $n \to +\infty$. We have discussed in Section 3 that Euler's discretization of the logistic differential equation $du/dt = Au(1-u)$ may produce chaotic behaviors in its orbit of the dynamical system (8.2) expressed as:

$$x_{n+1} - x_n = \Delta t A x_n(1 - x_n).$$

We can see why: Solving this equation with respect to x_{n+1},

$$x_{n+1} = \{(1 + \Delta t A) - \Delta t A x_n\} x_n,$$

and defining y_n and a as

$$\frac{\Delta t A x_n}{1+\Delta t A} = y_n, \qquad 1+\Delta t A = a,$$

we have $y_{n+1} = a y_n(1-y_n)$, as was already observed in Section 1.

Now let us generalize this fact for other functions. Suppose f in (8.1) satisfies the following three conditions:

$$\begin{cases} \text{(i)} & f(u) \text{ belongs to } C^2(\mathbb{R}),\\ \text{(ii)} & f(0)=f(1)=0,\\ \text{(iii)} & f''(u)<0 \text{ (for any } u\in\mathbb{R}). \end{cases} \tag{8.3}$$

We then rewrite (8.2) using $F_{\Delta t}(x) = x + \Delta t f(x)$ as

$$x_{n+1} = F_{\Delta t}(x_{n+1}). \tag{8.4}$$

THEOREM 8.1. *If we take Δt sufficiently large, then the dynamical system* (8.4) *under the assumptions* (8.3) *becomes chaotic in the sense of Li and Yorke.*

PROOF. First we indicate the meaning of the assumed conditions: The condition $f''(x)<0$ in the third assumption of (8.3) indicates that $f'(x)$ is always monotone decreasing; the second assumption means that $f(u)>0$ for $u\in(0,1)$ and $f'(0)>0$ (if not, $f(1)<0$); in addition, $f'(1)<0$.

Taking these facts into account, we have the proof of this theorem as follows. We differentiate the equation $F_{\Delta t}(x) = x+\Delta t f(x)$ with respect x. We then get

$$F'_{\Delta t}(x) = 1+\Delta t f'(x),$$
$$F'_{\Delta t}(0) = 1+\Delta t f'(0) > 0,$$
$$F''_{\Delta t}(x) = \Delta t f''(x) < 0.$$

Here again, the last inequality means that $F'_{\Delta t}(x)$ is monotone decreasing. Let us take Δt large enough so as to satisfy

$$\Delta t > \frac{-1}{f'(1)} > \frac{-1}{f'(+\infty)},$$

without excluding $f'(+\infty) = -\infty$. For this Δt, since $F'_{\Delta t}(+\infty)<0$ and $F'_{\Delta t}(0)>0$, we conclude $F'_{\Delta t}(x)$ has a unique vanishing point m $(0<m<1)$ such that $F'_{\Delta t}(m)=0$ and $F'_{\Delta t}(x)<0$ for $x>m$. It follows that $F_{\Delta t}(x)$ is monotone decreasing for x $(m<x<1)$ and $F_{\Delta t}(1)=1$. The point m is therefore a unique maximal point of $F_{\Delta t}(x)$. There exists $R>1$ which satisfies $F_{\Delta t}(R)=0$ and R is the unique vanishing point of $F_{\Delta t}(x)$.

Let us define M as the maximum of $F_{\Delta t}(x)$ for x such that $0<x<1$, namely, $M = F_{\Delta t}(m)$ (see Fig. 8.1). As M and R depend on Δt, let $M=M(\Delta t)$ and $R=R(\Delta t)$, and consider the behavior of $M(\Delta t)$ and $R(\Delta t)$ as functions of Δt. Since

$$F'_{\Delta t}(m) = 1+\Delta t f'(m) = 0$$

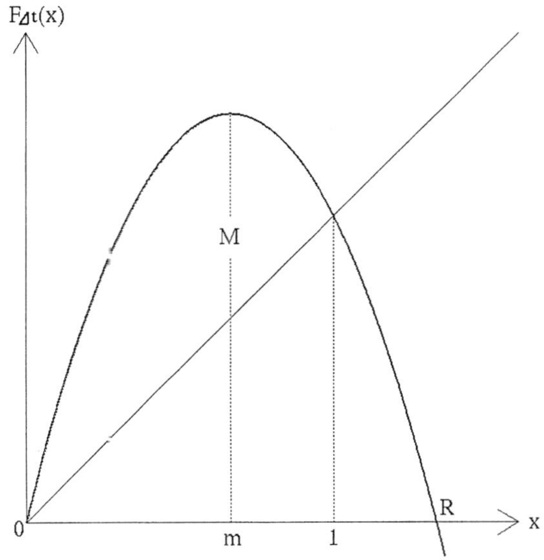

FIG. 8.1.

implies $f'(m) = -1/\Delta t$ (< 0), consequently $f(m)$ is increasing with respect to Δt. Then we get

$$M(\Delta t) = m + \Delta t f(m)$$

is increasing, and

$$M(+\infty) = +\infty.$$

On the other hand, the fact that for $R > 1$, $R(\Delta t) + \Delta t f(R(\Delta t)) = 0$ implies that

$$\frac{\mathrm{d}R}{\mathrm{d}\Delta t} = -\frac{f(R)}{1 + \Delta t f'(R)} = -\frac{f(R)}{F'_{\Delta t}(R)} < 0,$$

implies that $R(\Delta t)$ is decreasing from $R(0) = +\infty$ to $R(+\infty) = 1$. Thus, if we increase Δt starting from 0, we have a certain value of Δt such that

$$M(\Delta t) \geqslant R(\Delta t).$$

Now, we are prepared to apply the Li–Yorke theorem (see Section 2): Take $b = m$, $c = M(\Delta t) \geqslant R(\Delta t)$, and a as a unique positive value such that $F_{\Delta t}(a) = m$ (since $F'_{\Delta t}(x) > 0$ $(0 < x < m)$). Consequently, $d = F_{\Delta t}(M(\Delta t)) < 0$ and we obtain $d < 0 < a < b < c$ as was to be shown. □

This theorem has been proved in Section 3 under more general condition by YAMAGUTI and MATANO [1979].

8.2. *Modified Euler scheme*

Let us consider the new scheme called modified Euler scheme:

$$x_{n+1} = x_n + \frac{\Delta t}{2}\big(f(x_n) + f\big(x_n + \Delta t f(x_n)\big)\big). \tag{8.5}$$

This scheme is known to be a better approximation of the solution of O.D.E. than Euler scheme which have been studied in the previous sections. OSHIME [1987] showed that chaos may occur even in this scheme. Before we discuss the details, let us simplify the expression (8.5) using the notation $F_{\Delta t}(x) = x + \Delta t f(x)$ introduced in the previous section as

$$x_{n+1} = G_{\Delta t}(x_n),$$

where

$$G_{\Delta t}(x) = \tfrac{1}{2}\big(x + F_{\Delta t}^2(x)\big),$$

since

$$\begin{aligned} F_{\Delta t}^2(x) &= F_{\Delta t}\big(F_{\Delta t}(x)\big) \\ &= F_{\Delta t}(x) + \Delta t f\big(F_{\Delta t}(x)\big) \\ &= x + \Delta t f(x) + \Delta t f\big(x + \Delta t f(x)\big), \\ x + F_{\Delta t}^2(x) &= 2x + \Delta t\big(f(x) + f\big(x + \Delta t f(x)\big)\big). \end{aligned}$$

THEOREM 8.2. *If $f \in C^3(\mathbb{R})$ satisfies the following three conditions*:

$$\begin{aligned} &\text{(i)} \quad f(0) = f(1) = 0, \\ &\text{(ii)} \quad f''(u) < 0 \ (u \in \mathbb{R}), \\ &\text{(iii)} \quad f'''(u) \geqslant 0 \ (u \in \mathbb{R}). \end{aligned} \tag{8.6}$$

Then the discretization by modified Euler scheme shows chaotic phenomena for some large Δt.

We are going to sketch Oshime's proof of this theorem. It is more difficult than in the case of Euler scheme; We have to prove the following four theorems (8.3, 8.4, 8.5, and 8.6) before we prove this theorem.

THEOREM 8.3. *If $\Delta t > -1/f'(+\infty)$, then $F_{\Delta t}(x) = x + \Delta t f(x)$ has a unique maximal point x_m $(0 < x_m < 1)$ and a unique zero-point R (> 1) which satisfy*

$$x_m \leqslant \frac{R}{2}. \tag{8.7}$$

PROOF. The proof is quite analogous to that of Theorem 8.1 except for the last inequality (8.7). To begin with, we remark that condition (iii) of (8.6) implies that

$$F_{\Delta t}'''(x) = \Delta t f'''(x) \geqslant 0,$$

and therefore,

$$F''_{\Delta t}(x) \geqslant F''_{\Delta t}(x_m) \quad (x \geqslant x_m),$$
$$F''_{\Delta t}(x) \leqslant F''_{\Delta t}(x_m) \quad (x \leqslant x_m).$$

Next we integrate this inequality from x_m up to x taking in account $F'_{\Delta t}(x_m) = 0$:

$$F'_{\Delta t}(x) \geqslant F''_{\Delta t}(x_m)(x - x_m) \quad (\forall x \in \mathbb{R}).$$

Again, we integrate this inequality.

$$\begin{aligned} F_{\Delta t}(x) &\geqslant \tfrac{1}{2}F''_{\Delta t}(x_m)(x - x_m)^2 + F_{\Delta t}(x_m) \quad (x \geqslant x_m), \\ F_{\Delta t}(x) &\leqslant \tfrac{1}{2}F''_{\Delta t}(x_m)(x - x_m)^2 + F_{\Delta t}(x_m) \quad (x \leqslant x_m). \end{aligned} \tag{8.8}$$

Here the quadratic polynomial of x:

$$\tfrac{1}{2}F''_{\Delta t}(x_m)(x - x_m)^2 + F_{\Delta t}(x_m)$$

has the same maximum as $F_{\Delta t}(x)$ at x_m. The zeros of $F_{\Delta t}(x)$ are O and R. Then the inequalities (8.8) mean $x_m \leqslant R/2$ (see Fig. 8.2). □

THEOREM 8.4. *There exists $\Delta t > -1/f'(+\infty)$ such that $G_{\Delta t}(x)$ has a zero-point in $0 < x < 1$ and $G_{\Delta t}(x) \geqslant 0$ for x $(0 \leqslant x \leqslant 1)$.*

PROOF. We draw two graphs of $y = F_{\Delta t}(x)$ $(x \geqslant 0)$, and $x = -F_{\Delta t}(y)$ $(y \geqslant 0)$ in the same (x, y)-plane as we show in Figs. 8.3, 8.4 and 8.5. If Δt is sufficiently small, two graphs have no intersecting points in the first quadrant. To the contrary, if Δt is quite

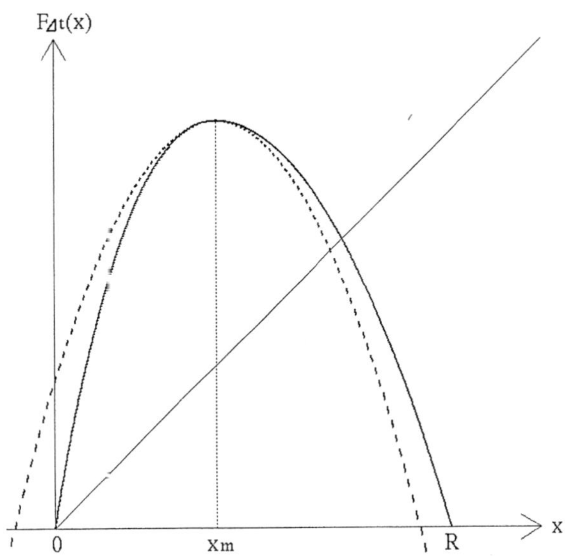

FIG. 8.2.

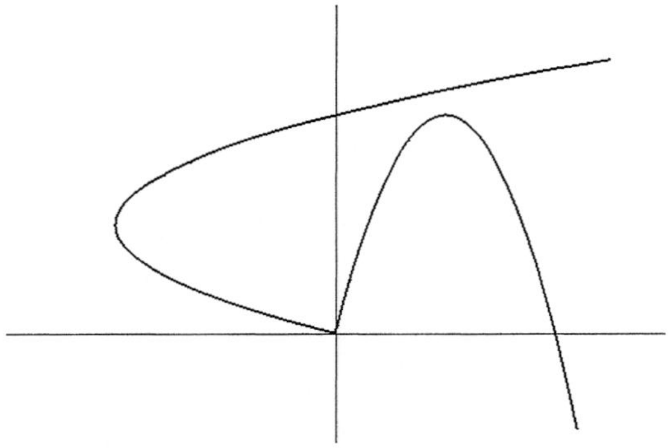

FIG. 8.3.

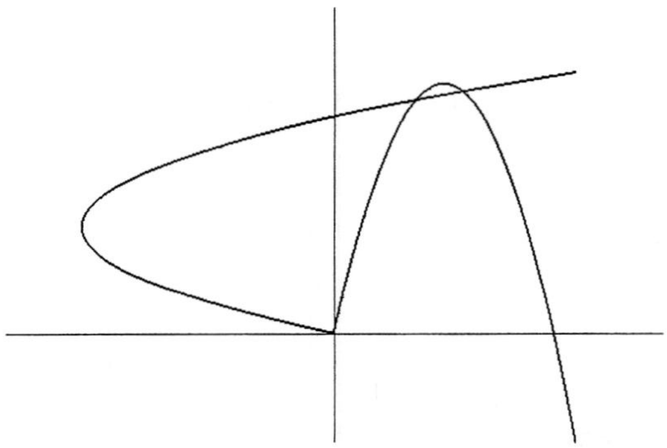

FIG. 8.4.

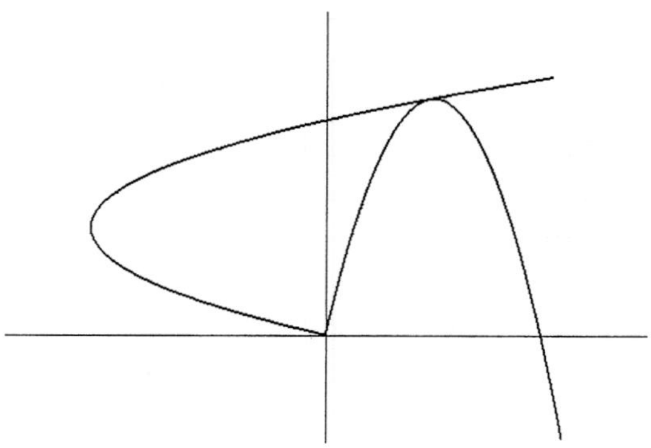

FIG. 8.5.

large, they have two points in common. There exists then certain Δt_c for which two graphs have one contact point in the first quadrant, and the graph of $x = -F_{\Delta t_c}(y)$ is above the graph of $y = F_{\Delta t_c}(x)$ (see Fig. 8.5).

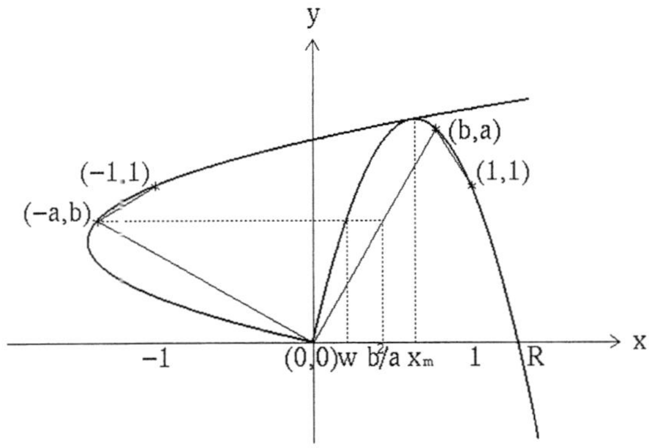

FIG. 8.6.

On the other hand, the quantity $x + F^2_{\Delta t_c}(x)$ is the horizontal distance from $A(-F^2_{\Delta t_c}(x), F_{\Delta t_c}(x))$ to $B(x, F_{\Delta t_c}(x))$. Fig. 8.6 shows this horizontal distance AB always satisfies $AB \geqslant 0$. Consequently

$$x + F^2_{\Delta t_c}(x) \geqslant 0 \quad (0 \leqslant x \leqslant 1),$$

and there exists m $(0 < m < 1)$ such that

$$m + F^2_{\Delta t_c}(m) = 0.$$

Thus,

$$G_{\Delta t_c}(x) = \tfrac{1}{2}\bigl(x + F^2_{\Delta t_c}(x)\bigr) \geqslant 0 \quad (0 \leqslant x \leqslant 1),$$

and $G_{\Delta t_c}(x)$ has a zero at m $(0 < m < 1)$. □

THEOREM 8.5. *If $\Delta t = \Delta t_c$, then we have*

$$G_{\Delta t}(x) \leqslant 1 \quad (0 \leqslant x \leqslant 1).$$

PROOF. Suppose the contrary, i.e. that

$$\max_{0 \leqslant x \leqslant 1} G_{\Delta t}(x) > 1.$$

Then there exists w such that $0 < w < 1$ and $G_{\Delta t}(w) = \frac{1}{2}(w + F^2_{\Delta t}(w)) > 1$, hence $F^2_{\Delta t}(w) > 1$. Here we take a and b such as

$$a = F^2_{\Delta t}(w) > 1,$$
$$b = F_{\Delta t}(w) < 1.$$

We draw two rectilinear polygons inscribed respectively in two graphs (see Fig. 8.6). Those are expressed as

$$\begin{cases} y = \dfrac{a}{b}x & (0 \leqslant x \leqslant b), \\ y = \dfrac{a-1}{b-1}(x-1) - 1 & (b \leqslant x), \end{cases}$$

and

$$\begin{cases} x = -\dfrac{a}{b}y & (0 \leqslant y \leqslant b), \\ x = -\dfrac{a-1}{b-1}(y-1) - 1 & (b \leqslant y). \end{cases}$$

Since the inside of the graph $y = F_{\Delta t}(x)$ is convex, we get

$$w + F_{\Delta t}^2(w) = w + a \leqslant \frac{b^2}{a} + a,$$

and $\Delta t = \Delta t_c$ implies that these polygons have no intersection. Comparing the ordinate of two polygons with the same height a,

$$-\frac{(a-1)^2}{b-1} - 1 < b,$$

(here $0 < b < 1$), it follows that

$$(a-1)^2 < 1 - b^2,$$

thus that

$$w + F_{\Delta}^2(w) \leqslant \frac{b^2}{a} + a < \frac{1-(a-1)^2}{a} + a = 2.$$

This means that

$$\tfrac{1}{2}\big(w + F_{\Delta t}^2(w)\big) \leqslant 1$$

a contradiction. □

Now we can consider a dynamical system $x_{n+1} = G_{\Delta t}(x_n)$ on $[0, 1]$ because $G_{\Delta t}([0, 1]) \subseteq [0, 1]$ is proved.

THEOREM 8.6. *We take $\Delta t = \Delta t_c$, then we get*

$$\max_{0 \leqslant x \leqslant m} G_{\Delta t}(x) \geqslant m.$$

PROOF. First, we show $F_{\Delta t}(x_m) > R$. If not, i.e. $F_{\Delta t}(x) \leqslant R$ $(0 < \forall x < R)$,

$$0 \leqslant F_{\Delta t}(x) \leqslant F_{\Delta t}(x_m) \leqslant R.$$

This means $0 \leqslant F_{\Delta t}^2(x) \leqslant R$ and it follows

$$G_{\Delta t}(x) = \tfrac{1}{2}\big(x + F_{\Delta t}^2(x)\big) \geqslant \tfrac{1}{2}x \quad (0 < x < 1).$$

This is a contradiction, because $G_{\Delta t}(x)$ has a zero m in $(0, 1)$.

Fig. 8.6 also shows that there exists x_1 such that $0 < x_1 < x_m$ and $F_{\Delta t}(x_1) = x_m$. Then,

$$\max_{0 \leqslant x \leqslant m} G_{\Delta t}(x) \geqslant G_{\Delta t}(x_1) = \tfrac{1}{2}\big(x_1 + F_{\Delta t}^2(x_1)\big) > \tfrac{1}{2}F_{\Delta t}(x_m) > \tfrac{1}{2}R.$$

In addition, Fig. 8.6 also shows that $x_m > m$. Therefore, we have $R/2 \geqslant x_m > m$ by Theorem 8.3. Finally, we get

$$\max_{0 \leqslant x \leqslant m} G_{\Delta t}(x) \geqslant m. \qquad \square$$

PROOF OF THEOREM 8.2. Now let us summarize the previous three theorems. If $\Delta t = \Delta t_c$,

$$\begin{cases} \text{(i)} & G_{\Delta t} : [0, 1] \to [0, 1], \\ \text{(ii)} & G_{\Delta t}(0) = G_{\Delta t}(m) = 0, \\ \text{(iii)} & \max_{0 \leqslant x \leqslant m} G_{\Delta t}(x) \geqslant m. \end{cases}$$

Then, there exists $\beta \in (0, m)$ such that $G_{\Delta t}(\beta) = m$. Moreover, there exists $\alpha \in (0, \beta)$ such that $G_{\Delta t}(\alpha) = \beta$. Hence

$$0 = G^3_{\Delta t}(\alpha) < \alpha < \beta = G_{\Delta t}(\alpha) < m = G^2_{\Delta t}(\alpha). \tag{8.9}$$

This is what is required to apply the Li–Yorke theorem. $\square$

REMARK 8.1. Inequalities (8.9) hold in an open neighborhood of $\Delta t = \Delta t_c$, hence chaos also occurs if Δt is in this neighborhood.

9. Central difference scheme and discretization of systems of O.D.E.'s

9.1. Central difference scheme

In numerical analysis, we use many kinds of finite difference schemes to approximate ordinary differential equations. Our principal consideration in this and the next sections is the central difference scheme for a special nonlinear equation:

$$\frac{\mathrm{d}x}{\mathrm{d}t} = x(1 - x). \tag{9.1}$$

By replacing $\mathrm{d}x/\mathrm{d}t$ with the difference quotient $(x_{n+1} - x_{n-1})/\Delta t$, we obtain the central difference scheme for this equation:

$$\frac{x_{n+1} - x_{n-1}}{\Delta t} = x_n(1 - x_n).$$

This scheme is known to be more accurate than Euler scheme; accurate here means that this scheme has smaller local truncation errors than those of Euler scheme. Yet, if we take Δt sufficiently small, this discretization also produces a chaotic dynamical system.

$$x_{n+1} = x_{n-1} + \Delta t x_n(1 - x_n). \tag{9.2}$$

Here, we remark that while the original differential equation has order one which means that the initial datum $x(0)$ determines a unique orbit $x(t)$, the order of this difference equation is two as it contains x_{n+1} and x_{n-1}, which requires two initial data x_0 and x_1 to determine its orbit $\{x_n\}$. If we consider this difference equation as a dynamical system for fixed value Δt, then we have to consider the following two-dimensional dynamical

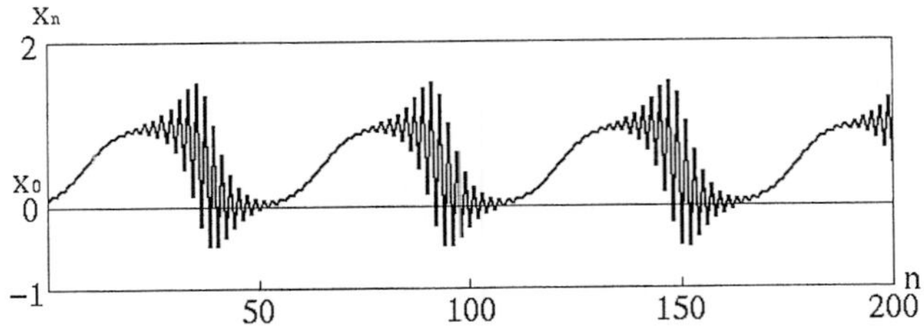

FIG. 9.1.

system:

$$\begin{cases} x_{n+1} = y_n + \Delta t x_n(1 - x_n), \\ y_{n+1} = x_n. \end{cases}$$

Many numerical experiments are solved by computers using this scheme, usually without a knowledge of x_1 because this is a numerical solution of (9.1). Then we take $x_1 = x_0 + \Delta t x_0(1 - x_0)$ as solution of the first step of Euler scheme for (9.1).

It is fairly popular among numerical analysts that when we use the above central scheme, there appears very strange phenomena: At the beginning, the numerical solution is very accurate but before long it starts to oscillate for a while and then the range of its amplitude diminishes and become accurate again, and the same cycle repeats in its process.

Fig. 9.1 illustrates the situation. This singular phenomena always appears for any small mesh size Δt. Moreover, it appears for any degree of accuracy. This phenomena called "ghost solution" was studied by USHIKI and one of the authors, YAMAGUTI [1981]. This phenomena can be explained as a kind of chaotic behavior of the dynamical system (9.2).

For the following argument, let us rewrite this system under the vector form, and introduce a map $\mathbf{F}_{\Delta t}$ defined as

$$\mathbf{F}_{\Delta t}\begin{pmatrix} x \\ y \end{pmatrix} = \begin{pmatrix} y + \Delta t x(1 - x) \\ x \end{pmatrix}.$$

Then (9.2) is

$$\begin{pmatrix} x_{n+1} \\ y_{n+1} \end{pmatrix} = \mathbf{F}_{\Delta t}\begin{pmatrix} x_n \\ y_n \end{pmatrix},$$

and using symbols x, y, X, Y with corresponds $x_n, y_n, x_{n+1}, y_{n+1}$, respectively, we can express (9.2) as,

$$\mathbf{F}_{\Delta t}\colon \begin{pmatrix} X \\ Y \end{pmatrix} = \mathbf{F}_{\Delta t}\begin{pmatrix} x \\ y \end{pmatrix}. \tag{9.3}$$

Putting $X = x$, $Y = y$, we can calculate the fixed points of (9.3), i.e. the system

$$\begin{cases} x = y + \Delta t x(1 - x), \\ y = x, \end{cases}$$

yields two points $A(0, 0)$ and $B(1, 1)$.

Meanwhile, the Jacobian matrix at these fixed points is

$$\partial \mathbf{F}_{\Delta t} = \frac{\partial(X, Y)}{\partial(x, y)} = \begin{pmatrix} 2\Delta t(1 - 2x) & 1 \\ 1 & 0 \end{pmatrix}.$$

Thus the Jacobian determinant is always -1 which means that the mapping (9.3) is a diffeomorphism on the plane onto the same plane and that $\mathbf{F}_{\Delta t}^{-1}$ exists.

9.2. *Eigenvalues and eigenvectors at the fixed points*

We can easily compute the eigenvalues and eigenvectors of the Jacobian matrix at the fixed points. At $A(0, 0)$, we have two eigenvalues λ_1 and λ_2:

$$\lambda_1 = \Delta t + \sqrt{\Delta t^2 + 1}, \qquad \lambda_2 = \Delta t - \sqrt{\Delta t^2 + 1}.$$

At $B(1, 1)$, we have two eigenvalues μ_1 and μ_2:

$$\mu_1 = -\Delta t + \sqrt{\Delta t^2 + 1}, \qquad \mu_2 = -\Delta t - \sqrt{\Delta t^2 + 1}.$$

We call all these fixed points hyperbolic fixed points of $\mathbf{F}_{\Delta t}$. The points $(0, 1)$ and $(1, 0)$ are 2-periodic points which are the fixed point of $\mathbf{F}_{\Delta t}^2$. Since eigenvalues of these periodic points are complex conjugate:

$$1 - 2\Delta t^2 \pm \sqrt{1 - \Delta t^2}\, \mathrm{i},$$

we call them elliptic fixed points of $\mathbf{F}_{\Delta t}^2$.

Let $W^u(B)$ denote the unstable manifold of B. Here, the unstable manifold is the set of points which repel from the fixed point B by the mapping (9.3). This unstable manifold is tangent to the eigenvector $(-\Delta t - \sqrt{\Delta t^2 + 1}, 1)$ at the fixed point B (see Fig. 9.2). On the other hand, let $W^s(A)$ denote the stable manifold of A, which is the set of points attracted to A by the mapping (9.3). This stable manifold is tangent to the eigenvector $(\Delta t - \sqrt{\Delta t^2 + 1}, 1)$ at the fixed point A.

The mapping $\mathbf{F}_{\Delta t}$ and $\mathbf{F}_{\Delta t}^{-1}$ are mutually topological conjugate by the symmetric transformation

$$S(x, y) = (1 - y, 1 - x),$$
$$\mathbf{F}_{\Delta t}^{-1} = S \circ \mathbf{F}_{\Delta t} \circ S^{-1}.$$

Thus $W^u(B)$ is one-to-one mapped by S on the manifold $W^s(A)$. Besides, $W^u(B)$ and $W^s(A)$ are symmetric with respect to the straight line $x + y = 1$. Numerical experiments show that for sufficiently small Δt, the graph of $W^u(B)$ forms a leaning-egg-shaped curve as seen in Fig. 9.2. This curve starting from B approaches to A while $W^s(A)$ spreads from A on both sides of the eigenvector at A. This $W^s(A)$ is symmetric for $W^u(B)$ with respect to $x + y = 1$. Although obtained through numerical computation, this stable manifold looks like almost identical to the unstable manifold. However, rigorous argument proves that these two manifolds are crossing each other infinite many times. This is the reason why we observe singular behaviors in the trajectory of (9.2). To prove this phenomena rigorously, USHIKI [1982] defined certain kind of chaos. What is surprising is that these chaotic behaviors arise however small the mesh size Δt is.

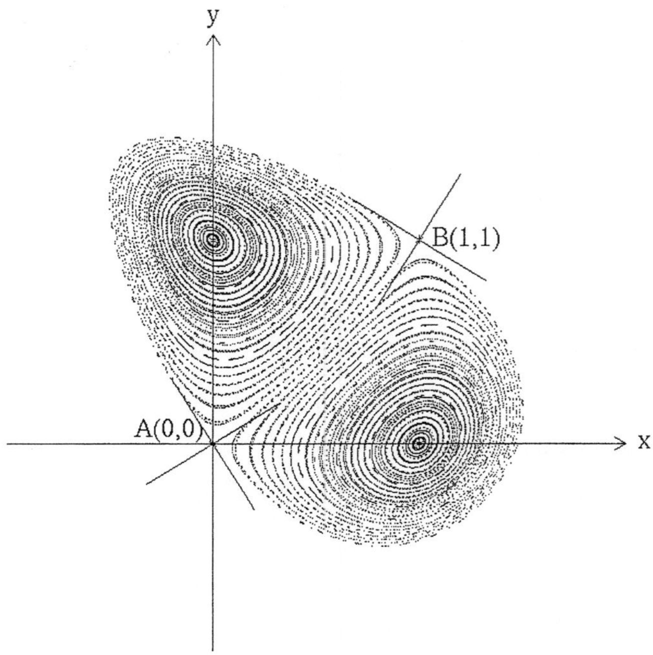

FIG. 9.2.

His proof was very deep and cannot be introduced here. The reader interested in it are desired to refer to his original paper.

9.3. Discretization of a system of ordinary differential equation

Previously, we have discussed one-dimensional dynamical system to explain the discretization of one-dimensional ordinary differential equations. Here, we consider a system of differential equations given as:

$$\begin{cases} \dfrac{\mathrm{d}u_1}{\mathrm{d}t} = f_1(u_1, u_2, \dots, u_n), \\ \dfrac{\mathrm{d}u_2}{\mathrm{d}t} = f_2(u_1, u_2, \dots, u_n), \\ \quad\vdots \\ \dfrac{\mathrm{d}u_n}{\mathrm{d}t} = f_n(u_1, u_2, \dots, u_n). \end{cases}$$

Euler's discretization of it is as follows:

$$\begin{cases} x_1^{k+1} = x_1^k + \Delta t f_1(x_1^k, x_2^k, \dots, x_n^k), \\ x_2^{k+1} = x_2^k + \Delta t f_2(x_1^k, x_2^k, \dots, x_n^k), \\ \quad\vdots \\ x_n^{k+1} = x_n^k + \Delta t f_n(x_1^k, x_2^k, \dots, x_n^k). \end{cases} \tag{9.4}$$

We rewrite these equations in a vector form:

$$\frac{\mathrm{d}\mathbf{u}}{\mathrm{d}t} = \mathbf{f}(\mathbf{u}),$$

$$\mathbf{x}^{k+1} = \mathbf{x}^k + \Delta t\, \mathbf{f}(\mathbf{x}^k).$$

Suppose that $\mathbf{f} \in C^1(\mathbb{R}^n)$. We rewrite (9.4) in more simple form:

$$\mathbf{x}^{k+1} = \mathbf{G}_{\Delta t}(\mathbf{x}^k),$$

$$\mathbf{G}_{\Delta t}(\mathbf{x}) = \mathbf{x} + \Delta t\, \mathbf{f}(\mathbf{x}).$$

Let E be the unit matrix. Then the Jacobian matrix of $\mathbf{G}_{\Delta t}$ is written as

$$\partial \mathbf{G}_{\Delta t} = E + \Delta t\, \partial \mathbf{f}.$$

Let $B(\mathbf{x}, r)$ denote the sphere of radius r whose center is $\mathbf{x}$ in $\mathbb{R}^n$. For a $\mathbf{x} \in \mathbb{R}^n$, $\|\mathbf{x}\|$ represents the Euclidean norm of $\mathbf{x}$. of A (that is a matrix whose entities are transposed and Further, for a matrix A, A^* means the adjoint matrix complex conjugate of the entities of A).

THEOREM 9.1 (HATA [1982]). *We assume that* $\mathbf{f}$ *is a continuously differentiable mapping from* $\mathbb{R}^n$ *to* $\mathbb{R}^n$*, and that this mapping has two equilibrium points, that is,* $\mathbf{f}(\bar{\mathbf{u}}) = \mathbf{f}(\bar{\mathbf{v}}) = \mathbf{O}$. *Moreover, we suppose the Jacobian of* $\mathbf{f}$ *does not vanishes at this two equilibrium points. Then the discretization* (9.4) *for* Δt *sufficiently large is a chaotic dynamical system in the sense of Marotto.*

This theorem is a generalization of Theorem 3.1 in Section 3 proved by Yamaguti and Matano under the assumption that there exist at least two equilibrium points. Chaos in the sense of Marotto means the existence of snap-back repeller which we discussed in Section 2.2. At the end of the proof, we will see that both equilibrium points $\bar{\mathbf{u}}$ and $\bar{\mathbf{v}}$ are snap-back repellers. We prepare beforehand the following three theorems: 9.2, 9.3, and 9.4.

THEOREM 9.2. *By our assumptions, there exist two positive constants* r_1 *and* c_1 *such that for* $\Delta t > c_1$,

$$\det \partial \mathbf{G}_{\Delta t}(\mathbf{x}) \neq 0$$

for all $\mathbf{x} \in B(\bar{\mathbf{u}}, r_1) \cup B(\bar{\mathbf{v}}, r_1)$.

PROOF. First, we can find $r_1 > 0$ such that

$$\det \partial \mathbf{f}(\mathbf{x}) \neq 0 \quad \forall \mathbf{x} \in B(\bar{\mathbf{u}}, r_1) \cup B(\bar{\mathbf{v}}, r_1).$$

We proceed to prove the existence of c_1 by contradiction: Suppose Theorem 9.2 is not true. Then there exists a sequence of $\Delta t_n \to \infty$ $(n \to +\infty)$, and corresponding x_n such that $\det \partial \mathbf{G}_{\Delta t_n}(\mathbf{x}_n) = 0$ for all $\mathbf{x}_n \in B(\bar{\mathbf{u}}, r_1) \cup B(\bar{\mathbf{v}}, r_1)$. On the other hand,

$$\partial \mathbf{G}_{\Delta t} = E + \Delta t\, \partial \mathbf{f},$$

therefore,

$$\det\left[\frac{E}{\Delta t_n} + \partial\mathbf{f}(\mathbf{x}_n)\right] = 0.$$

Hence we can say that $\mathbf{x}_n \to \mathbf{x}^*$ (taking a subsequence if necessary) with

$$\det \partial\mathbf{f}(\mathbf{x}^*) = 0,$$

where $\mathbf{x}^* \in B(\bar{\mathbf{u}}, r_1) \cup B(\bar{\mathbf{v}}, r_1)$. This is a contradiction. □

THEOREM 9.3. *Given a positive constant $\delta > 1$, there exist two positive constants r_2 and $c_2(\delta)$ such that*

$$\left\|\mathbf{G}_{\Delta t}(\mathbf{x}) - \mathbf{G}_{\Delta t}(\mathbf{y})\right\| \geqslant \delta\|\mathbf{x} - \mathbf{y}\|$$

for $\Delta t > c_2(\delta)$, and $\mathbf{x}, \mathbf{y} \in B(\bar{\mathbf{u}}, r_2)$.

PROOF. The assumption $\det \partial\mathbf{f}(\bar{\mathbf{u}}) \neq 0$ implies $\det \partial\mathbf{f}(\bar{\mathbf{u}})\partial\mathbf{f}^*(\bar{\mathbf{u}}) = (\det \partial\mathbf{f}(\bar{\mathbf{u}}))^2 \neq 0$. The matrix $\partial\mathbf{f}(\bar{\mathbf{u}})\partial\mathbf{f}^*(\bar{\mathbf{u}})$ is positive definite, then its minimal eigenvalue $\lambda_{\min}$ is positive. Therefore,

$$\left\|\partial\mathbf{f}(\bar{\mathbf{u}})\mathbf{x}\right\| \geqslant \sqrt{\lambda_{\min}}\|\mathbf{x}\| \quad (\text{for all } \mathbf{x} \in \mathbb{R}^n).$$

Thus, choosing $r_1 > 0$ by continuity,

$$\left\|\partial\mathbf{f}(\mathbf{x}) - \partial\mathbf{f}(\bar{\mathbf{u}})\right\| < \tfrac{1}{2}\sqrt{\lambda_{\min}} \quad (\mathbf{x} \in B(\bar{\mathbf{u}}, r_2)).$$

Consequently,

$$\begin{aligned}\left\|\mathbf{f}(\mathbf{x}) - \mathbf{f}(\mathbf{y})\right\| &= \left\|\int_0^1 \partial\mathbf{f}\big(\mathbf{y} + t(\mathbf{x} - \mathbf{y})\big)(\mathbf{x} - \mathbf{y})\,\mathrm{d}t\right\| \\ &\geqslant \left\|\partial\mathbf{f}(\bar{\mathbf{u}})(\mathbf{x} - \mathbf{y})\right\| - \tfrac{1}{2}\sqrt{\lambda_{\min}}\|\mathbf{x} - \mathbf{y}\| \\ &\geqslant \tfrac{1}{2}\sqrt{\lambda_{\min}}\|\mathbf{x} - \mathbf{y}\| \quad (\mathbf{x}, \mathbf{y} \in B(\bar{\mathbf{u}}, r_2)).\end{aligned}$$

Then, if we take $\Delta t > 2(1 + \delta)/\sqrt{\lambda_{\min}}$, we get

$$\begin{aligned}\left\|\mathbf{G}_{\Delta t}(\mathbf{x}) - \mathbf{G}_{\Delta t}(\mathbf{y})\right\| &> \Delta t\left\|\mathbf{f}(\mathbf{x}) - \mathbf{f}(\mathbf{y})\right\| - \|\mathbf{x} - \mathbf{y}\| \\ &\geqslant \left(\frac{\Delta t}{2}\sqrt{\lambda_{\min}} - 1\right)\|\mathbf{x} - \mathbf{y}\| \\ &\geqslant \delta\|\mathbf{x} - \mathbf{y}\|.\end{aligned}$$

□

THEOREM 9.4. *Let W be a bounded open set in $\mathbb{R}^n$. For a sufficiently small open neighborhood U around $\bar{\mathbf{u}}$, there exists a positive constant $c_3(U, W)$ such that for $\Delta t > c_3(U, W)$, the equation $\mathbf{G}_{\Delta t}(\mathbf{u}) = \mathbf{w}$ ($\mathbf{w} \in W$) has at least one root $\mathbf{u}$ in U.*

PROOF. We can suppose that $\mathbf{f}|_U$ is a topological isomorphism by taking a sufficiently small diameter of U. Let $\deg(\mathbf{O}, \mathbf{f}, U)$ be the degree of the mapping $\mathbf{f}$ in U with respect to $\mathbf{O}$. Then $\deg(\mathbf{O}, \mathbf{f}, U)$ is $+1$ or -1 since $\bar{\mathbf{u}}$ is an isolated zero.

Next, assume that $\mu_0\mathbf{u}_0 + \mathbf{f}(\mathbf{u}_0) = \mu_0\mathbf{w}_0$ for some $\mathbf{u}_0 \in \partial U$, $\mathbf{w}_0 \in W$ and $\mu_0 > 0$. Then

$$\mu_0 = \frac{\|\mathbf{f}(\mathbf{u}_0)\|}{\|\mathbf{u}_0 - \mathbf{w}_0\|} \geqslant \frac{\inf_{\mathbf{u}\in\partial U} \|\mathbf{f}(\mathbf{u})\|}{\sup_{\mathbf{u}\in\partial U, \mathbf{w}\in W} \|\mathbf{u} - \mathbf{w}\|} = \mu(U, W).$$

Taking μ such that $0 \leqslant \mu < \mu(U, W)$, we obtain $\mu\mathbf{w} \notin (\mu Id + \mathbf{f})(\partial U)$ for any $\mathbf{w} \in W$.

Now we consider the homotopy

$$(\nu Id + \mathbf{f})(\mathbf{x}), \quad (\nu, \mathbf{x}) \in [0, \mu] \times \overline{U}$$

and by the homotopy property of the degree, we get

$$\deg(\mu, \mathbf{w}, \mu Id + \mathbf{f}, U) = \deg(0, \mathbf{f}, U) \neq 0.$$

Therefore there exists $\mathbf{u} \in U$ such that

$$(\mu Id + \mathbf{f})(\mathbf{u}) = \mu\mathbf{w}.$$

This means that $\mathbf{G}_{\Delta t}(\mathbf{u}) = \mathbf{w}$ for $\Delta t = 1/\mu$. □

The same arguments of the above two theorems are true for $\bar{\mathbf{v}}$.

Now we are prepared to prove Theorem 9.1.

PROOF OF THEOREM 9.1 We choose sufficiently small open neighborhoods U, V of $\bar{\mathbf{u}}, \bar{\mathbf{v}}$, respectively such that $U \cap V = \emptyset$ and Theorem 9.4 holds for both $\bar{\mathbf{u}}$ and $\bar{\mathbf{v}}$. Let $r^* = \min(r_1, r_2)$, $c^* = \max(c_1, c_2(\delta), c_3(U, V), c_3(V, U))$. We can assume that

$$U \subset B(\bar{\mathbf{u}}, r^*), \qquad V \subset B(\bar{\mathbf{v}}, r^*).$$

By Theorem 9.4, for any $\Delta t > c^*$, there exist $\mathbf{v}_{\Delta t} \in V$ and $\mathbf{u}_{\Delta t} \in U$ such that $\mathbf{G}_{\Delta t}(\mathbf{v}_{\Delta t}) = \bar{\mathbf{u}}$, $\mathbf{G}_{\Delta t}(\mathbf{u}_{\Delta t}) = \mathbf{v}_{\Delta t}$. We also have $\mathbf{G}_{\Delta t}(\mathbf{u}_{\Delta t}) \neq 0$ and $\mathbf{G}_{\Delta t}(\mathbf{v}_{\Delta t}) \neq 0$ by Theorem 9.2. We can find $r_{\Delta t} > 0$ such that $B(\mathbf{u}_{\Delta t}, r_{\Delta t}) \subset B(\bar{\mathbf{u}}, r^*)$, $\mathbf{G}_{\Delta t}(B(\mathbf{u}_{\Delta t}, r_{\Delta t})) \subset U$, $\mathbf{G}^2_{\Delta t}(B(\mathbf{u}_{\Delta t}, r_{\Delta t})) \subset U$ and such that both $\mathbf{G}_{\Delta t}|_{B(\mathbf{u}_{\Delta t}, r_{\Delta t})}$ and $\mathbf{G}_{\Delta t}|_{\mathbf{G}_{\Delta t}(B(\mathbf{u}_{\Delta t}, r_{\Delta t}))}$ are homeomorphisms. Finally we can define a sequence of compact sets $\{B_k\}_{-\infty<k\leqslant 2}$ as follows:

$$\begin{aligned} B_1 &= \mathbf{G}_{\Delta t}\big(B(\mathbf{u}_{\Delta t}, r_{\Delta t})\big), \\ B_2 &= \mathbf{G}^2_{\Delta t}\big(B(\mathbf{u}_{\Delta t}, r_{\Delta t})\big), \quad \text{and} \\ B_{-k} &= \mathbf{G}^{-k}_{\Delta t}\big(B(\mathbf{u}_{\Delta t}, r_{\Delta t})\big) \quad \text{for all } k \geqslant 0, \end{aligned}$$

since $\mathbf{G}^{-k}_{\Delta t}$ is well defined by Theorem 9.3. This shows that $\bar{\mathbf{u}}$ is a snap-back repeller. Obviously the same argument holds for $\bar{\mathbf{v}}$. Thus we can apply Marotto theorem in Section 2.2. □

EXAMPLE 9.1. We can apply this theorem to the following example:

$$\begin{cases} \dfrac{du_1}{dt} = (a_1 - b_{11}u_1 - \cdots - b_{1n}u_n)u_1, \\ \dfrac{du_2}{dt} = (a_2 - b_{21}u_1 - \cdots - b_{2n}u_n)u_2, \\ \quad\vdots \\ \dfrac{du_n}{dt} = (a_n - b_{n1}u_1 - \cdots - b_{nn}u_n)u_n. \end{cases} \tag{9.5}$$

This system of differential equations appears in some problems of mathematical ecology proposed by USHIKI, YAMAGUTI and MATANO [1980]. Let us write $\mathbf{A} = (a_1, \dots, a_n)$, $B = (b_{ij})$, and $\mathbf{O} = (0, \dots, 0)$. If $\mathbf{A} \neq \mathbf{O}$ and $\det B \neq 0$, then we can easily see that the system (9.5) has at least two equilibrium points $\mathbf{O}$ and $B^{-1}\mathbf{A}$. Moreover, $\det \partial \mathbf{f}(\mathbf{O}) = a_1 \cdots a_n$ and $\det \partial \mathbf{f}(B^{-1}\mathbf{A}) = (-1)^n \bar{u}_1 \cdots \bar{u}_n \det B$. Therefore, the conclusion of the theorem above holds if $a_1 \cdots a_n \bar{u}_1 \cdots \bar{u}_n \neq 0$.

9.4. *Discretization of O.D.E. with one equilibrium*

In this section, we discuss an example of a two-dimensional ordinary differential equation which has only one equilibrium point, and prove that its Euler's discretization is chaotic for any time step. The fixed point of this difference equation is not exactly a snap-back repeller of Marotto, but this system is proved chaotic.

Consider the following two-dimensional differential equation:

$$\dot{x} = \frac{x(2y^2 - x^2)}{(x^2 + y^2)^{5/4}}, \qquad \dot{y} = \frac{y(y^2 - 2x^2)}{(x^2 + y^2)^{5/4}}. \tag{9.6}$$

Using polar coordinates, the above equations can be written in the form

$$\dot{r} = -\sqrt{r}\cos 2\theta, \qquad \dot{\theta} = -\frac{\sin 2\theta}{2\sqrt{r}}.$$

Note that this equation is not differentiable at the origin, but continuous in the whole space. We can get an exact figure of trajectories (see Fig. 9.3). Indeed, any solution of

$$\frac{dr}{d\theta} = \frac{\dot{r}}{\dot{\theta}} = 2r\cot 2\theta,$$

is given by the expression $r = C|\sin 2\theta|$ with certain positive constant C. Note that the x-axis and y-axis are also trajectories.

Let us check that the uniqueness of the solution is broken at the equilibrium point $(x, y) = (0, 0)$. If $0 \leqslant \theta \leqslant \pi/4$,

$$\dot{\theta} = -\frac{\sin 2\theta}{2\sqrt{C|\sin 2\theta|}} = -\frac{\sqrt{\sin 2\theta}}{2\sqrt{C}} \leqslant -\frac{\sqrt{\theta}}{2\sqrt{C}},$$

which means that in some finite time, θ becomes 0, and accordingly $r = C|\sin 2\theta|$ becomes 0, i.e. any trajectory arrives at $(0, 0)$ in a finite time. With the same argument in the domain of $\pi/4 \leqslant \theta \leqslant \pi/2$, any trajectory emerges from $(0, 0)$ at an arbitrary given

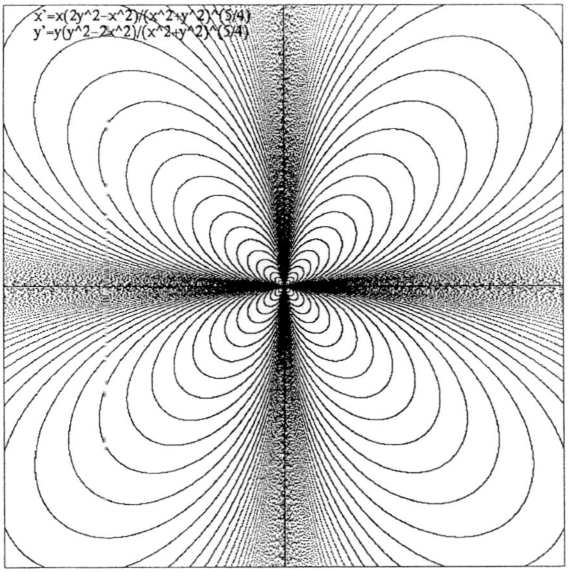

FIG. 9.3.

time. Since the equation has Lipschitz-continuity at any point except at the equilibrium point, the uniqueness of the solution breaks at only one point $(0,0)$ and it is broken to both directions of time. Recall that the one-dimensional differential equations treated in Section 6 have the breaking of the uniqueness only in the backward time direction.

Next, let us discretize the differential equation above with time step Δt. We obtain

$$\mathbf{F}\begin{pmatrix} x \\ y \end{pmatrix} = \begin{pmatrix} x + \Delta t \frac{x(2y^2-x^2)}{(x^2+y^2)^{5/4}} \\ y + \Delta t \frac{y(y^2-2x^2)}{(x^2+y^2)^{5/4}} \end{pmatrix} = \begin{pmatrix} \cos\theta(r + \Delta t\sqrt{r}(2\sin^2\theta - \cos^2\theta)) \\ \sin\theta(r + \Delta t\sqrt{r}(\sin^2\theta - 2\cos^2\theta)) \end{pmatrix}.$$

It is enough to consider the special case $\Delta t = 1$ without loss of generality, since as seen in Fig. 9.3, there is a scaling property between Δt and r. In fact, the multiplication of m to Δt corresponds to that of m^2 to r. Fig. 9.4 shows the image of map $\mathbf{F}$ ($\Delta t = 1$). We can see that $\mathbf{F}$ maps a disk to the dotted region.

From now on, we assume that $\Delta t = 1$ and we use occasionally the polar coordinates in the following way:

$$\mathbf{F}: (r, \theta) \to (R, \Theta).$$

The Jacobian matrix $\partial\mathbf{F}$ of $\mathbf{F}$ is

$$\begin{aligned} \partial\mathbf{F} &= \begin{pmatrix} 1 + \frac{(2y^2-3x^2)(x^2+y^2)-\frac{5}{2}x^2(2y^2-x^2)}{(x^2+y^2)^{9/4}} & \frac{\frac{13}{2}x^3y-xy^3}{(x^2+y^2)^{9/4}} \\ \frac{x^3y-\frac{13}{2}xy^3}{(x^2+y^2)^{9/4}} & 1 + \frac{(3y^2-2x^2)(x^2+y^2)-\frac{5}{2}y^2(y^2-2x^2)}{(x^2+y^2)^{9/4}} \end{pmatrix} \\ &= \begin{pmatrix} 1 + \frac{1}{\sqrt{r}}\left(2s^2 - 3c^2 - \frac{5}{2}c^2(2s^2 - c^2)\right) & \frac{1}{\sqrt{r}}cs\left(\frac{13}{2}c^2 - s^2\right) \\ \frac{1}{\sqrt{r}}cs\left(c^2 - \frac{13}{2}s^2\right) & 1 + \frac{1}{\sqrt{r}}\left(3s^2 - 2c^2 - \frac{5}{2}s^2(s^2 - 2c^2)\right) \end{pmatrix}, \end{aligned}$$

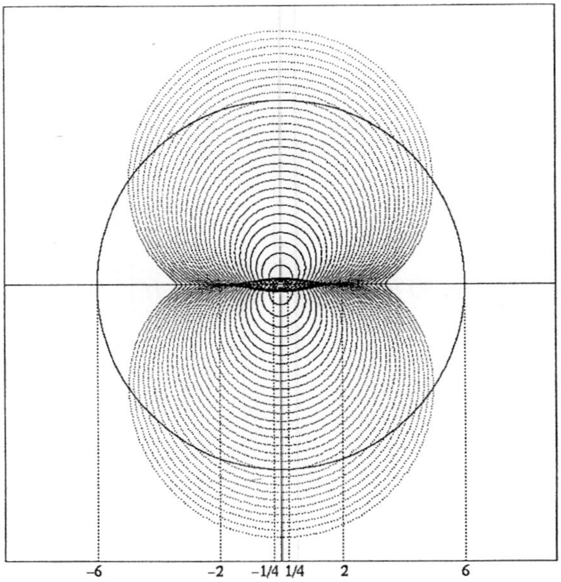

FIG. 9.4.

where $c = \cos\theta$ and $s = \sin\theta$.

$$\operatorname{tr}(\partial\mathbf{F}) = 2 - \frac{5}{2\sqrt{r}}\cos 2\theta,$$

$$\det(\partial\mathbf{F}) = 1 - \frac{5}{2\sqrt{r}}\cos 2\theta + \frac{1}{8r}(5 + 3\cos^2 2\theta),$$

$$\det(\partial\mathbf{F}) = 0 \quad \Leftrightarrow \quad \sqrt{r} = \frac{1}{4}\left(5\cos 2\theta \pm \sqrt{19\cos^2 2\theta - 10}\right).$$

Fig. 9.5 shows the curves of $\det(\partial\mathbf{F}) = 0$ and the degrees of the map $\mathbf{F}$.

Now, let us calculate the eigenvalues of $\mathbf{F}$:

$$\begin{aligned}\lambda_\pm &= \tfrac{1}{2}\left\{\operatorname{tr}(\partial\mathbf{F}) \pm \sqrt{\left(\operatorname{tr}(\partial\mathbf{F})\right)^2 - 4\left(\det(\partial\mathbf{F})\right)^2}\right\}\\ &= 1 + \frac{1}{4\sqrt{r}}\left(-5\cos 2\theta \pm \sqrt{19\cos^2 2\theta - 10}\right).\end{aligned}$$

Note that the eigenvalues are not equal to one for any r and θ. With $\theta_0 = \frac{1}{2}\arccos\sqrt{10/19}$, the eigenvalues in the first quadrant are classified as follows:

(i) $0 \leqslant \theta \leqslant \theta_0$ ($\lambda_\pm \in \mathbb{R}$): Both λ_- and λ_+ are monotone increasing with respect to r, and $\lambda_- \leqslant \lambda_+ < 1$, and

$$\lambda_\pm = -1 \quad \Leftrightarrow \quad \sqrt{r} = \frac{1}{8}\left(5\cos 2\theta \mp \sqrt{19\cos^2 2\theta - 10}\right).$$

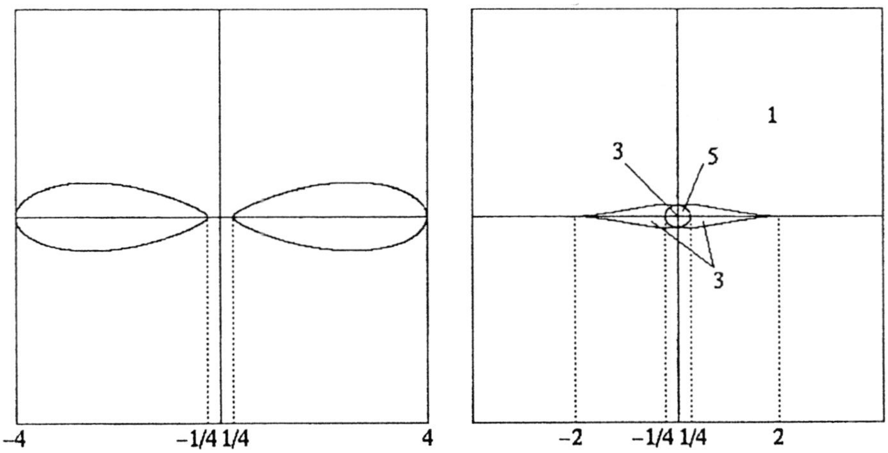

FIG. 9.5.

(ii) $\theta_0 \leqslant \theta \leqslant \frac{\pi}{2} - \theta_0$ ($\lambda_\pm \in \mathbb{C}$):

$$|\lambda_\pm|^2 = 1 - \frac{5}{2\sqrt{r}}\cos 2\theta + \frac{3\cos^2 2\theta + 5}{8r}.$$

In particular, $|\lambda_\pm| > 1$ when $\theta \geqslant \frac{\pi}{4}$. If $\theta < \frac{\pi}{4}$, then

$$|\lambda_\pm| = 1 \quad \Leftrightarrow \quad \sqrt{r} = \frac{3\cos^2 2\theta + 5}{20\cos 2\theta}.$$

(iii) $\pi/2 - \theta_0 \leqslant \theta \leqslant \pi/2$ ($\lambda_\pm \in \mathbb{R}$): Both λ_- and λ_+ are monotone decreasing with respect to r, and $1 < \lambda_- \leqslant \lambda_+$.

In Fig. 9.6, the letters R and C denote real and complex number of the eigenvalues respectively. The notation $+$ (resp. $-$) as superscript and subscript means that the modulus of the eigenvalue is greater (resp. less) than one.

Here is the statement (see Fig. 9.7):

THEOREM 9.5 (MAEDA [1998]). *The Euler's discretization* **F** *of* (9.6) *is chaotic in the sense of Marotto* (*i.e.* **F** *satisfies* (i), (ii), *and* (iii) *of Theorem* 2.3) *for any time step* Δt.

REMARK 9.1. By numerical computation, the number N of (i) in Theorem 2.3 is at most 7. There are 3 periodic orbits with period 2:

$$\left(\pm\frac{1}{4}, 0\right), \quad \left(\frac{\sqrt{6}}{12}, \pm\frac{\sqrt{3}}{12}\right), \quad \left(-\frac{\sqrt{6}}{12}, \pm\frac{\sqrt{3}}{12}\right).$$

REMARK 9.2. In particular, if the denominator of (9.6) is $(x^2 + y^2)^1$, chaotic phenomenon never occurs for sufficiently small time step.

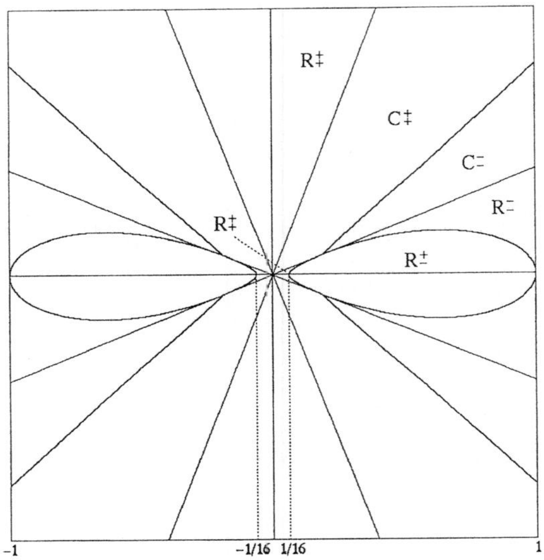

FIG. 9.6.

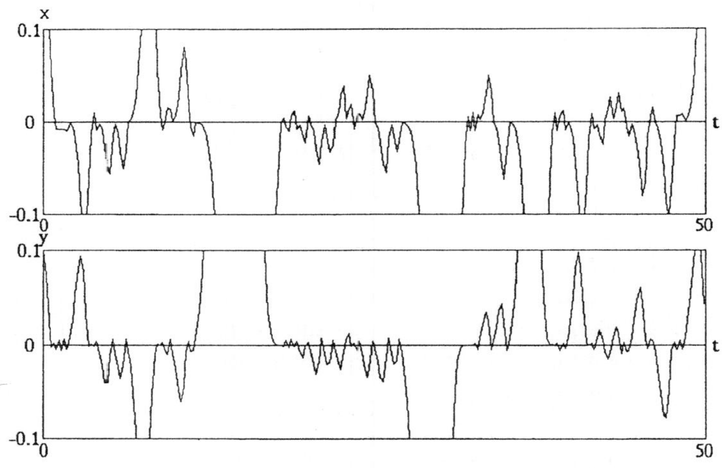

FIG. 9.7.

PROOF. This proof is quite long; we make use of following four theorems: From Theorems 9.6–9.9.

As already seen, we can assume $\Delta t = 1$ without loss of generality. Let $\mathbf{Z} = (0, 0)$ denote the unique fixed point of $\mathbf{F}$; $\mathbf{Z}_{-1} = (1, 0)$ be one of three inverse images of $\mathbf{Z}$, i.e. $\mathbf{F}(\mathbf{Z}_{-1}) = \mathbf{Z}$; $\mathbf{Z}_{-2} \approx (1.174, 0.723)$ be one of three inverse images of $\mathbf{Z}_{-1}$; and $\mathbf{Z}_{-3} \approx (0.627, 0.815)$ be a unique inverse image of $\mathbf{Z}_{-2}$. To define $\mathbf{Z}_{-n}$ $(n \in \mathbb{N})$ inductively, we prepare the next theorem, where $V_{(\pi/4,\pi/2)}(R_0)$ denotes the sector of the first quadrant formed by the circle $r = R_0$ and the rays $y = x$ and $x = 0$ (i.e. $0 \leqslant r \leqslant R_0, \pi/4 \leqslant \theta \leqslant \pi/2$).

THEOREM 9.6. *The restricted map* $\mathbf{F}: V_{(\pi/4,\pi/2)}(R_0) \to \mathbf{F}(V_{(\pi/4,\pi/2)}(R_0))$ *is injective and*

$$V_{(\pi/4,\pi/2)}(R_0) \subset \mathbf{F}\big(V_{(\pi/4,\pi/2)}(R_0)\big),$$

for any $R_0 > 0$. *Moreover, there exists a constant* $c > 1$ *which depends on* R_0 *and satisfies*

$$\|\mathbf{F}(\mathbf{X}) - \mathbf{Z}\| \geqslant c\|\mathbf{X} - \mathbf{Z}\| \quad (\forall \mathbf{X} \in V_{(\pi/4,\pi/2)}(R_0) - \{\mathbf{Z}\}).$$

PROOF. The boundary of $V_{(\pi/4,\pi/2)}(R_0)$ is composed of the following three curves:

$$l_{\pi/2}(R_0) = \left\{(r,\theta) \,\middle|\, \theta = \frac{\pi}{2},\ 0 \leqslant r \leqslant R_0\right\},$$

$$l_{\pi/4}(R_0) = \left\{(r,\theta) \,\middle|\, \theta = \frac{\pi}{4},\ 0 \leqslant r \leqslant R_0\right\},$$

$$C(R_0) = \left\{(r,\theta) \,\middle|\, r = R_0,\ \frac{\pi}{4} \leqslant \theta \leqslant \frac{\pi}{2}\right\}.$$

For any $r \in (0, R_0]$,

$$\begin{aligned} R^2 &= r\big(r + 2\sqrt{r}(\sin^2\theta - \cos^2\theta) + 1 - 3\sin^2\theta\cos^2\theta\big) \\ &\geqslant r\big(r + 1 - \tfrac{3}{4}\big) > r^2. \end{aligned}$$

Putting $r = R_0$, we get $\|\mathbf{F}(C(R_0)) - \mathbf{Z}\| > R_0$. Furthermore,

$$\frac{\|\mathbf{F}(\mathbf{X}) - \mathbf{Z}\|}{\|\mathbf{X} - \mathbf{Z}\|} = \frac{R}{r} \geqslant \sqrt{1 + \frac{1}{4R_0}},$$

then we can take $c = \sqrt{1 + 1/(4R_0)} (> 1)$. Note that

$$\mathbf{F}\big(l_{\pi/2}(R_0)\big) = \left\{(R,\Theta) \,\middle|\, \Theta = \frac{\pi}{2},\ 0 \leqslant R \leqslant R_0 + \sqrt{R_0}\right\}.$$

If $\mathbf{X} \in l_{\pi/4}(R_0)$,

$$\mathbf{F}(\mathbf{X})_x = \frac{\sqrt{2}}{2}\left(r + \frac{\sqrt{r}}{2}\right), \qquad \mathbf{F}(\mathbf{X})_y = \frac{\sqrt{2}}{2}\left(r - \frac{\sqrt{r}}{2}\right).$$

Hence $\mathbf{F}(\mathbf{X})_y < \mathbf{F}(\mathbf{X})_x$, that is, $\mathbf{F}(l_{\pi/4})$ is under the line $y = x$. Therefore, the image of the boundary of $V_{(\pi/4,\pi/2)}(R_0)$ includes $V_{(\pi/4,\pi/2)}(R_0)$ and does not have any self-intersection. Taking account $\det(\partial\mathbf{F}) \neq 0$ in $V_{(\pi/4,\pi/2)}(R_0)^\circ$ and the continuity of $\mathbf{F}$, we can conclude that $\mathbf{F}$ is injective and $V_{(\pi/4,\pi/2)}(R_0) \subset \mathbf{F}(V_{(\pi/4,\pi/2)}(R_0))$. □

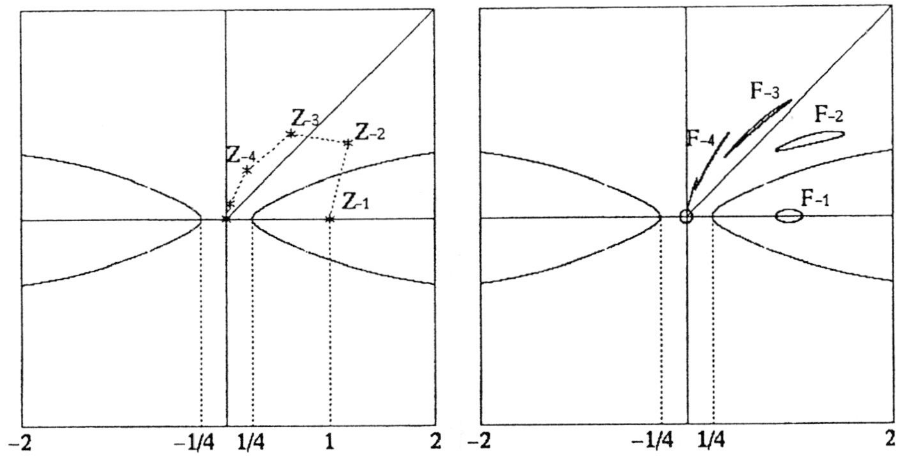

FIG. 9.8.

Let us come back to the proof of Theorem 9.5. We can uniquely define inverse images $\mathbf{Z}_{-n}$ $(n \geqslant 4)$ in the area $V_{(\pi/4,\pi/2)}(R_0)$, with

$$\mathbf{Z}_{-1} = (1, 0), \qquad \mathbf{Z}_{-2} \approx (1.174, 0.7237),$$
$$\mathbf{Z}_{-3} \approx (0.6271, 0.8157), \qquad \mathbf{Z}_{-4} \approx (0.1982, 0.4643),$$
$$\mathbf{Z}_{-5} \approx (0.0343, 0.1463), \qquad \mathbf{Z}_{-6} \approx (0.002228, 0.01814),$$
$$\mathbf{Z}_{-7} \approx (0.00002005, 0.0003254), \quad \ldots$$

Note that $\lim_{n\to+\infty} \mathbf{Z}_{-n} = \mathbf{Z}$.

The proof for (i) of Theorem 9.5 is as follows: Let us take $B_r(\mathbf{Z})$ and a sequence of inverse images $\{\mathbf{F}^{-n}(B_r(\mathbf{Z}))\}$ $(n \in \mathbb{N})$ as neighborhoods of $\mathbf{Z}_{-n}$. Fig. 9.8 (left) is the sequence $\{\mathbf{Z}_{-n}\}$ $(n \in \mathbb{N})$ and shows that $\det(\partial\mathbf{F}(\mathbf{Z}_{-n})) \neq 0$ $(n \in \mathbb{N})$. Hence for sufficiently small r,

$$\mathbf{F}^{-n} : B_r(\mathbf{Z}) \to \mathbf{F}^{-n}\big(B_r(\mathbf{Z})\big)$$

is an injective continuous map. Here, we take $r_0 = 1/16$ and fix it (see Fig. 9.8 (right)). Note that $\mathbf{F}^{-n}(B_{r_0}(\mathbf{Z}))$ must be included in $B_{r_0}(\mathbf{Z})$ for any n greater than certain integer N. In fact, we can show $N = 7$ by numerical computation:

$$\mathbf{F}^{-n}\big(B_{r_0}(\mathbf{Z})\big) \subset B_{r_0}(\mathbf{Z}) \quad (\forall n \geqslant 7).$$

Then, by the Brouwer fixed point theorem, there exists $\mathbf{Y}_n \in \mathbf{F}^{-n}(B_{r_0}(\mathbf{Z}))$ such that $\mathbf{F}^n(\mathbf{Y}_n) = \mathbf{Y}_n$. Since $\mathbf{F}^{-m}(B_{r_0}(\mathbf{Z})) \cap \mathbf{F}^{-n}(B_{r_0}(\mathbf{Z})) = \emptyset$ for any positive integers m, n $(m \neq n)$, it is clear that $\mathbf{Y}_n$ cannot have period less than n.

Before proving (ii) and (iii) of Theorem 9.5, we will prepare the next two theorems: 9.7 and 9.8; the latter one is a key theorem.

THEOREM 9.7. *The following inclusion holds*:

$$B_{r_0}(\mathbf{Z}) \subset \mathbf{F}\big(B_{r_0}(\mathbf{Z})\big).$$

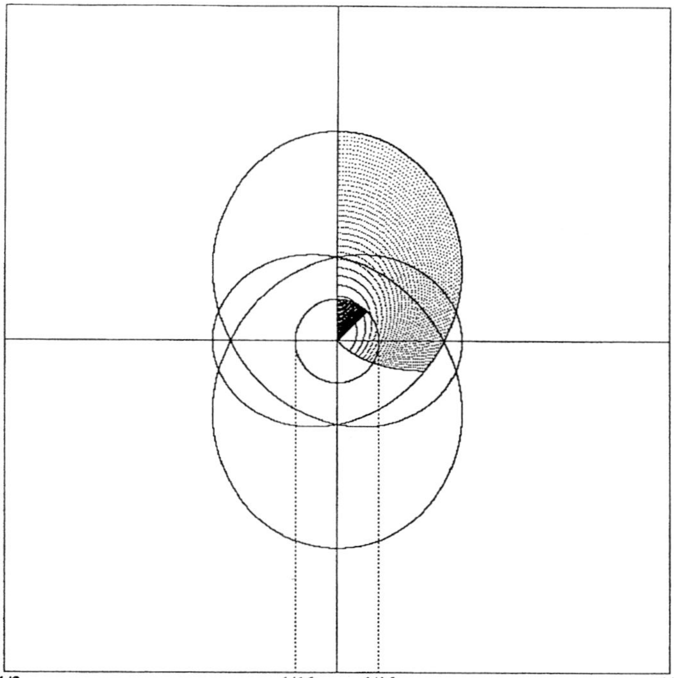

FIG. 9.9.

PROOF. To prove this, it suffices to show $V_{(0,\pi/2)}(r_0) \subset \mathbf{F}(V_{(\pi/4,\pi/2)}(r_0))$. If $\mathbf{X} \in l_{\pi/4}(r_0)$,

$$\mathbf{F}(\mathbf{X})_y = \frac{\sqrt{2}}{2}\left(r - \frac{\sqrt{r}}{2}\right) \leqslant 0,$$

for any $r \in [0, r_0]$, in other words, $\mathbf{F}(l_{\pi/4})$ is under the x-axis (see Fig. 9.9). The proof is completed through the same argument as in the proof of Theorem 9.6. □

Let U and V be compact neighborhoods of $\mathbf{Z}_{-1}$ and $\mathbf{Z}$ defined by

$$U = \mathbf{F}^{-1}\big(B_{r_0}(\mathbf{Z})\big), \qquad V = B_{r_0}(\mathbf{Z}),$$

and $\mathbf{H}$ be a function defined as $\mathbf{H}(\mathbf{X}) = \mathbf{F}^N(\mathbf{X})$. Note that $\mathbf{F}(U) = V$, $U \cap V = \emptyset$, and the distance δ between U and V is positive:

$$\delta = \inf\big\{\|\mathbf{X} - \mathbf{Y}\| \mid \mathbf{X} \in U,\ \mathbf{Y} \in V\big\} \geqslant \frac{3}{16} > 0.$$

THEOREM 9.8. *The following two inclusions hold*:

$$V \subset \mathbf{H}(U),$$
$$U \cup V \subset \mathbf{H}(V).$$

PROOF. From $V \subset \mathbf{F}(V)$,

$$V \subset \mathbf{F}(V) \subset \mathbf{F}\big(\mathbf{F}(V)\big) \subset \cdots \subset \mathbf{F}^n(V) \quad (\forall n \geqslant 1).$$

In particular, if $n = N$, $V \subset \mathbf{H}(V)$. Since $U = \mathbf{F}^{N-1}(\mathbf{F}^{-N}(B_{r_0}(\mathbf{Z}))$,

$$U \subset \mathbf{F}^{N-1}(V) \subset \mathbf{F}^N(V).$$

Hence $U \subset \mathbf{H}(V)$. To show $V \subset \mathbf{H}(U)$, take $n = N + 1$. Then,

$$V \subset \mathbf{F}^{N+1}(V) = \mathbf{F}^N(U). \qquad \square$$

The proof for (ii) of Theorem 9.5 is as follows: Let A be the set of sequences $E = \{E_n\}_{n=1}^{+\infty}$ where E_n equals either to U or to V, and $E_{n+1} = V$ if $E_n = U$. This restriction comes from the inclusions of Theorem 9.8. Let $R(E, n)$ be the number of E_i's which equal to U for $1 \leqslant i \leqslant n$. For each $w \in (0, 1)$, choose $E^w = \{E_n^w\}_{n=1}^{+\infty}$ from the sequences of A satisfying

$$\lim_{n \to +\infty} \frac{R(E^w, n^2)}{n} = w.$$

If B is defined by $B = \{E^w \mid w \in (0, 1)\} \subset A$, then B is uncountable. Then, for each $E^w \in B$, there exists a point $\mathbf{X}_w \in U \cup V$ with $\mathbf{H}^n(\mathbf{X}_w) \in E_n^w$ for all $n \geqslant 1$. Define S_H $(\subset \mathbb{R}^2)$ by

$$S_H = \big\{\mathbf{H}^n(\mathbf{X}_w) \mid E^w \in B,\ n \geqslant 0\big\}.$$

From the definition, $\mathbf{H}(S_H) \subset S_H$. S_H contains no periodic points of $\mathbf{H}$ (if not, $w = +\infty$). There exist infinite number of n's such that $\mathbf{H}^n(\mathbf{X}) \in U$ and $\mathbf{H}^n(\mathbf{Y}) \in V$ for any $\mathbf{X}, \mathbf{Y} \in S_H$ with $\mathbf{X} \neq \mathbf{Y}$. Therefore, for any $\mathbf{X}, \mathbf{Y} \in S_H$ with $\mathbf{X} \neq \mathbf{Y}$,

$$\limsup_{n \to +\infty} \big\|\mathbf{H}^n(\mathbf{X}) - \mathbf{H}^n(\mathbf{Y})\big\| \geqslant \delta.$$

Now let $S = \{\mathbf{F}^n(\mathbf{X}) \mid \mathbf{X} \in S_H,\ n \geqslant 0\}$. We can see that $\mathbf{F}(S) \subset S$, that S contains no periodic points of $\mathbf{F}$, and that for any $\mathbf{X}, \mathbf{Y} \in S$ with $\mathbf{X} \neq \mathbf{Y}$,

$$\limsup_{n \to +\infty} \big\|\mathbf{F}^n(\mathbf{X}) - \mathbf{F}^n(\mathbf{Y})\big\| \geqslant \delta > 0.$$

Thus we have proved (ii)(a) and (ii)(b).

For the proof of (ii)(c), it is also enough to show $\limsup_{n \to +\infty} \|\mathbf{H}^n(\mathbf{X}) - \mathbf{H}^n(\mathbf{Y})\| > 0$. Let $\mathbf{X} \in S$, and let $\mathbf{Y}$ be a periodic point with period k so that $\mathbf{Y}$ is also periodic with respect to $\mathbf{H}$ with period $k' \leqslant k$. Let $Q = \{\mathbf{H}^i(\mathbf{Y}) \mid 0 \leqslant i \leqslant k' - 1\}$ and for $\mathbf{x} \in \mathbb{R}^2$ let $R(\mathbf{x}, n)$ denote the number of i's in $\{1, 2, \ldots, n\}$ for which $\mathbf{H}^i(\mathbf{x}) \in U$. Then

$$\rho(\mathbf{Y}) = \lim_{n \to +\infty} \frac{R(\mathbf{Y}, n^2)}{n} = \begin{cases} 0 & (\text{if } Q \cap U = \emptyset), \\ +\infty & (\text{otherwise}). \end{cases}$$

If $\rho(\mathbf{Y}) = 0$, $\limsup_{n \to +\infty} \|\mathbf{H}^n(\mathbf{X}) - \mathbf{H}^n(\mathbf{Y})\| \geqslant \operatorname{dist}(Q, U)$. Since both Q and U are compact, $\operatorname{dist}(Q, U)$ is positive. Otherwise, $\rho(\mathbf{Y}) = +\infty$, there exists a subsequence $\{n_\nu\}_{\nu=1}^{+\infty}$ of $\mathbb{N}$ such that $\mathbf{H}^{n_\nu}(\mathbf{X}) \in V$ and $\mathbf{H}^{n_\nu}(\mathbf{Y}) \in U$ for any ν, hence $\limsup_{n \to +\infty} \|\mathbf{H}^n(\mathbf{X}) - \mathbf{H}^n(\mathbf{Y})\| \geqslant \delta$.

The proof of (iii) of Theorem 9.5 is as follows: First note that $\mathbf{Z}$ is expanding in $V_{(\pi/4,\pi/2)}(r_0)$. Let $D_n = \mathbf{H}^{-n}(B_{r_0}(\mathbf{Z}))$ for all $n \leqslant 0$. Given $\sigma > 0$ there exists $J = J(\sigma)$ such that $\|\mathbf{X} - \mathbf{Z}\| < \sigma$ for all $\mathbf{X} \in D_n$ and $n > J$.

For any sequence $E^w = \{E_n^w\}_{n=1}^{+\infty} \in A$, we further restrict the E^w in the following manner: If $E_n^w = U$, then $n = m^2$ for some integer m; if $E_n^w = U$ for both $n = m^2$ and $n = (m+1)^2$, then $E_n^w = D_{2m-k}$ for $n = m^2 + k$ where $k = 1, 2, \dots, 2m$; and for the remaining n's we shall assume $E_n^w = V$.

It can be easily checked that these sequences still satisfy $E_{n+1}^w \subset \mathbf{H}(E_n^w)$. Hence there exists a point $\mathbf{X}_w$ with $\mathbf{H}^n(\mathbf{X}_w) \in E_n^w$ for all $n \geqslant 0$. Now, define $S_0 = \{\mathbf{X}^w \mid w \in (3/4, 1)\}$, then S_0 is uncountable and $S_0 \subset S_H \subset S$. Finally, we prepare the next theorem for the completion of the proof.

THEOREM 9.9. *For any $s, t \in (3/4, 1)$, there exist infinitely many n's such that $E_k^s = E_k^t = U$ for both $k = n^2$ and $(n+1)^2$.*

PROOF. Let us proceed by contradiction: Under the negation of the original conclusion, we are to show $s + t \leqslant 3/2$. Without loss of generality, there is no n such that $E_k^s = E_k^t = U$ for both $k = n^2$ and $(n+1)^2$. Define a_n^w as:

$$a_n^w = \begin{cases} 1 & (\text{if } E_{n^2}^w = U), \\ 0 & (\text{if } E_{n^2}^w = V), \end{cases}$$

then,

$$\begin{aligned} s + t &= \lim_{n\to+\infty} \frac{R(E^s, n^2) + R(E^t, n^2)}{n} \\ &= \lim_{n\to+\infty} \frac{R(E^s, (2n)^2) + R(E^t, (2n)^2)}{2n} \\ &= \lim_{n\to+\infty} \frac{\sum_{i=1}^n (a_{2i-1}^s + a_{2i}^s) + \sum_{i=1}^n (a_{2i-1}^t + a_{2i}^t)}{2n} \\ &= \lim_{n\to+\infty} \frac{\sum_{i=1}^n (a_{2i-1}^s + a_{2i}^s + a_{2i-1}^t + a_{2i}^t)}{2n} \\ &\leqslant \lim_{n\to+\infty} \frac{\sum_{i=1}^n 3}{2n} = \frac{3}{2}. \end{aligned}$$

□

Now go back to the proof of (iii). By Theorem 9.9, for any $s, t \in (3/4, 1)$ there exist infinitely many m's such that $\mathbf{H}^n(\mathbf{X}_s) \in E_n^s = D_{2m-1}$ and $\mathbf{H}^n(\mathbf{X}_t) \in E_n^t = D_{2m-1}$ with $n = m^2 + 1$. But as already seen before, given $\sigma > 0$, $\|\mathbf{X} - \mathbf{Z}\| < \sigma/2$ for all $\mathbf{X} \in D_{2m-1}$ and sufficiently large m. Thus, for all $\sigma > 0$ there exists an integer m such that $\|\mathbf{H}^n(\mathbf{X}_s) - \mathbf{H}^n(\mathbf{X}_t)\| < \sigma$ with $n = m^2 + 1$. Since σ is arbitrary, we have

$$\liminf_{n\to+\infty} \left\| \mathbf{H}^n(\mathbf{X}_s) - \mathbf{H}^n(\mathbf{X}_t) \right\| = 0.$$

Therefore, for any $\mathbf{X}, \mathbf{Y} \in S_0$,

$$\liminf_{n\to+\infty} \left\| \mathbf{F}^n(\mathbf{X}_s) - \mathbf{F}^n(\mathbf{X}_t) \right\| = 0,$$

thus the proof of (iii) has been completed. □

10. Discrete models in mathematical sociology

10.1. Introduction

In Section 4, we focused on several applications of dynamical system to mathematical economics. Here, we examine an application to sociology in recent studies. This theory is fairly different from other social science such as Walrasian economics or Quetelet's social physics, in the sense that in these classical science, each person is treated as an atom in physics, namely, as an existence without his/her personality. But in the recent work by GRANOVETTER [1978], entitled "Threshold Models of Collective Behavior", behaviors in a collective are simulated as reflections of various attitudes of individuals in it. To make it clear, let us quote his riot model in a simple case.

Mathematical formulation of Granovetter's riot model is as follows: Assign to each member of the group a personal "threshold" (ratio) in choosing one of two decisions (e.g., to take part in a certain riot or not). We have then a distribution depending on these threshold values. Let $r\%$ be the ratio of individuals who have joined in the riot by the time t. With this setting, we are to compute the proportion of people in the collective who will have participated in the riot by the time $t+1$.

Based on the definition of the threshold, this proportion is determined by the individuals whose thresholds are less than $r\%$. Therefore, we have the following discrete dynamical system:

$$r(t+1) = F\bigl(r(t)\bigr). \tag{10.1}$$

Here, $F(r(t))$ is the cumulative distribution of the thresholds. We can figure out the temporal changes of $r(t)$ by (10.1). In some cases, we have an equilibrium value. The following two cases are examples of the possible types of riot.

Suppose there are ten persons in a collective, and each one has personal threshold. In the case of group A (see Table 10.1), if a persons with threshold 0 happened to start an action, then the cumulative number counts 1, which gives rise to the participation of the persons with threshold 1 to the riot. Successively, as the cumulative number increases to 2, the persons with threshold 2 participate in the riot. In this way, all the people in this collective will eventually participate in this riot.

On the other hand, in group B (Table 10.2), the maximum number of the participants is 4, for the absence of persons with threshold 3 and 4.

TABLE 10.1

Group A										
Threshold	0	1	2	3	4	5	6	7	8	9
Persons	1	1	1	1	1	1	1	1	1	1
Cumulation	1	2	3	4	5	6	7	8	9	10

TABLE 10.2

	Group B									
Threshold	0	1	2	3	4	5	6	7	8	9
Persons	2	1	1	0	0	1	1	1	1	2
Cumulation	2	3	4	4	4	4	4	4	4	4

10.2. A threshold model of collective behavior for a hairstyle fashion

Let us discuss another threshold model. We examine Granovetter and Ishii's "Two Threshold Model". Inspired specially by Ishii's paper, we elaborate a model for the propagation of a vogue, say, a hairstyle fashion (see YAMAGUTI [1994]). In this case, we suppose that each member has a pair of thresholds (L_w, U_w): Here, L_w stands for the threshold defined in the previous section and is regarded in this model as the "lower threshold" whereas U_w stands for the "upper threshold", namely, the maximum proportion $P(t)$ (= the proportion adopting this fashion/the total population) exceeding which, i.e. when $P(t) > U_w$, the person finds the fashion ceased to be novel and therefore he/she abandons it.

How can we compute the new proportion at $t+1$, that is, $P(t+1)$? We figure out $P(t+1)$ from $P(t)$ using two-dimensional distribution of (L_w, U_w). First, we remark that the new population accepting the fashion consists of persons whose pair of thresholds (L_w, U_w) satisfy

$$L_w < P(t) < U_w.$$

Now, we assume a certain two-dimensional distribution of the pair (L_w, U_w). The inequality $L_w < U_w$ means that the distribution occupies the upper half of the square above the diagonal. The following two cases show that this distribution plays an important role in this dynamics.

CASE 10.1 (*Uniform distribution of* (L_w, U_w)). Fig. 10.1 shows (L_w, U_w) is uniformly distributed in the whole upper triangle. The rectangle $ABCD$ represents the set of (L_w, U_w) which satisfies $L_w < P(t) < U_w$. Then we get,

$$\begin{aligned} P(t+1) &= \frac{\text{the area of } ABCD}{\text{the area of the whole triangle}} \\ &= \frac{P(t)(1-P(t))}{1/2} = 2P(t)\bigl(1-P(t)\bigr). \end{aligned}$$

The limit $P(t)$ for $t \to +\infty$ exists and is equal to $1/2$, which is the eventual proportion of the population adopting the fashion.

CASE 10.2 (*Concentrated distribution*). Fig. 10.2 shows that the total population is limited to the triangle ABC (point A and C are the middle points of each side), with the uniform density on this triangle. Then the proportion of the individuals who satisfy

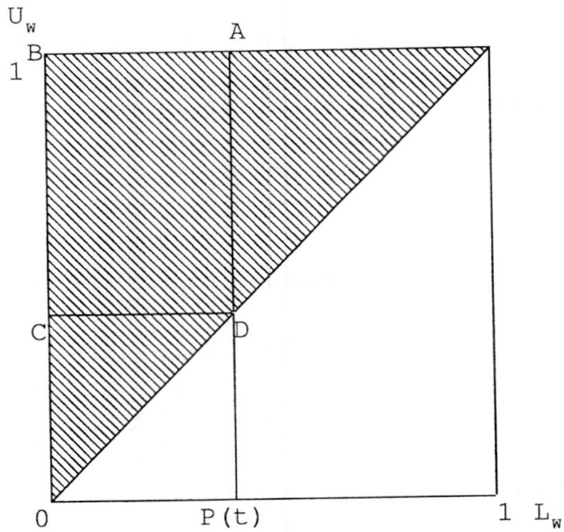

FIG. 10.1.

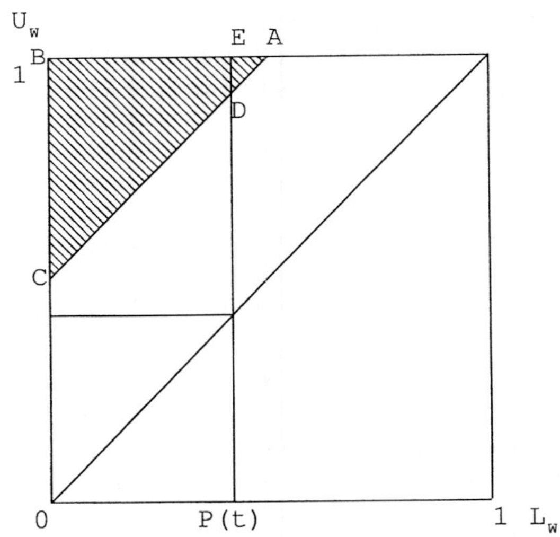

FIG. 10.2.

$L_w < P(t) < U_w$ is equal to the area $BCDE$. Then we get

$$\begin{aligned} P(t+1) &= \frac{\frac{1}{2}P(t)(1/2+1-P(t)-1/2)}{1/8} \\ &= 4P(t)\big(1-P(t)\big), \end{aligned}$$

which gives the famous chaotic dynamical system.

Finally, let us give a general formula for a given general distribution F of (L_w, U_w).

$$
\begin{aligned}
P(t+1) &= \frac{1}{A}\int_{P(t)}^{1}\int_{0}^{P(t)} F(L_w, U_w)\,\mathrm{d}L_w\,\mathrm{d}U_w \\
&= \frac{1}{A}\int_{0}^{1}\int_{0}^{P(t)} F(L_w, U_w)\,\mathrm{d}L_w\,\mathrm{d}U_w \\
&\quad - \frac{1}{A}\int_{0}^{P(t)}\int_{0}^{P(t)} F(L_w, U_w)\,\mathrm{d}L_w\,\mathrm{d}U_w,
\end{aligned}
$$

where

$$
\begin{aligned}
A &= \text{the total population} \\
&= \int_0^1\int_0^1 F(L_w, U_w)\,\mathrm{d}L_w\,\mathrm{d}U_w.
\end{aligned}
$$

Matsuda gave some examples of the cases of uniform distribution occupied in a triangle above the diagonal which is determined by the parameter a, as shown in Fig. 10.3. In this example, the total ration of the population is $\frac{1}{2}(1-a)^2$, and $P(t+1)$ is calculated by the following formulae with regard to the value of a:

(A) $0 < a < 1/2$.

$$
P(t+1) = \begin{cases}
\frac{P(t)(2-P(t)-2a)}{(1-a)^2} & (0 \leqslant P(t) \leqslant a), \\
\frac{2P(t)(1-P(t))-a^2}{(1-a)^2} & (a \leqslant P(t) \leqslant 1-a), \\
\frac{(P(t)-2a+1)(1-P(t))}{(1-a)^2} & (1-a \leqslant P(t) \leqslant 1).
\end{cases}
$$

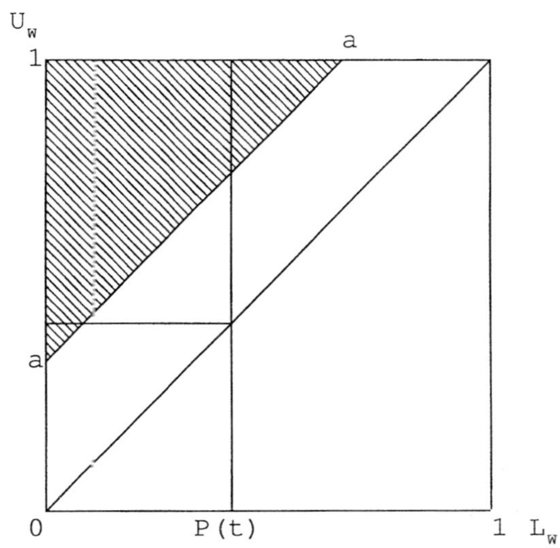

FIG. 10.3.

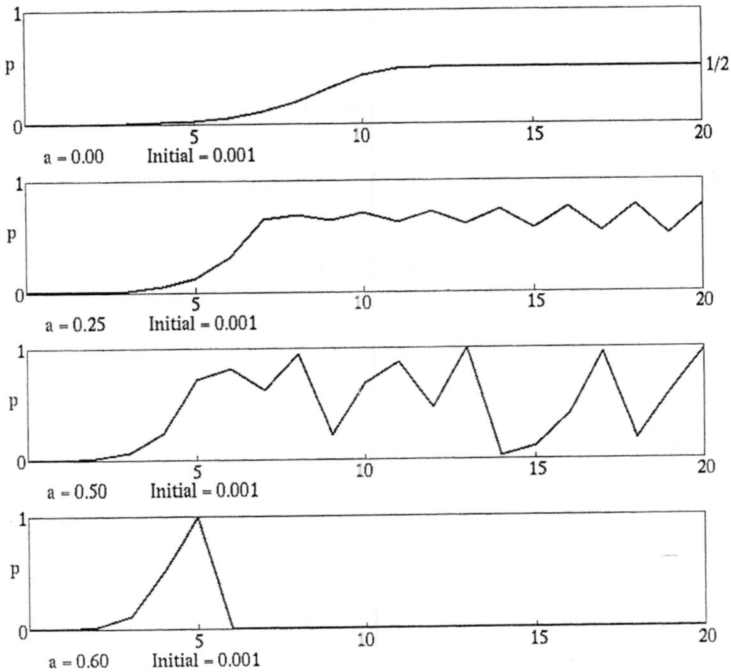

FIG. 10.4.

TABLE 10.3

Enquête Survey

If you are the member of a group of ten persons,
(1) *With how many others, you adopt the fashion?*
(2) *With how many others, you cease to follow this fashion?*

Number of person	(1) Adoption	(2) Stop of adoption
0		
1		
2		
3		
4		
5		
6		
7		
8		
9		

Please answer to the above questions by marking small circle.

(B) $1/2 \leqslant a < 1$.

$$P(t+1) = \begin{cases} \frac{P(t)(2-P(t)-2a)}{(1-a)^2} & (0 \leqslant P(t) \leqslant 1-a), \\ 1 & (1-a \leqslant P(t) \leqslant a), \\ \frac{(P(t)-2a+1)(1-P(t))}{(1-a)^2} & (a \leqslant P(t) \leqslant 1). \end{cases}$$

By changing the parameter a, we obtain various behaviors of the orbits of these dynamical systems. Fig. 10.4 shows some of them.

At the end of this section, let us explain a real experiment. We obtain a distribution of the pair of thresholds (L_w, U_w) through an enquête survey for a group of students. The following form given in Table 10.3 is a questionnaire table.

According to the results of this survey, we can obtain the discrete distribution and their dynamical systems.

11. Some fractal singular functions as the solutions of the boundary value problem

11.1. *Multigrid difference scheme*

Here we mean fractal singular function as a continuous function which has no finite derivative everywhere. For example, we have the Weierstrass function:

$$W_{a,b}(x) = \sum_{n=0}^{\infty} a^n \cos(b^n \pi x) \quad (ab \geqslant 1,\ 0 < a < 1,\ b > 0).$$

Also we have the Takagi function (see Fig. 11.1):

$$T(x) = \sum_{n=1}^{\infty} \frac{1}{2^n} \varphi^n(x),$$

where

$$\varphi(x) = \begin{cases} 2x & (0 \leqslant x \leqslant 1/2), \\ 2(1-x) & (1/2 \leqslant x \leqslant 1). \end{cases}$$

These functions are continuous and yet they have nowhere finite derivatives in $[0, 1]$. There exist many other similar functions.

Although in the first half of the twentieth century these functions were treated as pathological ones, some of these functions have recently been proved to be the solutions of boundary problems for certain multigrid difference scheme.

Let us explain this by a simple example. The Takagi function $T(x)$ is known as a unique bounded solution of the following functional equation:

$$T(x) = \tfrac{1}{2} T\big(\varphi(x)\big) + \frac{\varphi(x)}{2}.$$

Using the definition of $\varphi(x)$, we rewrite as

$$\begin{cases} T(x) = \frac{1}{2} T(2x) + x & (0 \leqslant x \leqslant 1/2), \\ T(x) = \frac{1}{2} T(2(1-x)) + 1 - x & (1/2 \leqslant x \leqslant 1). \end{cases}$$

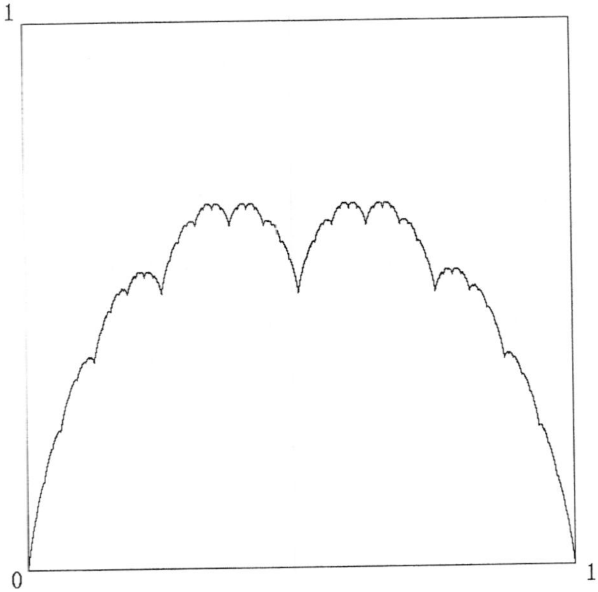

FIG. 11.1.

And successively using the Schauder expansion of continuous function, we get

$$T\left(\frac{2i+1}{2^{k+1}}\right)=\frac{1}{2}\left\{T\left(\frac{i}{2^k}\right)+T\left(\frac{i+1}{2^k}\right)\right\}+\frac{1}{2^{k+1}}, \tag{11.1}$$

for any non negative integer k and $0 \leqslant i \leqslant 2^k - 1$. Note that

$$T(0)=T(1)=0. \tag{11.2}$$

This series of equations are a generalization of the classical boundary value problem called the Dirichlet problem:

$$\frac{\mathrm{d}^2 y}{\mathrm{d}x^2}=-2 \quad (y(0)=y(1)=0),$$

whose solution is $y(x)=x(1-x)$. Through this process, we can say that the Takagi function $T(x)$ is the unique solution of the discrete Dirichlet problem of (11.1) with the boundary conditions (11.2). In fact, we can show that the solution is unique at the points $i/2^k$ by Eq. (11.1) with the boundary values. Then the uniqueness of the solution follows from the fact that the set of binary rational points is dense in [0, 1].

11.2. Lebesgue's singular function and the Takagi function

Another example is Lebesgue's singular function $L_\alpha(x)$ (see Fig. 11.2). That function is a continuous, strictly monotone increasing and has almost everywhere derivative zero.

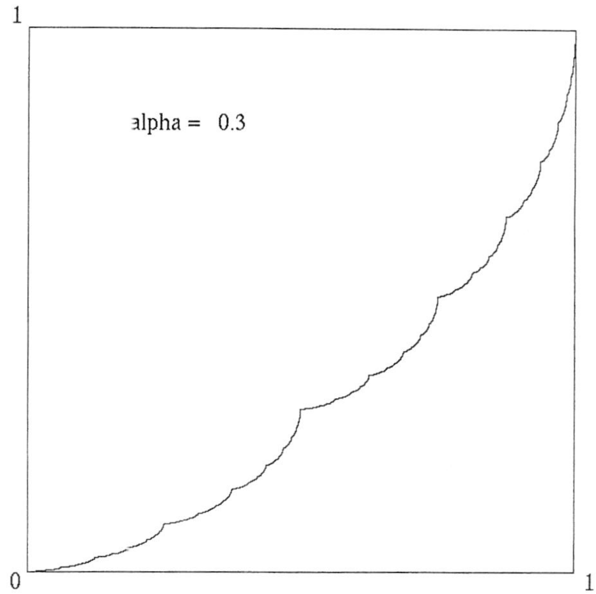

FIG. 11.2.

This $L_\alpha(x)$ $(\alpha \neq 1/2)$ satisfies

$$L_\alpha\left(\frac{2i+1}{2^{k+1}}\right) = (1-\alpha)L_\alpha\left(\frac{i}{2^k}\right) + \alpha L_\alpha\left(\frac{i+1}{2^k}\right),$$

for all non negative integer k and $0 \leqslant i \leqslant 2^k - 1$. The boundary conditions are

$$L_\alpha(0) = L_\alpha(1) = 1.$$

We find that there exists a beautiful relation between $L_\alpha(x)$ and $T(x)$. Namely, the next relation is shown:

$$\left.\frac{\partial L_\alpha(x)}{\partial \alpha}\right|_{\alpha=1/2} = 2T(x).$$

For the detail, see the article of YAMAGUTI and HATA [1984] and that of YAMAGUTI, HATA and KIGAMI [1997], pp. 33–46.

References

ARROW, K.J., HAHN, F.H. (1971). *General Competitive Analysis* (Holden Day).

BOOLE, G. (1860). *A Treatise on the Calculus of Finite Difference* (Macmillan, Cambridge).

FUJITA, H., UTIDA, S. (1953). The effect of population density on the growth of an animal population. *Ecology* **34** (3), 488–498.

GRANOVETTER, M. (1978). Threshold models of collective behavior. *Amer. J. Soci.* **83** (6), 1422–1442.

HATA, M. (1982). Euler's finite difference scheme and chaos in R^n. *Proc. Japan Acad. Ser. A* **58**, 178–180.

HICKS, J.R. (1946). *Value and Capital*, 2nd edn. (Oxford University Press, Oxford).

KAIZOUJI, T. (1994). Multiple equilibria and chaotic tâtonnement: Applications of the Yamaguti–Matano theorem. *J. Economic Behavior Organization* **24**, 357–362.

KEMENY, J.G., SNELL, J.L. (1963). *Mathematical Models in the Social Sciences* (Blaisdell Publishing Company, New York).

LI, T.Y., YORKE, J.A. (1975). Period three implies chaos. *Amer. Math. Monthly* **82**, 985–992.

LORENZ, E.N. (1963). Deterministic nonperiodic flow. *J. Atmospheric Sci.* **20**, 130.

MAEDA, Y. (1995). Euler's discretization revisited. *Proc. Japan Acad. Ser. A* **71** (3), 58–61.

MAEDA, Y. (1996). Euler's discretization of ordinary differential equation with symmetric non-linearity. *Japan J. Industrial Appl. Math.* **15** (1), 1–6.

MAEDA, Y. (1998). Euler's discretization of ordinary differential equation and chaos. Doctor Thesis of Ryukoku University.

MAEDA, Y., YAMAGUTI, M. (1996). Discretization of non-Lipschitz continuous O.D.E. and chaos. *Proc. Japan Acad. Ser. A* **72** (2), 43–45.

MAROTTO, F.R. (1978). Snap-back repellers imply chaos in R^n. *J. Math. Anal. Appl.* **63**, 199–223.

MAY, R. (1974). Biological populations with non overlapping generation: Stable points, stable cycles, and chaos. *Science* **186**, 645–647.

MYRBERG, P.J. (1962). Sur l'Iteration des Polynomes Reels Quadratiques. *J. Math. Pures Appl.* **41**, 339–351.

NEGISHI, T. (1962). The stability of a competitive economy: A survey article. *Econometrica* **30** (4), 635–669.

OSHIME, Y. (1987). On modified Euler's scheme, Wakayama–Daigaku keizaigakkai, Keizairiron (in Japanese).

SAARI, D.G. (1985). Iterative price mechanisms. *Econometrica* **53**, 1117–1131.

SHARKOVSKII, A.N. (1964). Coexistence of cycles of a continuous map of a line into itself. *Ukrainian Math. J.* **16**, 61–71.

USHIKI, S. (1982). Central difference scheme and chaos. *Physica D.*

USHIKI, S., YAMAGUTI, M., MATANO, H. (1980). Discrete population models and chaos. *Lecture Notes in Numerical Anal.* **2**, 1–25.

UTIDA, S. (1941). Studies on experimental population of the Azuki Bean Weevil, Callosobruchus Chinensis. I. The effect of population density on the progeny population. *Mem. Coll. Agr., Kyoto Imp. Univ.* **48**, 1–30.

VERHULST, P.F. (1838). Notice sur la loi que la population suit dans son accroisement. *Corr. Math. Phys.* **10**, 113.

WALRAS, L. (1926). Elements d'Economie Politique Pure, Paris et Lausanne.

YAMAGUTI, M. (1994). Discrete models in social sciences. *Comput. Math. Appl.* **28** (10–12), 263–267.

YAMAGUTI, M., HATA, M. (1984). Takagi function and its generalization. *Japan J. Appl. Math.* **1**, 186–199.

YAMAGUTI, M., HATA, M., KIGAMI, J. (1997). *Mathematics of Fractal* (Amer. Math. Soc., Providence, RI).

YAMAGUTI, M., MAEDA, Y. (1996). On the discretization of O.D.E.. *Z. Angew. Math. Mech.* **76** (4), 217–219.

YAMAGUTI, M., MATANO, H. (1979). Euler's finite difference scheme and chaos. *Proc. Japan Acad. Ser. A* **55**, 78–80.

YAMAGUTI, M., USHIKI, S. (1981). Chaos in numerical analysis of ordinary differential equation. *Physica D*.

Introduction to Partial Differential Equations and Variational Formulations in Image Processing

Guillermo Sapiro

Electrical and Computer Engineering, University of Minnesota,
Minneapolis, MN 55455, USA
e-mail: guille@ece.umn.edu

1. Introduction

The use of partial differential equations (PDEs) and curvature driven flows in image analysis has become an interest raising research topic in the past few years. The basic idea is to deform a given curve, surface, or image with a PDE, and obtain the desired result as the solution of this PDE. Sometimes, as in the case of color images, a system of coupled PDEs is used. The art behind this technique is in the design and analysis of these PDEs.

Partial differential equations can be obtained from variational problems. Assume a variational approach to an image processing problem formulated as

$$\arg\left\{\mathrm{Min}_I\, \mathcal{U}(u)\right\},$$

where $\mathcal{U}$ is a given energy computed over the image (or surface) I. Let $\mathcal{F}(\Phi)$ denote the Euler derivative (first variation) of $\mathcal{U}$. Since under general assumptions, a necessary condition for I to be a minimizer of $\mathcal{U}$ is that $\mathcal{F}(I) = 0$, the (local) minima may be computed via the steady state solution of the equation

$$\frac{\partial I}{\partial t} = \mathcal{F}(I),$$

Foundations of Computational Mathematics
Special Volume (F. Cucker, Guest Editor) of
HANDBOOK OF NUMERICAL ANALYSIS, VOL. XI
P.G. Ciarlet (Editor)

where t is an 'artificial' time marching parameter. PDEs obtained in this way have been used already for quite some time in computer vision and image processing, and the literature is large. The most classical example is the Dirichlet integral,

$$\mathcal{U}(I) = \int |\nabla I|^2(x)\,dx,$$

which is associated with the linear heat equation

$$\frac{\partial I}{\partial t}(t, x) = \Delta I(x).$$

More recently, extensive research is being done on the direct derivation of evolution equations which are not necessarily obtained from the energy approaches. Both types of PDEs are studied in this chapter.

Ideas on the use of PDEs in image processing go back at least to GABOR [1965], and a bit more recently, to JAIN [1977]. However, the field really took off thanks to the independent works by KOENDERINK [1984] and WITKIN [1983]. These researchers rigorously introduced the notion of *scale-space*, that is, the representation of images simultaneously at multiple scales. Their seminal contribution is to a large extent the basis of most of the research in PDEs for image processing. In their work, the multi-scale image representation is obtained by Gaussian filtering. This is equivalent to deforming the original image via the classical heat equation, obtaining in this way an isotropic diffusion flow. In the late 80s, HUMMEL [1986] noted that the heat flow is not the only parabolic PDE that can be used to create a scale-space, and indeed argued that an evolution equation which satisfies the maximum principle will define a scale-space as well. Maximum principle appears to be a natural mathematical translation of *causality*. Koenderink once again made a major contribution into the PDEs arena (this time probably involuntarily, since the consequences were not clear at all in his original formulation), when he suggested to add a thresholding operation to the process of Gaussian filtering. As later suggested by Osher and his colleagues, and proved by a number of groups, this leads to a geometric PDE, actually, one of the most famous ones, curvature motion.

PERONA and MALIK [1990] work on anisotropic diffusion has been one of the most influential papers in the area. They proposed to replace Gaussian smoothing, equivalent to isotropic diffusion via the heat flow, by a selective diffusion that preserves edges. Their work opened a number of theoretical and practical questions that continue to occupy the PDE image processing community, see, e.g., ALVAREZ, LIONS and MOREL [1992], ROMENY [1994]. In the same framework, the seminal works of OSHER and RUDIN [1990] on shock filters and RUDIN, OSHER and FATEMI [1992b] on Total Variation decreasing methods explicitly stated the importance and the need for understanding PDEs for image processing applications. We should also point out that about at the same time PRICE, WAMBACQ and OOSTERLINK published a very interesting paper [1990] on the use of Turing's reaction–diffusion theory for a number of image processing problems. Reaction–diffusion equations were also suggested to create patterns (TURK [1991], WITKIN and KASS [1991]).

Many of the PDEs used in image processing and computer vision are based on moving curves and surfaces with curvature based velocities. In this area, the level-set numerical method developed by OSHER and SETHIAN [1988] was very influential and crucial. Early developments on this idea were provided in OHTA, JASNOW and KAWASAKI [1982], and their equations were first suggested for shape analysis in computer vision in KIMIA, TANNENBAUM and ZUCKER [1990]. The basic idea is to represent the deforming curve, surface, or image, as the level-set of a higher dimensional hypersurface. This technique, not only provides more accurate numerical implementations, but also solves topological issues which were very difficult to treat before. The representation of objects as level-sets (zero-sets) is of course not completely new to the computer vision and image processing communities, since it is one of the fundamental techniques in mathematical morphology (SERRA [1988]). Considering the image itself as a collection of its level-sets, and not just as the level-set of a higher dimensional function, is a key concept in the PDEs community as shown in the very influential and seminal paper by ALVAREZ, GUICHARD, LIONS and MOREL [1993].

Other works, like the segmentation approach of MUMFORD and SHAH [1989] and the snakes of Kass, Witkin, and Terzopoulos have been very influential in the PDEs community as well.

It should be noted that a number of the above approaches rely quite heavily on a large number of mathematical advances in differential geometry for curve evolution (GRAYSON [1987]) and in viscosity solutions theory for curvature motion (see, e.g., EVANS and SPRUCK [1991].)

In this chapter we cover three applications of this framework: Segmentation, scalar image denoising, and contrast enhancement. These are just three classic and important applications, and much more has already been done in this area. We expect the interested reader to be inspired by this short description and search for additional material for these and other applications. Important sources of literature are the excellent collection of papers in the book edited by ROMENY [1994], the book by GUICHARD and MOREL [1995] that contains an outstanding description of the topic from the point of view of iterated infinitesimal filters, SETHIAN'S book [1996c] on level-sets covers in a very readable and comprehensive form level-sets, Osher's long-time expected book (until then see the review paper in OSHER [website]), LINDEBERG'S book [1994] on a classic in Scale-Space theory, WEICKERT'S book [1998] on anisotropic diffusion in image processing, KIMMEL'S lecture notes [1999], TOGA'S book [1998] on Brain Warping that includes a number of PDEs based algorithms for this, the special issue on the March 1998 issue of the IEEE Transactions on Image Processing (CASELLES, MOREL, SAPIRO and TANNENBAUM [1998]), the special issues in the Journal of Visual Communication and Image Representation, a series of Special Sessions at a number of IEEE International Conference on Image Processing (ICIP), and the Proceedings of the Scale Space Workshop. Finally, my recent published book also contains additional material (SAPIRO [2001]) and a considerably large bibliography.

2. Geodesic curves and minimal surfaces

In this section we show how a number of problems in image processing and computer vision can be formulated as the computation of paths or surfaces of minimal energy. We start with the basic formulation, connecting classical work on segmentation with the computation of geodesic curves in 2D. We then extend this work to three dimensions, and show the application of this framework to object tracking (stereo using this framework has been addressed in DERICHE, BOUVIN and FAUGERAS [1996], FAUGERAS and KERIVEN [1998]). The geodesic or minimal surface is computed via geometric PDEs, obtained from gradient descent flows. These flows are driven by intrinsic curvatures as well as forces that are derived from the image.

2.1. Basic two dimensional derivation

Since original work by KASS, WITKIN and TERZOPOULOS [1988], extensive research was done on "snakes" or active contour models for boundary detection. The classical approach is based on deforming an initial contour $\mathcal{C}_0$ towards the boundary of the object to be detected. The deformation is obtained by trying to minimize a functional designed so that its (local) minimum is obtained at the boundary of the object. These active contours are examples of the general technique of matching deformable models to image data by means of energy minimization (BLAKE and ZISSERMAN [1987], TERZOPOULOS, WITKIN and KASS [1988]). The energy functional is basically composed of two components, one controls the smoothness of the curve and another attracts the curve towards the boundary. This energy model is not capable of handling changes in the topology of the evolving contour when direct implementations are performed. Therefore, the topology of the final curve will be as the one of $\mathcal{C}_0$ (the initial curve), unless special procedures, many times heuristic, are implemented for detecting possible splitting and merging (MCINERNEY and TERZOPOULOS [1995], SZELISKI, TONNESEN and TERZOPOULOS [1993]). This is a problem when an unknown number of objects must be simultaneously detected. This approach is also nonintrinsic, since the energy depends on the parametrization of the curve and is not directly related to the objects geometry.

As we show in this section, a kind of "re-interpretation" of this model solves these problems and presents a new paradigm in image processing: The formulation of image processing problems as the search for geodesics curves or minimal surfaces. A particular case of the classical energy snakes model is proved to be equivalent to finding a geodesic curve in a Riemannian space with a metric derived from the image content. This means that in a certain framework, boundary detection can be considered equivalent to finding a curve of minimal weighted length. This interpretation gives a new approach for boundary detection via active contours, based on geodesic or local minimal distance computations. We also show that the solution to the geodesic flow exists in the viscosity framework, and is unique and stable. Consistency of the model can be obtained as well, showing that the geodesic curve converges to the right solution in the case of ideal objects. A number of examples of real images, showing the above properties, are presented.

Geodesic active contours

Energy based active contours. Let us briefly describe the classical energy based snakes. Let $\mathcal{C}(p):[0,1] \to \mathbb{R}^2$ be a parametrized planar curve and let $I:[0,a] \times [0,b] \to \mathbb{R}^+$ be a given image in which we want to detect the objects boundaries. The classical snakes approach (KASS, WITKIN and TERZOPOULOS [1988]) associates the curve $\mathcal{C}$ with an energy given by

$$E(\mathcal{C}) = \alpha \int_0^1 |\mathcal{C}'(p)|^2\, dp + \beta \int_0^1 |\mathcal{C}''(p)|^2\, dq - \lambda \int_0^1 |\nabla I(\mathcal{C}(p))|\, dp, \tag{2.1}$$

where α, β, and λ are real positive constants. The first two terms control the smoothness of the contours to be detected (internal energy), while the third term is responsible for attracting the contour towards the object in the image (external energy). Solving the problem of snakes amounts to finding, for a given set of constants α, β, and λ, the curve $\mathcal{C}$ that minimizes E. Note that when considering more than one object in the image, for instance for an initial prediction of $\mathcal{C}$ surrounding all of them, it is not possible to detect all the objects. Special topology-handling procedures must be added. Actually, the solution without those special procedures will be in most cases a curve which approaches a convex hull type figure of the objects in the image. In other words, the classical (energy) approach of snakes cannot directly deal with changes in topology. The topology of the initial curve will be the same as the one of the, possibly wrong, final curve. The model derived below, as well as the curve evolution models in CASELLES, CATTE, COLL and DIBOS [1993], MALLADI, SETHIAN and VEMURI [1994], MALLADI, SETHIAN and VEMURI [1995], MALLADI, SETHIAN and VEMURI [to appear], overcomes this problem.

Another possible problem of the energy based models is the need to select the parameters that control the trade-off between smoothness and proximity to the object. Let us consider a particular class of snakes model where the rigidity coefficient is set to zero, that is, $\beta = 0$. Two main reasons motivate this selections, which at least mathematically restricts the general model (2.1). First, this selection will allow us to derive the relation between this energy based active contours and geometric curve evolution ones. Second, the regularization effect on the geodesic active contours comes from curvature based curve flows, obtained only from the other terms in (2.1) (see Eq. (2.16) and its interpretation after it). This will allow to achieve smooth curves in the proposed approach without having the high order smoothness given by $\beta \neq 0$ in energy-based approaches. Moreover, the second order smoothness component in (2.1), assuming an arc-length parametrization, appears in order to minimize the total squared curvature (curve known as "elastica"). It is easy to prove that the curvature flow used in the new approach and presented below decreases the total curvature (ANGENENT [1991]). The use of the curvature driven curve motions as smoothing term was proved to be very efficient in previous literature (ALVAREZ, GUICHARD, LIONS and MOREL [1993], CASELLES, CATTE, COLL and DIBOS [1993], KIMIA, TANNENBAUM and ZUCKER [1995], MALLADI, SETHIAN and VEMURI [1994], MALLADI, SETHIAN and VEMURI [1995], MALLADI, SETHIAN and VEMURI [to appear], NIESSEN, TER HAAR ROMENY, FLORACK and SALDEN [1993], SAPIRO and TANNENBAUM [1994]), and is also supported by the experiments in this section. Therefore, curve smoothing will be obtained also with $\beta = 0$, having only the first regularization term. Assuming this (2.1) reduces to

$$E(\mathcal{C}) = \alpha \int_0^1 |\mathcal{C}'(p)|^2\, dp - \lambda \int_0^1 |\nabla I(\mathcal{C}(p))|\, dp. \tag{2.2}$$

Observe that by minimizing the functional (2.2), we are trying to locate the curve at the points of maxima $|\nabla I|$ (acting as "edge detector"), while keeping certain smoothness in the curve (object boundary). This is actually the goal in the general formulation (2.1) as well. The tradeoff between edge proximity and edge smoothness is played by the free parameters in the above equations.

Eq. (2.2) can be extended generalizing the edge detector part in the following way: Let $g : [0, +\infty[\to \mathbb{R}^+$ be a strictly decreasing function such that $g(r) \to 0$ as $r \to \infty$. Hence, $-|\nabla I|$ can be replaced by $g(|\nabla I|)^2$, obtaining a general energy functional given by

$$\begin{aligned} E(\mathcal{C}) &= \alpha \int_0^1 \left|\mathcal{C}'(p)\right|^2 dp + \lambda \int_0^1 g\big(\left|\nabla I\big(\mathcal{C}(p)\big)\right|\big)^2 dp \\ &= \int_0^1 \big(E_{\text{int}}\big(\mathcal{C}(p)\big) + E_{\text{ext}}\big(\mathcal{C}(p)\big)\big)\, dp. \end{aligned} \tag{2.3}$$

The goal now is to minimize E in (2.3) for $\mathcal{C}$ in a certain allowed space of curves. (In order to simplify the notation, we will sometimes write $g(I)$ or $g(\mathcal{X})$ ($\mathcal{X} \in \mathbb{R}^2$) instead of $g(|\nabla I|)$.) Note that in the above energy functional, only the ratio λ/α counts. The geodesic active contours will be derived from (2.3).

The functional in (2.3) is not intrinsic since it depends on the parametrization q that until now is arbitrary. This is an undesirable property, since parametrizations are not related to the geometry of the curve (or object boundary), but only to the velocity they are traveled. Therefore, it is not natural for an object detection problem to depend on the parametrization of the representation.

The geodesic curve flow. We now proceed and show that the solution of the particular energy snakes model (2.3) is given by a geodesic curve in a Riemannian space induced from the image I (a geodesic curve is a (local) minimal distance path between given points). To show this, we use the classical Maupertuis' Principle (DUBROVIN, FOMENKO and NOVIKOV [1984]) from dynamical systems. Giving all the background on this principle is beyond the scope of this chapter, so we will restrict the presentation to essential points and geometric interpretation. Let us define

$$\mathcal{U}(\mathcal{C}) := -\lambda g\big(\left|\nabla I(\mathcal{C})\right|\big)^2,$$

and write $\alpha = m/2$. Therefore,

$$E(\mathcal{C}) = \int_0^1 \mathcal{L}\big(\mathcal{C}(p)\big)\, dp,$$

where $\mathcal{L}$ is the Lagrangian given by

$$\mathcal{L}(\mathcal{C}) := \frac{m}{2}\left|\mathcal{C}'\right|^2 - \mathcal{U}(\mathcal{C}).$$

The Hamiltonian (DUBROVIN, FOMENKO and NOVIKOV [1984]) is then given by

$$H = \frac{q^2}{2m} + \mathcal{U}(\mathcal{C}),$$

where $q := mC'$. In order to show the relation between the energy minimization problem (2.3) and geodesic computations, we will need the following theorem (DUBROVIN, FOMENKO and NOVIKOV [1984]):

THEOREM 2.1 (Maupertuis' principle). *Curves $C(p)$ in Euclidean space which are extremal corresponding to the Hamiltonian $H = \frac{q^2}{2m} + \mathcal{U}(C)$, and have a fixed energy level E_0 (law of conservation of energy), are geodesics, with nonnatural parameter, with respect to the new metric* $(i, j = 1, 2)$

$$g_{ij} = 2m\big(E_0 - \mathcal{U}(C)\big)\delta_{ij}.$$

This classical theorem explains, among other things, when an energy minimization problem is equivalent to finding a geodesic curve in a Riemannian space. That means, when the solution to the energy problem is given by a curve of minimal "weighted distance" between given points. Distance is measured in the given Riemannian space with the first fundamental form g_{ij} (the first fundamental form defines the metric or distance measurement in the space). See the mentioned references (specially Section 3.3 in DUBROVIN, FOMENKO and NOVIKOV [1984]) for details on the theorem and the corresponding background in Riemannian geometry. According to the above result, minimizing $E(C)$ as in (2.3) with $H = E_0$ (conservation of energy) is equivalent to minimizing

$$\int_0^1 \sqrt{g_{ij}C'_iC'_j}\,dp \tag{2.4}$$

or $(i, j = 1, 2)$

$$\int_0^1 \sqrt{g_{11}C'^2_1 + 2g_{12}C'_1C'_2 + g_{22}C'^2_2}\,dp, \tag{2.5}$$

where $(C_1, C_2) = C$ (components of C), $g_{ij} = 2m(E_0 - \mathcal{U}(C))\delta_{ij}$.

We have just transformed the minimization (2.3) into the energy (2.5). As we see from the definition of g_{ij}, Eq. (2.5) has a free parameter, E_0. We deal now with this energy. Based on Fermat's Principle, we we will motivate the selection of the value of E_0. We then present an intuitive approach that brings us to the same selection.

Fixing the ratio λ/α, the search for the path minimizing (2.3) may be considered as a search for a path in the (x, y, p) space, indicating the nonintrinsic nature of this minimization problem. The Maupertuis Principle of least action used to derive (2.5) presents a purely geometric principle describing the orbits of the minimizing paths (BORN and WOLF [1986]). In other words, it is possible using the above theorem to find the projection of the minimizing path of (2.3) in the (x, y, p) space onto the (x, y) plane by solving an intrinsic problem. Observe that the parametrization along the path is yet to be determined after its orbit is tracked. The intrinsic problem of finding the projection of the minimizing path depends on a single free parameter E_0 incorporating the parametrization as well as λ and α ($E_0 = E_{\text{int}} - E_{\text{ext}} = \alpha|C'(p)|^2 - \lambda g(C(p))^2$).

The question to be asked is whether the problem in hand should be regarded as the behavior of springs and mass points leading to the nonintrinsic model (2.3). We shall take one step further, moving from springs to light rays, and use the following result from optics to motivate the proposed model (BORN and WOLF [1986], DUBROVIN, FOMENKO and NOVIKOV [1984]):

THEOREM 2.2 (Fermat's Principle). *In an isotropic medium the paths taken by light rays in passing from a point A to a point B are extrema corresponding to the traversal-time (as action). Such paths are geodesics with respect to the new metric* $(i, j = 1, 2)$

$$g_{ij} = \frac{1}{c^2(\mathcal{X})}\delta_{ij}.$$

$c(\mathcal{X})$ in the above equation corresponds to the speed of light at $\mathcal{X}$. Fermat's Principle defines the Riemannian metric for light waves. We define $c(\mathcal{X}) = 1/g(\mathcal{X})$ where "high speed of light" corresponds to the presence of an edge, while "low speed of light" corresponds to a nonedge area. The result is equivalent then to minimizing the intrinsic problem

$$\int_0^1 g\big(|\nabla I(\mathcal{C}(p))|\big)|\mathcal{C}'(p)|\,dp, \tag{2.6}$$

which is the same formulation as in (2.5), having selected $E_0 = 0$.

We shall return for a while to the energy model (2.3) to further explain the selection of $E_0 = 0$ from the point of view of object detection. As was explained before, in order to have a completely closed form for boundary detection via (2.5), we have to select E_0. It was shown that selecting E_0 is equivalent to fixing the free parameters in (2.3) (i.e., the parametrization and λ/α). Note that by Theorem 2.1, the interpretation of the snakes model (2.3) for object detection as a geodesic computation is valid for any value of E_0. The value of E_0 is selected to be zero from now on, which means that $E_{\text{int}} = E_{\text{ext}}$ in (2.3). This selection simplifies the notation and clarifies the relation of Theorem 2.1 and energy-snakes with (geometric) curve evolution active contours that results form Theorems 2.1 and 2.2. At an ideal edge, E_{ext} in (2.3) is expected to be zero, since $|\nabla I| = \infty$ and $g(r) \to 0$ as $r \to \infty$. Then, the ideal goal is to send the edges to the zeros of g. Ideally we should try as well to send the internal energy to zero. Since images are not formed by ideal edges, we choose to make equal contributions of both energy components. This choice, which coincides with the one obtained from Fermat's Principle and as said before allows to show the connection with curve evolution active contours, is also consistent with the fact that when looking for an edge, we may travel along the curve with arbitrarily slow velocity (given by the parametrization q, see equations obtained with the above change of parametrization). More comments on different selections of E_0, as well as formulas corresponding to $E_0 \neq 0$, are given in our extended papers.

Therefore, with $E_0 = 0$, and $g_{ij} = 2m\lambda g(|\nabla I(\mathcal{C})|)^2\delta_{ij}$, Eq. (2.4) becomes

$$\min \int_0^1 \sqrt{2m\lambda}\, g\big(|\nabla I(\mathcal{C}(p))|\big)|\mathcal{C}'(p)|\,dp. \tag{2.7}$$

Since the parameters above are constants, without loss of generality we can set now $2\lambda m = 1$ to obtain

$$\min \int_0^1 g\big(\big|\nabla I\big(\mathcal{C}(p)\big)\big|\big)\big|\mathcal{C}'(p)\big|\,dp. \tag{2.8}$$

We have transformed the problem of minimizing (2.3) into a problem of geodesic computation in a Riemannian space, according to a new metric.

Let us, based on the above theory, give the above expression a further geodesic curve interpretation from a slightly different perspective. The Euclidean length of the curve $\mathcal{C}$ is given by

$$L := \oint \big|\mathcal{C}'(p)\big|\,dp = \oint ds, \tag{2.9}$$

where ds is the Euclidean arc-length (or Euclidean metric). The flow

$$\mathcal{C}_t = \kappa\vec{\mathcal{N}}, \tag{2.10}$$

where again κ is the Euclidean curvature, gives the fastest way to reduce L, that is, moves the curve in the direction of the gradient of the functional L. Looking now at (2.8), a new length definition in a different Riemannian space is given,

$$L_R := \int_0^1 g\big(\big|\nabla I\big(\mathcal{C}(p)\big)\big|\big)\big|\mathcal{C}'(p)\big|\,dp. \tag{2.11}$$

Since $|\mathcal{C}'(p)|\,dp = ds$, we obtain

$$L_R := \int_0^{L(\mathcal{C})} g\big(\big|\nabla I\big(\mathcal{C}(s)\big)\big|\big)\,ds. \tag{2.12}$$

Comparing this with the classical length definition as given in (2.9), we observe that the new length is obtained by weighting the Euclidean element of length ds by $g(|\nabla I(\mathcal{C}(s)|)$, which contains information regarding the boundary of the object. Therefore, when trying to detect an object, we are not just interested in finding the path of minimal classical length ($\oint ds$) but the one that minimizes a new length definition which takes into account image characteristics. Note that (2.8) is general, besides being a positive decreasing function, no assumptions on g were made. Therefore, the theory of boundary detection based on geodesic computations given above can be applied to any general "edge detector" functions g. Recall that (2.8) was obtained from the particular case of energy based snakes (2.3) using Maupertuis' Principle, which helps to identify variational approaches that are equivalent to computing paths of minimal length in a new metric space.

In order to minimize (2.8) (or L_R) we search for the gradient descent direction of (2.8), which is a way of minimizing L_R via the steepest-descent method. Therefore, we need to compute the Euler–Lagrange of (2.8). Details on this computation are given in our papers. Thus, according to the steepest-descent method, to deform the initial curve $\mathcal{C}(0) = \mathcal{C}_0$ towards a (local) minima of L_R, we should follow the curve evolution equation (compare with (2.10))

$$\frac{\partial \mathcal{C}(t)}{\partial t} = g(I)\,\kappa\vec{\mathcal{N}} - (\nabla g \cdot \vec{\mathcal{N}})\vec{\mathcal{N}}, \tag{2.13}$$

where κ is the Euclidean curvature as before, $\vec{\mathcal{N}}$ is the unit inward normal, and the right hand side of the equation is given by the Euler–Lagrange of (2.8) as derived in our extended reports. This equation shows how each point in the active contour $\mathcal{C}$ should move in order to decrease the length L_R. The detected object is then given by the steady state solution of (2.13), that is $\mathcal{C}_t = 0$.

To summarize, Eq. (2.13) presents a curve evolution flow that minimizes the weighted length L_R, which was derived from the classical snakes case (2.3) via Maupertuis' Principle of least action. This is the basic geodesic curve flow we propose for object detection (the full model is presented below). In the following section we embed this flow in a level-set formulation in order to complete the model and show its connection with previous curve evolution active contours. This embedding will also help to present theoretical results regarding the existence of the solution of (2.13). We note that minimization of a normalized version of (2.12) was proposed in FUA and LECLERC [1990] from a different perspective, leading to a different geometric method.

The level-sets geodesic flow: Derivation. In order to find the geodesic curve, we computed the corresponding steepest-descent flow of (2.8), Eq. (2.13). Eq. (2.13) is represented using the level-sets approach (OSHER and SETHIAN [1988]). Assume that the curve $\mathcal{C}$ is a level-set of a function $u : [0, a] \times [0, b] \to \mathbb{R}$. That is, $\mathcal{C}$ coincides with the set of points $u =$ constant (e.g., $u = 0$). u is therefore an implicit representation of the curve $\mathcal{C}$. This representation is parameter free, then intrinsic. The parametrization is also topology free since different topologies of the zero level-set do not imply different topologies of u. If a curve $\mathcal{C}$ evolves according to

$$\mathcal{C}_t = \beta \vec{\mathcal{N}}$$

for a given function β, then the embedding function u should deform according to

$$u_t = \beta |\nabla u|,$$

where β is computed on the level-sets. Embedding the evolution of $\mathcal{C}$ in that of u, topological changes of $\mathcal{C}(t)$ are handled automatically and accuracy and stability are achieved using the proper numerical algorithm (OSHER and SETHIAN [1988]). This level-set representation was formally analyzed in CHEN, GIGA and GOTO [1991], EVANS and SPRUCK [1991], SONER [1993], proving for example that in the viscosity framework, the solution is independent of the embedding function u for a number of velocities β (see Section 2.1 as well as Theorem 5.6 and Theorem 7.1 in CHEN, GIGA and GOTO [1991]). In our case u is initiated to be the signed distance function. Therefore, based on (2.8) and embedding (2.13) in u, we obtain that solving the geodesic problem is equivalent to searching for the steady state solution ($\frac{\partial u}{\partial t} = 0$) of the following evolution equation ($u(0, \mathcal{C}) = u_0(\mathcal{C})$):

$$\begin{aligned}\frac{\partial u}{\partial t} &= |\nabla u| \operatorname{div}\left(g(I)\frac{\nabla u}{|\nabla u|}\right)\\ &= g(I)|\nabla u| \operatorname{div}\left(\frac{\nabla u}{|\nabla u|}\right) + \nabla g(I) \cdot \nabla u\\ &= g(I)|\nabla u|\kappa + \nabla g(I) \cdot \nabla u, \end{aligned} \tag{2.14}$$

where the right hand of the flow is the Euler–Lagrange of (2.8) with $\mathcal{C}$ represented by a level-set of u, and the curvature κ is computed on the level-sets of u. That means, (2.14) is obtained by embedding (2.13) into u for $\beta = g(I)\kappa - \nabla g \cdot \vec{\mathcal{N}}$. On the equation above we made use of the fact that

$$\kappa = \operatorname{div}\left(\frac{\nabla u}{|\nabla u|}\right).$$

Eq. (2.14) is the main part of the proposed active contour model.

The level-sets geodesic flow: Boundary detection. Let us proceed and explore the geometric interpretation of the geodesic active contours Eq. (2.14) from the point of view of object segmentation, as well as its relation to other geometric curve evolution approaches to active contours. In CASELLES, CATTE, COLL and DIBOS [1993], MALLADI, SETHIAN and VEMURI [1994], MALLADI, SETHIAN and VEMURI [1995], MALLADI, SETHIAN and VEMURI [to appear], the authors proposed the following model for boundary detection:

$$\begin{aligned}\frac{\partial u}{\partial t} &= g(I)|\nabla u| \operatorname{div}\left(\frac{\nabla u}{|\nabla u|}\right) + cg(I)|\nabla u| \\ &= g(I)(c+\kappa)|\nabla u|, \end{aligned} \tag{2.15}$$

where c is a positive real constant. Following CASELLES, CATTE, COLL and DIBOS [1993], MALLADI, SETHIAN and VEMURI [1994], MALLADI, SETHIAN and VEMURI [1995], MALLADI, SETHIAN and VEMURI [to appear], Eq. (2.15) can be interpreted as follows: First, the flow

$$u_t = (c+\kappa)|\nabla u|$$

means that each one of the level-sets $\mathcal{C}$ of u is evolving according to

$$\mathcal{C}_t = (c+\kappa)\vec{\mathcal{N}},$$

where $\vec{\mathcal{N}}$ is the inward normal to the curve. This equation was first proposed in OSHER and SETHIAN [1988], were extensive numerical research on it was performed and then studied in KIMIA, TANNENBAUM and ZUCKER [1995] for shape analysis. The Euclidean shortening flow

$$\mathcal{C}_t = \kappa\vec{\mathcal{N}}, \tag{2.16}$$

denoted also as *Euclidean heat flow*, is well-known for its very satisfactory geometric smoothing properties (ANGENENT [1991], GAGE and HAMILTON [1986], GRAYSON [1987]). The flow decreases the total curvature as well as the number of zero-crossings and the value of maxima/minima curvature. Recall that this flow also moves the curve in the gradient direction of its length functional. Therefore, it has the properties of "shortening" as well as "smoothing". This shows that having only the first regularization component in (2.1), $\alpha \neq 0$ and $\beta = 0$, is enough to obtain smooth active contours as argued in when the selection $\beta = 0$ was done. The constant velocity $c\vec{\mathcal{N}}$, is related to classical mathematical morphology (SAPIRO, KIMMEL, SHAKED, KIMIA and BRUCKSTEIN

[1993]) and shape offsetting in CAD (KIMMEL and BRUCKSTEIN [1993]), is similar to the balloon force introduced in COHEN [1991]. Actually this velocity pushes the curve inwards (or outward) and it is crucial in the above model in order to allow convex initial curves to capture nonconvex shapes. Of course, the c parameter must be specified a priori in order to make the object detection algorithm automatic. This is not a trivial issue as pointed out in CASELLES, CATTE, COLL and DIBOS [1993] where possible ways of estimating this parameter are considered. Summarizing, the "force" $(c+\kappa)$ acts as the internal force in the classical energy based snakes model, smoothness being provided by the curvature part of the flow. The Euclidean heat flow $\mathcal{C}_t = \kappa\vec{\mathcal{N}}$ is exactly the regularization curvature flow that "replaces" the high order smoothness term in (2.1).

The external image dependent force is given by the stopping function $g(I)$. The main goal of $g(I)$ is actually to stop the evolving curve when it arrives to the objects boundaries. In CASELLES, CATTE, COLL and DIBOS [1993], MALLADI, SETHIAN and VEMURI [1994], MALLADI, SETHIAN and VEMURI [1995], MALLADI, SETHIAN and VEMURI [to appear], the authors choose

$$g = \frac{1}{1+|\nabla\hat{I}|^p}, \tag{2.17}$$

where $\hat{I}$ is a smoothed version of I and $p=1$ or 2. $\hat{I}$ was computed using Gaussian filtering, but more effective geometric smoothers can be used as well. Note that other decreasing functions of the gradient may be selected as well. For an ideal edge, $\nabla\hat{I}=\delta$, $g=0$, and the curve stops ($u_t=0$). The boundary is then given by the set $u=0$.

In contrast with classical energy models of snakes, the curve evolution model given by (2.15) is topology independent. That is, there is no need to know a priori the topology of the solution. This allows it to detect any number of objects in the image, without knowing their exact number. This is achieved with the help of the mentioned level-set numerical algorithm.

Let us return to the full model. Comparing Eq. (2.14) to (2.15), we see that the term $\nabla g\cdot\nabla u$, naturally incorporated via the geodesic framework, is missing in the old model. This term attracts the curve to the boundaries of the objects (∇g points toward the middle of the boundaries). Note that in the old model, the curve stops when $g=0$. This only happens at an ideal edge. In cases where there are different gradient values along the edge, as often happens in real images, g gets different values at different locations along the boundaries. It is necessary to restrict the g values, as well as possible gaps in the boundary, so that the propagating curve is guaranteed to stop. This makes the geometric model (2.15) inappropriate for the detection of boundaries with (unknown) high variations of the gradients. In the proposed model, the curve is attracted towards the boundary by the new gradient term. The gradient vectors are all directed towards the middle of the boundary. Those vectors direct the propagating curve into the "valley" of the g function. In the 2D case, $\nabla g\cdot\vec{\mathcal{N}}$ is effective in case the gradient vectors coincide with normal direction of the propagating curve. Otherwise, it will lead the propagating curve into the boundary and eventually force it to stay there. To summarize, this new force increases the attraction of the deforming contour towards the boundary, being of special help when this boundary has high variations on its gradient values. Thereby, it is also possible to detect boundaries with high differences in their gradient values, as well as small gaps. The second advantage of this new term is that we partially remove the necessity of the constant velocity given by c. This constant velocity, that mainly allows the detection of nonconvex objects, introduces an extra parameter to the model, that in most cases is an undesirable property. In the full model, the new term will allow the detection of nonconvex objects as well. This constant motion term may help to avoid certain local minima (as the balloon force), and is also of importance when starting from curves inside the object as we will see in Section 2.1. In case we wish to add this constant velocity, in order, for example, to increase the speed of convergence, we can consider the term $cg(I)|\nabla u|$ like an "area constraint" to the geodesic problem (2.8) (c being the Lagrange multiplier), obtaining

$$\frac{\partial u}{\partial t} = |\nabla u| \operatorname{div}\left(g(I) \frac{\nabla u}{|\nabla u|} \right) + cg(I)|\nabla u|. \tag{2.18}$$

Before proceeding, note that constant velocity is derived from an energy involving area. That is, $\mathcal{C} = c\vec{\mathcal{N}}$ minimizes the area enclosed by $\mathcal{C}$. Therefore, adding constant velocity is like solving $L_R + c$ area($\mathcal{C}$) (CASELLES, KIMMEL, SAPIRO and SBERT [1997b], EVANS [1998], SIDDIQI, BERUBE, TANNENBAUM and ZUCKER [1998]).

Eq. (2.18) is, of course, equivalent to

$$\frac{\partial u}{\partial t} = g(c+\kappa)|\nabla u| + \nabla u \cdot \nabla g \tag{2.19}$$

and means that the level-sets move according to

$$\mathcal{C}_t = g(I)(c+\kappa)\vec{\mathcal{N}} - (\nabla g \cdot \vec{\mathcal{N}})\vec{\mathcal{N}}. \tag{2.20}$$

Eq. (2.18), which is the level-sets representation of the modified solution of the geodesic problem (2.8) derived from the energy (2.3), constitutes the general *geodesic active contour* model we propose. The solution to the object detection problem is then given by the zero level-set of the steady state ($u_t = 0$) of this flow.

An important issue of the proposed model is the selection of the stopping function g in the model. According to the results in CASELLES, CATTE, COLL and DIBOS [1993] and in Section 2.1, in the case of ideal edges the described approach of object detection via geodesic computation is independent of the choice of g, as long as g is a positive strictly decreasing function and $g(r) \to 0$ as $r \to \infty$. Since real images do not contain ideal edges, g must be specified. In the following experimental results we use g as in Malladi et al. and Caselles et al., given by (2.17). This is a very simple "edge detector", similar to the ones used in previous active contours models, both curve evolution and energy based ones, and suffers from the well known problems of gradient based edge detectors. In spite of this, and as we can appreciate from the following examples, accurate results are obtained using this simple function. The use of better edge detectors, as for example energy ones (FREEMAN and ADELSON [1991], PERONA and MALIK [1991]), will immediately improve the results.

In the derivations above we have followed the formulation in CASELLES, KIMMEL and SAPIRO [1995], CASELLES, KIMMEL and SAPIRO [1997]. The obtained geodesic equation, as well as its 3D extension (see next section), was independently proposed by

KICHENASSAMY, KUMAR, OLVER, TANNENBAUM and YEZZI [1995], KICHENASSAMY, KUMAR, OLVER, TANNENBAUM and YEZZI [1996] and SHAH [1996] based on a different initial approaches In the case of SHAH [1996], g is obtained from an elaborated segmentation procedure obtained from MUMFORD and SHAH [1989] approach. In KICHENASSAMY, KUMAR, OLVER, TANNENBAUM and YEZZI [1996] the authors give a number of important theoretical results as well. The formal mathematical connections between energy models and curve evolution ones was done in CASELLES, KIMMEL and SAPIRO [1997], before this work the two approaches are considered independent. Three dimensional examples are given in WHITAKER [1995], where similar equations as the presented in the next section are proposed. The equations there are obtained by extending the flows in CASELLES, CATTE, COLL and DIBOS [1993], MALLADI, SETHIAN and VEMURI [1994]. In TEK and KIMIA [1995], the authors extend the models in CASELLES, CATTE, COLL and DIBOS [1993], MALLADI, SETHIAN and VEMURI [1994], motivated by work reported in KIMIA, TANNENBAUM and ZUCKER [1990], KIMIA, TANNENBAUM and ZUCKER [1995]. One of the key ideas, motivated by the shape theory of shocks developed by Kimia et al., is to perform multiple initializations, while using the same equations as in CASELLES, CATTE, COLL and DIBOS [1993], MALLADI, SETHIAN and VEMURI [1994]. Possible advantages of this are reported in the mentioned manuscript. That paper (TEK and KIMIA [1995]) uses the same equations as in CASELLES, CATTE, COLL and DIBOS [1993], MALLADI, SETHIAN and VEMURI [1994] and not the new ones described in this chapter (and in KICHENASSAMY, KUMAR, OLVER, TANNENBAUM and YEZZI [1995], KICHENASSAMY, KUMAR, OLVER, TANNENBAUM and YEZZI [1996]), also without showing its connection with classical snakes. A normalized version of (2.8) was derived in FUA and LECLERC [1990] from a different point of view, giving as well different flows for 2D active contours.

Existence, uniqueness, stability, and consistency of the geodesic model

Before proceeding with the experimental results, we want to present results regarding existence and uniqueness of the solution to (2.18). Based on the theory of viscosity solutions (CRANDALL, ISHII and LIONS [1992]), the Euclidean heat flow as well as the geometric model (2.15), are well defined for nonsmooth images as well (CASELLES, CATTE, COLL and DIBOS [1993], CHEN, GIGA and GOTO [1991], EVANS and SPRUCK [1991]). We now present similar results for our model (2.18). Note that besides the work in CASELLES, CATTE, COLL and DIBOS [1993], there is not much formal analysis for active contours approaches in the literature. The results presented in this section, together with the results on numerical analysis of viscosity solutions, ensures the existence and uniqueness of the solution of the geodesic active contours model.

Let us first recall the notion of viscosity solutions for this specific equation; see CRANDALL, ISHII and LIONS [1992] for details. We re-write Eq. (2.18) in the form

$$\begin{cases} \dfrac{\partial u}{\partial t} - g(\mathcal{X})a_{ij}(\nabla u)\partial_{ij}u - \nabla g \cdot \nabla u - cg(\mathcal{X})|\nabla u| = 0, \\ [t, \mathcal{X}) \in [0, \infty) \times \mathbb{R}^2, \\ u(0, \mathcal{X}) = u_0(\mathcal{X}), \end{cases} \tag{2.21}$$

where $a_{ij}(q) = \delta_{ij} - \frac{p_i, p_j}{|p|^2}$ if $p \neq 0$. We used in (2.21) and we shall use below the usual notations $\partial_i = \frac{\partial}{\partial x_i}$ and $\partial_{ij} = \frac{\partial^2}{\partial x_i \partial x_j}$, together with the classical Einstein summation convention. The terms $g(\mathcal{X})$ and ∇g are assumed to be continuous.

Eq. (2.21) should be solved in $D = [0,1]^2$ with Neumann boundary conditions. In order to simplify the notation and as usual in the literature, we extend the images by reflection to $\mathbb{R}^2$ and we look for solutions verifying $u(\mathcal{X} + 2h) = u(\mathcal{X})$ for all $\mathcal{X} \in \mathbb{R}^2$ and $h \in \mathbb{Z}^2$. The initial condition u_0 as well as the data $g(\mathcal{X})$ are taken extended to $\mathbb{R}^2$ with the same periodicity.

Let $u \in C([0,T] \times \mathbb{R}^2)$ for some $T \in]0,\infty[$. We say that u is a viscosity sub-solution of (2.21) if for any function $\phi \in C(\mathbb{R} \times \mathbb{R}^2)$ and any local maxima $(t_0, \mathcal{X}_0) \in]0,T] \times \mathbb{R}^2$ of $u - \phi$ we have if $\nabla\phi(t_0, \mathcal{X}_0) \neq 0$, then

$$\begin{aligned}&\frac{\partial \phi}{\partial t}(t_0, \mathcal{X}_0) - g(\mathcal{X}_0) a_{ij}\big(\nabla\phi(t_0, \mathcal{X}_0)\big)\partial_{ij}\phi(t_0, \mathcal{X}_0)\\ &\quad - \nabla g(\mathcal{X}_0) \cdot \nabla\phi(t_0, \mathcal{X}_0) - c g(\mathcal{X}_0)\big|\nabla\phi(t_0, \mathcal{X}_0)\big| \leqslant 0,\end{aligned}$$

and if $\nabla\phi(t_0, \mathcal{X}_0) = 0$, then

$$\frac{\partial \phi}{\partial t}(t_0, \mathcal{X}_0) - g(\mathcal{X}_0) \lim_{q \to 0} \sup a_{ij}(q)\partial_{ij}\phi(t_0, \mathcal{X}_0) \leqslant 0$$

and

$$u(0, \mathcal{X}) \leqslant u_0(\mathcal{X}).$$

In the same way, a viscosity super-solution is defined by changing in the expressions above "local maxima" by "local minima", "$\leqslant$" by "$\geqslant$", and "lim sup" by "lim inf". A viscosity solution is a functions which is both a viscosity sub-solution and a viscosity super-solution. Viscosity solutions is one of the most popular frameworks for the analysis of nonsmooth solutions of PDEs, having physical relevance as well. The viscosity solution coincides with the classical one if this exists.

With the notion of viscosity solutions, we can now present the following result regarding the geodesic model:

THEOREM 2.3. *Let $W^{1,\infty}$ denote the space of bounded Lipschitz functions in $\mathbb{R}^2$. Assume that $g \geqslant 0$ is such that $\sup_{\mathcal{X} \in \mathbb{R}^2} |Dg^{1/2}(\mathcal{X})| < \infty$ and $\sup_{\mathcal{X} \in \mathbb{R}^2} |D^2 g(\mathcal{X})| < \infty$. Let $u_0 \in \mathrm{BUC}(\mathbb{R}^2) \cap W^{1,\infty}(\mathbb{R}^2)$. (In the experimental results, the initial function u_0 will be the distance function, with $u_0 = 0$ at the boundary of the image.) Then*

1. *Eq.* (2.21) *admits a unique viscosity solution*

$$u \in C\big([0,\infty) \times \mathbb{R}^2\big) \cap L^\infty\big(0,T; W^{1,\infty}(\mathbb{R}^2)\big) \quad \textit{for all } T < \infty.$$

Moreover, u satisfies

$$\inf u_0 \leqslant u(t, \mathcal{X}) \leqslant \sup u_0.$$

2. *Let $v \in C([0,\infty) \times \mathbb{R}^2)$ be the viscosity solution of* (2.21) *corresponding to the initial data $v_0 \in C(\mathbb{R}^2) \cap W^{1,\infty}(\mathbb{R}^2)$. Then*

$$\big\|u(t,\cdot) - v(t,\cdot)\big\|_\infty \leqslant \|u_0 - v_0\|_\infty \quad \textit{for all } t \geqslant 0.$$

This shows that the unique solution is stable.

The assumptions of the theorem above are just technical. They imply the smoothness of the coefficients of (2.21) is required to prove the result using the method in ALVAREZ, LIONS and MOREL [1992], CASELLES, CATTE, COLL and DIBOS [1993]. In particular, Lipschitz continuity in $\mathcal{X}$ is required. This implies a well defined trajectory of the flow $\mathcal{X}_t = \nabla g(\mathcal{X})$, going to every point $\mathcal{X}_0 \in \mathbb{R}^2$, which is reasonable in our context. The proof of this theorem follows the same steps of the corresponding proofs for the model (2.15); see CASELLES, CATTE, COLL and DIBOS [1993], Theorem 3.1, and we shall omit the details (see also ALVAREZ, LIONS and MOREL [1992]).

In the next theorem, we recall results on the independence of the generalized evolution with respect to the embedding function u_0. Let Γ_0 be the initial active contour, oriented such that it contains the object. In this case the initial condition u_0 is selected to be the signed distance function, such that it is negative in the interior of Γ_0 and positive in the exterior. Then, we have

THEOREM 2.4 (Theorem 7.1, CHEN, GIGA and GOTO [1991]). *Let $u_0 \in W^{1,\infty}(\mathbb{R}^2) \cap \mathrm{BUC}(\mathbb{R}^2)$. Let $u(t,x)$ be the solution of the proposed geodesic evolution equation as in previous theorem. Let $\Gamma(t) := \{\mathcal{X}\colon u(t,\mathcal{X}) = 0\}$ and $\mathcal{D}(t) := \{\mathcal{X}\colon u(t,\mathcal{X}) < 0\}$. Then, $(\Gamma(t), \mathcal{D}(t))$ are uniquely determined by $(\Gamma(0), \mathcal{D}(0))$.*

This theorem is adopted from CHEN, GIGA and GOTO [1991], where a slightly different formulation is given. The techniques there can be applied to the present model.

Let us present some further remarks on the proposed geodesic flows (2.14) and (2.18), as well as the previous geometric model (2.15). First note that these equations are invariant under increasing re-arrangements of contrast (morphology invariant ALVAREZ, GUICHARD, LIONS and MOREL [1993]). This means that $\Theta(u)$ is a viscosity solution of the flow if u is and $\Theta : \mathbb{R} \to \mathbb{R}$ is an increasing function. On the other hand, while (2.14) is also contrast invariant, i.e., invariant to the transformation $u \leftarrow -u$ (remember that u is the embedding function used by the level-set approach), Eqs. (2.15) and (2.18) are not due to the presence of the "constant velocity" component $cg(I)|\nabla u|$. This has a double effect. First, for Eq. (2.14), it can be shown that the generalized evolution of the level-sets $\Gamma(t)$ only depends on Γ_0 (EVANS and SPRUCK [1991], Theorem 2.8), while for (2.18), the result in Theorem 3 is given. Second, for Eq. (2.14) one can show that if a smooth classical solution of the curve flow (2.13) exists and is unique, then it coincides with the generalized solution obtained via the level-sets representation (2.14) during the lifetime of the classical solution (EVANS and SPRUCK [1991], Theorem 6.1). The same result can then be proved for the general curve flow (2.20) and its level-set representation (2.18), although a more delicate proof, on the lines of Corollary 11.2 in SONER [1993], is required.

We have just presented results concerning the existence, uniqueness, and stability of the solution of the geodesic active contours. Moreover, we have observed that the evolution of the curve is independent of the embedding function, at least as long as we precise its interior and exterior regions. These results are presented in the viscosity framework. To conclude this section, let us mention that, in the case of a smooth ideal edge $\widehat{\Gamma}$, one can prove that the generalized motion $\Gamma(t)$ converges to $\widehat{\Gamma}$ as $t \to \infty$, making the proposed approach consistent:

THEOREM 2.5. *Let* $\widehat{\Gamma} = \{\mathcal{X} \in \mathbb{R}^2\colon\ g(\mathcal{X}) = 0\}$ *be a simple Jordan curve of class* C^2 *and* $Dg(\mathcal{X}) = 0$ *in* $\widehat{\Gamma}$. *Furthermore, assume* $u_0 \in W^{1,\infty}(\mathbb{R}^2) \cap \mathrm{BUC}(\mathbb{R}^2)$ *is of class* C^2 *and such that the set* $\{\mathcal{X} \in \mathbb{R}^2\colon\ u_0(\mathcal{X}) \leqslant 0\}$ *contains* $\widehat{\Gamma}$ *and its interior. Let* $u(t, \mathcal{X})$ *be the solution of* (2.18) *and* $\Gamma(t) = \{\mathcal{X} \in \mathbb{R}^2\colon\ u(t, \mathcal{X}) = 0\}$. *Then, if c, the constant component of the velocity, is sufficiently large,* $\Gamma(t) \to \widehat{\Gamma}$ *as* $t \to \infty$ *in the Hausdorff distance.*

A similar result can be proved for the basic geodesic model, that is for $c = 0$, assuming the maximal distance between $\widehat{\Gamma}$ and the initial curve $\Gamma(0)$ is given and bounded (to avoid local minima).

Experimental results

Let us present some examples of the proposed geodesic active contours model (2.18). In the numerical implementation of Eq. (2.18) we have chosen central difference approximation in space and forward difference approximation in time. This simple selection is possible due to the stable nature of the equation, however, when the coefficient c is taken to be of high value, more sophisticated approximations are required (OSHER and SETHIAN [1988]). See the mentioned references for details on the numerics.

Fig. 2.1 presents two wrenches with inward flow. Note that this is a difficult image, not only for the existence of 2 objects, separated by only a few pixels, but also for the existence of many artifacts, like the shadows, which can derive the edge detection to a wrong solution. We applied the geodesic model (2.18) to the image, and indeed, both objects are detected. The original connected curve splits in order to detect both objects. The geodesic contours also manages not to be stopped by the shadows (false contours), due to the stronger attraction force provided by the term $\nabla g \cdot \nabla u$ towards the real boundaries. Observe that the process of preferring the real edge over the shadow one, starts at their connection points, and the contour is pulled to the real edge, "like closing a zipper". We run the model also with $c = 0$, obtaining practically the same results with slower convergence.

We continue the geodesic experiments with an ultrasound image, to show the flexibility of the approach. This is given in Fig. 2.2, where the fetus is detected. In this case, the image was smoothed with a Gaussian-type kernel (2–3 iterations of a 3×3 window filter

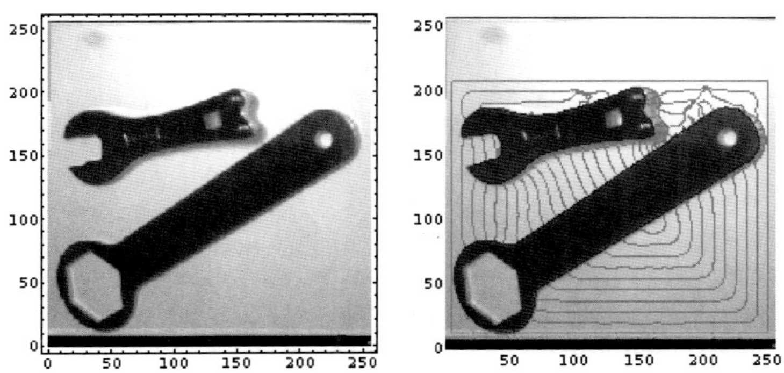

FIG. 2.1. Detecting two wrenches with the geodesic flow, moving inwards.

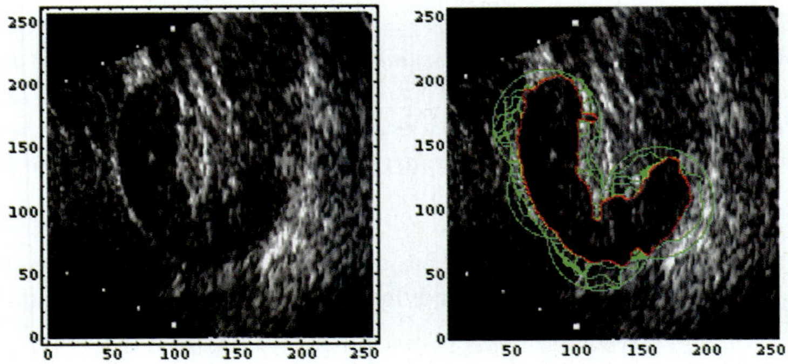

FIG. 2.2. Inward geodesic flow for ultrasound detection.

are usually applied) before the detection was performed. This avoids possible local minima, and together with the attraction force provided by the new term, allowed to detect an object with gaps in its boundary. In general, gaps in the boundary (flat gradient) can be detected if they are of the order of magnitude of $1/(2c)$ (after smoothing). Note also that the initial curve is closer to the object to be detected to avoid further possible detection of false contours (local minima). Although this problem is significantly reduced by the new term incorporated in the geodesic model, is not completely solved. In many applications, as interactive segmentation of medical data, this is not a problem, since the user can provide a rough initial contour as the one in Fig. 2.2 (or remove false contours). This problem might be automatically solved using better stopping function g, as explained in the previous sections, or by higher values of c, the constant velocity, imitating the balloon force of Cohen et al. Another classical technique for avoiding some local minima is to solve the geodesic flow in a multiscale fashion. Starting from a contour surrounding all the image, and a low resolution of it, the algorithm is applied. Then, the result of it (steady state) is used as initial contour for the next higher resolution, and the process continues up to the original resolution. Multiresolution can help as well to reduce the computational complexity (GEIGER, GUPTA, COSTA and VLONTZOS [1995]).

Fig. 2.3 shows results for the segmentation of skin lesions. Three examples are given in the first row. The second row shows the combination of the geodesic active contours with the continuous scale morphology introduced in the previous chapter. The hairy image on the left is first pre-processed with a directional morphology PDE to remove the hairs (middle figure), and the segmentation is performed on this filtered image, right.

2.2. *Three-dimensional minimal surfaces*

The 2D geodesic active contours presented above can be easily extended to 3D replacing the length element by an area element, thereby computing minimal surfaces (CASELLES, KIMMEL, SAPIRO and SBERT [1997a], CASELLES, KIMMEL, SAPIRO and SBERT [1997b], YEZZI, KICHENASSAMY, OLVER and TANNENBAUM [1997]). A few examples are presented below.

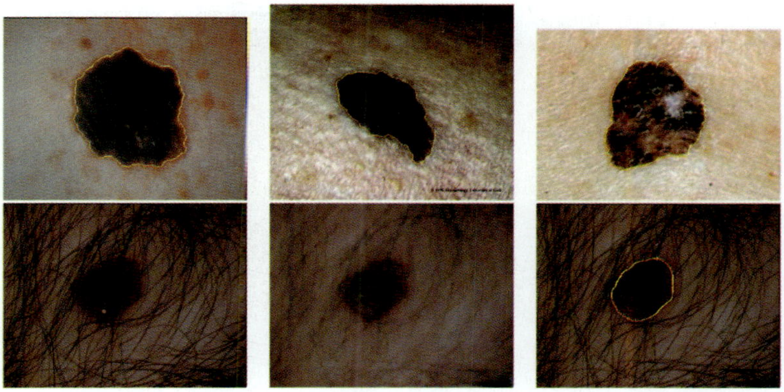

FIG. 2.3. Geodesic flow for skin lesion segmentation. Three examples are given in the first row, while the second row shows the original hairy image, result of hair removal via continuous scale morphological PDEs, and the segmented lesion.

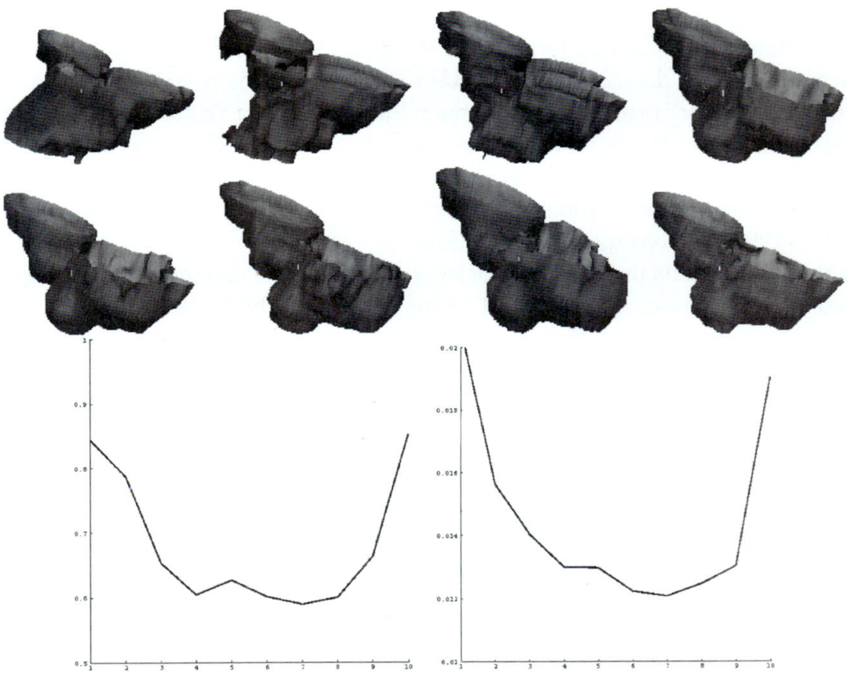

FIG. 2.4. Detection of a heart with the minimal surfaces model. Several consecutive time stages are shown, together with the plot of the area and volume (vertical axes) against time (horizontal axes).

Fig. 2.4 shows a number of sequences of an active hart. This figure is reproduced from MALLADI, KIMMEL, ADALSTEINSSON, SAPIRO, CASELLES and SETHIAN [1996], and was obtained combining the minimal surfaces model with fast numerics developed by Malladi and Sethian. Fig. 2.5 shows the detection of a head from MRI data.

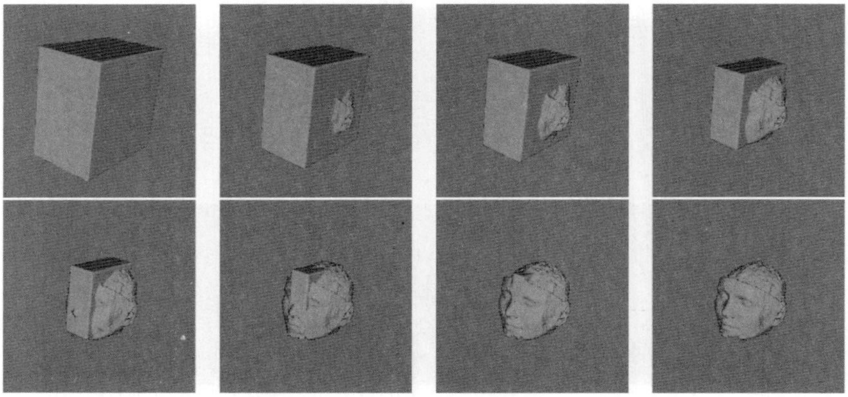

FIG. 2.5. Detecting a head from 3D MRI. Eight steps of the evolution of the minimal surface are shown.

2.3. *Geodesics in vector-valued images*

We now extend the geodesic formulation to object detection in vector-valued images, presenting what we denote as *color snakes* (*color active contours*). (In this section we use the word "color" to refer to general multi-valued images.) Vector-valued images are not just obtained in image modalities were the data is recorded in a vector fashion, as in color, medical, and LANDSAT applications. The vector-valued data can be obtained also from scale and orientation decompositions very popular in texture analysis; see SAPIRO [1997] for corresponding references.

We should mention that a number of results on vector-valued segmentation were reported in the literature, e.g., in ZHU, LEE and YUILLE [1995]. See SAPIRO [1997] for the relevant references and further comparisons. Here we address the geodesic active contours approach with vector-image metrics, a simple and general approach. Other algorithms can also be extended to vector-valued images following the framework described in this section.

Vector-valued edges

We present a definition of edges in vector-valued images based on classical Riemannian geometry (KREYSZIG [1959]). Early approaches to detecting discontinuities in multi-valued images attempted to combine the response of single-valued edge detectors applied separately to each of the image components (see, for example, NEVATIA [1977]). The way the responses for each component are combined is in general heuristic, and has no theoretical basis. A principled way to look at gradients in multivalued images, and the one we adopt in this chapter, has been described in DI ZENZO [1986].

The idea is the following. Let $I(x_1, x_2): \mathbb{R}^2 \to \mathbb{R}^m$ be a multi-valued image with components $I_i(x_1, x_2): \mathbb{R}^2 \to \mathbb{R}$, $i = 1, 2, \ldots, m$. For color images, for example, we have $m = 3$ components. The value of the image at a given point (x_1^0, x_2^0) is a vector in $\mathbb{R}^m$, and the difference of image values at two points $P = (x_1^0, x_2^0)$ and $Q = (x_1^1, x_2^1)$ is given by $\Delta I = I(P) - I(Q)$. When the (Euclidean) distance $d(P, Q)$ between P

and Q tends to zero, the difference becomes the arc element

$$dI = \sum_{i=1}^{2} \frac{\partial I}{\partial x_i} dx_i, \tag{2.22}$$

and its squared norm is given by

$$dI^2 = \sum_{i=1}^{2} \sum_{j=1}^{2} \frac{\partial I}{\partial x_i} \frac{\partial I}{\partial x_j} dx_i \, dx_j. \tag{2.23}$$

This quadratic form is called the *first fundamental form* (KREYSZIG [1959]). Let us denote $g_{ij} := \frac{\partial I}{\partial x_i} \cdot \frac{\partial I}{\partial x_j}$, then

$$dI^2 = \sum_{i=1}^{2} \sum_{j=1}^{2} g_{ij} \, dx_i \, dx_j = \begin{bmatrix} dx_1 \\ dx_2 \end{bmatrix}^T \begin{bmatrix} g_{11} g_{12} \\ g_{21} g_{22} \end{bmatrix} \begin{bmatrix} dx_1 \\ dx_2 \end{bmatrix}. \tag{2.24}$$

The first fundamental form allows the measurement of changes in the image. For a unit vector $\hat{v} = (v_1, v_2) = (\cos\theta, \sin\theta)$, $dI^2(\hat{v})$ is a measure of the rate of change of the image in the $\hat{v}$ direction. The extrema of the quadratic form (2.24) are obtained in the directions of the eigenvectors of the matrix $[g_{ij}]$, and the values attained there are the corresponding eigenvalues. Simple algebra shows that the eigenvalues are

$$\lambda_{\pm} = \frac{g_{11} + g_{22} \pm \sqrt{(g_{11} - g_{22})^2 + 4g_{12}^2}}{2}, \tag{2.25}$$

and the eigenvectors are $(\cos\theta_{\pm}, \sin\theta_{\pm})$, where the angles $\theta_{\pm}$ are given (modulo π) by

$$\theta_+ = \frac{1}{2} \arctan \frac{2g_{12}}{g_{11} - g_{22}}, \qquad \theta_- = \theta_+ + \pi/2. \tag{2.26}$$

Thus, the eigenvectors provide the direction of maximal and minimal changes at a given point in the image, and the eigenvalues are the corresponding rates of change. We call θ_+ the *direction of maximal change* and λ_+ the *maximal rate of change*. Similarly, θ_- and λ_- are the *direction of minimal change* and the *minimal rate of change* respectively. Note that for $m = 1$, $\lambda_+ \equiv \|\nabla I\|^2$, $\lambda_- \equiv 0$, and $(\cos\theta_+, \sin\theta_+) = \nabla I / \|\nabla I\|$.

In contrast to grey-level images ($m = 1$), the minimal rate of change λ_- may be different than zero. In the single-valued case, the gradient is always perpendicular to the level-sets, and $\lambda_- \equiv 0$. As a consequence, the "strength" of an edge in the multi-valued case is not simply given by the rate of maximal change, λ_+, but by how λ_+ compares to λ_-. For example, if $\lambda_+ = \lambda_-$, we know that the image changes at an equal rate in all directions. Image discontinuities can be detected by defining a function $f = f(\lambda_+, \lambda_-)$ that measures the dissimilarity between λ_+ and λ_-. A possible choice is $f = f(\lambda_+ - \lambda_-)$, which has the nice property of reducing to $f = f(\|\nabla I\|^2)$ for the one dimensional case, $m = 1$.

CUMANI [1991] extended some of the above ideas. He analyzed the directional derivative of λ_+ in the direction of its corresponding eigenvector, and looked for edges by localizing the zero crossings of this function. Cumani's work attempts to generalize the ideas of TORRE and POGGIO [1986] to multivalued images. Note that this approach entirely neglects the behavior of λ_-. As we already mentioned, it is the relationship between λ_- and λ_+ that is important.

A noise analysis for the above vector-valued edge detector has been presented in LEE and COK [1991]. It was shown that, for correlated data, this scheme is more robust to noise than the simple combination of the gradient components.

Before concluding this section we should point out that based on the theory above, improved edge detectors for vector-valued images can be obtained following, for example, the developments on energy-based edge detectors (FREEMAN and ADELSON [1991]). In order to present the color snakes, the theory developed above is sufficient.

Color snakes

Let $f_{\text{color}} = f(\lambda_+, \lambda_-)$ be the edge Detector defined above. The edge stopping function g_{color} is then defined such that $g_{\text{color}} \to 0$ when $f \to \max(\infty)$, as in the grey-scale case. For example, $f_{\text{color}} := (\lambda_+ - \lambda_-)^{1/p}$, $p > 0$, or $f_{\text{color}} := \sqrt{\lambda_+}$, and $g_{\text{color}} := \frac{1}{1+f}$ or $g_{\text{color}} := \exp\{-f\}$. The function (metric) g_{color} defines the space on which we compute the geodesic curve. Defining

$$L_{\text{color}} := \int_0^{\text{length}} g_{\text{color}}\, dv, \tag{2.27}$$

the object detection problem in vector-valued images is then associated with minimizing L_{color}. We have formulated the problem of object segmentation in vector-valued images as a problem on finding a geodesic curve in a space defined by a metric induced from the whole vector image.

In order to minimize L_{color}, that is the *color length*, we compute as before the gradient descent flow. The equations developed for the geodesic active contours are independent of the specific selection of the function g. Replacing g_{grey} by g_{color} and embedding the evolving curve $\mathcal{C}$ in the function $u : \mathbb{R}^2 \to \mathbb{R}$, we obtain the general flow, with additional unit speed, for the *color snakes*,

$$\frac{\partial u}{\partial t} = g_{\text{color}}(\nu + \kappa)|\nabla u| + \nabla u \cdot \nabla g_{\text{color}}. \tag{2.28}$$

Recapping, Eq. (2.28) is the modified level-sets flow corresponding to the gradient descent of L_{color}. Its solution (steady-state) is a geodesic curve in the space define by the metric $g_{\text{color}}(\lambda_\pm)$ of the vector-valued image. This solution gives the boundaries of objects in the scene. (Note that $\lambda_\pm$ can be computed on a smooth image obtained from vector-valued anisotropic diffusion (SAPIRO and RINGACH [1996]).) Following work presented previously in this chapter, theoretical results regarding the existence, uniqueness, stability, and correctness of the solutions to the color active contours can be obtained.

Fig. 2.6 presents an example of the vector snakes model for a medical image. Fig. 2.7 shows an example for texture segmentation. The original image is filtered with Gabor filters tuned to frequency and orientation as proposed in LEE, MUMFORD and YUILLE [1992] for texture segmentation (see SAPIRO [1997] for additional related references). From this set of frequency/orientation images, g_{color} is computed according to the formulas above, and the vector-valued snakes flow is applied. Four frequencies and four

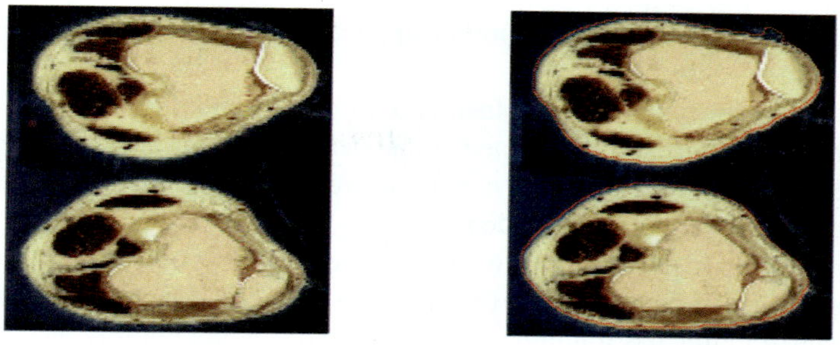

FIG. 2.6. Example of the color snakes. The original image is on the left, and the one with the segmented objects (red lines) on the right. The original curve contained both objects. The computations were done on the $L^*a^*b^*$ space.

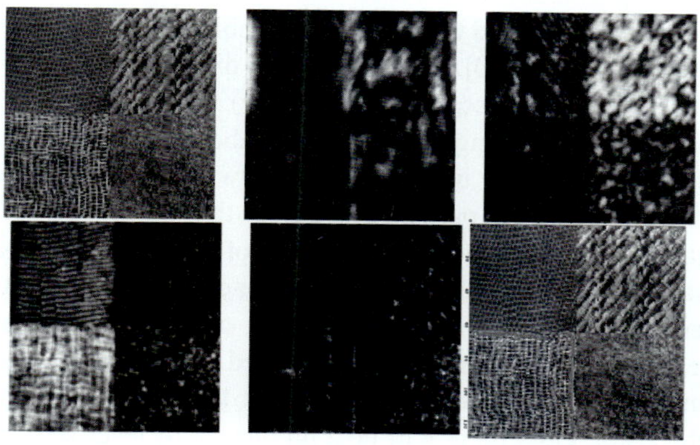

FIG. 2.7. Example of the vector snakes for a texture image. The original texture (top-left) is decomposed into frequency/orientation (four frequencies and four orientations) components via Gabor filters and this collection of images is used to compute the metric g_{color} for the snakes flow. A subset of the different components are shown, followed (bottom-right) by the result of the evolving vector snakes (green), segmenting one of the texture boundaries (red).

orientations are used, obtaining sixteen images. More examples can be found in SAPIRO [1997].

Remark on level-lines of vector valued images

As we have seen, and we will continue to develop later in this chapter, level-sets provide a fundamental concept and representation for scalar images. The basic idea is that a scalar image $I(x, y): \mathbb{R}^2 \to \mathbb{R}$ is represented as a collection of sets

$$\Lambda_h := \{(x, y): I(x, y) = h\},$$

or

$$\hat{\Lambda}_h := \{(x, y): I(x, y) \leqslant h\}.$$

This representation not only brings state of the art image processing algorithms, it is also the source of the connected components concept that we will develop later in this chapter and that has given light to important applications like contrast enhancement and image registration.

One of the fundamental questions then is if we can extend the concept of level-sets (and after that, the concept of connected components), to multivalued data, that is, images of the form $I(x, y):\mathbb{R}^2 \to \mathbb{R}^n$, $n > 1$. As we have seen, this data includes color images, multispectral images, and video.

One straightforward possibility is to consider the collection of classical level-sets for each one of the image components $I_i(x, y):\mathbb{R}^2 \to \mathbb{R}$, $1 \leqslant i \leqslant n$. Although this is a interesting approach, this has a number of caveats and it is not entirely analogue to the scalar level-sets. For example, it is not clear how to combine the level-sets from the different components. In contrast with the scalar case, a point on the plane belongs to more than one level-set $\Lambda_i(x, y)$, and therefore in might belong to more than one connected component. A possible solution to this is to consider lexicographic orders. That means that some arbitrary decisions need to be taken to combine the multiple level-sets.

In CHUNG and SAPIRO [2000] we argued for a different approach. Basically, we redefine the level-set lines of a scalar image $I(x, y):\mathbb{R}^2 \to \mathbb{R}$ as the integral curves of the directions of minimal change θ_- as defined above. In other words, we select a given pixel (x, y) and travel the image plane $\mathbb{R}^2$ always in the direction of minimal change. Scalar images have the particular property that the minimal change is zero, and therefore, the integral curves of the directions of minimal change are exactly the classical level-sets lines defined above. The advantage of this definition is that it is dimension independent, and as we have seen before, minimal change and direction of minimal change can also be defined for multivalued images following classical Riemannian geometry, we have obtained a definition of level-sets for multivalued images as well. As in the scalar case, there will be a unique set of level-sets, in contrast with the case when we treat each component separately. This concept is fully developed and used to represent vectorial data with scalar images in CHUNG and SAPIRO [2000].

2.4. *Finding the minimal geodesic*

Fig. 2.8 shows an image of a neuron from the central nervous system. This image was obtained via electronic microscopy (EM). After the neuron is identified, it is marked via the injection of a color fluid. Then, a portion of the tissue is extracted, and after some processing, it is cut into thin slices and observed and captured via the EM system. The figure shows the output of the EM after some simple post-processing, mainly composed by contrast enhancement. The goal of the biologist is to obtain a three dimensional reconstruction of this neuron. As we observe from the example in Fig. 2.8, the image is very noisy, and the boundaries of the neuron are difficult to identify. Segmenting the neuron is then a difficult task.

One of the most commonly used approaches to segment objects as the neuron in Fig. 2.8 are *active contours* as those described earlier in this chapter. As shown before, this reduces to the minimization of a weighted length given by

$$\int_C g(\|\nabla(I)\|)\,ds, \tag{2.29}$$

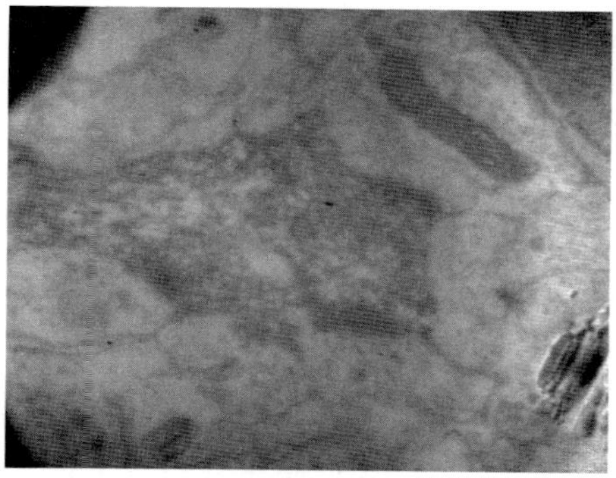

FIG. 2.8. Example of an EM image of a neuron (one slice).

where $\mathcal{C}:\mathbb{R}\to\mathbb{R}^2$ is the deforming curve, $I:\mathbb{R}^2\to\mathbb{R}$ the image, ds stands for the curve arc-length ($\|\partial\mathcal{C}/\partial s\|=1$), $\nabla(\cdot)$ stands for the gradient, and $g(\cdot)$ is such that $g(r)\to 0$ while $r\to\infty$ (the "edge detector").

There are two main techniques to find the geodesic curve, that is, the minimizer of (2.29):

(1) Compute the gradient descent of (2.29), and starting from a closed curve either inside or outside the object, deform it toward the (possibly local) minima, finding a geodesic curve. This approach gives a curve evolution flow, based on curvature motion, leading to very efficient solutions for a large number of applications. This was the approach followed in previous sections in this chapter. This models, which computes a *minimal geodesic*, gives a completely automatic segmentation procedure. When tested with images like the one in Fig. 2.8, we find two major drawbacks (BERTALMIO, SAPIRO and RANDALL [1998]). First, due to the large amount of noise, spurious objects are detected, and is left to the user to manually eliminate them. Second, due to the fact that the boundary of the real neuron is very weak, this is not always detected.

(2) Connect between a few points marked by the user on the neuron's boundary, while keeping the weighted length (2.29) to a minimum. This was developed in COHEN and KIMMEL [to appear], and it is the approach we present now (see VAZQUEZ, SAPIRO and RANDALL [1998]). In contrast with the technique described above, this approach always needs user intervention to mark the initial points. On the other hand, for images like the one in Fig. 2.8, it permits a better handling of the noise. In the rest of this section we will briefly describe this technique and the additions incorporated to address our specific problem.

Computing the minimal geodesic

We now describe the algorithm used to compute the minimal weighted path between points on the objects boundary. That is, given a set of boundary points $\{\mathcal{P}\}_{i=1}^{N}$, and following (2.29), we have to find the N curves that minimize ($\mathcal{P}_{N+1}\equiv\mathcal{P}_1$)

$$d\big(I(\mathcal{P}_i), I(\mathcal{P}_{i+1})\big) := \int_{\mathcal{P}_i}^{\mathcal{P}_{i+1}} g\big(\|\nabla I\|\big)\,ds. \tag{2.30}$$

The algorithm is composed of three main steps: (1) Image regularization, (2) Computation of equal distance contours, (3) Back propagation. We briefly describe each one of these steps now. For details on the first step, see BLACK, SAPIRO, MARIMONT and HEEGER [1998]. For details on the other steps, see COHEN and KIMMEL [to appear].

Image regularization. In order to reduce the noise on the images obtained from EM, we perform the following two steps (these are just examples of common approaches used to reduce noise before segmenting):

(1) Subsampling.
We use a 4-taps filter, approximating a Gaussian function, to smooth the image before a 2×2 subsampling is performed. This not only removes noise but also gives a smaller image to work with, thereby accelerating the algorithm by a factor of 4. That is, we will work on the subsampled image (although the user marks the end points on the original image), and only after the segmentation is computed, the result is extrapolated to the original size image. Further subsampling was found to already produce unaccurate results. The result from this step is then an image $I_{2\times 2}$ which is one quarter of the original image I.

(2) Smoothing.
In order to further reduce noise in the image, we smooth the image either with a Gaussian filter or with one of the anisotropic diffusion flows presented in the next section.

At the end of the pre-processing stage we then obtain an image $\hat{I}_{2\times 2}$ which is the result of the subsampling of I followed by noise removal. Although the user marks the points $\{\mathcal{P}\}_{i=1}^{N}$ on the original image I, the algorithm makes all the computations on $\hat{I}_{2\times 2}$ and then extrapolates and displays them on I.

Equal distance contours computation. After the image $\hat{I}_{2\times 2}$ is computed, we have to compute, for every point $\widehat{\mathcal{P}}_i$, where $\widehat{\mathcal{P}}_i$ is the point in $\hat{I}_{2\times 2}$ corresponding to the point $\mathcal{P}_i$ in I (coordinates divided by two), the weighted distance map, according to the weighted distance d. That is, we have to compute the function

$$\mathcal{D}_i(x, y) := d\big(\hat{I}_{2\times 2}(\widehat{\mathcal{P}}_i), \hat{I}_{2\times 2}(x, y)\big),$$

or in words, the weighted distance between the pair of image points $\widehat{\mathcal{P}}_i$ and (x, y).

There are basically two ways of making this computation, computing equal distance contours, or directly computing $\mathcal{D}_i$. We briefly describe each one of these now.

Equal distance contours $\mathcal{C}_i$ are curves such that every point on them has the same distance d to $\widehat{\mathcal{P}}_i$. That is, the curves $\mathcal{C}_i$ are the level-sets or isophotes of $\mathcal{D}_i$. It is easy to see (COHEN and KIMMEL [to appear]), that following the definition of d, these contours are obtained as the solution of the curve evolution flow

$$\frac{\partial \mathcal{C}_i(x, y, t)}{\partial t} = \frac{1}{g(\|\nabla \hat{I}_{2\times 2}\|)}\vec{\mathcal{N}},$$

where $\vec{\mathcal{N}}$ is the outer unit normal to $\mathcal{C}_i(x, y, t)$. This type of flow should be implemented using the standard level-sets method (OSHER and SETHIAN [1988]).

A different approach is the one presented is based on the fact that the distance function $\mathcal{D}_i$ holds the following Hamilton–Jacobi equation:

$$\frac{1}{g(\|\nabla \hat{I}_{2\times 2}\|)}\|\nabla \mathcal{D}_i\| = 1.$$

Optimal numerical techniques have been proposed to solve this static Hamilton–Jacobi equation by TSITSIKLIS [1995] motivated by optimal control, and later independently obtained by HELMSEN, PUCKETT, COLLELA and DORR [1996] and SETHIAN [1996a], and extended and applied to numerous fields by SETHIAN [1996b], SETHIAN [1996c] (see also OSHER and HELMSEN [in preparation], KIMMEL [1999], KIMMEL and SETHIAN [1998], POLYMENAKOS, BERTSEKAS and TSITSIKLIS [1988]). Due to this optimality, this is the approach we follow in the examples below. At the end of this step, we have $\mathcal{D}_i$ for each point $\widehat{\mathcal{P}}_i$. We should note that we do not need to compute $\mathcal{D}_i$ for all the image plane. It is actually enough to stop the computations when the value at $\widehat{\mathcal{P}}_{i+1}$ is obtained.

Back propagation. After the distance functions $\mathcal{D}_i$ are computed, we have to trace the actual minimal path between $\widehat{\mathcal{P}}_i$ and $\widehat{\mathcal{P}}_{i+1}$ that minimizes d. Once again it is easy to show (see, for example, KIMMEL and SETHIAN [1996], SETHIAN [1996c]), that this path should be perpendicular to the level-curves $\mathcal{C}_i$ of $\mathcal{D}_i$, and therefore tangent to $\nabla\mathcal{D}_i$. The path is then computed backing from $\widehat{\mathcal{P}}_{i+1}$, in the gradient direction, until we return to the point $\widehat{\mathcal{P}}_i$. This back propagation is of course guaranteed to converge to the point $\widehat{\mathcal{P}}_i$, and then gives the path of minimal weighted distance.

Examples

We present now a number of examples of the algorithm described above. We compare the results with those obtained with *PictureIt*, a commercially available general purpose image processing package developed by Microsoft (to the best of our knowledge, the exact algorithm used by this product was not published). As in the tracing algorithm, this software allows for the user to click a few points on the object's boundary, while the program automatically completes the rest of it. Three to five points are used for each one of the examples. The points are usually marked at extrema of curvature or at areas where the user, after some experience, predicts possible segmentation difficulties. The same points were marked in the tracing algorithm and in *PictureIt*. The results are shown in Fig. 2.9. We observe that the tracing technique outperforms *PictureIt*. Moreover, we found *PictureIt* to be extremely sensible to the exact position of the marked points, a difference of one or two pixels can cause a very large difference in the segmentation results. Our algorithm is very robust to the exact position of the points marked by the user. Additional examples can be found in SAPIRO [2001].

2.5. *Affine invariant active contours*

The geodesic formulation allows us to perform affine invariant segmentation as well. We now describe affine invariant active contours. In order to obtain them, we must replace the classical gradient based edge detector by an affine invariant one, and we

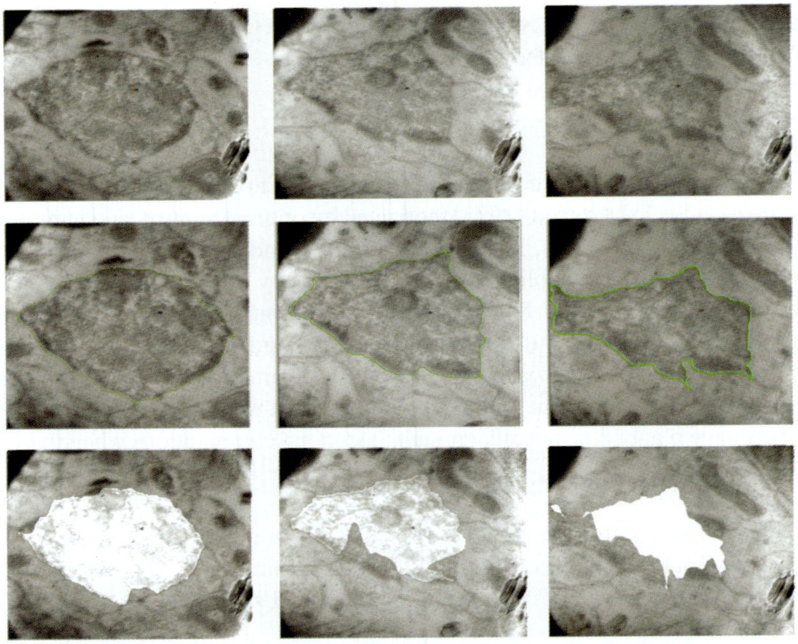

FIG. 2.9. Comparison of the minimal geodesic results with those obtained with the commercial software *PictureIt*. For each column, the original image is shown on the top, the result of the tracing algorithm (green line) on the middle, and the result of *PictureIt* on the bottom. For this last, the area segmented by the algorithm is shown brighter.

should also compute affine invariant gradient descent flows. Both extensions are given now, following OLVER, SAPIRO and TANNENBAUM [1996] and OLVER, SAPIRO and TANNENBAUM [1999].

Affine invariant gradient

Let $I : \mathbb{R}^2 \to \mathbb{R}^+$ be a given image in the continuous domain. In order to detect edges in an affine invariant form, a possible approach is to replace the classical gradient magnitude $\|\nabla I\| = \sqrt{I_x^2 + I_y^2}$, which is only Euclidean invariant, by an *affine invariant gradient*. By this we mean that we search for an affine invariant function from $\mathbb{R}^2$ to $\mathbb{R}$ that has, at image edges, values significantly different from those at flat areas, and such that this values are preserved, at corresponding image points, under affine transformations. In order to accomplish this, we have to verify that we can use basic affine invariant descriptors which can be computed from I in order to find an expression that (qualitatively) behaves like $\|\nabla I\|$. Using the classification developed in OLVER [1993], OLVER [1995], we found that the two basic independent affine invariant descriptors are (note that the simplest Euclidean invariant differential descriptor is exactly $\|\nabla I\|$, which is enough to formulate a basic Euclidean invariant edge detector)

$$H := I_{xx} I_{yy} - I_{xy}^2, \qquad J := I_{xx} I_y^2 - 2 I_x I_y I_{xy} + I_x^2 I_{yy}.$$

We should point out that there is no (nontrivial) first order affine invariant descriptor, and that all other second order differential invariants are functions of H and J. Therefore, the simplest possible affine gradient must be expressible as a function $\mathcal{F} = \mathcal{F}(H, J)$ of these two invariant descriptors.

The differential invariant J is related to the Euclidean curvature of the level-sets of the image. Indeed, if a curve $\mathcal{C}$ is defined as the level-set of I, then the curvature of $\mathcal{C}$ is given by $\kappa = \frac{J}{\|\nabla I\|^3}$. LINDEBERG [1994] used J to compute corners and edges, in an affine invariant form, that is,

$$\mathcal{F} := J = \kappa \|\nabla I\|^3.$$

This singles out image structures with a combination of high gradient (edges) and high curvature of the level-sets (corners). Note that in general edges and corners do not have to lie on a unique level-set. Here, by combining both H and J, we present a more general affine gradient approach. Since both H and J are second order derivatives of the image, the order of the affine gradient is not increased while using both invariants.

DEFINITION 2.1. The (basic) affine invariant gradient of a function I is defined by the equation

$$\widehat{\nabla}_{\text{aff}} I := \left| \frac{H}{J} \right|. \tag{2.31}$$

Technically, since $\widehat{\nabla}_{\text{aff}} I$ is a scalar (a map from $\mathbb{R}^2$ to $\mathbb{R}$), it measures just the magnitude of the affine gradient, so our definition may be slightly misleading. However, an affine invariant gradient direction does not exist, since directions (angles) are not affine invariant, and so we are justified in omitting "magnitude" for simplicity.

Note also that if photometric transformations are allowed, then $\widehat{\nabla}_{\text{aff}} I$ becomes only a relative invariant. In order to obtain an absolute invariant, we can use, for example, the combination $\frac{H^{3/2}}{J}$. Since in this case, going from relative to absolute invariants is straightforward, we proceed with the development of the simpler function $\widehat{\nabla}_{\text{aff}} I$.

The justification for our definition is based on a (simplified) analysis of the behavior of $\widehat{\nabla}_{\text{aff}} I$ near edges in the image defined by I. Near the edge of an object, the gray-level values of the image can be (ideally) represented via $I(x, y) = f(y - h(x))$, where $y = h(x)$ is the edge, and $f(t)$ is a slightly smoothed step function with a jump near $t = 0$. Straightforward computations show that, in this case,

$$H = -h'' f' f'', \qquad J = -h'' f'^3.$$

Therefore

$$H/J = f''/f'^2 = \left(-1/f'\right)'.$$

Clearly H/J is large (positive or negative) on either side of the object $y = f(x)$, creating an approximation of a zero crossing at the edge. (Note that the Euclidean gradient is the opposite, high at the ideal edge and zero everywhere else. Of course, this does not make any fundamental difference, since the important part is to differentiate between edges and flat regions. In the affine case, edges are given by doublets.) This is due to the fact that $f(x) = \text{step}(x)$, $f'(x) = \delta(x)$, and $f''(x) = \delta'(x)$. (We are omitting the points where $f' = 0$.) Therefore, $\widehat{\nabla}_{\text{aff}} I$ behaves like the classical Euclidean gradient magnitude.

In order to avoid possible difficulties when the affine invariants H or J are zero, we replace $\widehat{\nabla}_{\text{aff}}$ by a slight modification. Indeed, other combinations of H and J can provide similar behavior, and hence be used to define affine gradients. Here we present the general technique as well as a few examples.

We will be more interested in edge stopping functions than in edge maps. These are functions which are as close as possible to zero at edges, and close to the maximal possible value at flat regions. We then proceed to make modifications on $\widehat{\nabla}_{\text{aff}}$ which allow us to compute well-defined stopping functions instead of just edge maps.

In Euclidean invariant edge detection algorithms based on active contours, as well as in anisotropic diffusion, the stopping term is usually taken in the form $(1 + \|\nabla I\|^2)^{-1}$, the extra 1 being taken to avoid singularities where the Euclidean gradient vanishes. Thus, in analogy, the corresponding affine invariant stopping term should have the form

$$\frac{1}{1 + (\widehat{\nabla}_{\text{aff}} I)^2} = \frac{J^2}{H^2 + J^2}.$$

However, this can still present difficulties when both H and J vanish, so we propose a second modification.

DEFINITION 2.2. The *normalized affine invariant gradient* is given by:

$$\nabla_{\text{aff}} I = \sqrt{\frac{H^2}{J^2 + 1}}. \tag{2.32}$$

The motivation comes from the form of the *affine invariant stopping term*, which is now given by

$$\frac{1}{1 + (\nabla_{\text{aff}} I)^2} = \frac{J^2 + 1}{H^2 + J^2 + 1}. \tag{2.33}$$

Formula (2.33) avoids all difficulties where either H or J vanishes, and hence is a proper candidate for affine invariant edge detection. Indeed, in the neighborhood of an edge we obtain

$$\frac{J^2 + 1}{H^2 + J^2 + 1} = \frac{f'^6 h''^2 + 1}{h''^2 f'^2 (f'^4 + f''^2) + 1},$$

which, assuming h'' is moderate, gives an explanation of why it serves as a barrier for the edge. Barriers, that is, functions that go to zero at (salient) edges, will be important for the affine active contours presented in the following sections.

Examples of the affine invariant edge detector (2.33) are given in Fig. 2.10. As with the affine invariant edge detection scheme introduced in previous section, this algorithm might produce gaps in the objects boundaries, as a result of the existence of perfectly straight segments with the same gray value. In this case, an edge integration algorithm is

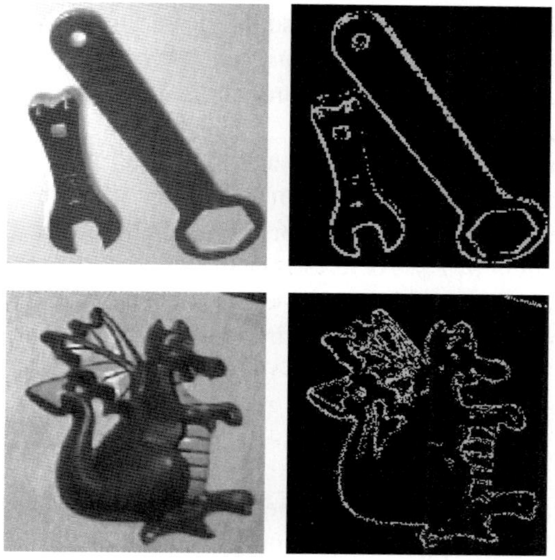

FIG. 2.10. Examples of the affine invariant edge detector (after thresholding).

needed to complete the object boundary. The affine invariant active contours presented later is a possible remedy of this problem.

Affine invariant gradient snakes

Based on the gradient active contours and affine invariant edge detectors above, it is almost straightforward to define affine invariant gradient active contours. In order to carry this program out, we will first have to define the proper norm. Since affine geometry is defined only for convex curves, we will initially have to restrict ourselves to the (Fréchet) space of thrice-differentiable convex closed curves in the plane, i.e.,

$$\mathbf{C}_0 := \{\mathcal{C} : [0, 1] \to \mathbb{R}^2 \colon \mathcal{C} \text{ is convex, closed and } C^3\}.$$

Let ds denote the affine arc-length. Then, letting $L_{\mathrm{aff}} := \oint ds$ be the *affine length*, we proceed to define the *affine norm* on the space $\mathbf{C}_0$

$$\|\mathcal{C}\|_{\mathrm{aff}} := \int_0^1 \|\mathcal{C}(p)\|_{\mathrm{a}}\, dp = \int_0^{L_{\mathrm{aff}}} \|\mathcal{C}(s)\|_{\mathrm{a}}\, ds,$$

where

$$\|\mathcal{C}(p)\|_{\mathrm{a}} := [\mathcal{C}(p), \mathcal{C}_p(p)].$$

Note that the area enclosed by $\mathcal{C}$ is just

$$A = \frac{1}{2}\int_0^1 \|\mathcal{C}(p)\|_{\mathrm{a}}\, dp = \frac{1}{2}\int_0^1 [\mathcal{C}, \mathcal{C}_p]\, dp = \frac{1}{2}\|\mathcal{C}\|_{\mathrm{aff}}. \tag{2.34}$$

Observe that

$$\|\mathcal{C}_s\|_{\mathrm{a}} = [\mathcal{C}_s, \mathcal{C}_{ss}] = 1, \qquad \|\mathcal{C}_{ss}\|_{\mathrm{a}} = [\mathcal{C}_{ss}, \mathcal{C}_{sss}] = \mu,$$

where μ is the *affine curvature*, i.e., the simplest nontrivial differential affine invariant. This makes the affine norm $\|\cdot\|_{\text{aff}}$ consistent with the properties of the Euclidean norm on curves relative to the Euclidean arc-length dv. (Here we have that $\|\mathcal{C}_v\| = 1$, $\|\mathcal{C}_{vv}\| = \kappa$.)

We can now formulate the functionals that will be used to define the affine invariant snakes. Accordingly, assume that $\phi_{\text{aff}} = \phi(w_{\text{aff}})$ is an affine invariant stopping term, based on the affine invariant edge detectors considered before. Therefore, ϕ_{aff} plays the role of the weight ϕ in L_ϕ. As in the Euclidean case, we regard ϕ_{aff} as an affine invariant conformal factor, and replace the affine arc length element ds by a conformal counterpart $ds_{\phi_{\text{aff}}} = \phi_{\text{aff}}\, ds$ to obtain the first possible functional for the affine active contours

$$L_{\phi_{\text{aff}}} := \int_0^{L_{\text{aff}}(t)} \phi_{\text{aff}}\, ds, \tag{2.35}$$

where as above L_{aff} is the affine length. The obvious next step is to compute the gradient flow corresponding to $L_{\phi_{\text{aff}}}$ in order to produce the affine invariant model. Unfortunately, as we will see, this will lead to an impractically complicated geometric contour model which involves four spatial derivatives. In the meantime, using the connection between the affine and Euclidean arc lengths, note that the above equation can be re-written in Euclidean space as

$$L_{\phi_{\text{aff}}} = \int_0^{L(t)} \phi_{\text{aff}} \kappa^{1/3}\, dv, \tag{2.36}$$

where $L(t)$ denotes the ordinary Euclidean length of the curve $\mathcal{C}(t)$ and dv stands for the Euclidean arc-length.

The snake model which we will use comes from another (special) affine invariant, namely *area*, cf. (2.34). Let $\mathcal{C}(p,t)$ be a family of curves in $\mathbf{C}_0$. A straightforward computation reveals that the first variation of the area functional

$$A(t) = \frac{1}{2}\int_0^1 [\mathcal{C}, \mathcal{C}_p]\, dp$$

is

$$A'(t) = -\int_0^{L_{\text{aff}}(t)} [\mathcal{C}_t, \mathcal{C}_s]\, ds.$$

Therefore the gradient flow which will decrease the area as quickly as possible relative to $\|\cdot\|_{\text{aff}}$ is exactly

$$\mathcal{C}_t = \mathcal{C}_{ss},$$

which, modulo tangential terms, is equivalent to

$$\mathcal{C}_t = \kappa^{1/3}\vec{\mathcal{N}},$$

which is precisely the affine invariant heat equation studied by SAPIRO and TANNENBAUM [1994]. It is this functional that we will proceed to modify with the conformal factor ϕ_{aff}. Therefore, we define the conformal area functional to be

$$A_{\phi_{\text{aff}}} := \int_0^1 [\mathcal{C}, \mathcal{C}_p]\phi_{\text{aff}}\,dp = \int_0^{L_{\text{aff}}(t)} [\mathcal{C}, \mathcal{C}_s]\phi_{\text{aff}}\,ds.$$

(An alternative definition could be to use ϕ_{aff}, to define the affine analogue of the weighted area that produces the Euclidean weighted constant motion $\mathcal{C}_t = \phi\vec{\mathcal{N}}$.) The first variation of $A_{\phi_{\text{aff}}}$ will turn out to be much simpler than that of $L_{\phi_{\text{aff}}}$ and will lead to an implementable geometric snake model.

The precise formulas for the variations of these two functionals are given in the following result. They use the definition of $Y^{\perp}$, which is the unit vector perpendicular to Y. The proof follows by an integration by parts argument and some simple manipulations (CASELLES, KIMMEL and SAPIRO [1997], CASELLES, KIMMEL, SAPIRO and SBERT [1997a], KICHENASSAMY, KUMAR, OLVER, TANNENBAUM and YEZZI [1995], KICHENASSAMY, KUMAR, OLVER, TANNENBAUM and YEZZI [1996]).

LEMMA 2.1. *Let $L_{\phi_{\text{aff}}}$ and $A_{\phi_{\text{aff}}}$ denote the conformal affine length and area functionals respectively.*

(1) *The first variation of $L_{\phi_{\text{aff}}}$ is given by*

$$\frac{dL_{\phi_{\text{aff}}}(t)}{dt} = -\int_0^{L_{\text{aff}}(t)} \big[\mathcal{C}_t, (\nabla\phi_{\text{aff}})^{\perp}\big]\,ds + \int_0^{L_a(t)} \phi_{\text{aff}}\mu[\mathcal{C}_t, \mathcal{C}_s]\,ds. \tag{2.37}$$

(2) *The first variation of $A_{\phi_{\text{aff}}}$ is given by*

$$\frac{dA_{\phi_{\text{aff}}}(t)}{dt} = -\int_0^{L_{\text{aff}}(t)} \left[\mathcal{C}_t, \left(\phi_{\text{aff}}\mathcal{C}_s + \frac{1}{2}\big[\mathcal{C}, (\nabla\phi)^{\perp}\mathcal{C}_s\big]\right)\right]ds. \tag{2.38}$$

The affine invariance of the resulting variational derivatives follows from a general result governing invariant variational problems having volume preserving symmetry groups (OLVER, SAPIRO and TANNENBAUM [1997]):

THEOREM 2.6. *Suppose G is a connected transformation group, and $\mathcal{I}[\mathcal{C}]$ is a G-invariant variational problem. Then the variational derivative (or gradient) $\delta\mathcal{I}$ of $\mathcal{I}$ is a differential invariant if and only if G is a group of volume-preserving transformations.*

We now consider the corresponding gradient flows computed with respect to $\|\cdot\|_{\text{aff}}$. First, the flow corresponding to the functional $L_{\phi_{\text{aff}}}$ is

$$\mathcal{C}_t = \big\{(\nabla\phi_{\text{aff}})^{\perp} + \phi_{\text{aff}}\mu\mathcal{C}_s\big\}_s = \big((\nabla\phi_{\text{aff}})^{\perp}\big)_s + (\phi_{\text{aff}}\mu)_s\mathcal{C}_s + \phi_{\text{aff}}\mu\mathcal{C}_{ss}.$$

As before, we ignore the tangential components, which do not affect the geometry of the evolving curve, and so obtain the following possible model for geometric affine invariant active contours:

$$\mathcal{C}_t = \phi_{\text{aff}}\mu\kappa^{1/3}\vec{\mathcal{N}} + \big\langle\big((\nabla\phi_{\text{aff}})^{\perp}\big)_s, \vec{\mathcal{N}}\big\rangle\vec{\mathcal{N}}. \tag{2.39}$$

The geometric interpretation of the affine gradient flow (2.39) minimizing $L_{\phi_{\text{aff}}}$ is analogous to that of the corresponding Euclidean geodesic active contours. The term

$\phi_{\text{aff}}\mu\kappa^{1/3}$ minimizes the affine length L_{aff} while smoothing the curve according to the results in SAPIRO and TANNENBAUM [1994], being stopped by the affine invariant stopping function ϕ_{aff}. The term associated with $((\nabla\phi_{\text{aff}})^{\perp})_s$ creates a potential valley, attracting the evolving curve to the affine edges. Unfortunately, this flow involves μ which makes it difficult to implement. (Possible techniques to compute μ numerically were recently reported in CALABI, OLVER and TANNENBAUM [1996], CALABI, OLVER, SHAKIBAN, TANNENBAUM and HAKER [1988], FAUGERAS and KERIVEN [1995].)

The gradient flow coming from the first variation of the modified area functional on the other hand is much simpler:

$$\mathcal{C}_t = \left(\phi_{\text{aff}}\mathcal{C}_s + \frac{1}{2}[\mathcal{C}, (\nabla\phi_{\text{aff}})^{\perp}]\mathcal{C}_s\right)_s. \tag{2.40}$$

Ignoring tangential terms (those involving $\mathcal{C}_s$) this flow is equivalent to

$$\mathcal{C}_t = \phi_{\text{aff}}\mathcal{C}_{ss} + \frac{1}{2}[\mathcal{C}, (\nabla\phi_{\text{aff}})^{\perp}]\mathcal{C}_{ss}, \tag{2.41}$$

which in Euclidean form gives the second possible affine contour snake model:

$$\mathcal{C}_t = \phi_{\text{aff}}\kappa^{1/3}\vec{\mathcal{N}} + \frac{1}{2}\langle\mathcal{C}, \nabla\phi_{\text{aff}}\rangle\kappa^{1/3}\vec{\mathcal{N}}. \tag{2.42}$$

Notice that although both models (2.39) and (2.42) were derived for *convex curves*, the flow (2.42) makes sense in the nonconvex case as well, which makes this the only candidate for a practical affine invariant geometric contour method. Thus we will concentrate on (2.42) from now on, and just consider (2.39) as a model with some theoretical interest.

In order to better capture concavities, to speed up the evolution, as well as to be able to define outward evolutions, a constant inflationary force of the type $\nu\phi\vec{\mathcal{N}}$ may be added to (2.42). This can be obtained from the affine gradient descent of $A_{\phi_{\text{aff}}} := \iint \phi_{\text{aff}}\,dx\,dy$. Note that the inflationary force $\mathcal{C}_t = \phi\vec{\mathcal{N}}$ in the Euclidean case of active contours if obtained from the (Euclidean) gradient descent of $A := \iint \phi\,dx\,dy$, where ϕ is a regular edge detector. We should note that although this constant speed term is not always needed to capture a given contour, for real-world images it is certainly very helpful. We have not found an affine invariant "inflationary" type term, and given the fact that the affine normal involves higher derivatives, we doubt that there is such an expression. Formal results regarding existence and uniqueness of solutions to (2.42) can be derived following the same techniques used for the Euclidean case.

2.6. *Additional extensions and modifications*

A number of fundamental extensions to the theory of geodesic active contours exist. We briefly present a few of them now in order to encourage and motivate the reader to refer to these very important contributions.

In LORIGO, FAUGERAS, GRIMSON, KERIVEN and KIKINIS [1998] the authors showed how to combine the geodesic active contours framework with the high codimension level-sets theory of AMBROSIO and SONER [1996] in order to segment thin tubes in 3D.

One of the fundamental assumptions of the geodesic active contours in all its variants described so far is the presence of significant edges at the boundary of the objects of interests. Although this is significantly alleviated due to the attraction term $\nabla g \cdot \vec{\mathcal{N}}$, and can be further alleviated with the use of advanced edge detection functions g, the existence of edges is an intrinsic assumption of the schemes (note that as explained before, those edges do not have to have the same gradient value, and can have gaps). A number of works have been proposed in the literature to further address this problem. Paragios and Deriche proposed the *geodesic active regions*, where the basic idea is to have the geodesic contours not only driven by edges but also by regions. In other words, the goal is not only to have the geodesic contours converge to regions of high gradients, as in the original models described above, but also to have them separate "uniform" regions. The segmentation is then driven both by the search of uniform regions and the search of jumps in uniformity (edges). Uniform regions are defined using statistical measurements and texture analysis. The reader is encouraged to read their work (PARAGIOS and DERICHE [1999b], PARAGIOS and DERICHE [1999c]), in order to obtain details and to see how the geodesic active contours can be combined with statistical analysis, texture analysis, and unsupervised learning capabilities in order to produce a state-of-the-art framework for image segmentation.

A related approach was developed by CHAN and VESE [1998], CHAN, SANDBERG and VESE [1999]. (Another related approach was developed by Yezzi, Tsai, and Willsky). The authors have connected active contours, level-sets, variational level-sets, and the Mumford–Shah segmentation algorithm, and developed a framework for image segmentation "without-edges". The basic idea is to formulate a variational problem, related to Mumford–Shah, that searches for uniform regions while penalizing for the number of distinct regions. This energy is then embedded in the variational level-sets framework described in the previous section, and then solved using the appropriate numerical analysis machinery. A bit more detailed, the basic idea is to find two regions defined by a closed curve $\mathcal{C}$, each region with (unknown) gray-value averages A_1 and A_2, such that the set $(\mathcal{C}, A_1, A_2)$ minimizes

$$\begin{aligned} E(A_1, A_2, \mathcal{C}) = {} & \mu_1 \operatorname{length}(\mathcal{C}) + \mu_2 \text{ area inside}(\mathcal{C}) \\ & + \mu_3 \int_{\text{inside}(\mathcal{C})} \big(I(x, y) - A_1\big)^2 \, dx\, dy \\ & + \mu_4 \int_{\text{outside}(\mathcal{C})} \big(I(x, y) - A_2\big)^2 \, dx\, dy, \end{aligned}$$

where μ_i are parameters and $I(x, y)$ is the image. As mentioned above, this is embedded in the variational level-sets framework described before.

Even without any basic modifications in the geodesic active contours, a number of improvements and extensions can be obtained playing with the g function, that is, with the "edge" detection component of the framework. An example of this was presented above, where we showed how to re-define g for vector-valued images, permitting the segmentation of color and texture data. Paragios and Deriche pioneered the use of the geodesic active contours framework for tracking and the detection of moving objects, e.g., PARAGIOS and DERICHE [1998b], PARAGIOS and DERICHE [1999d]. The basic idea is to add to g a temporal component, that is, edges are not only defined by spatial gradients (as in still images), but also by temporal gradients. An object that is not moving will not be detected since its temporal gradient is zero.

This is still an active area of research, and more important contributions continue to appear by these and other authors (see also TERZOPOULOS and SZELISKI [1992]).

2.7. *Tracking and morphing active contours*

In the previous sections, the metric g used to detect the objects in the scene was primarily based on the spatial gradient or spatial vector gradient. That is, the metric favors areas of high gradient. In PARAGIOS and DERICHE [1997], the authors propose a different metric that allows for the detection, not just of the scene objects, but of the scene objects that are moving. In other words, given two consecutive frames of a video sequence, the goal is now to detect and track objects that are moving in the scene. The idea is then to define a new function g that not only includes spatial gradient information, which will direct the geodesic curve to all the objects in the scene, but also temporal gradient, driving the geodesic to only the objects that are moving. Having this concept in mind, many different g functions can be proposed.

We now describe a related approach for tracking. The basic idea, inspired in part by the work on geodesic active contours and the work of Paragios and Deriche mentioned above, is to use information from one or more images to perform some operation on an additional image. Examples of this are given in Fig. 2.11. On the top row we have two consecutive slices of a 3D image obtained from electronic microscopy. The image on the left has, superimposed, the contour of an object (a slice of a neuron). We can use this information to drive the segmentation of the next slice, the image on the right. On the bottom row we see two consecutive frames of a video sequence. The image on the left shows a marked object that we want to track. Once again, we can use the image on the left to perform the tracking operation in the image on the right. These are the type of problems we address in this section.

Our approach is based on deforming the contours of interest from the first image toward the desired place in the second one. More specifically, we use a system of coupled partial differential equations (PDEs) to achieve this (coupled PDEs have already been used in the past to address other image processing tasks, see ROMENY [1994], PROESMANS, PAUWELS and VAN GOOL [1994] and references therein). The first partial differential equation deforms the first image, or features of it, toward the second one. The additional PDE is driven by the deformation velocity of the first one, and it deforms the curves of interest in the first image toward the desired position in the second one. This last deformation is implemented using the level-sets numerical scheme, allowing for changes in the topology of the deforming curve. That is, if the objects of interest split or merge from the first image to the second one, these topology changes are automatically handled by the algorithm. This means that we will be able to track scenes with dynamic occlusions and to segment 3D medical data where the slices contain cuts with different topologies.

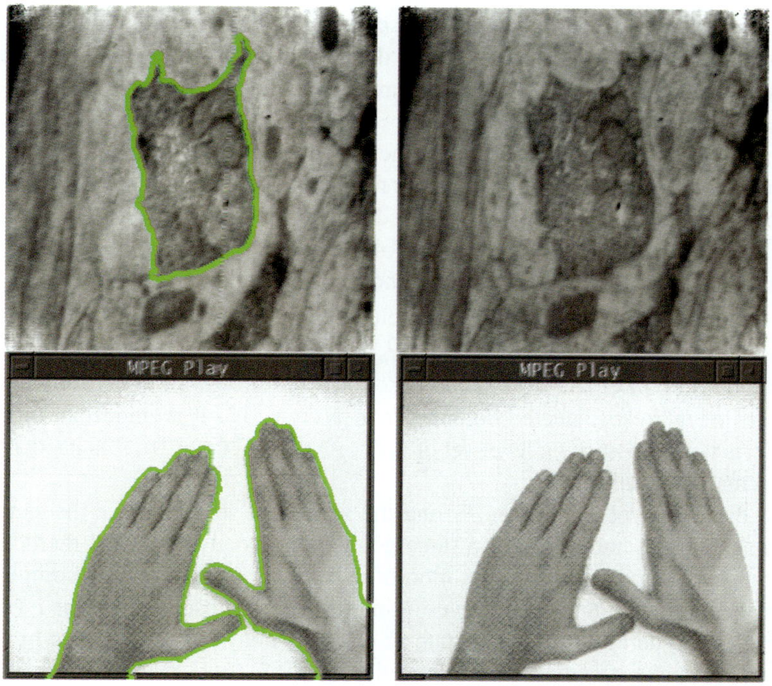

FIG. 2.11. Examples of the problems addressed in this section. See text.

Let $I_1(x, y, 0): \mathbb{R}^2 \to \mathbb{R}$ be the current frame (or slice), where we have already segmented the object of interest. The boundary of this object is given by $\mathcal{C}_{I_1}(p, 0): \mathbb{R} \to \mathbb{R}^2$. Let $I_2(x, y): \mathbb{R}^2 \to \mathbb{R}$ be the image of the next frame, where we have to detect the new position of the object originally given by $\mathcal{C}_{I_1}(p, 0)$ in $I_1(x, y, 0)$. Let us define a continuous and Lipschitz function $u(x, y, 0): \mathbb{R}^2 \to \mathbb{R}$, such that its zero level-set is the curve $\mathcal{C}_{I_1}(p, 0)$. This function can be for example the signed distance function from $\mathcal{C}_{I_1}(p, 0)$. Finally, let's also define $\mathcal{F}_1(x, y, 0): \mathbb{R}^2 \to \mathbb{R}$ and $\mathcal{F}_2(x, y): \mathbb{R}^2 \to \mathbb{R}$ to be images representing features of $I_1(x, y, 0)$ and $I_2(x, y)$ respectively (e.g., $\mathcal{F}_i \equiv I_i$, or $\mathcal{F}_i$ equals the edge maps of I_i, $i = 1, 2$). With these functions as initial conditions, we define the following system of coupled evolution equations (t stands for the marching variable):

$$
\begin{aligned}
\frac{\partial \mathcal{F}_1(x, y, t)}{\partial t} &= \beta(x, y, t) \left\| \nabla \mathcal{F}_1(x, y, t) \right\|, \\
\frac{\partial u(x, y, t)}{\partial t} &= \hat{\beta}(x, y, t) \left\| \nabla u(x, y, t) \right\|,
\end{aligned}
\tag{2.43}
$$

where the velocity $\hat{\beta}(x, y, t)$ is given by

$$
\hat{\beta}(x, y, t) := \beta(x, y, t) \frac{\nabla \mathcal{F}_1(x, y, t)}{\|\nabla \mathcal{F}_1(x, y, t)\|} \cdot \frac{\nabla u(x, y, t)}{\|\nabla u(x, y, t)\|}. \tag{2.44}
$$

The first equation of this system is the *morphing* equation, where $\beta(x, y, t): \mathbb{R}^2 \times [0, \tau) \to \mathbb{R}$ is a function measuring the 'discrepancy' between the selected features $\mathcal{F}_1(x, y, t)$ and $\mathcal{F}_2(x, y)$. This equation is morphing $\mathcal{F}_1(x, y, t)$ into $\mathcal{F}_2(x, y, t)$, so that $\beta(x, y, \infty) = 0$.

The second equation of this system is the *tracking* equation. The velocity in the second equation, $\hat{\beta}$, is just the velocity of the first one projected into the normal direction of the level-sets of u. Since tangential velocities do not affect the geometry of the evolution, both the level-sets of $\mathcal{F}_1$ and u are following exactly the same geometric flow. In other words, being $\vec{\mathcal{N}}_{\mathcal{F}_1}$ and $\vec{\mathcal{N}}_u$ the inner normals of the level-sets of $\mathcal{F}_1$ and u respectively, these level-sets are moving with velocities $\beta\vec{\mathcal{N}}_{\mathcal{F}_1}$ and $\hat{\beta}\vec{\mathcal{N}}_u$ respectively. Since $\hat{\beta}\vec{\mathcal{N}}_u$ is just the projection of $\beta\vec{\mathcal{N}}_{\mathcal{F}_1}$ into $\vec{\mathcal{N}}_u$, both level sets follow the same geometric deformation. In particular, the zero level-set of u is following the deformation of $\mathcal{C}_{I_1}$, the curves of interest (detected boundaries in $I_1(x, y, 0)$). It is important to note that since $\mathcal{C}_{I_1}$ is not necessarily a level-set of $I_1(x, y, 0)$ or $\mathcal{F}_1(x, y, 0)$, u is needed to track the deformation of this curve.

Since the curves of interest in $\mathcal{F}_1$ and the zero level-set of u have the same initial conditions and they move with the same geometric velocity, they will then deform in the same way. Therefore, when the morphing of $\mathcal{F}_1$ into $\mathcal{F}_2$ has been completed, the zero level-set of u should be the curves of interest in the subsequent frame $I_2(x, y)$.

One could argue that the steady state of (2.43) is not necessarily given by the condition $\beta = 0$, since it can also be achieved with $\|\nabla\mathcal{F}_1(x, y, t)\| = 0$. This is correct, but it should not affect the tracking since we are assuming that the boundaries to track are not placed over regions where there is no information and then the gradient is flat. Therefore, for a certain band around the boundaries the evolution will only stop when $\beta = 0$, thus allowing for the tracking operation.

For the examples below, we have opted for a very simple selection of the functions in the tracking system, namely

$$\mathcal{F}_i = \mathcal{L}(I_i), \quad i = 1, 2, \tag{2.45}$$

and

$$\beta(x, y, t) = \mathcal{F}_2(x, y) - \mathcal{F}_1(x, y, t), \tag{2.46}$$

where $\mathcal{L}(\cdot)$ indicates a band around $\mathcal{C}_{I_1}$. That is, for the evolving curve $\mathcal{C}_{I_1}$ we have an evolving band B of width w around it, and $\mathcal{L}(f(x, y, t)) = f(x, y, t)$ if (x, y) is in B, and it is zero otherwise. This particular *morphing* term is a local measure of the difference between $I_1(t)$ and I_2. It works increasing the grey value of $I_1(x_0, y_0, t)$ if it is smaller than $I_2(x_0, y_0)$, and decreasing it otherwise. Therefore, the steady state is obtained when both values are equal $\forall x_0, y_0$ in B, with $\|\nabla I_1\| \neq 0$. Note that this is a *local* measure, and that no hypothesis concerning the shape of the object to be tracked has been made. Having no model of the boundaries to track, the algorithm becomes very flexible. Being so simple, the main drawback of this particular selection is that it requires an important degree of similarity among the images for the algorithm to track the curves of interest and not to detect spurious objects. If the set of curves $\mathcal{C}_{I_1}$ isolates an almost uniform interior from an almost uniform exterior, then there is no need for

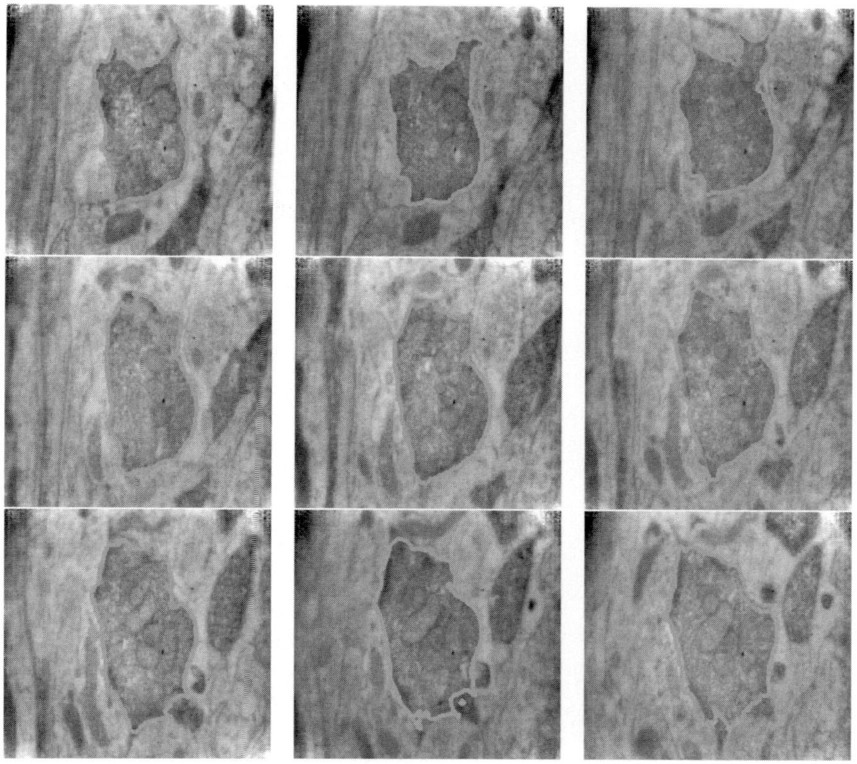

FIG. 2.12. Nine consecutive slices of neural tissue. The first image has been segmented manually. The segmentation over the sequence has been performed using the algorithm described in this section.

a high similarity among consecutive images. On the other hand, when working with more cluttered images, if $\mathcal{C}_{I_1}(0)$ is too far away from the expected limit $\lim_{t\to\infty}\mathcal{C}_{I_1}(t)$, then the above-mentioned errors in the tracking procedure may occur. This *similarity* requirement concerns not only the shapes of the objects depicted in the image but especially their grey levels, since this β function measures grey-level differences. Therefore, histogram equalization is always performed as a pre-processing operation.

We should also note that this particular selection of β involves information of the two present images. Better results are expected if information from additional images in the sequence are taken into account to perform the *morphing* among these two.

The first example of the tracking algorithm is presented in Fig. 2.12. This figure shows nine consecutive slices of neural tissue obtained via electronic microscopy (EM). The goal of the biologist is to obtain a three dimensional reconstruction of this neuron. As we observe from these examples, the EM images are very noisy, and the boundaries of the neuron are not easy to identify or to tell apart from other similar objects. Segmenting the neuron is then a difficult task. Before processing for segmentation, the images are regularized using anisotropic diffusion, see next chapter. Active contours techniques as those in described before will normally fail with this type of images. Since the variation

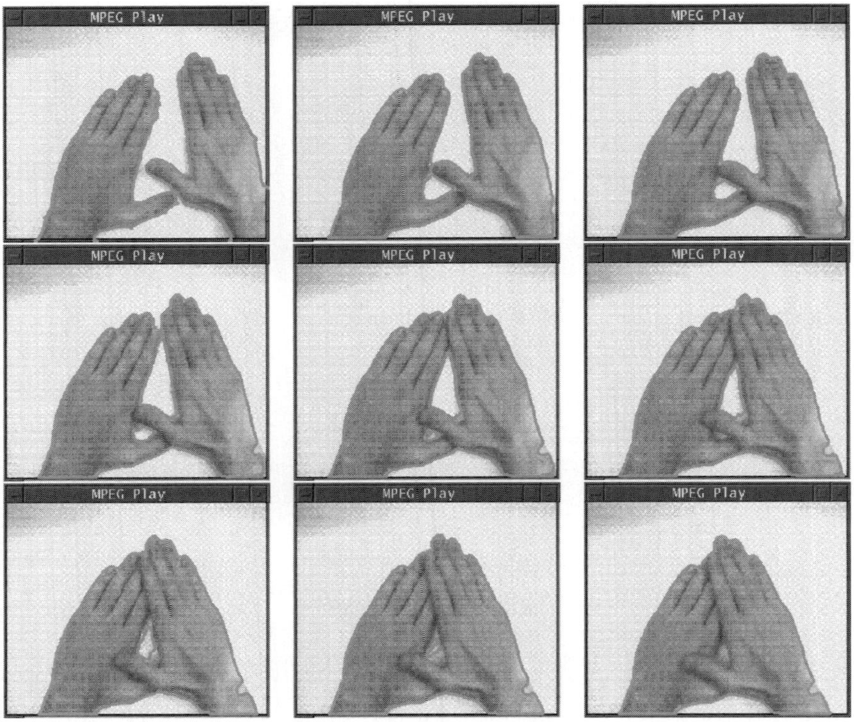

FIG. 2.13. Nine frames of a movie. The first image has been segmented manually. The segmentation over the sequence has been performed using the algorithm described in this section. Notice the automatic handling of topology changes.

between consecutive slices is not too large, we can use the segmentation obtained for the first slice (segmentation obtained either manually or with the edge tracing technique described before), to drive the segmentation of the next one, and then automatically proceed to find the segmentation in the following images. In this figure, the top left image shows the manual or semi-automatic segmentation superimposed, while the following ones show the boundaries found by the algorithm. Due to the particular choice of the β function, dissimilarities among the images cause the algorithm to mark as part of the boundary small objects which are too close to the object of interest. These can be removed by simple morphological operations. Cumulative errors might cause the algorithm to lose track of the boundaries after several slices, and re-initialization would be required.

One could argue that we could also use the segmentation of the first frame to initialize the active contours techniques mentioned above for the next frame. We still encounter a number of difficulties with this approach: (1) The deforming curve gets attracted to local minima, and often fails to detect the neuron; (2) Those algorithms normally deform either inwards or outwards (mainly due to the presence of balloon-type forces), while the boundary curve corresponding to the first image is in general neither inside nor outside the object in the second image. To solve this, more elaborated techniques, e.g., PARAGIOS and DERICHE [1999b], have to be used. Therefore, even if the image is not noisy, special techniques need to be developed and implemented to direct different points of the curve toward different directions.

Fig. 2.13 shows an example of object tracking. The top left image has, superimposed, the contours of the objects to track. The following images show the contours found by the algorithm. For sake of space, only one every three frames is shown. Notice how topological changes are handled automatically. As mentioned before, a pioneering topology independent algorithm for tracking in video sequences, based on the general geodesic framework can be found in PARAGIOS and DERICHE [1998a], PARAGIOS and DERICHE [1999b]. In contrast with this approach, that scheme is based on a unique PDE (no morphing flow), deforming the curve toward a (local) geodesic curve, and it is very sensible to spatial and temporal noisy gradients. Due to the similarity between frames, the algorithm just described converges very fast. The CONDENSATION algorithm described in BLAKE and ISARD [1998] can also achieve, in theory, topology-free tracking, though to the best of our knowledge real examples showing this capability have not been yet reported. In addition, this algorithm requires having a model of the object to track and a model of the possible deformations, even for simple and useful examples as the ones shown below (note that the algorithm here proposed requires no previous or learned information). On the other hand, the outstanding tracking capabilities for cluttered scenes shown with the CONDENSATION scheme can not be obtained with the simple selections for $\mathcal{F}_i$ and β used for the examples below, and more advanced selections must be investigated. Additional tracking examples are given in SAPIRO [2001].

3. Geometric diffusion of scalar images

In the previous section we presented PDEs that deformed curves and surfaces according to combinations of their intrinsic geometry and external, image-dependent, velocities. In this and following sections we will deal with PDEs applied to full images. We first show how linear PDEs are related to linear, Gaussian, filtering, and then extend these models introducing nonlinear PDEs for image enhancement.

3.1. Gaussian filtering and linear scale-spaces

The simplest, and possible one of the most popular ways to smooth an image $I(x, y):\mathbb{R}^2 \to \mathbb{R}$ is to filter it with a (radial) Gaussian filter (BABAUD, WITKIN, BAUDIN and DUDA [1986]), centered at 0, and with isotropic variance t. We then obtain

$$I(x, y, t) = I(x, t)G_{0,t}.$$

For different variances t we obtain different levels of smoothing; see Fig. 3.1. This defines a *scale-space* for the image. That is, we get copies of the image at different scales. Note of course that any scale t_1 can be obtained from a scale t_0, where $t_0 < t_1$, as well as from the original images. This is what we denoted before as the causality criteria for scale-spaces.

FIG. 3.1. Smoothing an image with a series of Gaussian filters. Note how the information is gradually being lost when the variance of the Gaussian filter increases from left to right, top to bottom.

Filtering with a Gaussian has many properties, and this *scale-space* can be obtained following a series of intuitive physically based axioms. One of the fundamental properties of Gaussian filtering, which we have already seen when dealing with curves, is that $I(x, y, t)$ satisfies the *linear heat flow* or *Laplace equation*:

$$\frac{\partial I(x, y, t)}{\partial t} = \Delta I(x, y, t) = \frac{\partial^2 I(x, y, t)}{\partial x^2} + \frac{\partial^2 I(x, y, t)}{\partial y^2}. \tag{3.1}$$

This is very easy to show, first showing that the Gaussian function itself satisfies the Laplace equation (it is the *kernel*), and then adding the fact that derivatives and filtering are linear operations.

The flow (3.1) is also denoted as *isotropic diffusion*, since it is diffusing the information equally in all directions. It is well know that the "heat" (gray-values) will spread, and at the end, a uniform image, equal to the average of the initial "heat" (gray-values) is obtained. This can be observed in Fig. 3.1. Although this is very good for local reducing noise (averaging is optimal for additive noise), this filter also destroys the image content, that is, its boundaries. The goal of the rest of this section is to replace the isotropic diffusion by anisotropic ones, that remove noise while preserving edges.

3.2. *Edge stopping diffusion*

We have just seen that we can smooth an image using the Laplace equation or heat flow, with the image as initial condition. This flow though not only removes noise, but also blurs the image. HUMMEL [1986] suggested that the heat flow is not the only PDE that can be used to enhance an image, and that in order to keep the scale-space property we only need to make sure that the flow we use holds the maximum principle. Many approaches have been taken in the literature to implement this idea. Although not all of the them really hold the maximum principle, they do share the concept of replacing the linear heat flow by a nonlinear PDEs that does not diffuse the image in a uniform way. These flows are normally denoted as *anisotropic diffusion*, to contrast with the heat flow, that diffuses the image *isotropically* (equal in all directions).

The first elegant formulation of anisotropic diffusion was introduced by PERONA and MALIK [1990] (see GABOR [1965] for very early work in this topic and also RUDIN, OSHER and FATEMI [1992a] for additional pioneer work), and since then a considerable amount of research has been devoted to the theoretical and practical understanding of this and related methods for image enhancement. Research in this area has been oriented toward understanding the mathematical properties of anisotropic diffusion and related variational formulations (AUBERT and VESE [to appear], CATTE, LIONS, MOREL and COLL [1992], KICHENASSAMY [1996], PERONA and MALIK [1990], YOU, XU, TANNENBAUM and KAVEH [1996]), developing related well-posed and stable equations (ALVAREZ, GUICHARD, LIONS and MOREL [1993], ALVAREZ, LIONS and MOREL [1992], CATTE, LIONS, MOREL and COLL [1992], GUICHARD [1993], NITZBERG and SHIOTA [1992], RUDIN, OSHER and FATEMI [1992a], YOU, XU, TANNENBAUM and KAVEH [1996]), extending and modifying anisotropic diffusion for fast and accurate implementations, modifying the diffusion equations for specific applications (GERIG, KUBLER, KIKINIS and JOLESZ [1992]), and studying the relations between anisotropic diffusion and other image processing operations (SAPIRO [1996], SHAH [1996]). In this section we develop a statistical interpretation of anisotropic diffusion, specifically, from the point of view of robust statistics. We will present the Perona–Malik diffusion equation and show that it is equivalent to a robust procedure that estimates a piecewise constant image from a noisy input image.

The robust statistical interpretation also provides a means for detecting the boundaries (edges) between the piecewise constant regions in an image that has been smoothed with anisotropic diffusion. The boundaries between the piecewise constant regions are considered to be "outliers" in the robust estimation framework. Edges in a smoothed image are, therefore, very simply detected as those points that are treated as outliers.

We will also show (following BLACK and RANGARAJAN [1996]) that, for a particular class of robust error norms, anisotropic diffusion is equivalent to regularization with an explicit line process. The advantage of the line-process formulation is that we can add constraints on the spatial organization of the edges. We demonstrate that adding such constraints to the Perona–Malik diffusion equation results in a qualitative improvement in the continuity of edges.

Perona–Malik formulation

As we saw in previous section, diffusion algorithms remove noise from an image by modifying the image via a partial differential equation (PDE). For example, consider applying the isotropic diffusion equation (the heat equation) discussed in previous section, given by

$$\frac{\partial I(x,y,t)}{\partial t} = \operatorname{div}(\nabla I),$$

using of course the original (degraded/noisy) image $I(x,y,0)$ as the initial condition, where once again, $I(x,y,0):\mathbb{R}^2 \to \mathbb{R}^+$ is an image in the continuous domain, (x,y) specifies spatial position, t is an artificial time parameter, and where ∇I is the image gradient.

PERONA and MALIK [1990] replaced the classical isotropic diffusion equation with

$$\frac{\partial I(x,y,t)}{\partial t} = \operatorname{div}\bigl(g(\|\nabla I\|)\nabla I\bigr), \tag{3.2}$$

where $\|\nabla I\|$ is the gradient magnitude, and $g(\|\nabla I\|)$ is an "edge-stopping" function. This function is chosen to satisfy $g(x) \to 0$ when $x \to \infty$ so that the diffusion is "stopped" across edges.

As mentioned before, (3.2) motivated a large number of researchers to study the mathematical properties of this type of equation, as well as its numerical implementation and adaptation to specific applications. The stability of the equation was the particular concern of extensive research, e.g., ALVAREZ, LIONS and MOREL [1992], CATTE, LIONS, MOREL and COLL [1992], KICHENASSAMY [1996], PERONA and MALIK [1990], YOU, XU, TANNENBAUM and KAVEH [1996]. We should note that the mathematical study of that equation is not straightforward, although it can be shown that if $g(\cdot)$ is computed over a smoothed version of I, the flow is well posed and has a unique solution (CATTE, LIONS, MOREL and COLL [1992]). Note of course that a reasonable numerical implementation intrinsically smooths the gradient, and then it is expected to be stable.

In what follow we present equations that are modifications of (3.2). We do not discuss the stability of these modified equations because the stability results can be obtained from the mentioned references. Note again that possible stability problems will typically be solved, or at least moderated, by the spatial regularization and temporal delays introduced by the numerical methods for computing the gradient in $g(\|\nabla I\|)$ (CATTE, LIONS, MOREL and COLL [1992], KICHENASSAMY [1996], OSHER and RUDIN [1990]).

Perona–Malik discrete formulation. Perona and Malik discretized their anisotropic diffusion equation as follows:

$$I_s^{t+1} = I_s^t + \frac{\lambda}{|\eta_s|} \sum_{p \in \eta_s} g(\nabla I_{s,p}) \nabla I_{s,p}, \tag{3.3}$$

where I_s^t is a discretely-sampled image, s denotes the pixel position in a discrete, two-dimensional grid, and t now denotes discrete time steps (iterations). The constant $\lambda \in \mathbb{R}^+$ is a scalar that determines the rate of diffusion, η_s represents the spatial neighborhood of pixel s, and $|\eta_s|$ is the number of neighbors (usually 4, except at the image boundaries). Perona and Malik linearly approximated the image gradient (magnitude) in a particular direction as

$$\nabla I_{s,p} = I_p - I_s^t, \quad p \in \eta_s. \tag{3.4}$$

Fig. 3.9 shows examples of applying this equation to an image, using two different choices for the edge-stopping function, $g(\cdot)$. Qualitatively, the effect of anisotropic diffusion is to smooth the original image while preserving brightness discontinuities. As we will see, the choice of $g(\cdot)$ can greatly affect the extent to which discontinuities are preserved. Understanding this is one of the main goals of this chapter.

A statistical view. Our goal is to develop a statistical interpretation of the Perona–Malik anisotropic diffusion equation. Toward that end, we adopt an oversimplified statistical model of an image. In particular, we assume that a given input image is a piecewise constant function that has been corrupted by zero-mean Gaussian noise with small variance. In YOU, XU, TANNENBAUM and KAVEH [1996], the authors presented interesting theoretical (and practical) analysis of the behavior of anisotropic diffusion for piecewise constant images. We will return later to comment on their results.

Consider the image intensity differences, $I_p - I_s$, between pixel s and its neighboring pixels p. Within one of the piecewise constant image regions, these neighbor differences will be small, zero-mean, and normally distributed. Hence, an optimal estimator for the "true" value of the image intensity I_s at pixel s minimizes the square of the neighbor differences. This is equivalent to choosing I_s to be the mean of the neighboring intensity values.

The neighbor differences will not be normally distributed, however, for an image region that includes a boundary (intensity discontinuity). Consider, for example, the image region illustrated in Fig. 3.2. The intensity values of the neighbors of pixel s are drawn from two different populations, and in estimating the "true" intensity value at s we want to include only those neighbors that belong to the same population. In particular, the pixel labeled p is on the wrong side of the boundary so I_p will skew the estimate of I_s significantly. With respect to our assumption of Gaussian noise within each constant region, the neighbor difference $I_p - I_s$ can be viewed as an *outlier* because it does not conform to the statistical assumptions.

Robust estimation. The field of robust statistics (HAMPEL, RONCHETTI, ROUSSEEUW and STAHEL [1986], HUBER [1981]) is concerned with estimation problems in which the data contains gross errors, or outliers.

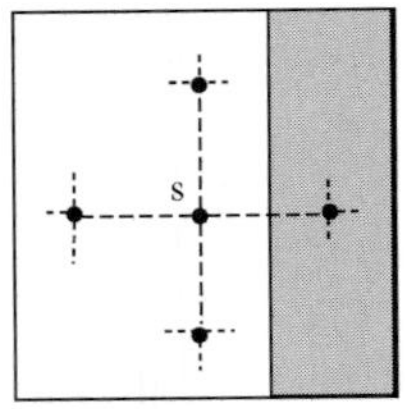

FIG. 3.2. Local neighborhood of pixels at a boundary (intensity discontinuity).

Many robust statistical techniques have been applied to standard problems in computer vision (PROC. INT. WORKSHOP ON ROBUST COMPUTER VISION [1990], MEER, MINTZ, ROSENFELD and KIM [1991], SCHUNCK [1990]). There are robust approaches for performing local image smoothing (BESL, BIRCH and WATSON [1988]), image reconstruction (GEMAN and YANG [1995], GEMAN and GEMAN [1984]), blur classification (CHEN and SCHUNCK [1990]), surface reconstruction (SINHA and SCHUNCK [1992]), segmentation (MEER, MINTZ and ROSENFELD [1990]), pose estimation (KUMAR and HANSON [1990]), edge detection (LUI, SCHUNCK and MEYER [1990]), structure from motion or stereo (TIRUMALAI, SCHUNCK and JAIN [1990], WENG and COHEN [1990]), optical flow estimation (BLACK and ANANDAN [1991], BLACK and ANANDAN [1993], SCHUNCK [1989]), and regularization with line processes (BLACK and RANGARAJAN [1996]). For further details see HAMPEL, RONCHETTI, ROUSSEEUW and STAHEL [1986] or, for a review of the applications of robust statistics in computer vision, see MEER, MINTZ, ROSENFELD and KIM [1991].

The problem of estimating a piecewise constant (or smooth) image from noisy data can also be posed using the tools of robust statistics. We wish to find an image I that satisfies the following optimization criterion:

$$\min_I \sum_{s\in I} \sum_{p\in\eta_s} \rho(I_p - I_s, \sigma), \tag{3.5}$$

where $\rho(\cdot)$ is a robust error norm and σ is a "scale" parameter that will be discussed further below. To minimize (3.5), the intensity at each pixel must be "close" to those of its neighbors. As we shall see, an appropriate choice of the ρ-function allows us to minimize the effect of the outliers, $(I_p - I_s)$, at the boundaries between piecewise constant image regions.

Eq. (3.5) can be solved by gradient descent

$$I_s^{t+1} = I_s^t + \frac{\lambda}{|\eta_s|} \sum_{p\in\eta_s} \psi\big(I_p - I_s^t, \sigma\big), \tag{3.6}$$

where $\psi(\cdot) = \rho'(\cdot)$, and t again denotes the iteration. The update is carried out simultaneously at every pixel s.

The specific choice of the robust error norm or ρ-function in (3.5) is critical. To analyze the behavior of a given ρ-function, we consider its derivative (denoted ψ), which is proportional to the *influence function* (HAMPEL, RONCHETTI, ROUSSEEUW and STAHEL [1986]). This function characterizes the bias that a particular measurement has on the solution. For example, the quadratic ρ-function has a linear ψ-function.

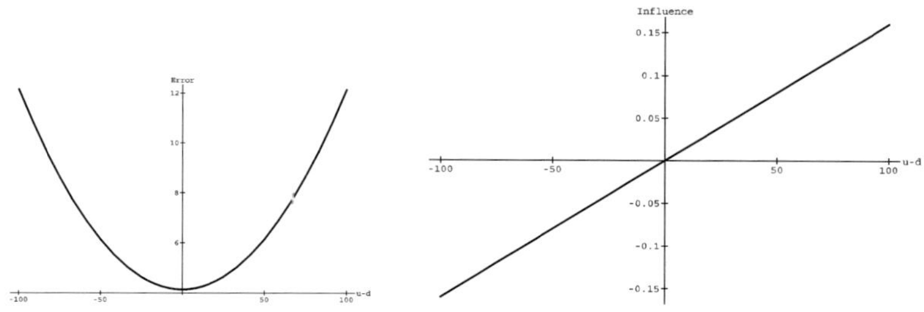

$\rho(x,\sigma)$ $\psi(x,\sigma)$

FIG. 3.3. Least-squares (quadratic) error norm.

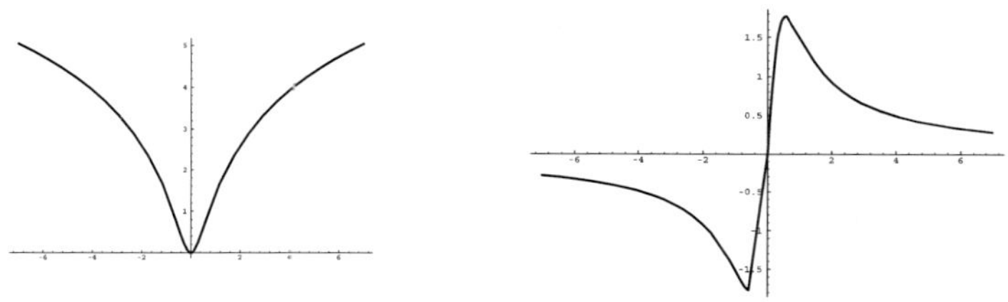

$\rho(x,\sigma)$ $\psi(x,\sigma)$

FIG. 3.4. Lorentzian error norm.

If the distribution of values $(I_p - I_s^t)$ in every neighborhood is a zero-mean Gaussian, then $\rho(x,\sigma) = x^2/\sigma^2$ provides an optimal local estimate of I_s^t. This *least-squares* estimate of I_s^t is, however, very sensitive to outliers because the influence function increases linearly and without bound (see Fig. 3.3). For a quadratic ρ, I_s^{t+1} is assigned to be the mean of the neighboring intensity values I_p.This selection leads to the isotropic diffusion flow. When these values come from different populations (across a boundary) the mean is not representative of either population, and the image is blurred too much. Hence, the quadratic gives outliers (large values of $|\nabla I_{s,p}|$) too much *influence*.

To increase robustness and *reject* outliers, the ρ-function must be more forgiving about outliers; that is, it should increase less rapidly than x^2. For example, consider the following *Lorentzian* error norm plotted in Fig. 3.4:

$$\rho(x,\sigma) = \log\left(1 + \frac{1}{2}\left(\frac{x}{\sigma}\right)^2\right), \qquad \psi(x,\sigma) = \frac{2x}{2\sigma^2 + x^2}. \tag{3.7}$$

Examination of the ψ-function reveals that, when the absolute value of the gradient magnitude increases beyond a fixed point determined by the scale parameter σ, its influence is reduced. We refer to this as a *redescending* influence function (HAMPEL, RONCHETTI, ROUSSEEUW and STAHEL [1986]). If a particular local difference, $\nabla I_{s,p} = I_p - I_s^t$, has a large magnitude then the value of $\psi(\nabla I_{s,p})$ will be small and therefore that measurement will have little effect on the update of I_s^{t+1} in (3.6).

Robust statistics and anisotropic diffusion

We now explore the relationship between robust statistics and anisotropic diffusion by showing how to convert back and forth between the formulations. Recall the continuous anisotropic diffusion equation:

$$\frac{\partial I(x, y, t)}{\partial t} = \operatorname{div}\big(g\big(\|\nabla I\|\big)\nabla I\big). \tag{3.8}$$

The continuous form of the robust estimation problem in (3.5) can be posed as:

$$\min_I \int_\Omega \rho\big(\|\nabla I\|\big)\, d\Omega, \tag{3.9}$$

where Ω is the domain of the image and where we have omitted σ for notational convenience. One way to minimize (3.9) is via gradient descent using the calculus of variations theory (see, for example, GUICHARD [1993], NORDSTRÖM [1990], PERONA and MALIK [1990], YOU, XU, TANNENBAUM and KAVEH [1996] for the use of this formulation):

$$\frac{\partial I(x, y, t)}{\partial t} = \operatorname{div}\left(\rho'\big(\|\nabla I\|\big)\frac{\nabla I}{\|\nabla I\|}\right). \tag{3.10}$$

By defining

$$g(x) := \frac{\rho'(x)}{x}, \tag{3.11}$$

we obtain the straightforward relation between image reconstruction via robust estimation (3.9) and image reconstruction via anisotropic diffusion (3.8). YOU, XU, TANNENBAUM and KAVEH [1996] show and make extensive use of this important relation in their analysis.

The same relationship holds for the discrete formulation; compare (3.3) and (3.6) with $\psi(x) = \rho'(x) = g(x)\,x$. Note that additional terms will appear in the gradient descent equation if the magnitude of the image gradient is discretized in a nonlinear fashion. We proceed with the discrete formulation as given in previous section. The basic results we present hold for the continuous domain as well.

Perona and Malik suggested two different edge stopping $g(\cdot)$ functions in their anisotropic diffusion equation. Each of these can be viewed in the robust statistical framework by converting the $g(\cdot)$ functions into the related ρ-functions.

Perona and Malik first suggested

$$g(x) = \frac{1}{1 + \frac{x^2}{K^2}} \tag{3.12}$$

for a positive constant K. We want to find a ρ-function such that the iterative solution of the diffusion equation and the robust statistical equation are equivalent. Letting $K^2 = 2\sigma^2$, we have

$$g(x)x = \frac{2x}{2 + \frac{x^2}{\sigma^2}} = \psi(x, \sigma), \tag{3.13}$$

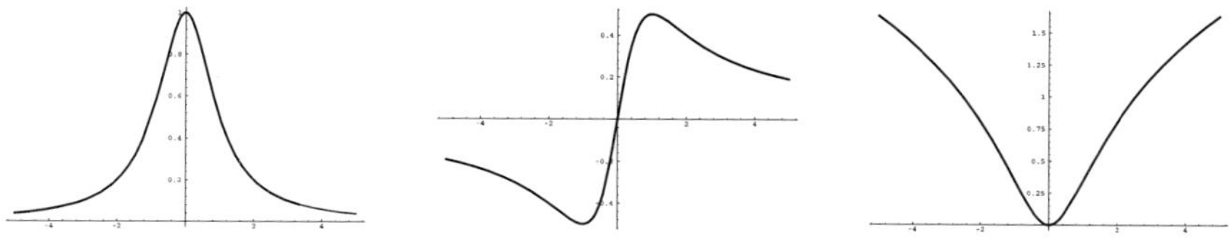

$g(x)$ $g(x)x = \psi(x)$ $\rho(x)$

FIG. 3.5. Lorentzian error norm and the Perona–Malik g stopping function.

where $\psi(x, \sigma) = \rho'(x, \sigma)$. Integrating $g(x)x$ with respect to x gives:

$$\int g(x)x\,dx = \sigma^2 \log\left(1 + \frac{1}{2}\left(\frac{x^2}{\sigma^2}\right)\right) = \rho(x). \tag{3.14}$$

This function $\rho(x)$ is proportional to the Lorentzian error norm introduced in the previous section, and $g(x)x = \rho'(x) = \psi(x)$ is proportional to the influence function of the error norm; see Fig. 3.5. Iteratively solving (3.6) with a Lorentzian for ρ is therefore equivalent to the discrete Perona–Malik formulation of anisotropic diffusion. This relation was previously pointed out in YOU, XU, TANNENBAUM and KAVEH [1996] (see also BLACK and RANGARAJAN [1996], NORDSTRÖM [1990]).

The same treatment can be used to recover a ρ-function for the other g-function proposed by Perona and Malik

$$g(x) = e^{-\frac{x^2}{k^2}}. \tag{3.15}$$

The resulting ρ-function is related to the robust error norm proposed by LECLERC [1989]. The derivation is straightforward and is omitted here.

Exploiting the relationship

The above derivations demonstrate that anisotropic diffusion is the gradient descent of an estimation problem with a familiar robust error norm. What's the advantage of knowing this connection? We argue that the robust statistical interpretation gives us a broader context within which to evaluate, compare, and choose between alternative diffusion equations. It also provides tools for automatically determining what should be considered an outlier (an "edge"). In this section we illustrate these connections with an example.

While the Lorentzian is more robust than the L_2 (quadratic) norm, its influence does not descend all the way to zero. We can choose a more "robust" norm from the robust statistics literature which does descend to zero. The *Tukey's biweight*, for example, is plotted along with its influence function in Fig. 3.6:

$$\rho(x, \sigma) = \begin{cases} \frac{x^2}{\sigma^2} - \frac{x^4}{\sigma^4} + \frac{x^6}{3\sigma^6} & |x| \leqslant \sigma, \\ \frac{1}{3} & \text{otherwise,} \end{cases} \tag{3.16}$$

$$\psi(x, \sigma) = \begin{cases} x(1 - (x/\sigma)^2)^2 & |x| \leqslant \sigma, \\ 0 & \text{otherwise,} \end{cases} \tag{3.17}$$

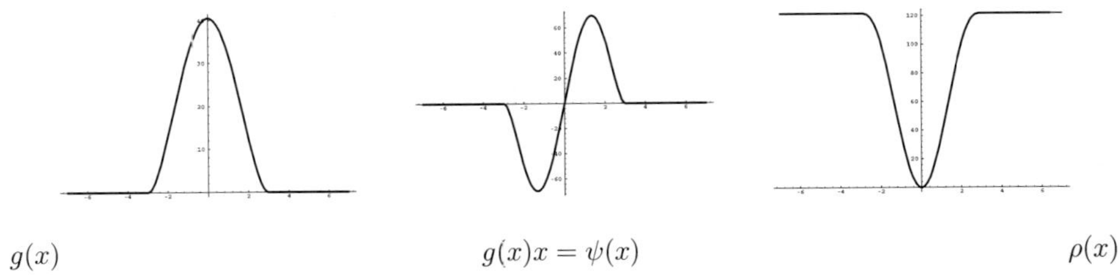

FIG. 3.6. Tukey's biweight.

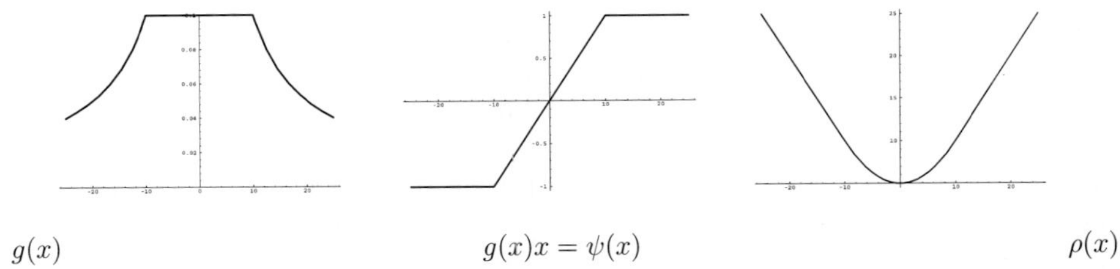

FIG. 3.7. Huber's minmax estimator (modification of the L_1 norm).

$$g(x,\sigma) = \begin{cases} \frac{1}{2}(1-(x/\sigma)^2)^2 & |x| \leqslant \sigma, \\ 0 & \text{otherwise}. \end{cases} \tag{3.18}$$

Another error norm from the robust statistics literature, Huber's [1981] *minimax* norm (see also RUDIN, OSHER and FATEMI [1992a], YOU, XU, TANNENBAUM and KAVEH [1996]), is plotted along with its influence function in Fig. 3.7. Huber's minmax norm is equivalent to the L_1 norm for large values. But, for normally distributed data, the L_1 norm produces estimates with higher variance than the optimal L_2 (quadratic) norm, so Huber's minmax norm is designed to be quadratic for small values:

$$\rho(x,\sigma) = \begin{cases} x^2/2\sigma + \sigma/2, & |x| \leqslant \sigma, \\ |x|, & |x| > \sigma, \end{cases} \tag{3.19}$$

$$\psi(x,\sigma) = \begin{cases} x/\sigma, & |x| \leqslant \sigma, \\ \operatorname{sign}(x), & |x| > \sigma, \end{cases} \tag{3.20}$$

$$g(x,\sigma) = \begin{cases} 1/\sigma, & |x| \leqslant \sigma, \\ \operatorname{sign}(x)/x, & |x| > \sigma. \end{cases} \tag{3.21}$$

We would like to compare the influence (ψ-function) of these three norms, but a direct comparison requires that we dilate and scale the functions to make them as similar as possible.

First, we need to determine how large the image gradient can be before we consider it to be an outlier. We appeal to tools from robust statistics to automatically estimate the "robust scale," σ_e, of the image as (ROUSSEEUW and LEROY [1987])

$$\begin{aligned} \sigma_e &= 1.4826\,\mathrm{MAD}(\nabla I) \\ &= 1.4826\,\mathrm{median}_I\big(\big\|\nabla I - \mathrm{median}_I\big(\|\nabla I\|\big)\big\|\big), \end{aligned} \tag{3.22}$$

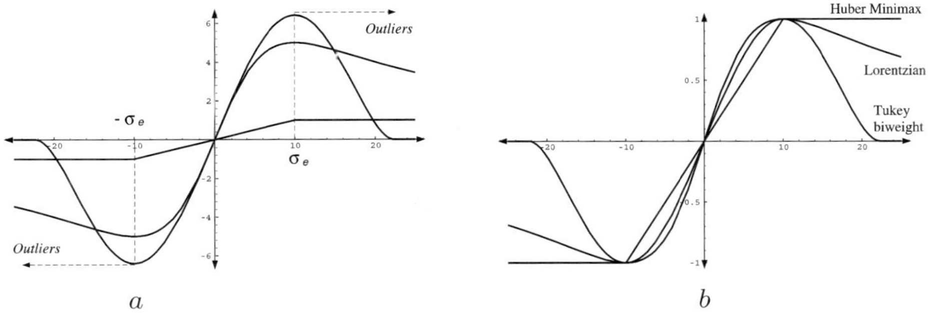

FIG. 3.8. Lorentzian, Tukey, and Huber ψ-functions. (a) values of σ chosen as a function of σ_e so that outlier "rejection" begins at the same value for each function; (b) the functions aligned and scaled.

where "MAD" denotes the median absolute deviation and the constant is derived from the fact that the MAD of a zero-mean normal distribution with unit variance is $0.6745 = 1/1.4826$. For a discrete image, the robust scale, σ_e, is computed using the gradient magnitude approximation introduced before.

Second, we choose values for the scale parameters σ to dilate each of the three influence functions so that they begin rejecting outliers at the same value: σ_e. The point where the influence of outliers first begins to decrease occurs when the derivative of the ψ-function is zero. For the modified L_1 norm this occurs at $\sigma_e = \sigma$. For the Lorentzian norm it occurs at $\sigma_e = \sqrt{2}\,\sigma$ and for the Tukey norm it occurs at $\sigma_e = \sigma/\sqrt{5}$. Defining σ with respect to σ_e in this way we plot the influence functions for a range of values of x in Fig. 3.8a. Note how each function now begins reducing the influence of measurements at the same point.

Third, we scale the three influence functions so that they return values in the same range. To do this we take λ in (3.3) to be one over the value of $\psi(\sigma_e, \sigma)$. The scaled ψ-functions are plotted in Fig. 3.8b.

Now we can compare the three error norms directly. The modified L_1 norm gives all outliers a constant weight of one while the Tukey norm gives *zero* weight to outliers whose magnitude is above a certain value. The Lorentzian (or Perona–Malik) norm is in between the other two. Based on the shape of $\psi(\cdot)$ we would correctly predict that diffusing with the Tukey norm produces sharper boundaries than diffusing with the Lorentzian (standard Perona–Malik) norm, and that both produce sharper boundaries than the modified L_1 norm. We can also see how the choice of function affects the "stopping" behavior of the diffusion; given a piecewise constant image where all discontinuities are above a threshold, the Tukey function will leave the image unchanged whereas the other two functions will not.

These predictions are born out experimentally, as can be seen in Fig. 3.9. The figure compares the results of diffusing with the Lorentzian $g(\cdot)$ function and the Tukey g function. The value of $\sigma_e = 10.278$ was estimated automatically using (3.22) and the values of σ and λ for each function were defined with respect to σ_e as described above. The figure shows the diffused image after 100 iterations of each method. Observe how the Tukey function results in sharper discontinuities.

FIG. 3.9. Comparison of the Perona–Malik (Lorentzian) function (left) and the Tukey function (right) after 100 iterations. Top: original image. Middle: diffused images. Bottom: magnified regions of diffused images.

We can detect edges in the smoothed images very simply by detecting those points that are treated as outliers by the given ρ-function. Fig. 3.10 shows the outliers (edge points) in each of the images, where $|\nabla I_{s,p}| > \sigma_e$.

Finally, Fig. 3.11 illustrates the behavior of the two functions in the limit (shown for 500 iterations). The Perona–Malik formulation continues to smooth the image while the Tukey version has effectively "stopped."

These examples illustrate how ideas from robust statistics can be used to evaluate and compare different g-functions and how new functions can be chosen in a principled way. See BLACK and RANGARAJAN [1996] for other robust ρ-functions which could be used for anisotropic diffusion. See also GEIGER and YUILLE [1991] for related work connecting anisotropic diffusion, the mean-field ρ-function, and binary line processes.

FIG. 3.10. Comparison of edges (outliers) for the Perona–Malik (Lorentzian) function (left) and the Tukey function (right) after 100 iterations. Bottom row shows a magnified region.

It is interesting to note that common robust error norms have frequently been proposed in the literature without mentioning the motivation from robust statistics. For example, RUDIN, OSHER and FATEMI [1992a] proposed a formulation that is equivalent to using the L_1 norm (TV or Total Variation). As expected from the robust formulation here described, and further analyzed in detail by Caselles and colleagues, this norm will have a flat image as unique steady state solution. YOU, XU, TANNENBAUM and KAVEH [1996] explored a variety of anisotropic diffusion equations and reported better results for some than for others. In addition to their own explanation for this, their results are predicted, following the development presented here, by the robustness of the various error norms they use. Moreover, some of their theoretical results, e.g., Theorem 1 and Theorem 3, are easily interpreted based on the concept of influence functions. Finally, Mead and colleagues (HARRIS, KOCH, STAATS and LUO [1990], MEAD [1989]) have used analog VLSI (aVLSI) technology to build hardware devices that perform regularization. The aVLSI circuits behave much like a resistive grid, except that the resistors are replaced with "robust" resistors made up of several transistors. Each such resistive grid circuit is equivalent to using a different robust error norm.

Robust estimation and line processes

This section derives the relationship between anisotropic diffusion and regularization with line processes. The connection between robust statistics and line processes has been explored elsewhere; see BLACK and RANGARAJAN [1996] for details and exam-

FIG. 3.11. Comparison of the Perona–Malik (Lorentzian) function (left) and the Tukey function (right) after 500 iterations.

ples as well as BLAKE and ZISSERMAN [1987], CHARBONNIER, BLANC-FERAUD, AUBERT and BARLAUD [to appear], GEIGER and YUILLE [1991], GEMAN and YANG [1995], GEMAN and REYNOLDS [1992] for recent related results. While we work with the discrete formulation here, it is easy to verify that the connections hold for the continuous formulation as well.

Recall that the robust formulation of the smoothing problem was posed as the minimization of

$$E(I) = \sum_s E(I_s), \tag{3.23}$$

where

$$E(I_s) = \sum_{p \in \eta_s} \rho(I_p - I_s, \sigma). \tag{3.24}$$

There is an alternative, equivalent, formulation of this problem that makes use of an explicit *line process* in the minimization:

$$E(I, \mathbf{l}) = \sum_s E(I_s, \mathbf{l}), \tag{3.25}$$

where

$$E(I_s, \mathbf{l}) = \sum_{p\in\eta_s}\left[\frac{1}{2\sigma^2}(I_p - I_s)^2 l_{s,p} + P(l_{s,p})\right] \tag{3.26}$$

and where $l_{s,p} \in \mathbf{l}$ are analog line processes ($0 \leqslant l_{s,p} \leqslant 1$) (GEIGER and YUILLE [1991], GEMAN and GEMAN [1984]). The line process indicates the presence (l close to 0) or absence (l close to 1) of discontinuities or *outliers*. The last term, $P(l_{s,p})$, *penalizes* the introduction of line processes between pixels s and p. This penalty term goes to zero when $l_{s,p} \to 1$ and is large (usually approaching 1) when $l_{s,p} \to 0$.

One benefit of the line-process approach is that the "outliers" are made explicit and therefore can be manipulated. For example, we can add constraints on these variables which encourage specific types of spatial organizations of the line processes.

Numerous authors have shown how to convert a line-process formulation into the robust formulation with a ρ-function by minimizing over the line variables (BLAKE and ZISSERMAN [1987], GEIGER and YUILLE [1991], GEMAN and REYNOLDS [1992]). That is

$$\rho(x) = \min_{0\leqslant l\leqslant 1} E(x, l),$$

where

$$E(x, l) = \left(x^2 l + P(l)\right).$$

For our purposes here it is more interesting to consider the other direction: can we convert a robust estimation problem into an equivalent line-process problem? We have already shown how to convert a diffusion problem with a $g(\cdot)$ function into a robust estimation problem. If we can make the connection between robust ρ-functions and line processes then we will be able to take a diffusion formulation like the Perona–Malik equation and construct an equivalent line process formulation.

Then, our goal is to take a function $\rho(x)$ and construct a new function, $E(x, l) = (x^2 l + P(l))$, such that the solution at the minimum is unchanged. Clearly the penalty term $P(\cdot)$ will have to depend in some way on $\rho(\cdot)$. By taking derivatives with respect to x and l, it can be shown that the condition on $P(l)$ for the two minimization problems to be equivalent is given by

$$-x^2 = P'\left(\frac{\psi(x)}{2x}\right).$$

By integrating this equation we obtain the desired line process penalty function $P(l)$. See BLACK and RANGARAJAN [1996] for details on the explicit computation of this integral. There are a number of conditions on the form of ρ that must be satisfied in order to recover the line process, but as described in BLACK and RANGARAJAN [1996], these conditions do in fact hold for many of the redescending ρ-functions of interest.

In the case of the Lorentzian norm, it can be shown that $P(l) = l - 1 - \log l$; see Fig. 3.12. Hence, the equivalent line-process formulation of the Perona–Malik equation is:

$$E(I_s, \mathbf{l}) = \sum_{p\in\eta_s}\left[\frac{1}{2\sigma^2}(I_p - I_s)^2 l_{s,p} + l_{s,p} - 1 - \log l_{s,p}\right]. \tag{3.27}$$

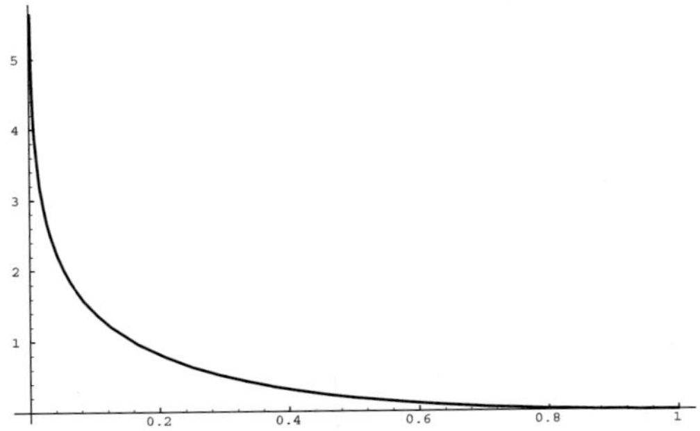

FIG. 3.12. Lorentzian (Perona–Malik) penalty function, $P(l)$, $0 \leqslant l \leqslant 1$.

Differentiating with respect to I_s and l gives the following iterative equations for minimizing $E(I_s, \mathbf{l})$:

$$I_s^{t+1} = I_s^t + \frac{\lambda}{|\eta_s|} \sum_{p \in \eta_s} l_{s,p} \nabla I_{s,p}, \tag{3.28}$$

$$l_{s,p} = \frac{2\sigma^2}{2\sigma^2 + \nabla I_{s,p}^2}. \tag{3.29}$$

Note that these equations are equivalent to the discrete Perona–Malik diffusion equations. In particular, $l_{s,p}$ is precisely equal to $g(\|\nabla I_{s,p}\|)$.

Spatial organization of outliers. One advantage of the connection between anisotropic diffusion and line processes, obtained through the connection of both techniques to robust statistics, is the possibility of improving anisotropic flows by the explicit design of line processes with spatial coherence. In the classical Perona–Malik flow, which relies on the Lorentzian error norm, there is no spatial coherence imposed on the detected outliers; see Fig. 3.10. Since the outlier process is explicit in the formulation of the line processing (Eq. (3.26)), we can add additional constraints on its spatial organization. While numerous authors have proposed spatial coherence constraints for discrete line processes (CHOU and BROWN [1990], GEMAN and GEMAN [1984], GEMAN and REYNOLDS [1992], MURRAY and BUXTON [1987]), we need to generalize these results to the case of analog line processes (BLACK and RANGARAJAN [1996]). This is fully addressed in BLACK, SAPIRO, MARIMONT and HEEGER [1998].

3.3. Directional diffusion

We have seen that the heat flow diffuses the image in all directions in an equal amount. One possible way to correct this is to "stop" the diffusion across edges, and this was the technique presented above. An alternative way, introduced in ALVAREZ, LIONS and MOREL [1992], is to direct the diffusion in the desired direction. If ξ indicates the direction perpendicular the ∇I ($\xi = \frac{1}{\|\nabla I\|}(-I_y, I_x)$), that is, parallel to the image jumps (edges), the we are interested in diffusing the image I only in this direction:

$$\frac{\partial I}{\partial t} = \frac{\partial^2 I}{\partial \xi^2}.$$

Using the classical formulas for directional derivatives, we can write $\frac{\partial^2 I}{\partial \xi^2}$ as a function of the derivatives of I in the x and y directions, obtaining

$$\frac{\partial I(x,y,t)}{\partial t} = \frac{I_{xx}I_y^2 - 2I_xI_yI_{xy} + I_{yy}I_x^2}{(\|\nabla u\|)^2}.$$

Recalling that the curvature of the level sets of I is given by

$$\kappa = \frac{I_{xx}I_y^2 - 2I_xI_yI_{xy} + I_{yy}I_x^2}{(\|\nabla u\|)^3},$$

the anisotropic flow is equivalent to

$$\frac{\partial I(x,y,t)}{\partial t} = \kappa\|\nabla u\|, \tag{3.30}$$

which means that the level-sets $\mathcal{C}$ of I are moving according to the Euclidean geometric heat flow

$$\frac{\partial \mathcal{C}}{\partial t} = \kappa\vec{\mathcal{N}}.$$

Directional diffusion is then equivalent to smoothing each one of the level sets according to the geometric heat flow.

In order to further stop diffusion across edges, it is common to modify the flow (3.30) to obtain

$$\frac{\partial I(x,y,t)}{\partial t} = g(\|\nabla u\|)\kappa\|\nabla u\|, \tag{3.31}$$

where $g(r)$ is such that it goes to zero when $r \to \infty$. Fig. 3.13 shows an example of this flow.

3.4. *Introducing prior knowledge*

Anisotropic diffusion applied to the raw image data is well motivated only when the noise is additive and signal independent. For example, if two objects in the scene have the same mean and differ only in variance, anisotropic diffusion of the data is not effective. In addition to this problem, anisotropic diffusion does not take into consideration the special content of the image. That is, in a large number of applications, like magnetic resonance imaging of the cortex and SAR data, it is known in advance the number of different types of objects in the scene, and directly applying anisotropic diffusion to the raw data does not take into consideration this important information given a priori.

A possible solution to the problems described above is presented now. The proposed scheme constitutes one of the steps of a complete system for the segmentation of MRI volumes of human cortex. The technique comprises three steps. First, the posterior probability of each pixel is computed from its likelihood and a homogeneous prior; i.e., a prior that reflects the relative frequency of each class (white matter, gray matter, and nonbrain in the case of MRI of the cortex), but is the same across all pixels. Next, the posterior probabilities for each class are anisotropically smoothed. Finally, each

FIG. 3.13. Example of anisotropic diffusion (original on the left and processed on the right).

pixel is classified independently using the MAP rule. Fig. 3.14 compares the classification of cortical white matter with and without the anisotropic smoothing step. The anisotropic smoothing produces classifications that are qualitatively smoother within regions while preserving detail along region boundaries. The intuition behind the method is straightforward. Anisotropic smoothing of the posterior probabilities results in piecewise constant posterior probabilities which, in turn, yield piecewise "constant" MAP classifications.

This technique, originally developed for MRI segmentation, is quite general, and can be applied to any given (or learned) probability distribution functions. We now first describe the technique in its general formulation. Then, in SAPIRO [2001] we discuss the mathematical theory underlying the technique. We demonstrated that anisotropic smoothing of the posterior probabilities yields the MAP solution of a discrete MRF with a noninteracting, analog discontinuity field. In contrast, isotropic smoothing of the posterior probabilities is equivalent to computing the MAP solution of a single, discrete MRF using continuous relaxation labeling. Combining a discontinuity field with a discrete MRF is important as it allows the disabling of clique potentials across discontinuities. Furthermore, explicit representation of the discontinuity field suggests new algorithms that incorporate hysteresis and nonmaximal suppression as shown when presenting the robust diffusion filter before.

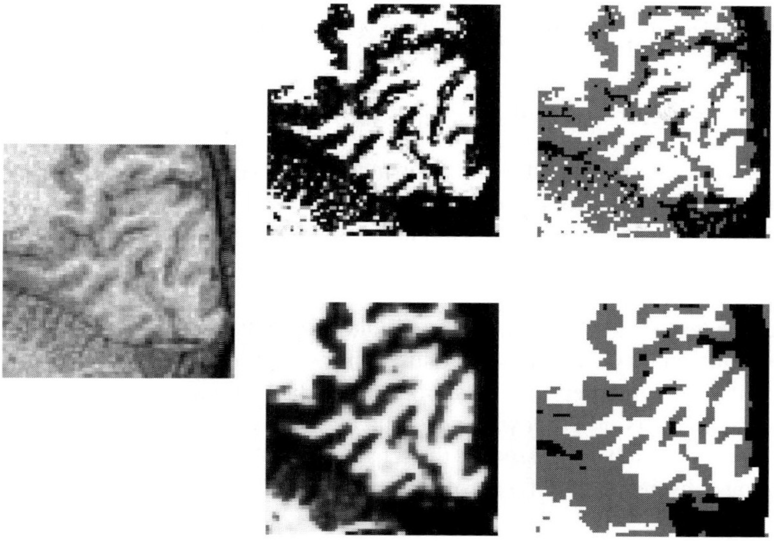

FIG. 3.14. (Top-row) Left: Intensity image of MRI data. Middle: Image of posterior probabilities corresponding to white matter class. Right: Image of corresponding MAP classification. Brighter regions in the posterior image correspond to areas with higher probability. White regions in the classification image correspond to areas classified as white matter; black regions correspond to areas classified as CSF. (Bottom-row) Left: Image of white matter posterior probabilities after being anisotropically smoothed. Right: Image of MAP classification computed using smoothed posteriors.

We now deal with the scalar case, while in PARDO and SAPIRO [to appear] we showed how to diffuse the whole probability vector with coupled PDEs.

The general technique

Let's assume that it is given to us, in the form of a priori information, the number k of classes (different objects) in the image. In MRI of the cortex for example, these classes would be white matter, gray matter, and CSF (nonbrain). In the case of SAR images, the classes can be "object" and "background." Similar classifications can be obtained for a large number of additional applications and image modalities.

In the first stage, the pixel or voxel intensities within each class are modeled as independent random variables with given or learned distributions. Thus, the likelihood $\Pr(V_i = v | C_i = c)$ of a particular pixel (or voxel in the case of 3D MRI), V_i, belonging to a certain class, C_i, is given. For example, in the case of normally distributed likelihood, we have

$$\Pr(V_i = v | C_i = c) = \frac{1}{\sqrt{2\pi}\sigma_c} \exp\left(-\frac{1}{2}\frac{(v-\mu_c)^2}{\sigma_c^2}\right). \tag{3.32}$$

Here, i is a spatial index ranging over all pixels or voxels in the image, and the index c stands for one of the k classes. V_i and C_i correspond to the intensity and classification of voxel i respectively. The mean and variance (μ_c and σ_c) are given, learned from examples, or adjusted by the user.

The posterior probabilities of each voxel belonging to each class are computed using Bayes' rule:

$$\Pr(C_i = c \mid V_i = v) = \frac{1}{K} \Pr(V_i = v \mid C_i = c) \Pr(C_i = c), \tag{3.33}$$

where K is a normalizing constant independent of c. As in the case of the likelihood, the prior distribution, $\Pr(C_i = c)$, is not restricted, and can be arbitrarily complex. In a large number of applications, we can adopt a homogeneous prior, which implies that $\Pr(C_i = c)$ is the same over all spatial indices i. The prior probability typically reflects the relative frequency of each class.

After the posterior probabilities are computed (note that we will have now k images), the posterior images are smoothed anisotropically (in two or three dimensions), but preserving discontinuities. The anisotropic smoothing technique applied can be based on the original version proposed by Perona et al. or any other of the extensions later proposed and already discussed in this chapter. As we have seen, this step involves simulating a discretization of a partial differential equation:

$$\frac{\partial P_c}{\partial t} = \operatorname{div}\bigl(g(\|\nabla P_c\|)\nabla P_c\bigr), \tag{3.34}$$

where $P_c = \Pr(C = c \mid V)$ stand for the posterior probabilities for class c, the stopping term $g(\|\nabla P_c\|) = \exp(-(\|\nabla P_c\|/\eta_c)^2)$, and η_c represents the rate of diffusion for class c. The function $g(\cdot)$ controls the local amount of diffusion such that diffusion across discontinuities in the volume is suppressed. Since we are now smoothing probabilities, to be completely formal, these evolving probabilities should be normalized each step of the iteration to add to one. This problem is formally addressed and solved later in PARDO and SAPIRO [to appear].

Finally, the classifications are obtained using the maximum a posteriori probability (MAP) estimate after anisotropic diffusion. That is,

$$C_i^* = \arg\max_k \Pr{}^*(C_i = c \mid V_i = v), \tag{3.35}$$

where $\Pr^*(C_i = c \mid V_i = v)$ corresponds to the posterior following anisotropic diffusion.

Recapping, the proposed algorithm has the following steps:

(1) Compute the priors and likelihood functions for each one of the classes in the images.
(2) Using Bayes rule, compute the posterior for each class.
(3) Apply anisotropic diffusion (combined with normalization) to the posterior images.
(4) Use MAP to obtain the classification.

This techniques solves both problems mentioned in the introduction. That is, it can handle nonadditive noise, and more important, introduces prior information about the type of images being processed. As stated before, in SAPIRO [2001] we discuss the relations of this technique with other algorithms proposed in the literature.

The anisotropic posterior smoothing scheme was first proposed and used to segment white matter from MRI data of human cortex. Pixels at a given distance from the boundaries of the white matter classification were then automatically classified as gray matter. Thus, gray matter segmentation relied heavily on the white matter segmentation being accurate. Fig. 3.15 shows comparisons between gray matter segmentations produced automatically by the proposed method and those obtained manually. More examples can be found in TEO, SAPIRO and WANDELL [1997]. Note that when applied to MRI, the technique being proposed bears some superficial resemblance to schemes that anisotropically smooth the raw image before classification (GERIG, KUBLER, KIKINIS and JOLESZ [1992]). Besides the connection between our technique and MAP estimation of Markov random fields, which is absent in schemes that smooth the image

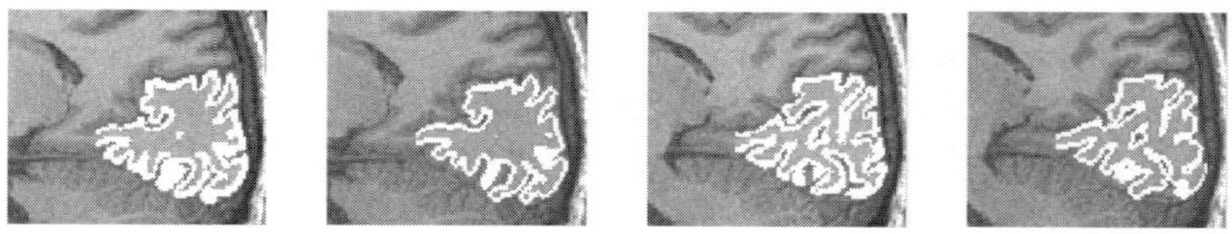

FIG. 3.15. The two left images show manual gray matter segmentation results; the two right images show the automatically computed gray matter segmentation (same slices shown).

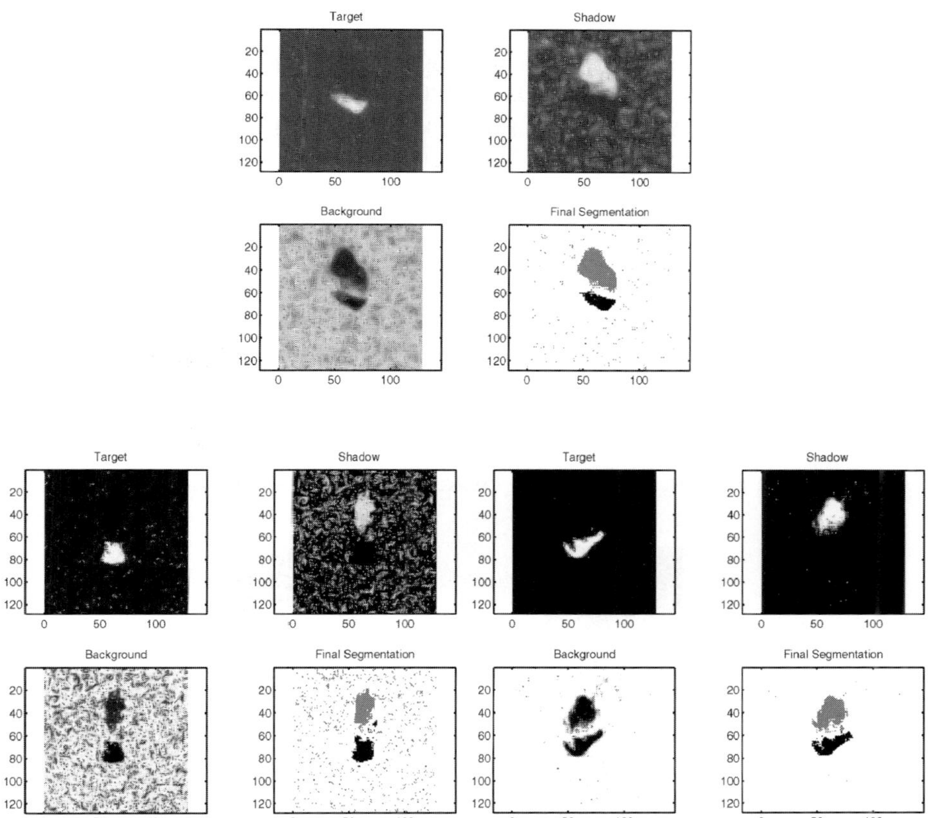

FIG. 3.16. Segmentation of SAR data (scalar and video), see HAKER, SAPIRO and TANNENBAUM [2000] and HAKER, SAPIRO and TANNENBAUM [1998].

directly, there are other important differences, since GERIG, KUBLER, KIKINIS and JOLESZ [1992] suffers from the common problems of diffusion raw data detailed in the introduction. We should note again that applying anisotropic smoothing on the posterior probabilities is feasible even when the class likelihoods are described by general probability mass functions (and even multi-variate distributions!).

Fig. 3.16 shows the result of the algorithm applied to SAR data. Here the three classes are object, shadow, and background. The first row shows the result of segmenting a single frame. Uniform priors and Gaussian distributions are used. The next rows show the results for the segmentation of video data with learned priors. The second row shows the segmentation of the first frame of a video sequence (left). Uniform priors and Gaussian distributions are used for this frame as well. To illustrate the learning importance, only 2 steps of posterior diffusion were applied. After this frame, smooth posteriors of frame i are used as priors of frame $i+1$. The figure on the right shows the result, once again with only 2 posterior smoothing steps, for the eight frame in the sequence.

3.5. Diffusion on general manifolds

In this section we have covered the diffusion of scalar images. This has been extended to vectorial data, as well as to data defined on nonflat manifolds; see PERONA [1998], BERTALMIO, SAPIRO, CHENG and OSHER [2000], CHAN and SHEN [1999], SOCHEN, KIMMEL and MALLADI [1998], TANG, SAPIRO and CASELLES [2000a], TANG, SAPIRO and CASELLES [2000b] and MEMOLI, SAPIRO and OSHER [2002] for details.

4. Contrast enhancement

Images are captured at low contrast in a number of different scenarios. The main reason for this problem is poor lighting conditions (e.g., pictures taken at night or against the sun rays). As a result, the image is too dark or too bright, and is inappropriate for visual inspection or simple observation. The most common way to improve the contrast of an image is to modify its pixel value distribution, or *histogram*. A schematic example of the contrast enhancement problem and its solution via histogram modification is given in Fig. 4.1. On the left, we see a low contrast image with two different squares, one

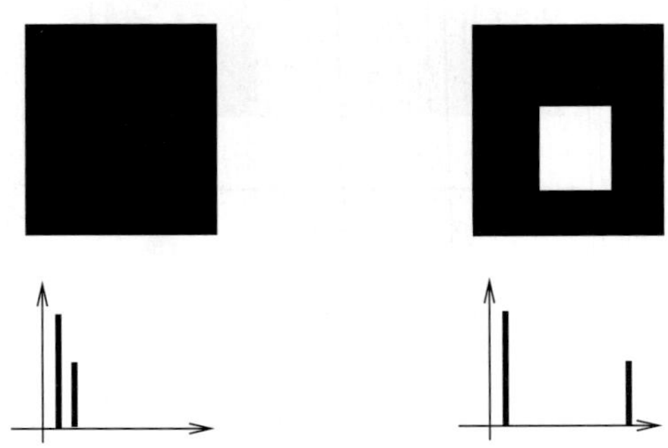

FIG. 4.1. Schematic explanation of the use of histogram modification to improve image contrast.

FIG. 4.2. Example of contrast enhancement. Note how objects that are not visible on the original image on the left (e.g., the 2nd chair and the objects through the window), are now detectable in the processed one (right).

inside the other, and its corresponding histogram. We can observe that the image has low contrast, and the different objects can not be identified, since the two regions have almost identical grey values. On the right we see what happens when we modify the histogram in such a way that the grey values corresponding to the two regions are separated. The contrast is improved immediately. An additional example, this time for a real image, is given in Fig. 4.2.

In this section, we first follow SAPIRO and CASELLES [1997a], SAPIRO and CASELLES [1997b] and show how to obtain any gray-level distribution as the steady state of an ODE, and present examples for different pixel value distributions. Uniform distributions are usually used in most contrast enhancement applications. On the other hand, for specific tasks, the exact desirable distribution can be dictated by the application, and the technique here presented applies as well. After this basic equation is presented and analyzed, we can combine it with the smoothing operators proposed in as those previously discussed, obtaining contrast normalization and denoising at the same time. We also extend the flow to local contrast enhancement both in the image plane and in the gray-value space. Local contrast enhancement in the gray-value space is performed for example to improve the visual appearance of the image (the reason for this will be explained latter). We should mention that the straightforward extension of this formulation to local contrast enhancement in the image domain, that is, in the neighborhood of each pixel, does not achieves good results. Indeed, in this case, fronts parallel to the edges are created (this is common to most contrast enhancement techniques). At the experimental level, one can avoid this by combining local and global techniques in the framework here presented, or using the approach described in the second half of this section, which follows CASELLES, LISANI, MOREL and SAPIRO [1997], CASELLES, LISANI, MOREL and SAPIRO [1999].

After giving the basic image flows for histogram modification, a variational interpretation of the histogram modification flow and theoretical results regarding existence of solutions to the proposed equations are presented.

In the second part of this section, we discuss *local* histogram modification operations which preserve the family of connected components of the level-sets of the image, that is, following the morphology school, preserve shape. Local contrast enhancement is mainly used to further improve the image contrast and facilitate the visual inspection of the data. As we will later see, global histogram modification not always produces good contrast, and specially small regions, are hardly visible after such a global operation. On the other hand, local histogram modification improves the contrast of small regions as well, but since the level-sets are not preserved, artificial objects are created. The theory developed will enjoy the best of both words: The shape-preservation property of global techniques and the contrast improvement quality of local ones.

Before proceeding, we should point out that in PERONA and TARTAGNI [1994] the authors presented a diffusion network for image normalization. In their work, the image $I(x, y)$ is normalized via $\frac{I-I_a}{I_M-I_m}$, where I_a, I_M, and I_m are the average, maximum, and minimum of I over local areas. These values are computed using a diffusion flow, which minimizes a cost functional. The method was generalized computing a full local frame of reference for the gray level values of the image. This is achieved changing the variables in the flow. A number of properties, including existence of the solution of the diffusion flow, were presented as well.

4.1. Global PDEs based approach

Histogram modification

As explained in the Introduction, the most common way to improve the contrast of an image is to modify the pixel value distribution, i.e., the histogram. We shall do this by means of an evolution equation. We start with the equation for histogram equalization, and then we extend it for any given distribution. In histogram equalization, the goal is to achieve an uniform distribution of the image values (PRATT [1991]). That means, given the gray-level distribution p of the original image, the image values are mapped into new ones such that the new distribution $\hat{p}$ is uniform. In the case of digital images, $p(i), 0 \leqslant i \leqslant M$, is computed as

$$p(i) = \frac{\text{Number of pixels with value } i}{\text{Total number of pixels in the image}},$$

and uniformity can be obtained only approximately.

We proceed to show an image evolution equation that achieves this uniform distribution when it arrives to steady state. Assume that the continuous image $I(x, y, t): [0, N]^2 \times [0, T) \to [0, M]$ evolves according to

$$\frac{\partial I(x, y, t)}{\partial t} = \left(N^2 - N^2/MI(x, y, t)\right) - \mathcal{A}\left[(v, w)\colon\ I(v, w, t) \geqslant I(x, y, t)\right], \quad (4.1)$$

where $\mathcal{A}[\cdot]$ represents area (or number of pixels in the discrete case). For the steady state solution ($I_t = 0$) we have

$$\mathcal{A}\left[(v, w)\colon\ I(v, w) \geqslant I(x, y)\right] = \left(N^2 - N^2/MI(x, y)\right).$$

Then, for $a, b \in [0, M]$, $b > a$, we have

$$\mathcal{A}\left[(v, w)\colon\ b \geqslant I(v, w) \geqslant a\right] = \left(N^2/M\right)(b - a),$$

which means that the histogram is constant. Therefore, the steady state solution of (4.1), if it exists (see below), gives the image after normalization via histogram equalization.

From (4.1) we can extend the algorithm to obtain any given gray-value distribution $h:[0, M] \to \mathbb{R}^+$. Let $H(s) := \int_0^s h(\xi)\,d\xi$. That is, $H(s)$ gives the density of points between 0 and s. Then, if the image evolves according to

$$\frac{\partial I(x,y,t)}{\partial t} = \left(N^2 - H\left[I(x,y,t)\right]\right) - \mathcal{A}\left[(v,w)\colon\ I(v,w,t) \geqslant I(x,y,t)\right], \tag{4.2}$$

the steady state solution is given by

$$\mathcal{A}\left[(v,w)\colon\ I(v,w) \geqslant I(x,y)\right] = \left(N^2 - H\left[I(x,y)\right]\right).$$

Therefore,

$$\mathcal{A}\left[(v,w)\colon\ I(x,y) \leqslant I(v,w) \leqslant I(x,y)+\delta\right] = H\left[I(x,y)+\delta\right] - H\left[I(x,y)\right],$$

and taking Taylor expansion when $\delta \to 0$ we obtain the desired result. Note that of course (4.1) is a particular case of (4.2), with $h = \text{constant}$.

Existence and uniqueness of the flow. We present now results related to the existence and uniqueness of the proposed flow for histogram equalization. We will see that the flow for histogram modification has an explicit solution, and its steady state is straightforward to compute. This is due to the fact, as we will prove below, that the value of $\mathcal{A}$ is constant in the evolution. This is not unexpected, since it is well known that histogram modification can be performed with look-up tables. In spite of this, it is important, and not only from the theoretical point of view, to first present the basic flow for histogram modification, in order to arrive to the energy based interpretation and to derive the extensions later presented. These extensions do not have explicit solutions.

Let I_0 be an image, i.e., a bounded measurable function, defined in $[0, N]^2$ with values in the range $[a, b]$, $0 \leqslant a < b \leqslant M$. We assume that the distribution function of I_0 is continuous, that is

$$\mathcal{A}\left[X\colon\ I_0(X) = \lambda\right] = 0 \tag{4.3}$$

for all $X \in [0, N]^2$ and all $\lambda \in [a, b]$. To equalize the histogram of I_0 we look for solutions of

$$I_t(t,X) = \mathcal{A}\left[Z\colon\ I(t,Z) < I(t,X)\right] - \frac{N^2}{b-a}\left(I(t,X) - a\right), \tag{4.4}$$

which also satisfy

$$\mathcal{A}\left[X\colon\ I(t,X) = \lambda\right] = 0. \tag{4.5}$$

Hence the distribution function of $I(t,X)$ is also continuous. This requirement, mainly technical, avoids the possible ambiguity of changing the sign "$<$" by "$\leqslant$" in the computation of $\mathcal{A}$ (see also the remarks at the end of this section).

Let us recall the definition of $\operatorname{sign}^-(\cdot)$:

$$\operatorname{sign}^-(r) = \begin{cases} 1 & \text{if } r < 0, \\ [0,1] & \text{if } r = 0, \\ 0 & \text{if } r > 0, \end{cases}$$

With this notation, I satisfying (4.4) and (4.5) may be written as

$$I_t(t,X)=\int_{[0,N]^2}\operatorname{sign}^-\bigl(I(t,Z)-I(t,X)\bigr)\,dZ-\frac{N^2}{b-a}\bigl(I(t,X)-a\bigr). \tag{4.6}$$

Observe that as a consequence of (4.5), the real value of sign^- at zero is unimportant, avoiding possible ambiguities. In order to simplify the notation, let us normalize I such that it is defined on $[0,1]^2$ and takes values in the range $[0,1]$. This is done just by the change of variables given by $I(t,X)\leftarrow\frac{I(\mu t,NX)-a}{b-a}$, where $\mu=\frac{b-a}{N^2}$. Then, I satisfies the equation

$$I_t(t,X)=\int_{[0,1]^2}\operatorname{sign}^-\bigl(I(t,Z)-I(t,X)\bigr)\,dZ-I(t,X). \tag{4.7}$$

Therefore, without loss of generality we can assume $N=1$, $a=0$, and $b=1$, and analyze (4.7). Let us make precise our notion of solution for (4.7).

DEFINITION 4.1. A bounded measurable function $I:[0,\infty)\times[0,1]^2\to[0,1]$ will be called a solution of (4.7) if, for almost all $X\in[0,1]^2$, $I(.,X)$ is continuous in $[0,\infty)$, $I_t(.,X)$ exists a.e. with respect to t and (4.7) holds a.e. in $[0,\infty)\times[0,1]^2$.

Now we may state the following result:

THEOREM 4.1. *For any bounded measurable function $I_0:[0,1]^2\to[0,1]$ such that $\mathcal{A}[Z\colon I_0(Z)=\lambda]=0$ for all $\lambda\in[0,1]$, there exists a unique solution $I(t,X)$ in $[0,\infty)\times[0,1]^2$ with range in $[0,1]$ satisfying the flow (4.7) with initial condition given by I_0, and such that $\mathcal{A}[Z\colon I(t,Z)=\lambda]=0$ for all $\lambda\in[0,1]$. Moreover, as $t\to\infty$, $I(t,X)$ converges to the histogram equalization of $I_0(X)$.*

The above theorem can be adapted to any required gray-value distribution h. As pointed out before, the specific distribution depends on the application. Uniform distributions are the most common choice. If it is known in advance for example that the most important image information is between certain gray-value region, h can be such that it allows this region to expand further, increasing the detail there. Another possible way of finding h is to equalize between local minima of the original histogram, preserving certain type of structure in the image.

Variational interpretation of the histogram flow. The formulation given by Eqs. (4.6) and (4.7) not only helps to prove the theorem above, also gives a variational interpretation of the histogram modification flow. Variational approaches are frequently used in image processing. They give explicit solutions for a number of problems, and very often help to give an intuitive interpretation of this solution, interpretation which is many times not so easy to achieve from the corresponding Euler–Lagrange or PDE. Variational formulations help to derive new approaches to solve the problem as well.

Let us consider the following functional

$$\mathcal{U}(I)=\frac{1}{2}\int\left(I(X)-\frac{1}{2}\right)^2dX-\frac{1}{4}\iint\bigl|I(X)-I(Z)\bigr|\,dX\,dZ, \tag{4.8}$$

where $I\in L^2[0,1]^2$, $0\leqslant I(X)\leqslant 1$. $\mathcal{U}$ is a Lyapunov functional for Eq. (4.7):

LEMMA 4.1. *Let I be the solution of* (4.7) *with initial data I_0 as in Theorem* 4.1. *Then*

$$\frac{d\mathcal{U}(I)}{dt} \leqslant 0.$$

Therefore, when solving (4.7) we are indeed minimizing the functional $\mathcal{U}$ given by (4.8) restricted to the condition that the minimizer satisfies the constraint.

This variational formulation gives a new interpretation to histogram modification and contrast enhancement in general. It is important to note that in contrast with classical techniques of histogram modification, it is completely formulated in the image domain and not in the probability one (although the spatial relation of the image values is still not important, until the formulations below). The first term in $\mathcal{U}$ stands for the "variance" of the signal, while the second one gives the contrast between values at different positions. To the best of our knowledge, this is the first time a formal image based interpretation to histogram equalization is given, showing the effect of the operation to the image contrast.

From this formulation, other functionals can be proposed to achieve contrast modification while including image and perception models. One possibility is to change the metric which measures contrast, second term in the equation above, by metrics which better model for example visual perception. It is well known that the total absolute difference is not a good perceptual measurement of contrast in natural images. At least, this absolute difference should be normalized by the local average. This also explains why ad-hoc techniques that segment the grey-value domain and perform (independent) local histogram modification in each one of the segments perform better than global modifications. This is due to the fact that the normalization term is less important when only pixels of the same range value are considered. Note that doing this is straightforward in our framework, $\mathcal{A}$ should only consider pixels in between certain range in relation with the value of the current pixel being updated.

From the variational formulation is also straightforward to extend the model to local (in image space) enhancement by changing the limits in the integral from global to local neighborhoods. In the differential form, the corresponding equations is

$$\frac{\partial I}{\partial t} = \big(N^2 - H\big[I(x,y,t)\big]\big) - \mathcal{A}\big[(v,w) \in \mathrm{B}(v,w,\delta)\colon\ I(v,w,t) \geqslant I(x,y,t)\big],$$

where $\mathrm{B}(v,w,\delta)$ is a ball of center (v,w) and radius δ ($\mathrm{B}(v,w)$ can also be any other surrounding neighborhood, obtained from example from previously performed segmentation). The main goal of this type of local contrast enhancement is to enhance the image for object detection. This formulation does not work perfectly in practice when the goal of the contrast enhancement algorithm is to obtain a visually pleasant image. Fronts parallel to the edges are created as we can see in the examples below. (Further experiments with this model are presented in Section 4.1.) Effects of the local model can be moderated by combining it with the global model.

Based on the same approach, it is straightforward to derive models for contrast enhancement of movies, integrating over corresponding zones in different frames, when correspondence could be given, for example, by optical flow. Another advantage is the possibility to combine it with other operations. As an example, in the next section, an

FIG. 4.3. Original image (top-left) and results of the histogram equalization process with the software package *xv* (top-right), the proposed image flow for histogram equalization (bottom-left), and the histogram modification flow for a piece-wise linear distribution (bottom-right).

additional smoothing term will be added to the model (4.7). It is also possible to combine contrast enhancement with other variational algorithms like image denoising, and this was developed as well in our original paper.

Experimental results

Before presenting experimental results, let us make some remarks on the complexity of the algorithm. Each iteration of the flow requires $O(N^2)$ operations. In our examples we observed that no more than 5 iterations are usually required to converge. Therefore, the complexity of the proposed algorithm is $O(N^2)$, which is the minimal expected for any image processing procedure that operates on the whole image.

The first example is given in Fig. 4.3. The original image is presented on the top-left. On the right we present the image after histogram equalization performed using the popular software *xv* (copyright 1993 by John Bradley), and on the bottom-left the one obtained from the steady state solution of (4.1). On the bottom-right we give an example of Eq. (4.2) for $h(I)$ being a piece-wise linear function of the form $-\alpha(|I - M/2| - M/2)$, where α is a normalization constant, I the image value, and M the maximal image value.

Fig. 4.4 presents an example of combining local and global histogram modification. The original image is given on the top left. The result of global histogram equalization on the top right, and the one for local contrast enhancement (16×16 neighborhood) on the bottom left. We see fronts appearing parallel to the edges. Finally, on the bottom right we show the combination of local and global contrast modification: we apply one

FIG. 4.4. Result of the combination of local and global contrast enhancement. The original image is given on the top left. The result of global histogram equalization on the top right, and the one for local contrast enhancement (16×16 neighborhood) on the bottom left. Finally, on the bottom right we show the combination of local and global contrast modification: The image on the left is further processed by the global histogram modification flow.

(or several) global steps after k successive local steps. Note that the algorithm described is natural for this kind of combination, since all what is needed is the area $\mathcal{A}$ to be computed in a "time" dependent neighborhood.

4.2. *Shape preserving contrast enhancement*

We have extended the results above and designed *local* histogram modification operations which preserve the family of connected components of the level-sets of the image, that is, following the morphology school, preserve shape. Local contrast enhancement is mainly used to further improve the image contrast and facilitate the visual inspection of the data. We have already seen, and it will be further exemplified later, that global histogram modification not always produces good contrast, and specially small regions, are hardly visible after such a global operation. On the other hand, local histogram modification improves the contrast of small regions as well, but since the level-sets are not preserved, artificial objects are created. The theory developed in CASELLES, LISANI, MOREL and SAPIRO [1997] and exemplified below enjoys the best of both words: The shape-preservation property of global techniques and the contrast improvement qual-

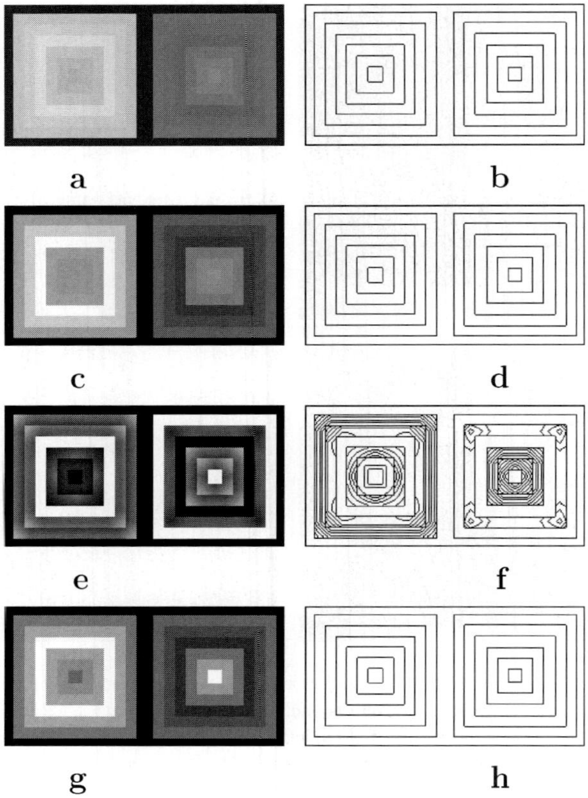

FIG. 4.5. Example of the level-sets preservation. The top row shows the original image and its level-sets. The second row shows the result of global histogram modification and the corresponding level-sets. Results of classical local contrast enhancement and its corresponding level-sets are shown in the third row. The last row shows the result of the algorithm. Note how the level-sets are preserved, in contrast with the result on the 3rd row, while the contrast is much better than the global modification.

ity of local ones. Examples are presented below, while details on this technique can be found in CASELLES, LISANI, MOREL and SAPIRO [1997].

In Fig. 4.5 we compare our local technique with classical ones. In the classical algorithm the procedure is to define an $n \times m$ neighborhood and move the center of this area from pixel to pixel. At each location we compute the histogram of the $n \times m$ points in the neighborhood and obtain a histogram equalization (or histogram specification) transformation function. This function is used to map the level of the pixel centered in the neighborhood. The center of the $n \times m$ region is then moved to an adjacent pixel location and the procedure is repeated. In practice one updates the histogram obtained in the previous location with the new data introduced at each motion step. Fig. 4.5a shows the original image whose level-lines are displayed in Fig. 4.5b. In Fig. 4.5c we show the result of the global histogram equalization of Fig. 4.5a. Its level-lines are displayed in Fig. 4.5d. Note how the level-sets lines are preserved, while

FIG. 4.6. Additional example of shape preserving local histogram modification for real data. Figure a is the original image. Figures b–d are the results of global histogram equalization, classical local scheme (61 $\times$ 61 neighborhood), and shape preserving algorithm, respectively.

the contrast of small objects is reduced. Fig. 4.5e shows the result of the classical local histogram equalization described above (31 $\times$ 31 neighborhood), with level-lines displayed in Fig. 4.5f. All the level sets for grey-level images are displayed at intervals of 20 grey-values. We see that new level-lines appear thus modifying the topographic map (the set of level-lines) of the original image, introducing new objects. Fig. 4.5g shows the result of the algorithm for local histogram equalization. Its corresponding level-lines are displayed in Fig. 4.5h. We see that they coincide with the level-lines of the original image, Fig. 4.5b.

An additional example is given in Fig. 4.6. Fig. 4.6a is the original image. Figs. 4.6b–4.6d are the results of global histogram equalization, classical local scheme (61 $\times$ 61 neighborhood), and shape preserving algorithm, respectively.

Experiments with a color image are given in our extended report on this subject.

Acknowledgements

The work on geodesic active contours is joint with V. Caselles and R. Kimmel. The work on edges for vector-valued images is joint with D. Ringach and D.H. Chung. The work on computing the minimal geodesic is joint with L. Vazquez and D.H. Chung, while its extensions for skin segmentation were developed with D.H. Chung. The work on affine invariant detection is joint with P. Olver and A. Tannenbaum. The work on tracking is joint with M. Bertalmio and G. Randall. The work on robust diffusion is joint with M. Black, D. Marimont, and D. Heeger. The work on posterior diffusion is the result of collaborations with P. Teo, B. Wandell, S. Haker, A. Tannenbaum, and A. Pardo. The work on contrast enhancement is the result of collaborations with V. Caselles, J.L. Lisani, and J.M. Morel. The work summarized in this chapter is supported by a grant from the Office of Naval Research ONR-N00014-97-1-0509, the Office of Naval Research Young Investigator Award, the Presidential Early Career Awards for Scientists and Engineers (PECASE), a National Science Foundation CAREER Award, and by the National Science Foundation Learning and Intelligent Systems Program (LIS).

References

ALVAREZ, L., GUICHARD, F., LIONS, P.L., MOREL, J.M. (1993). Axioms and fundamental equations of image processing. *Arch. Rational Mech.* **123**, 199–257.

ALVAREZ, L., LIONS, P.L., MOREL, J.M. (1992). Image selective smoothing and edge detection by nonlinear diffusion. *SIAM J. Numer. Anal.* **29**, 845–866.

AMBROSIO, L., SONER, M. (1996) Level set approach to mean curvature flow in arbitrary codimension. *J. Differential Geom.* **43**, 693–737.

ANGENENT, S. (1991). Parabolic equations for curves on surfaces, Part II. Intersections, blow-up, and generalized solutions. *Ann. of Math.* **133**, 171–215.

AUBERT, G., VESE, L. (to appear). A variational method for image recovery. *SIAM J. Appl. Math.*, to appear.

BABAUD, J., WITKIN, A.P., BAUDIN, B., DUDA, R.O. (1986). Uniqueness of the Gaussian kernel for scale-space filtering. *IEEE-PAMI* **8**, 26–33.

BERTALMIO, M., SAPIRO, G., CHENG, L.-T., OSHER, S. (2000). A framework for solving surface PDEs for computer graphics applications, UCLA CAM TR 00-43.

BERTALMIO, M., SAPIRO, G., RANDALL, G. (1998). Morphing active contours: A geometric, topology-free, technique for image segmentation and tracking. In: *Proc. IEEE Int. Conference Image Processing, Chicago.*

BESL, P.J., BIRCH, J.B., WATSON, L.T. (1988). Robust window operators. In: *Proc. Int. Conf. on Comp. Vision, ICCV-88*, pp. 591–600.

BLACK, M.J., ANANDAN, P. (1991). Robust dynamic motion estimation over time. In: *Proc. Computer Vision and Pattern Recognition, CVPR-91, Maui, Hawaii*, pp. 296–302.

BLACK, M., ANANDAN, P. (1993). A framework for the robust estimation of optical flow. In: *Fourth International Conf. on Computer Vision, Berlin, Germany*, pp. 231–236.

BLACK, M., RANGARAJAN, A. (1996). On the unification of line processes, outlier rejection, and robust statistics with applications in early vision. *Internat. J. Computer Vision* **19**, 57–92.

BLACK, M., SAPIRO, G., MARIMONT, D., HEEGER, D. (1998). Robust anisotropic diffusion. *IEEE Trans. Image Processing* **7** (3), 421–432.

BLAKE, A., ISARD, M. (1998). *Active Contours* (Springer-Verlag, New York).

BLAKE, A., ZISSERMAN, A. (1987). *Visual Reconstruction* (MIT Press, Cambridge).

BORN, M., WOLF, W. (1986). *Principles of Optics*, 6th (corrected) Edition (Pergamon Press).

CALABI, E., OLVER, P.J., SHAKIBAN, C., TANNENBAUM, A., HAKER, S. (1988). Differential and numerical invariant signature curves applied to object recognition. *Internat. J. Computer Vision* **26**, 107–135.

CALABI, E., OLVER, P.J., TANNENBAUM, A. (1996). Affine geometry, curve flows, and invariant numerical approximations. *Adv. Math.* **124**, 154–196.

CASELLES, V., CATTE, F., COLL, T., DIBOS, F. (1993). A geometric model for active contours. *Numerische Mathematik* **66**, 1–31.

CASELLES, V., KIMMEL, R., SAPIRO, G. (1995). Geodesic active contours. In: *Proc. Int. Conf. Comp. Vision '95, Cambridge.*

CASELLES, V., KIMMEL, R., SAPIRO, G. (1997). Geodesic active contours. *Internat. J. Computer Vision* **22** (1), 61–79.

CASELLES, V., KIMMEL, R., SAPIRO, G., SBERT, C. (1997a). Minimal surfaces: A geometric three-dimensional segmentation approach. *Numer. Math.* **77**, 423–451.

CASELLES, V., KIMMEL, R., SAPIRO, G., SBERT, C. (1997b). Minimal surfaces based object segmentation. *IEEE-PAMI* **19** (4), 394–398.

CASELLES, V., LISANI, J.-L., MOREL, J.-M., SAPIRO, G. (1997). Shape-preserving local contrast enhancement. In: *Proc. IEEE–International Conference on Image Processing, Santa Barbara, CA*.

CASELLES, V., LISANI, J.-L., MOREL, J.-M., SAPIRO, G. (1999). Shape-preserving local contrast enhancement. *IEEE Trans. Image Processing* **8**, 220–230.

CASELLES, V., MOREL, J.-M., SAPIRO, G., TANNENBAUM, A. (1998). Introduction to the special issue on PDEs and geometry driven diffusion in image processing and analysis. *IEEE Trans. Image Processing* **7** (3), 269–273.

CATTE, F., LIONS, P.-L., MOREL, J.-M., COLL, T. (1992). Image selective smoothing and edge detection by nonlinear diffusion. *SIAM J. Numer. Anal.* **29**, 182–193.

CHAN, T., SHEN, J. (1999). Variational restoration of nonflat image features: Models and algorithms, UCLA CAM-TR 99-20.

CHAN, T.F., SANDBERG, B.Y., VESE, L.A. (1999). Active contours without edges for vector-valued images, UCLA CAM Report 99-35.

CHAN, T.F., VESE, L.A. (1998). Active contours without edges, UCLA CAM Report 98-53.

CHARBONNIER, P., BLANC-FERAUD, L., AUBERT, G., BARLAUD, M. (to appear). Deterministic edge-preserving regularization in computed imaging. *IEEE Trans. Image Processing*, to appear.

CHEN, Y.G., GIGA, Y., GOTO, S. (1991). Uniqueness and existence of viscosity solutions of generalized mean curvature flow equations. *J. Differential Geom.* **33**.

CHEN, D.S., SCHUNCK, B.G. (1990). Robust statistical methods for building classification procedures. In: *Proc. Int. Workshop on Robust Computer Vision, Seattle, WA*, pp. 72–85.

CHOU, P.B., BROWN, C.M. (1990). The theory and practice of Bayesian image labeling. *Internat. J. Computer Vision* **4** (3), 185–210.

CHUNG, D.H., SAPIRO, G. (2000). On the level-lines and geometry of vector-valued images. *IEEE Signal Processing Letters* **7**, 241–243.

COHEN, L.D. (1991). On active contour models and balloons. *CVGIP* **53**.

COHEN, L.D., KIMMEL, R. (to appear). Global minimum for active contours models: A minimal path approach, *International Journal of Computer Vision*, to appear. (A short version appeared in: *Proc. of CVPR'96, San Francisco, CA*, pp. 666–673 (1996)).

CRANDALL, M.G., ISHII, H., LIONS, P.L. (1992). User's guide to viscosity solutions of second order partial linear differential equations. *Bull. Amer. Math. Soc.* **27**, 1–67.

CUMANI, A. (1991). Edge detection in multispectral images. *CVGIP: Graphical Models and Image Processing* **53**, 40–51.

DERICHE, R., BOUVIN, C., FAUGERAS, O. (1996). A level-set approach for stereo, INRIA Technical Report (Sophia-Antipolis).

DI ZENZO, S. (1986). A note on the gradient of a multi-image. *CVGIP* **33**, 116–125.

DUBROVIN, B.A., FOMENKO, A.T., NOVIKOV, S.P. (1984). *Modern Geometry – Methods and Applications I* (Springer-Verlag, New York).

EVANS, L.C. (1998). *Partial Differential Equations* (American Mathematical Society, Providence, RI).

EVANS, L.C., SPRUCK, J. (1991). Motion of level sets by mean curvature, I. *J. Differential Geom.* **33**, 635–681.

FAUGERAS, O.D., KERIVEN, R. (1995). Scale-spaces and affine curvature. In: Mohr, R., Wu, C. (eds.), *Proc. Europe–China Workshop on Geometrical modeling and Invariants for Computer Vision*, pp. 17–24.

FAUGERAS, O.D., KERIVEN, R. (1998). Variational principles, surface evolution, PDEs, level-set methods, and the stereo problem. *IEEE Trans. Image Processing* **73**, 336–344.

FUA, P., LECLERC, Y.G. (1990). Model driven edge detection. *Machine Vision and Applications* **3**, 45–56.

FREEMAN, W.T., ADELSON, E.H. (1991). The design and use of steerable filters. *IEEE-PAMI* **9**, 891–906.

GABOR, D. (1965). Information theory in electron microscopy. *Laboratory Investigation* **14**, 801–807.

GAGE, M., HAMILTON, R.S. (1986). The heat equation shrinking convex plane curves. *J. Differential Geom.* **23**.

GEIGER, D., GUPTA, A., COSTA, L.A., VLONTZOS, J. (1995). Dynamic programming for detecting, tracking, and matching deformable contours. *IEEE-PAMI* **17** (3).

GEIGER, D., YUILLE, A. (1991). A common framework for image segmentation. *Internat. J. Computer Vision* **6**, 227–243.

GEMAN, S., GEMAN, D. (1984). Stochastic relaxation, Gibbs distributions, and the Bayesian restoration of images. *IEEE-PAMI* **6** (6), 721–742.

GEMAN, D., REYNOLDS, G. (1992). Constrained restoration and the recovery of discontinuities. *IEEE-PAMI* **14**, 367–383.

GEMAN, D., YANG, C. (1995). Nonlinear image recovery with half-quadratic regularization. *IEEE Trans. Image Processing* **4** (7), 932–946.

GERIG, G., KUBLER, O., KIKINIS, R., JOLESZ, F.A. (1992). Nonlinear anisotropic filtering of MRI data. *IEEE Trans. Medical Imaging* **11**, 221–232.

GRAYSON, M. (1987). The heat equation shrinks embedded plane curves to round points. *J. Differential Geom.* **26**.

GUICHARD, F. (1993). Multiscale analysis of movies: Theory and algorithms, PhD Dissertation, CEREMADE (Paris).

GUICHARD, F., MOREL, J.M. (1995). Introduction to partial differential equations in image processing. In: *IEEE Int. Conf. Image Proc., Washington, DC*. Tutorial notes.

HAKER, S., SAPIRO, G., TANNENBAUM, A. (1998). Knowledge based segmentation of SAR data. In: *Proc. IEEE-ICIP '98, Chicago*.

HAKER, S., SAPIRO, G., TANNENBAUM, A. (2000). Knowledge-based segmentation of SAR data with learned priors. *IEEE Trans. Image Processing* **9**, 299–301.

HAMPEL, F.R., RONCHETTI, E.M., ROUSSEEUW, P.J., STAHEL, W.A. (1986). *Robust Statistics: The Approach Based on Influence Functions* (John Wiley & Sons, New York).

HARRIS, J.G., KOCH, C., STAATS, E., LUO, J. (1990). Analog hardware for detecting discontinuities in early vision. *Internat. J. Computer Vision* **4**, 211–223.

HELMSEN, J., PUCKETT, E.G., COLLELA, P., DORR, M. (1996). Two new methods for simulating photolithography development in 3D. In: Proc. SPIE Microlithography **IX**, p. 253.

HUBER, P.J. (1981). *Robust Statistics* (John Wiley and Sons, New York).

HUMMEL, R.A. (1986). Representations based on zero-crossings in scale-space. In: *Proc. IEEE CVPR*, pp. 204–209.

JAIN, A.K. (1977). Partial differential equations and finite-difference methods in image processing, part 1: Image representation. *J. of Optimization Theory and Applications* **23**, 65–91.

KASS, M., WITKIN, A., TERZOPOULOS, D. (1988). Snakes: Active contour models. *Internat. J. Computer Vision* **1**, 321–331.

KICHENASSAMY, S. (1996). Edge localization via backward parabolic and hyperbolic PDE, Preprint (University of Minnesota).

KICHENASSAMY, S., KUMAR, A., OLVER, P., TANNENBAUM, A., YEZZI, A. (1995). Gradient flows and geometric active contour models. In: *Proc. Internat. Conf. Computer Vision '95, Cambridge*, pp. 810–815.

KICHENASSAMY, S., KUMAR, A., OLVER, P., TANNENBAUM, A., YEZZI, A. (1996). Conformal curvature flows: from phase transitions to active vision. *Arch. Rational Mech. Anal.* **134**, 275–301.

KIMIA, B.B., TANNENBAUM, A., ZUCKER, S.W. (1990). Toward a computational theory of shape: An overview. In: Lecture Notes in Computer Science **427** (Springer-Verlag, New York), pp. 402–407.

KIMIA, B.B., TANNENBAUM, A., ZUCKER, S.W. (1995). Shapes, shocks, and deformations, I. *Internat. J. Computer Vision* **15**, 189–224.

KIMMEL, R. (1999). Numerical geometry of images: Theory, algorithms, and applications, Technion CIS Report 9910.

KIMMEL, R., BRUCKSTEIN, A.M. (1993). Shape offsets via level sets. *CAD* **25** (5), 154–162.

KIMMEL, R., SETHIAN, J.A. (1996). Fast marching method for computation of distance maps, LBNL Report 38451 (UC Berkeley).

KIMMEL, R., SETHIAN, J.A. (1998). Computing geodesic paths on manifolds. *Proc. Nat. Acad. Sci.* **95** (15), 8431–8435.

KOENDERINK, J.J. (1984). The structure of images. *Biological Cybernetics* **50**, 363–370.

KREYSZIG, E. (1959). *Differential Geometry* (University of Toronto Press, Toronto).

KUMAR, R., HANSON, A.R. (1990). Analysis of different robust methods for pose refinement. In: *Proc. Int. Workshop on Robust Computer Vision, Seattle, WA*, pp. 167–182.

LECLERC, Y.G. (1989). Constructing simple stable descriptions for image partitioning. *Internat. J. Computer Vision* **3** (1), 73–102.

LEE, H.-C., COK, D.R. (1991). Detecting boundaries in a vector field. *IEEE Trans. Signal Proc.* **39**, 1181–1194.

LEE, T.S., MUMFORD, D., YUILLE, A.L. (1992). Texture segmentation by minimizing vector-valued energy functionals: The coupled-membrane model. In: *ECCV'92*. In: Lecture Notes in Computer Science **588** (Springer-Verlag), pp. 165–173.

LINDEBERG, T. (1994). *Scale-Space Theory in Computer Vision* (Kluwer).

LORIGO, L.M., FAUGERAS, O., GRIMSON, W.E.L., KERIVEN, R., KIKINIS, R. (1998). Segmentation of bone in clinical knee MRI using texture-based geodesic active contours. In: *Proceedings Medical Image Computing and Computer-Assisted Intervention, MICCAI '98* (Springer, Cambridge, MA), pp. 1195–1204.

LUI, L., SCHUNCK, B.G., MEYER, C.C. (1990). On robust edge detection. In: *Proc. Int. Workshop on Robust Computer Vision, Seattle, WA*, pp. 261–286.

MALLADI, R., KIMMEL, R., ADALSTEINSSON, D., SAPIRO, G., CASELLES, V., SETHIAN, J.A. (1996). A geometric approach to segmentation and analysis of 3D medical images. In: *Proc. Mathematical Methods in Biomedical Image Analysis Workshop, San Francisco*, pp. 21–22.

MALLADI, R., SETHIAN, J.A., VEMURI, B.C. (1994). Evolutionary fronts for topology independent shape modeling and recovery. In: *Proc. of the 3rd ECCV, Stockholm, Sweden*, pp. 3–13.

MALLADI, R., SETHIAN, J.A., VEMURI, B.C. (1995). Shape modeling with front propagation: A level set approach. *IEEE Trans. on PAMI* **17**, 158–175.

MALLADI, R., SETHIAN, J.A., VEMURI, B.C. (to appear). A fast level set based algorithm for topology independent shape modeling. In: ROSENFELD, A., KONG, Y. (eds.), *J. Mathematical Imaging and Vision*, special issue on Topology and Geometry, to appear.

MCINERNEY, T., TERZOPOULOS, D. (1995). Topologically adaptable snakes. In: *Proc. ICCV, Cambridge, MA*.

MEAD, C. (1989). *Analog VLSI and Neural Systems* (Addison-Wesley, New York).

MEER, P., MINTZ, D., ROSENFELD, A. (1990). Robust recovery of piecewise polynomial image structure. In: *Proc. Int. Workshop on Robust Computer Vision, Seattle, WA*, pp. 109–126.

MEER, P., MINTZ, D., ROSENFELD, A., KIM, D.Y. (1991). Robust regression methods for computer vision: A review. *Internat. J. Computer Vision* **6** (1), 59–70.

MEMOLI, F., SAPIRO, G., OSHER, S. (2002). Harmonic maps into implicit manifolds, Preprint.

MUMFORD, D., SHAH, J. (1989). Optimal approximations by piecewise smooth functions and variational problems. *Comm. Pure and Appl. Math.* **42**.

MURRAY, D.W., BUXTON, B.F. (1987). Scene segmentation from visual motion using global optimization. *IEEE Trans. on Pattern Analysis and Machine Intelligence* **9** (2), 220–228.

NEVATIA, R. (1977). A color edge detector and its use in scene segmentation. *IEEE Trans. Syst. Man, Cybern.* **7**, 820–826.

NIESSEN, W.J., TER HAAR ROMENY, B.M., FLORACK, L.M.J., SALDEN, A.H. (1993). Nonlinear diffusion of scalar images using well-posed differential operators, Technical Report (Utrecht University, The Netherlands).

NITZBERG, M., SHIOTA, T. (1992). Nonlinear image filtering with edge and corner enhancement. *IEEE-PAMI* **14**, 826–833.

NORDSTRÖM, N. (1990). Biased anisotropic diffusion: A unified regularization and diffusion approach to edge detection. *Image and Vision Computing* **8**, 318–327.

OHTA, T., JASNOW, D., KAWASAKI, K. (1982). Universal scaling in the motion of random interfaces. *Physical Review Letters* **47**, 1223–1226.

OLVER, P.J. (1993). *Applications of Lie Groups to Differential Equations*, 2nd Edition (Springer-Verlag).

OLVER, P. (1995). *Equivalence, Invariants, and Symmetry* (Cambridge University Press).

OLVER, P., SAPIRO, G., TANNENBAUM, A. (1996). Affine invariant detection: Edges, active contours, and segments. In: *Proc. Computer Vision Pattern Recognition, San Francisco*.

OLVER, P., SAPIRO, G., TANNENBAUM, A. (1997). Invariant geometric evolutions of surfaces and volumetric smoothing. *SIAM J. of Applied Math.* **57** (1), 176–194.

OLVER, P., SAPIRO, G., TANNENBAUM, A. (1999). Affine invariant edge maps and active contours, Affine invariant detection: Edge maps, anisotropic diffusion, and active contour. *Acta Appl. Math.* **59**, 45–77.

OSHER, S. (website). UCLA Technical Reports, located at http://www.math.ucla.edu/applied/cam/index.html.

OSHER, S., HELMSEN, J. (in prep.). A generalized fast algorithm with applications to ion etching, in preparation.

OSHER, S., RUDIN, L.I. (1990). Feature-oriented image enhancement using shock filters. *SIAM J. Numer. Anal.* **27**, 919–940.

OSHER, S.J., SETHIAN, J.A. (1988). Fronts propagation with curvature dependent speed: Algorithms based on Hamilton–Jacobi formulations. *J. Comput. Physics* **79**, 12–49.

PARAGIOS, N., DERICHE, R. (1997). A PDE-based level-set approach for detection and tracking of moving objects. INRIA Technical Report 3173 (Sophia-Antipolis).

PARAGIOS, N., DERICHE, R. (1998a). A PDE-based level-set approach for detection and tracking of moving objects. In: *Proc. Int. Conf. Comp. Vision '98, Bombay, India.*

PARAGIOS, N., DERICHE, R. (1998b). Geodesic active regions for tracking. In: *European Symposium on Computer Vision and Mobile Robotics CVMR'98, Santorini, Greece.*

PARAGIOS, N., DERICHE, R. (1999a). Geodesic active regions for motion estimation and tracking Technical Report (INRIA–Sophia Antipolis) 3631.

PARAGIOS, N., DERICHE, R. (1999b). Geodesic active contours for supervised texture segmentation. In: *Proc. Computer Vision Pattern Recognition, Colorado.*

PARAGIOS, N., DERICHE, R. (1999c). Geodesic active regions for supervised texture segmentation. In: *Proc. International Conference Computer Vision, Corfu, Geece.*

PARAGIOS, N., DERICHE, R. (1999d). Geodesic active contours and level sets for detection and tracking of moving objects, *IEEE transactions on Pattern Analysis Machine Intelligence*, to appear.

PARDO, A., SAPIRO, G. (to appear). Vector probability diffusion, *IEEE Signal Processing Letters*, to appear.

PERONA, P. (1998). Orientation diffusion. *IEEE Trans. Image Processing* **7**, 457–467.

PERONA, P., MALIK, J. (1990). Scale-space and edge detection using anisotropic diffusion. *IEEE-PAMI* **12**, 629–639.

PERONA, P., MALIK, J. (1991). Detecting and localizing edges composed of steps, peaks, and roofs, MIT–CICS Technical Report.

PERONA, P., TARTAGNI, M. (1994). Diffusion network for on-chip image contrast normalization. In: *Proc. IEEE – International Conference on Image Proc., Austin, Texas, Vol. 1*, pp. 1–5.

POLYMENAKOS, L.C., BERTSEKAS, D.P., TSITSIKLIS, J.N. (1988). Implementation of efficient algorithms for globally optimal trajectories. *IEEE Trans. on Automatic Control* **43** (2), 278–283.

PRATT, W.K. (1991). *Digital Image Processing* (John Wiley & Sons, New York).

PRICE, C.B., WAMBACQ, P., OOSTERLINK, A. (1990). Image enhancement and analysis with reaction-diffusion paradigm. *IEE Proc.* **137**, 136–145.

Proc. Int. Workshop on Robust Computer Vision, Seattle, WA, 1990.

PROESMANS, M., PAUWELS, E., VAN GOOL, J. (1994). Coupled geometry-driven diffusion equations for low-level vision. In: Romeny, B. (ed.), *Geometry Driven Diffusion in Computer Vision* (Kluwer).

ROMENY, B. (ed.) (1994). *Geometry Driven Diffusion in Computer Vision* (Kluwer).

ROUSSEEUW, P.J., LEROY, A.M. (1987). *Robust Regression and Outlier Detection* (John Wiley & Sons, New York).

RUDIN, L.I., OSHER, S., FATEMI, E. (1992a). Nonlinear total variation based noise removal algorithms. *Physica D* **60**, 259–268.

RUDIN, L., OSHER, S., FATEMI, E. (1992b). Nonlinear total variation based noise removal algorithms. In: *Proc. Modélisations Matématiques pour le traitement d'images* (INRIA), pp. 149–179.

SAPIRO, G. (1996). From active contours to anisotropic diffusion: Connections between the basic PDEs in image processing. In: *Proc. IEEE – International Conference on Image Processing, Lausanne.*

SAPIRO, G. (1997). Color snakes. *Computer Vision and Image Understanding* **68** (2), 247–253.

SAPIRO, G. (2001). *Geometric Partial Differential Equations and Image Analysis* (Cambridge University Press, New York).

SAPIRO, G., CASELLES, V. (1997a). Histogram modification via differential equations. *J. Differential Equations* **135** (2), 238–268.

SAPIRO, G., CASELLES, V. (1997b). Contrast enhancement via image evolution flows. *Graphical Models and Image Processing* **59** (6), 407–416.

SAPIRO, G., KIMMEL, R., SHAKED, D., KIMIA, B.B., BRUCKSTEIN, A.M. (1993). Implementing continuous-scale morphology via curve evolution. *Pattern Recog.* **26** (9).

SAPIRO, G., RINGACH, D. (1996). Anisotropic diffusion of multivalued images with applications to color filtering. *IEEE Trans. on Image Processing* **5**, 1582–1586.

SAPIRO, G., TANNENBAUM, A. (1994). On affine plane curve evolution. *J. Funct. Anal.* **119** (1), 79–120.

SCHUNCK, B.G. (1989). Image flow segmentation and estimation by constraint line clustering. *IEEE Transactions on Pattern Analysis and Machine Intelligence* **11** (10), 1010–1027.

SCHUNCK, B.G. (1990). Robust computational vision. In: *Proc. Int. Workshop on Robust Computer Vision, Seattle, WA*, pp. 1–18.

SERRA, J. (1988). *Image Analysis and Mathematical Morphology, Vol. 2: Theoretical Advances* (Academic Press).

SETHIAN, J. (1996a). Fast marching level set methods for three-dimensional photolithography development. In: *Proc. SPIE International Symposium on Microlithography, Santa Clara, CA*.

SETHIAN, J.A. (1996b). A fast marching level-set method for monotonically advancing fronts. *Proc. Nat. Acad. Sci.* **93** (4), 1591–1595.

SETHIAN, J.A. (1996c). *Level Set Methods: Evolving Interfaces in Geometry, Fluid Mechanics, Computer Vision and Materials Sciences* (Cambridge University Press, Cambridge, UK).

SHAH, J. (1996). A common framework for curve evolution, segmentation, and anisotropic diffusion. In: *Proc. CVPR, San Francisco*.

SIDDIQI, K., BERUBE, TANNENBAUM, A., ZUCKER, S. (1998). Area and length minimizing flows for shape segmentation. *IEEE Trans. Image Processing* **7** (3), 433–443.

SINHA, S.S., SCHUNCK, B.G. (1992). A two-stage algorithm for discontinuity-preserving surface reconstruction. *IEEE Transactions on Pattern Analysis and Machine Intelligence* **14** (1), 36–55.

SOCHEN, N., KIMMEL, R., MALLADI, R. (1998). A general framework for low-level vision. *IEEE Trans. Image Processing* **7**, 310–318.

SONER, H.M. (1993). Motion of a set by the curvature of its boundary. *J. Differential Equations* **101**.

SZELISKI, R., TONNESEN, D., TERZOPOULOS, D. (1993). Modeling surfaces of arbitrary topology with dynamic particles. In: *Proc. CVPR*, pp. 82–87.

TANG, B., SAPIRO, G., CASELLES, V. (2000a). Diffusion of general data on nonflat manifolds via harmonic maps theory: The direction diffusion case. *Internat. J. Computer Vision* **36** (2), 149–161.

TANG, B., SAPIRO, G., CASELLES, V. (2000b). Chromaticity diffusion. In: *Proc. IEEE – International Conference on Image Processing, Vancouver, Canada*.

TEK, H., KIMIA, B.B. (1995). Image segmentation by reaction–diffusion bubbles. In: *Proc. ICCV'95, Cambridge*, pp. 156–162.

TEO, P., SAPIRO, G., WANDELL, B. (1997). Creating connected representations of cortical gray matter for functional MRI visualization. *IEEE Trans. Medical Imaging* **16** (06), 852–863.

TERZOPOULOS, D., SZELISKI, R. (1992). Tracking with Kalman snakes. In: Blake, A., Zisserman, A. (eds.), *Active Vision* (MIT Press, Cambridge, MA).

TERZOPOULOS, D., WITKIN, A., KASS, M. (1988). Constraints on deformable models: Recovering 3D shape and non-rigid motions. *AI* **36**.

TIRUMALAI, A.P., SCHUNCK, B.G., JAIN, R.C. (1990). Robust dynamic stereo for incremental disparity map refinement. In: *Proc. Int. Workshop on Robust Computer Vision, Seattle, WA*, pp. 412–434.

TOGA, A.W. (1998). *Brain Warping* (Academic Press, New York).

TORRE, V., POGGIO, T. (1986). On edge detection. *IEEE-PAMI* **8**, 147–163.

TSITSIKLIS, J.N. (1995). Efficient algorithms for globally optimal trajectories. *IEEE Trans. on Automatic Control* **40**, 1528–1538.

TURK, G. (1991). Generating synthetic textures using reaction-diffusion. *Computer Graphics* **25** (3).

VAZQUEZ, L., SAPIRO, G., RANDALL, G. (1998). Segmenting neurons in electronic microscopy via geometric tracing. In: *Proc. IEEE ICIP, Chicago*.

WEICKERT, J. (1998). *Anisotropic Diffusion in Image Processing*, ECMI Series (Teubner-Verlag, Stuttgart).

WENG, J., COHEN, P. (1990). Robust motion and structure estimation using stereo vision. In: *Proc. Int. Workshop on Robust Computer Vision, Seattle, WA*, pp. 367–388.

WHITAKER, R.T. (1995). Algorithms for implicit deformable models. In: *Proc. ICCV'95, Cambridge*, pp. 822–827.

WITKIN, A.P. (1983). Scale-space filtering. In: *Int. Joint Conf. Artificial Intelligence*, pp. 1019–1021.

WITKIN, A.P., KASS, M. (1991). Reaction-diffusion textures. *Computer Graphics* **25** (3).

YEZZI, A., KICHENASSAMY, S., OLVER, P., TANNENBAUM, A. (1997). Geometric active contours for segmentation of medical imagery. *IEEE Trans. Medical Imaging* **16**, 199–210.

YOU, Y.L., XU, W., TANNENBAUM, A., KAVEH, M. (1996). Behavioral analysis of anisotropic diffusion in image processing. *IEEE Trans. Image Processing* **5**, 1539–1553.

ZHU, S.C., LEE, T.S., YUILLE, A.L. (1995). Region competition: Unifying snakes, region growing, energy/Bayes/MDL for multi-band image segmentation. In: *Proc. Int. Conf. Comp. Vision '95, Cambridge*, pp. 416–423.

Further reading

BERTALMIO, M. (1998). Morphing active contours, M.Sc. Thesis I.I.E. (Universidad de la Republica, Uruguay).

BESAG, J. (1986). On the statistical analysis of dirty pictures. *J. Royal Statistical Society* **48**, 259–302.

CROMARTIE, R., PIZER, S.M. (1991). Edge-affected context for adaptive contrast enhancement. In: *Proceedings Information Processing in Medical Imaging, Wye, UK*. In: Lecture Notes in Comp. Science **511**, pp. 474–485.

FAUGERAS, O.D., BERTHOD, M. (1981). Improving consistency and reducing ambiguity in stochastic labeling: an optimization approach. *IEEE-PAMI* **3**, 412–423.

HUMMEL, R.A., ZUCKER, S.W. (1983). On the foundations of relaxation labeling processes. *IEEE Trans. Pattern Analysis and Machine Intelligence* **5** (2), 267–286.

KIRKPATRICK, S., GELLATT, C.D., VECCHI, M.P. (1983). Optimization by simulated annealing. *Science* **220**, 671–680.

LI, S.Z., WANG, H., PETROU, M. (1994). Relaxation labeling of Markov random fields. In: *Int'l Conf. Pattern Recognition*, pp. 488–492.

PERONA, P., SHIOTA, T., MALIK, J. (1994). Anisotropic diffusion. In: Romeny, B. (ed.), *Geometry Driven Diffusion in Computer Vision* (Kluwer).

ROSENFELD, A., HUMMEL, R., ZUCKER, S. (1976). Scene labeling by relaxation operations. *IEEE Trans. Systems, Man, and Cybernetics* **6** (6), 420–433.

SCHWARTZ, L. (1991). *Analyse I. Theorie des Ensembles et Topologie* (Hermann).

SERRA, J. (1982). *Image Analysis and Mathematical Morphology* (Academic Press, New York).

WEISS, Y., ADELSON, E. (1996). Perceptually organized EM: a framework for motion segmentation that combines information about form and motion. In: *Int'l Conf. Computer Vision and Pattern Recognition*, pp. 312–326.

YOU, Y.L., KAVEH, M. (1996). A regularization approach to joint blur identification and image restoration. *IEEE Trans. Image Processing* **5**, 416–428.

ZHU, S.C., MUMFORD, D. (1997). GRADE: Gibbs reaction and diffusion equations: A framework for pattern synthesis, image denoising, and removing clutter, Preprint.

Subject Index